Application of Basic Neuroscience to Child Psychiatry

This book is dedicated to the sacred memory of the one and one-half million Jewish children who were killed during the long and terrible Night. They are not forgotten.

Application of Basic Neuroscience to Child Psychiatry

Edited by

Stephen I. Deutsch

Veterans Administration Medical Center
and Georgetown University School of Medicine
Washington, D.C.

Abraham Weizman and Ronit Weizman

Tel Aviv University
Tel Aviv, Israel

Plenum Medical Book Company
New York and London

Library of Congress Cataloging-in-Publication Data

Application of basic neuroscience to child psychiatry / edited by
Stephen I. Deutsch, Abraham Weizman, and Ronit Weizman.
p. cm.
Includes bibliographical references.
Includes index.
ISBN-13: 978-1-4612-7849-8 e-ISBN-13: 978-1-4613-0525-5
DOI: 10.1007/978-1-4613-0525-5
1. Biological child psychiatry. 2. Neuroregulators--physiology.
3. Synaptic Receptors--physiology. I. Deutsch, Stephen I.
II. Weizman, Abraham. III. Weizman, Ronit.
[DNLM: 1. Mental Disorders--etiology. 2. Mental Disorders--in
fancy & childhood. 3. Mental Disorders--physiopathology. WS 350
A6515]
RJ486.5.A67 1990
618.92'8907--dc20
DNLM/DLC
for Library of Congress 90-7203
CIP

The work described in "Brain Recognition Sites for Methylphenidate and the Amphetamines: Their Relationship to the Dopamine Transport Complex, Glucoreceptors, and Serotonergic Neurotransmission in the Central Nervous System" was done as part of the authors' employment with the federal government and is therefore in the public domain.

233 Spring Street, New York, N.Y. 10013
Softcover reprint of the hardcover 1st edition 1990

Plenum Medical Book Company is an imprint of Plenum Publishing Corporation

Contributors

George M. Anderson • Yale Child Study Center, Yale University School of Medicine, New Haven, Connecticut 06510

Itzchak Angel • Department of Biology, Laboratoires d'Etudes de Recherches Synthelabo, 75013 Paris, France

Paul Berger • Clinical Neuroscience Branch, National Institute of Mental Health, National Institutes of Health, Bethesda, Maryland 20892

James R. Brasic • Division of Child and Adolescent Psychiatry, Department of Psychiatry, New York University School of Medicine, New York, New York 10016

Boris Birmaher • Department of Child Psychiatry, College of Physicians and Surgeons of Columbia University, New York, New York 10032

Phillip B. Chappell • Yale Child Study Center, Yale University School of Medicine, New Haven, Connecticut 06510

Roland D. Ciaranello • Department of Psychiatry and Behavioral Sciences, Division of Child Psychiatry, Stanford University School of Medicine, Stanford, California 94305

Donald J. Cohen • Yale Child Study Center, Yale University School of Medicine, New Haven, Connecticut 06510

Lynn H. Deutsch • Psychiatry Service, Veterans Administration Medical Center, and Department of Psychiatry, Georgetown University School of Medicine, Washington, D.C. 20007

Stephen I. Deutsch • Psychiatry Service, Veterans Administration Medical Center, and Department of Psychiatry, Georgetown University School of Medicine, Washington, D.C. 20007

Nelson Freimer • College of Physicians and Surgeons of Columbia University, New York, New York 10032

Wayne H. Green • Department of Psychiatry, New York University School of Medicine, and Child and Adolescent Psychiatric Clinic, Bellevue Hospital Center, New York, New York 10016

Laurence L. Greenhill • Department of Clinical Psychiatry, College of Physicians and Surgeons of Columbia University, New York, New York 10032

Richard L. Hauger • San Diego Veterans Administration Medical Center, and Department of Psychiatry, University of California–San Diego, School of Medicine, La Jolla, California 92093

Bridget Hulihan-Giblin • Clinical Neuroscience Branch, National Institute of Mental Health, National Institutes of Health, Bethesda, Maryland 20892

Aaron Janowsky • Research Service, Department of Psychiatry, Veterans Administration Medical Center, and Oregon Health Sciences University, Portland, Oregon 97207

Abba J. Kastin • Veterans Administration Medical Center, Tulane University School of Medicine, New Orleans, Louisiana 70146

Elizabeth V. Koby • Department of Child Psychiatry, Georgetown University School of Medicine, Washington, D.C. 20007

Arthur F. Kohrman • Department of Pediatrics, Pritzker School of Medicine, The University of Chicago, and La Rabida Children's Hospital and Research Center, Chicago, Illinois 60649

Karin A. Kook • Division of Neuropharmacological Drug Products, Food and Drug Administration, Rockville, Maryland 20857

James F. Leckman • Yale Child Study Center, Yale University School of Medicine, New Haven, Connecticut 06510

Len Leven • Division of Child and Adolescent Psychiatry, Department of Psychiatry, New York University School of Medicine, New York, New York 10016

John Mastropaolo • Psychiatry Service, Veterans Administration Medical Center, Washington, D.C. 20422

William L. Nyhan • Department of Pediatrics, University of California–San Diego, La Jolla, California 92093

Theodore Page • Department of Pediatrics, University of California–San Diego, La Jolla, California 92093

David Pauls • Yale Child Study Center, Yale University School of Medicine, New Haven, Connecticut 06510

Jonathan H. Pincus • Department of Neurology, Georgetown University Hospital, Washington, D.C. 20007

Ausma Rabe • New York State Institute for Basic Research in Developmental Disabilities, Staten Island, New York 10314

Judith L. Rapoport • Child Psychiatry Branch, National Institute of Mental Health, National Institutes of Health, Bethesda, Maryland 20892

Ann M. Rasmusson • Yale Child Study Center, Yale University School of Medicine, New Haven, Connecticut 06510

Moshe Rehavi • Department of Physiology and Pharmacology, Sackler Faculty of Medicine, Tel Aviv University, Tel Aviv 69978, Israel

Mark A. Riddle • Yale Child Study Center, Yale University School of Medicine, New Haven, Connecticut 06510

Anthony L. Riley • Psychopharmacology Laboratory, Psychology Department, The American University, Washington, D.C. 20016

Wilma G. Rosen • Department of Psychiatry, Columbia University, and The Presbyterian Hospital, Neurological Institute, Columbia-Presbyterian Medical Center, New York, New York 10032

John L. R. Rubenstein • Department of Psychiatry and Behavioral Sciences, Division of Child Psychiatry, Stanford University School of Medicine, Stanford, California 94305

Curt A. Sandman • State Developmental Research Institute, Fairview, Costa Mesa, California 92626; and Department of Psychiatry and Human Behavior, University of California–Irvine Medical Center, Orange, California 92668

Nicole Schupf • New York State Institute for Basic Research in Developmental Disabilities, Staten Island, New York 10314

Michael Stanley • Departments of Psychiatry and Pharmacology, College of Physicians and Surgeons of Columbia University, and Department of Neurochemistry, New York State Psychiatric Institute, New York, New York 10032

Susan E. Swedo • Child Psychiatry Branch, National Institute of Mental Health, National Institutes of Health, Bethesda, Maryland 20892

Susan K. Theut • Psychiatry Service, Veterans Administration Medical Center, and Department of Psychiatry, Georgetown University School of Medicine, Washington, D.C. 20007

Frank J. Vocci, Jr. • Medications Development Program, Division of Preclinical Research, National Institute on Drug Abuse, Rockville, Maryland 20857

Myrna M. Weissman • Department of Psychiatry, College of Physicians and Surgeons of Columbia University, and Division of Clinical–Genetic Epidemiology, New York State Psychiatric Institute, New York, New York 10032

Abraham Weizman • Geha Psychiatric Hospital, Beilinson Medical Center, Petah-Tiqva 49100; and Sackler Faculty of Medicine, Tel Aviv University, Tel Aviv 69978, Israel

Ronit Weizman • Pediatric Department, Hasharon Hospital, Petah Tiqva 49372; and Sackler Faculty of Medicine, Tel Aviv University, Tel Aviv 69978, Israel

Henryk M. Wisniewski • New York State Institute for Basic Research in Developmental Disabilities, Staten Island, New York 10314

Krystyna E. Wisniewski • New York State Institute for Basic Research in Developmental Disabilities, Staten Island, New York 10314

Dona Lee Wong • Department of Psychiatry and Behavioral Sciences, Division of Child Psychiatry, Stanford University School of Medicine, Stanford, California 94305

J. Gerald Young • Division of Child and Adolescent Psychiatry, Department of Psychiatry, New York University School of Medicine, New York, New York, 10016

Preface

The idea for this book developed during the course of several discussions among the editors while we were working together as staff scientists in the laboratories of the Clinical Neuroscience Branch of the National Institute of Mental Health. It was a happy coincidence that the three of us, child psychiatrists with predominantly clinical interests, selected a collaborative bench research project involving neurotransmitter receptor characterization and regulation. We appreciated the relevance of our work to child psychiatry and wished for a forum to share the excitement we enjoyed in the laboratory with our clinical colleagues. Moreover, it seemed to us that much of the pharmacological research in child psychiatry proceeded on an empirical basis, often without a compelling neurochemical rationale. This could reflect the paucity of neurochemical data that exists in child psychiatry and the very limited understanding of the pathophysiology in most psychiatric disorders that occur in childhood. Also, we bemoaned the fact that there was a virtual absence of meaningful interchange between clinical investigators in child psychiatry and their colleagues in the neurosciences. We believed that an edited book appealing to clinicians and basic scientists could serve as an initial effort to foster interchange between them. The editors wish to emphasize that this book is viewed as only a beginning in the process of interchange that must take place. It is our hope that this book will stimulate new ideas that will be pursued collaboratively in studies that flow to and from the bench and the bedside. These studies must be undertaken because the severity of the problems confronting children—which interfere with their healthy growth and development, and their capacity to enjoy life—demands it.

The editors would like to express their sincerest appreciation to Ms. Monica Novitzki, who assumed many of the administrative and secretarial functions associated with the execution of this project. She performed these difficult and often frustrating tasks with celerity and professionalism. We would also like to acknowledge the patience and support of Mr. Eliot Werner, who served as the sponsoring editor of this project for Plenum Publishing Corporation. Mr. Werner's kindness and understanding contributed to the pleasantness of the task. Finally, we would like to thank our patients and children, from whom we have so much to learn.

Stephen I. Deutsch
Abraham Weizman
Ronit Weizman

Washington and Tel Aviv

Contents

Introduction

Stephen I. Deutsch, Abraham Weizman, and Ronit Weizman

1. INTRODUCTION

The editors were impressed with the encyclopedic scope and thoroughness of the contributors as they discussed issues in neuroscience with an immediate or potential relevance to child psychiatry. The review of these submissions suggested many potentially fruitful avenues of future investigation. The stimulation of new research ideas and the exposure to problems that could be pursued most effectively collaboratively in the laboratory and at the bedside were essential objectives in proposing and developing the idea for this book. The most optimistic goals in editing this book were that it would serve as a catalyst for interaction between investigators from seemingly unrelated areas and would lead to novel approaches to the understanding and treatment of major disorders in child psychiatry. There is a compelling need to apply the developments in the neurosciences to child psychiatry as rapidly as they emerge in order to relieve children of life-long disabilities. The editors hope that this book will introduce clinicians to developments in the neurosciences, and basic neuroscientists to clinical problems.

2. BRAIN IMAGING STUDIES OF AUTISM

Despite the thoroughness of this edited volume, several important areas were not formally considered in depth or presented as chapters. Often, these omissions reflected deliberate editorial decisions necessitated by an attempt to keep the final volume within its recommended page limit. However, the book would be less than complete if the application of developments in brain imaging to an understanding of autism and the potential relevance of recent findings in the regulation of axonal outgrowth and neural grafting to child psychiatry were not acknowledged.

Stephen I. Deutsch • Psychiatry Service, Veterans Administration Medical Center, and Department of Psychiatry, Georgetown University School of Medicine, Washington, D.C. 20007. *Abraham Weizman* • Geha Psychiatric Hospital, Beilinson Medical Center, Petah Tiqva 49100, Israel; and Sackler Faculty of Medicine, Tel Aviv University, Tel Aviv 69978, Israel. *Ronit Weizman* • Pediatric Department, Hasharon Hospital, Petah Tiqva 49372; and Sackler Faculty of Medicine, Tel Aviv University, Tel Aviv 69978, Israel.

The newly developed techniques for imaging the living brain in vivo have a great potential application to child psychiatry; to date, they have not been as extensively explored in child psychiatry as they have been in adult psychiatry. For example, in vivo magnetic resonance scanning (MRI) revealed evidence of developmental hypoplasia of the neocerebellar hemispheres and vermal lobules VI and VII in 14 of 18 autistic patients whose disorders were not complicated by known genetic or neurologic abnormalities, including severe mental retardation, cerebral palsy, and epilepsy.[1] Moreover, the interpretation of this neocerebellar abnormality was not confounded by a history of anticonvulsant or antipsychotic medication. The existence of this anatomic abnormality suggests that neural circuits connecting deep cerebellar nuclei with the reticular activating system could be disrupted in autistic patients. Furthermore, the in vivo anatomic findings may be relevant to the degeneration of cerebellar Purkinje cells observed in postmortem histopathological studies of four autistic patients.[2] However, in an MRI investigation of the posterior fossae of 15 autistic patients and 15 age- and sex-matched normal children, no differences in the midsagittal area and volume of the fourth ventricle were detected.[3] Interestingly, seven adult autistic patients did not show functional evidence of cerebellar hypofunction or diminished glucose metabolic rate in a positron-emission tomography (PET) study of the cerebellum and vermal lobes VI and VII.[4] These patients were normally intelligent and had no past or current history of a metabolic or neurological disorder nor use of antipsychotic drugs. Heh et al. speculated that the failure to detect diminished glucose metabolic rate in the cerebella of these patients could be due to an inefficiency of metabolism. Also, glucose uptake was measured during a task of visual vigilance (i.e., continuous performance test) that could have obscured differences between patients and controls. In another PET study of the cerebral metabolic rate of 10 autistic men, each with well-documented histories of infantile autism, there was no evidence of diminished glucose metabolic rate.[5] In fact, the autistic group displayed elevated glucose utilization in various brain regions, although there was significant overlap with the control group. This study did not find any evidence of a focal metabolic abnormality. However, the absence of a focal abnormality could reflect the fact that eight of the 10 patients had updated diagnoses of autism, residual state, and glucose utilization was measured at rest without the benefit of an activation procedure. In any event, the assessment of brain metabolic activity with PET using (F-18) 2-fluoro-2-deoxy-D-glucose in the autistic patients studied to date has not contributed to a greater understanding of the pathophysiology of this disorder. Gross anatomic abnormalities could not be detected in a computed tomographic (CT) study of nine autistic boys (ages 9 years, 8 months to 16 years, 3 months) with evidence of left hemispheric dysfunction on neuropsychological testing.[6] The study assumes added significance in view of the fact that these autistic boys were of normal or near-normal intelligence and none had a history of neurologic insult. These data argue against the presence of gross anatomic abnormalities in classic presentations of autism. However, reliance on linear measurements, potential artifacts introduced by head tilt, and bias associated with the selection of CT slices for analysis were inherent limitations of the CT study. In order to obviate many of these difficulties, volumetric measures of cerebrospinal fluid (CSF), white matter, gray matter, the third ventricle, the lateral ventricles, the caudate nuclei, lenticular nuclei, and thalami were determined in CT scans of 12 autistic men without histories of neurologic diseases or seizures.[7] These measures were compared with those obtained in an appropriately selected control group of 16 healthy men. The autistic men could not be distinguished from the control group by any of the volumetric measures or by abnormal ventricular or nuclear asymmetries. These data suggest that gross anatomic defects are not uniformly seen in autistic patients. Early CT reports of ventriculomegaly and asymmetries of posterior cerebral structures and ventricles in autism may be confined to small subgroups of patients and not specific to the disorder.[8–12] There may, however, be greater variance in several of the volumetric measures among autistic patients.[9]

In summary, the possible existence of neocerebellar hypoplasia has been the most provocative finding to result from the application of imaging techniques to the study of autism; this finding would be most consistent with the neuropathological reports of cerebellar Purkinje cell degeneration. Future directions in this avenue of research could include imaging of neurotransmitter receptors with recently developed PET ligands, e.g., [^{11}C]-Ro15-1788, and the implication of specific neurotransmitter abnormalities in the pathophysiology of this disorder.

3. REGULATION OF NEURITIC EXTENSION

The physiology of the growth cone and regulation of the directional outgrowth of axons have been investigated extensively over the past decade.[13,14] The filamentous protein actin seems to be responsible for the motility of growth cones, i.e., the tip of advancing axons; motility involves the polymerization and depolymerization of actin monomers. The process of actin polymerization—depolymerization is dependent on ATP and its hydrolysis and is regulated in vivo by so-called actin-binding proteins. Axonal guidance or regulation of growth cone extension is influenced by many factors, including the embryological stage of development. For example, in earliest development, general cell adhesion molecules, e.g., neural cell adhesion molecule (N-CAM), provide a permissive substrate for axons to extend through environments composed of neuroepithelial and mesenchymal cells. These molecules are on the surfaces of axons and neural epithelium; they promote adhesion by their homophilic interaction on apposing cell surfaces.

A more restricted pathway for axonal extension may be provided by laminin, a glycoprotein on the surface of the extracellular matrix that is less uniformly expressed in the developing nervous system than general cell adhesion molecules. Laminin directs axonal outgrowth and is recognized by a class of axonal glycoprotein receptors referred to as *integrins*.

Other classes of glycoproteins may mediate axonal fasciculation or bundling of axons; i.e., axons destined to innervate the same region may form bundles and travel together. The molecules responsible for fasciculation may guide developing axons along preexisting fiber tracts. There are also molecules in the developing and adult nervous systems, especially on the surface of oligodendrocytes, that inhibit axonal extension.

Finally, there are molecules that exert a chemotropic influence on growth cones and that play an essential role in the maintenance and survival of axons that have reached their final destinations. Nerve growth factor is the prototype of this final class of molecule. It is conceivable that abnormalities in growth cone motility and/or regulation of the directional outgrowth of axons will be implicated in the etiopathogenesis of some of the major neuropsychiatric disorders of childhood. The basic science relevant to these topics was recently reviewed.[13,14]

4. TRANSPLANTATION AND CELLULAR REPLACEMENT THERAPY

The combined application of molecular biological and neural grafting techniques may have important therapeutic implications for the treatment of degenerative brain diseases and traumatic brain injury.[15] The retrograde degeneration of cholinergic neurons in the basal forebrain of rats after transection of the fimbria-fornix was prevented by neural implantation of rat fibroblasts from an established cell line that secreted nerve growth factor (NGF) constitutively. NGF exerts a trophic influence on cholinergic cell bodies and is transported in a retrograde fashion from cholinergic nerve terminals in the hippocampus to cell bodies in the septum. In fact, the highest levels of messenger RNA (mRNA) for NGF are found in the target areas of cholinergic neurons, i.e., hippocampus and cortex, consistent with its proposed role in the maintenance and survival of cholinergic neurons.[16] The implanted fibroblasts were infected with a retroviral vector that contained the genetic insert encoding the precursor to mouse NGF. Before their injection into the cavity formed as a result of transection of the fimbria-fornix, these cells were shown to secrete high levels of biologically active NGF in vitro. Presumably, the high rate of NGF secretion continued in vivo after implantation. That cholinergic cell bodies survived after implantation was reflected in cell counts and immunohistochemical staining for choline acetyltransferase (ChAT) activity, a specific marker of cholinergic neurons. The implications of this research are profound, although not immediately relevant for application to child psychiatry. For example, it is possible that a neural implantation of autologous cells from a patient, e.g., primary skin fibroblasts, that were genetically modified to secrete growth factors could attenuate or prevent degeneration of specific neurons. The degeneration of specific neurons essential to the etiopathogenesis of a given disorder and the growth factor(s) necessary for their maintenance would have to be identified. This approach may hold great promise for the treatment of Alzheimer's disease, a disorder characterized by degeneration of cholinergic projections from the basal forebrain to the cortex and hippocampus. Clearly, a therapeutic application of this approach in child psychiatry is remote and seems unlikely at this time; however, there are some data to suggest that cerebellar Purkinje cells may be degenerating in at least a few patients with autism.[2]

Fetal neural tissue may also be considered for cellular replacement therapy in degenerative brain disease and traumatic brain injury. Viable mesencephalic neural tissue from first-trimester fetal cadavers can be implanted into the caudate nuclei of normal monkeys after cryopreservation in liquid nitrogen for periods of about 2 months.[17] The implants showed dense neuritic profiles and stained immunohistochemically positive for tyrosine hydroxylase, a marker of catecholaminergic neurons. The neuritic arborization of the grafts was reminiscent of the morphological appearance of the zona reticulata region of the substantia nigra in situ. Moreover, after cryopreservation, cell cultures could be derived from this mesencephalic fetal tissue, and maintained in vitro for 14 days. Thus, a thorough screening of potential cellular implants for immunological, bacteriological, and virological safety could be performed before surgery. The ability to maintain tissue banks of viable fetal neural tissue holds its most immediate promise for the treatment of parkinsonian patients. The application of cellular replacement therapy to child psychiatry remains only a theoretical possibility. The editors are aware of the many ethical issues associated with this area of inquiry that must be resolved prior to an investigation of its therapeutic application in children.

The final chapter of the book discusses some of the unique ethical issues facing clinical investigators in child psychiatry. Several investigators contend that the special and appropriate sensitivities aroused in reviewing research proposals involving children, especially when protocols include procedures without obvious diagnostic or therapeutic indications, can interfere with progress toward increased understanding and improved treatment. Some of these investigators would argue that the ethical constraints imposed on them have been counterproductive delaying the development of newer and more effective interventions. In any event, a consideration of ethical issues by investigators in child psychiatry is often unavoidable, and the final chapter is intended to introduce the reader to some of these issues.

REFERENCES

1. Courchesne E, Yeung-Courchesne R, Press GA, et al: Hypoplasia of cerebellar vermal lobules VI and VII in autism. N Engl J Med 318:1349–1354, 1988
2. Ritvo ER, Freeman BJ, Scheibel AB, et al: Lower Purkinje cell counts in the cerebella of four autistic subjects: Initial findings of the UCLA—NSAC autopsy research report. Am J Psychiatry 143:862–866, 1986

3. Garber HJ, Ritvo ER, Chiu LC, et al: A magnetic resonance imaging study of autism: Normal fourth ventricle size and absence of pathology. Am J Psychiatry 146:532–534, 1989
4. Heh CWC, Smith R, Wu J, et al: Positron emission tomography of the cerebellum in autism. Am J Psychiatry 146:242–245, 1989
5. Rumsey JM, Duara R, Grady C, et al: Brain metabolism in autism. Resting cerebral glucose utilization rates as measured with positron emission tomography. Arch Gen Psychiatry 42:448–455, 1985
6. Prior MR, Tress B, Hoffman WL, et al: Computed tomographic study of children with classic autism. Arch Neurol 41:482–484, 1984
7. Creasey H, Rumsey JM, Schwartz M, et al: Brain morphometry in autistic men as measured by volumetric computed tomography. Arch Neurol 43;669–672, 1986
8. Damasio H, Maurer RG, Damasio AR, et al: Computerized tomographic scan findings in patients with autistic behavior. Arch Neurol 37:504–510, 1980
9. Rosenbloom S, Campbell M, George AE, et al: High resolution CT scanning in infantile autism: A quantitative approach. J Am Acad Child Psychiatry 23:72–77, 1984
10. Campbell M, Rosenbloom S, Perry R, et al: Computerized axial tomographic scans in young autistic children. Am J Psychiatry 139:510–512, 1982
11. Caparulo BK, Cohen DJ, Rothman SL, et al: Computed tomographic brain scanning in children with developmental neuropsychiatric disorders. J Am Acad Child Psychiatry 20:338–357, 1981
12. Hier DE, LeMay M, Rosenberger PB: Autism and unfavorable left-right asymmetries of the brain. J Autism Dev Disord 9:153–159, 1979
13. Dodd J, Jessell TM: Axon guidance and the patterning of neuronal projections in vertebrates. Science 242:692–699, 1988
14. Smith SJ: Neuronal cytomechanics: The actin-based motility of growth cones. Science 242:708–715, 1988
15. Rosenberg MB, Friedmann T, Robertson RC, et al: Grafting genetically modified cells to the damaged brain: Restorative effects of NGF expression. Science 242:1575–1578, 1988
16. Ayer-LeLievre C, Olson L, Ebendal T, et al: Expression of the *B*-nerve growth factor gene in hippocampal neurons. Science 240:1339–1341, 1988
17. Redmond DE, Jr, Naftolin F, Collier TJ, et al: Cryopreservation, culture, and transplantation of human fetal mesencephalic tissue into monkeys. Science 242:768–771, 1988

I

Basic Neuroscience Considerations

The contributions in this section describe recent developments in the basic sciences that have an immediate or potential relevance to clinicians and investigators concerned with neuropsychiatric disorders of childhood. Molecular biological concepts are introduced in the first chapter. A knowledge of gene expression and its regulation is essential to an understanding of the etiology of many disorders and recent approaches to their treatment. Clinicians and molecular geneticists are collaborating closely in attempts to identify abnormal genes with precise chromosomal assignment and localization. Curiously, in some instances, it may be possible to identify precisely the chromosomal location of an abnormal gene with only limited insight into the functional role of the primary gene product. The next chapter discusses the current status of our ability to assess the "firing" of specific neurons by measuring neurotransmitter-specific metabolites in accessible body fluids. These measures may serve as clinically useful indices of the "presynaptic" activity of specific neurons. The measurement of neurotransmitter metabolites may clarify pathophysiological mechanisms in neuropsychiatric disorders, and prove useful to assessing the impact of pharmacological interventions.

Several of the chapters in this section focus on a description of neurotransmitter receptor complexes, especially approaches to their characterization and elucidation of their regulation and physiological roles. A neurotransmitter receptor complex is more than an "agonist recognition site" or "acceptor" with which a ligand interacts reversibly and with high affinity.

Often, clinical investigators are limited in terms of their access to tissues and biological fluids whose study is essential to a complete understanding of the biology of neuropsychiatric disorders. Therefore, tissues and fluids "outside" of the central nervous system are frequently chosen for study. The relevance of some of these "peripheral" indices to central nervous system activity is considered in this section. Hopefully, the developments in functional brain imaging (e.g., positron emission tomography) will eliminate some of the existing limitations to the study of central neurochemistry in childhood disorders. The intimate relationship between the immune system and nervous system and its potential relevance to an understanding of pathogenetic mechanisms, and newer approaches to therapy are also discussed.

1

Molecular Neurobiology and Disorders of Brain Development

Roland D. Ciaranello, Dona Lee Wong, and John L. R. Rubenstein

1. INTRODUCTION

Developmental neurobiology and molecular neurobiology are potentially areas of great promise for scientists and clinicians concerned with developmental disorders in children. Child psychiatric researchers have struggled to solve complex developmental problems without adequate biologic or neurologic tools; as a consequence, the biologic database of the field is quite meager. That situation promises to change, however. With the advent of recombinant DNA technology and the more recent application of these techniques to molecular neurobiology, we now have at hand tools that enable us to learn more about the role played by genes in regulating neuronal development and about how environmental interactions influence gene expression. With this new knowledge, we can begin to ask more focused questions about human central nervous system (CNS) development and seek to relate our findings to developmental disorders in children.

This chapter focuses on some basic principles of genetics and molecular biology. We illustrate the key steps from gene transcription to protein synthesis. We then describe some well-known diseases of the CNS whose pathophysiology has been greatly clarified by the use of molecular biologic approaches. We conclude by pointing out where, in our opinion, these approaches will be of benefit to child psychiatry in the future and how they may be applied to clinical conditions the child psychiatrist is likely to encounter.

2. MENDELIAN GENETICS

The genetic information used by cells to replicate themselves and carry out their biologic activity is contained on chromosomes. These are highly compact bodies residing within the

---

Roland D. Ciaranello, Dona Lee Wong, and John L. R. Rubenstein • Department of Psychiatry and Behavioral Sciences, Division of Child Psychiatry, Stanford University School of Medicine, Stanford, California 94305.

cell nucleus made up of nucleic acids and proteins. They consist of a pair of sex chromosomes (XX or XY) and paired homologous autosomes. The total number of chromosomes is unique for each species. Genes, the actual units of heredity, are arranged in a linear array on each chromosome. The existence of genes was predicted by the Austrian monk Gregor Mendel, who postulated the role of "particulate factors," which determined heritable traits. On the basis of his breeding experiments with peas, Mendel recognized that genetic factors occurred in pairs and that some factors were dominant (the organism's phenotype could be determined by the presence of a single gene copy), while others were recessive (two copies are necessary to determine phenotype).

Mendel also recognized that the parents in a particular mating could have two identical copies of the same factor or have two different copies. Mendel's experiments led to the formulation of two laws, which for many years formed the basis for much of what was known in genetics. Mendel's first law states that alleles, which are variant forms of a particular gene, segregate each generation. Thus, in the formation of sperm or ova, the germ cells will contain one copy of each allele. If the parent is homozygous at a particular locus, so that both alleles are identical, the germ cell will contain only one form of that allele. If the parent is heterozygous at that locus and thus has two different alleles, each germ cell will contain one allele or the other.

Mendel's second law states that individual genes segregate independently. Thus, for two traits, each determined by a single gene, the ratios of phenotypes arising from the first mating (the F_1 generation) and from subsequent crosses ($F_1 \times F_1 = F_2$) can be predicted by supposing an entirely random association between alleles determining each phenotype. We now know that this is strictly true only for unlinked genes, e.g., genes residing on separate chromosomes, or sufficiently distant from each other on the same chromosome to behave independently. The practical consequences of Mendel's laws are much more familiar to us than the laws themselves: When two parents mate, each bearing an autosomal-recessive gene for a particular trait, the odds of producing an offspring with the trait are 1 in 4. If one parent carries an autosomal-dominant gene, all the offspring will express the phenotype (assuming complete penetrance), regardless of their gender and independent of the genotype of the other parent. The same laws allow us to predict the distribution of phenotypes for sex-linked recessive and dominant traits, and to track a particular gene through multiple generations of breeding.

3. DNA, RNA, AND PROTEIN

3.1. DNA Structure

Genes are made up of deoxyribonucleic acid (DNA). DNA is a polymer of deoxyribonucleotides (usually shortened to nucleotides). Each nucleotide consists of a purine or pyrimidine base attached to deoxypentose phosphate, a modified sugar. When nucleotides are polymerized through linkages between their phosphate groups, they become nucleic acids. These relationships are pictured in Figs. 1 and 2.

In the cell nucleus, DNA exists as a paired structure consisting of two complementary strands intertwined to form a helical duplex. The four nucleotides that make up DNA are adenylic acid (A), guanylic acid (G), cytidylic acid (C), and thymidylic acid (T). A and G are purine nucleotides, while C and T are pyrimidine nucleotides. Each DNA strand has a defined polarity; at one end of each strand there is a free 5′ phosphate group, while at the other end there is a free 3′ hydroxyl group (Fig. 2). Since the polarity of each strand in a nucleic acid duplex runs in opposite directions, the two strands are said to be antiparallel.

The two DNA strands pair by hydrogen bonding to form a double-helical structure with

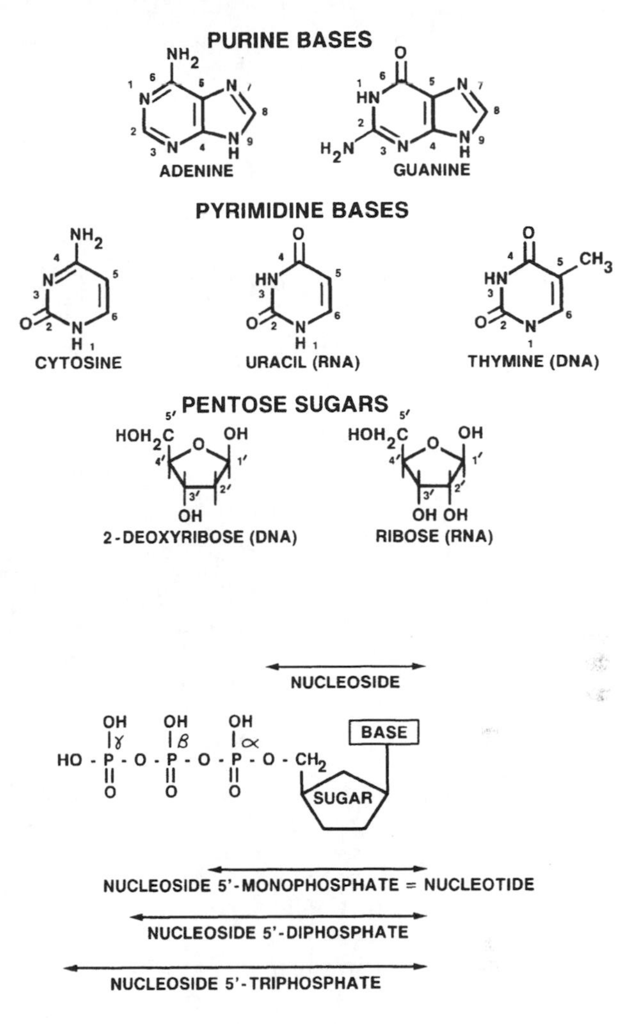

Figure 1. Chemical constituents of nucleic acids. (Top) The chemical structures of the purine and pyrimidine bases in DNA and RNA and the two forms of ribose sugars are illustrated. (Bottom) Nomenclature for various components associated with nucleic acids. The combination of a base and sugar is known as a nucleoside. Addition of a phosphate group forms a nucleoside monophosphate; this is typically called a deoxyribonucleotide in DNA or ribonucleotide in RNA, although both are frequently referred to as nucleotides.

their free 5′ phosphates and 3′ hydroxyls at opposite ends. The hydrogen bonds that form between opposing base pairs are essential to maintain the paired structure of the DNA molecule. Base pairing is strictly governed: the most stable hydrogen bonding occurs between A–T and G–C pairs. As a result, A always pairs with T, and G always pairs with C. This is termed the principle of complementarity, and it has a number of important consequences. Complementarity explains why single-stranded nucleic acids can combine to form stable duplexes, and it is a critical element in nucleic acid synthesis, where one strand acts as a template for the synthesis of the complementary strand. The principles of polarity and base pairing provide a very powerful predictive tool: Knowledge of the sequence of bases in one strand automatically reveals the sequence of bases in the complementary strand.

3.2. DNA Replication

If cells are to replicate themselves, they must pass on their genetic information without copying error. The capability for precise self-replication is an essential property of DNA. During DNA replication, the two strands unwind. Specific enzymes, known as DNA poly-

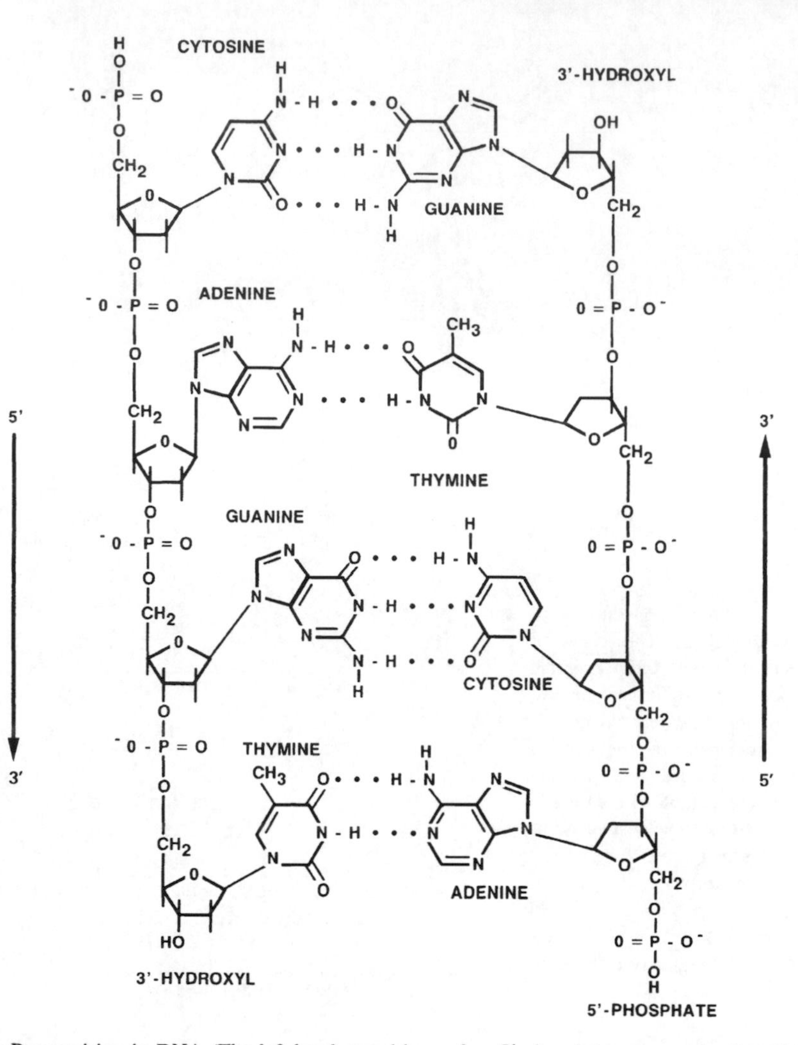

Figure 2. Base pairing in DNA. The left-hand strand has a free 5′-phosphate group and a free 3′-hydroxyl group on the deoxyribose moiety. The right-hand strand is identically constructed, but the two strands are oriented in opposite directions and are said to be antiparallel.

merases, travel along the unwinding double helix. In DNA replication, each strand serves as the template for synthesis of the complementary strand (Fig. 3). Synthesis of the daughter strands takes place by addition of the 5′-phosphate of a deoxyribonucleotide to the free 3′-hydroxyl on the growing strand. Chain growth thus takes place in a 5′→3′ direction. In this way, each parent strand gives rise to a complementary daughter strand with which it immediately forms a new duplex structure joined by hydrogen bonding. At the end of the replication cycle, two new DNA duplexes have been formed, each consisting of one parental strand and one daughter strand. This process is termed semiconservative replication.

3.3. Gene Structure

Although mammalian genes can be extremely complex units, they frequently share common structural features. Figure 4 illustrates our conception of a "generic" gene. Starting from

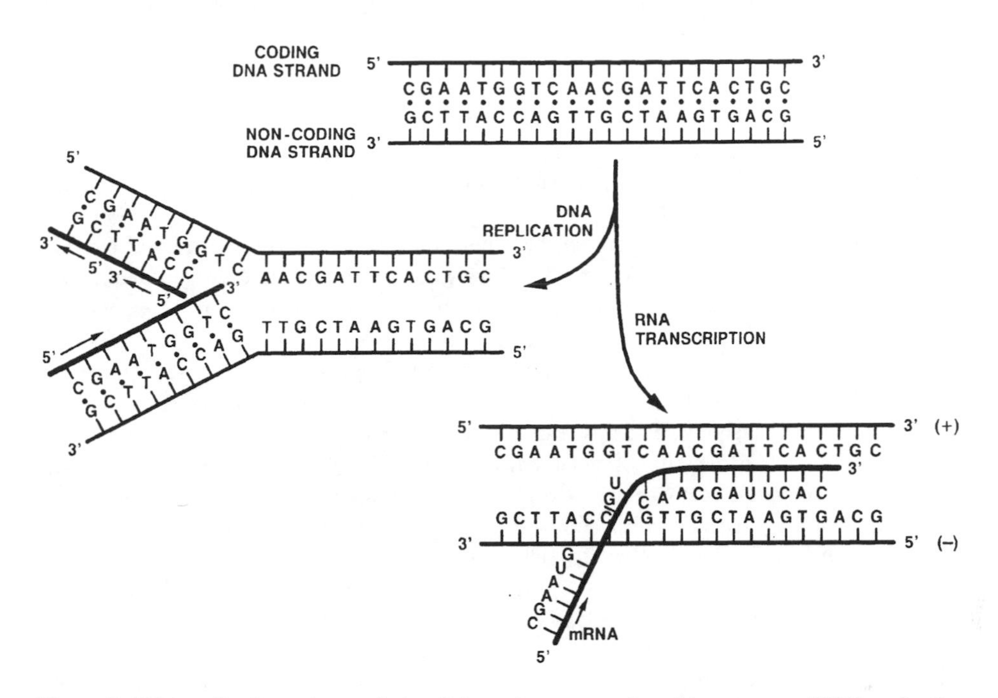

Figure 3. DNA replication and transcription. Schematic representation of the processes of DNA replication and transcription. Synthesis of the daughter strands takes place as the parental strands unwind. Both daughter strands are synthesized in a 5′→3′ direction. One strand is therefore synthesized continuously from the 3′ parental end in the direction of strand separation. Directional synthesis of the other strand is identical. However, this strand must be synthesized discontinuously from the replication fork outward, after which the segments are ligated. During transcription, the DNA strands again separate to permit one of them to be transcribed. In this figure, the primary mRNA transcript is a copy of the uppermost (positive) DNA strand, except that a U has been substituted for every T.

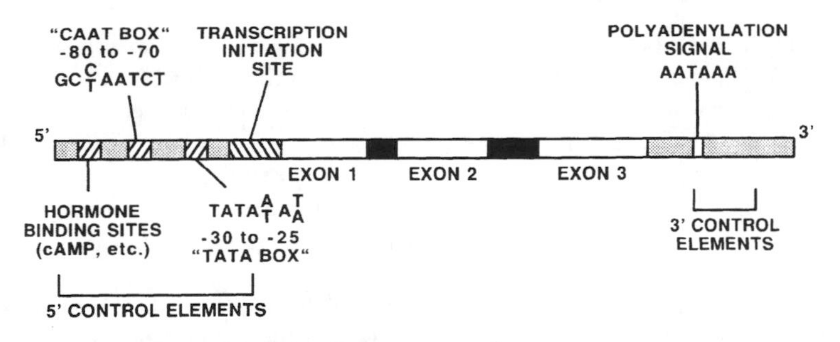

Figure 4. A "generic" gene. Important components of a "generic" gene. The promoter region of the gene lies 5′ or upstream to the transcription initiation site. It consists of upstream promoter elements (UPEs), including the CAAT box and one or more TATA boxes. Hormone binding sites may also be present in this region. They may be part of the promoter or act at a distance from it. Our generic gene consists of three exons (white) interrupted by two introns (black). The third exon is followed by a region of 3′ control elements. These include the polyadenylation signal as well as transcriptional enhancer elements. The transcription termination site is not shown.

the 5′ end, we find short clusters of bases that, as a group, are termed the 5′ regulatory elements. These are involved in activating or inhibiting gene transcription. One important group of regulatory elements are the upstream promoter elements (UPEs). UPEs consist of short consensus sequences of about 8–12 base pairs (bp). Groups of UPEs constitute the promoter region of the gene, a stretch of about 100 bp located a variable distance upstream from the transcription initiation site. The promoter region is necessary for activation of gene transcription; its absence or mutation may result in loss or severe attenuation of gene transcription.

Frequently the promoter region can be identified by the presence of a short stretch of repeating TA sequences (the TATA box). The TATA box is thought to be a signal sequence for RNA polymerase to initiate gene transcription about 25–30 bp downstream. It appears to be important in initiating transcription accurately.

The CAAT box is an upstream promoter element found in many different genes. The identity and mechanism of action of UPEs are a subject of intense activity (for review see Maniatis et al.[1] and Ptashne[2]). The upstream promoter elements bind transcriptional activating proteins that modulate the rate of gene transcription. The precise mechanisms by which transcriptional activators regulate gene transcription are not entirely understood, but they are believed to interact physically with both the DNA promoter elements and with RNA polymerase through specific binding sites for each. Other 5′ regulatory elements include consensus sequences for binding of hormones and cyclic adenosine monophosphate (cAMP) to the gene; these compounds also play critical roles in the modulation of gene transcription.

The coding region of the gene comprises that portion that will be transcribed into the primary messenger RNA (mRNA) transcript. It consists of sequences between the transcription initiation and termination sites, sites that remain imprecisely defined for mammalian genes. The coding regions of eukaryotic genes vary greatly in length from a few hundred bases to as many as two million. The coding region is organized into sequences that will appear in the mature mRNA transcript (exons) and sequences that will be spliced out (introns). A few genes, e.g., those encoding some neurotransmitter receptors, consist of a single uninterrupted exon. But in most cases, the exon domains are broken up by intervening introns; these may account for the majority of bases within the transcribed region of the gene.

3.4. Exon–Intron Boundaries

Although the two ends of an intron show little homology, they contain short consensus sequences which define the site at which splicing of the primary mRNA transcript is to take place. On the 5′ (left-hand) end of the intron is the dinucleotide sequence -GT-. On the right-hand (3′) end, an -AG- sequences appears. These are referred to as the donor and acceptor sites, respectively. The primary mRNA transcript is a copy of the coding region of the gene, including both exon and intron domains. Excision of the intron sequences from the primary transcript is carried out by specific splicing enzymes, which use the donor and acceptor splice sequences as excision markers.

The sequences downstream of the coding region also contain regulatory elements. One of these is the sequence AATAAA, which functions as a polyadenylation signal. The polyadenylation signal is believed to direct a nuclease to cut the mRNA chain at a region 10–15 bases further downstream. A second enzyme then adds a run of 100–200 bases of A (AAAAAA......A) to the mRNA chain. The function of the poly(A) tail is not entirely clear. A particular mRNA may be represented by a mixed population containing poly(A) tails of different length or have no poly(A) tail at all. There appears to be no difference in the functionality or stability of these various mRNA forms. Other regulatory elements in the 3′ downstream regions include hormone-binding sites and transcriptional enhancer elements, sequences of bases that modulate the rate of gene transcription in a manner similar to the upstream elements. Since mammalian genes commonly span many kilobases in length, the 3′ regulatory elements have the capability to modulate gene transcription over very long distances.

3.5. RNA Synthesis

RNA, or ribonucleic acid, is a single-stranded polymer of ribonucleotides. RNA shares some important structural and chemical properties with DNA, but the two molecules differ in important ways, and carry out very different functions. Like DNA, RNA consists of four nucleotides: adenylic acid (A), guanylic acid (G), and cytidylic acid (C) are common to both DNA and RNA. However, RNA contains uridylic acid (U) instead of thymidylic acid. The two bases, uracil and thymine, differ in the presence of a methyl group at the 5-carbon (Fig. 1). In contrast to deoxyribose, the pentose in RNA has a hydroxyl group at the 2′ position (ribose). Polymerization of the ribose phosphate during RNA synthesis takes place in the same way as in DNA synthesis. RNA thus has the same 5′→3′ polarity as DNA.

RNA exists in the cell as a single stranded molecule. Through pairing of complementary bases, RNA strands can form hybrid duplexes with single-stranded DNA or RNA molecules. In addition, complementary sequences within a RNA strand may self-hybridize, forming intrastrand loops, as in transfer RNA (tRNA). Hybridization in all these examples takes place according to the conventional base-pairing rules.

Although numerous types of RNA are known, for purposes of this chapter we are only concerned with two, mRNA and tRNA. mRNA serves as the information carrier for the genetic instructions encoded in DNA. In the cell nucleus, specific enzymes, called RNA polymerases, can faithfully transcribe an mRNA copy from the DNA template.

Figure 3 illustrates the process of mRNA synthesis from DNA and conveys a number of important principles. The two DNA strands separate at the site at which mRNA synthesis begins. mRNA synthesis proceeds by addition of a 5′-ribonucleotide to the terminal 3′-hydroxyl on the growing mRNA chain. Growth of the chain thus proceeds from the 5′ to the 3′ end, a constant feature of nucleic acid synthesis. Figure 3 also shows that only one DNA strand will be transcribed into a primary mRNA transcript. By convention, we refer to that DNA strand as the positive strand and to the other as the negative strand. Nucleic acid strands are not copied directly but rather serve as templates for synthesis of their complementary strands. Therefore, it follows that the negative strand of the DNA duplex serves as the template for mRNA synthesis. In this way, the mRNA product is an exact copy of the positive strand of the DNA duplex, except that U appears in mRNA wherever T is specified in genomic DNA.

3.6. mRNA Processing

The primary mRNA transcript derived from intron-containing genes must first be processed before it becomes a functional message (mature transcript). Splicing enzymes cut the transcript at specified splice sites and rejoin the free ends. The mature mRNA transcript is thus a continuous structure made up entirely of exons. mRNA transcripts of all higher eukaryotic organisms contain consensus splice junctions, providing a common mechanism for intron excision and generation of a mature transcript. This process is illustrated in Fig. 5.

3.7. The Genetic Code

During the 1950s, scientists were baffled by the problem of deciphering the information in DNA. It was known by then that chromosomes consisted of DNA and protein, so one or the

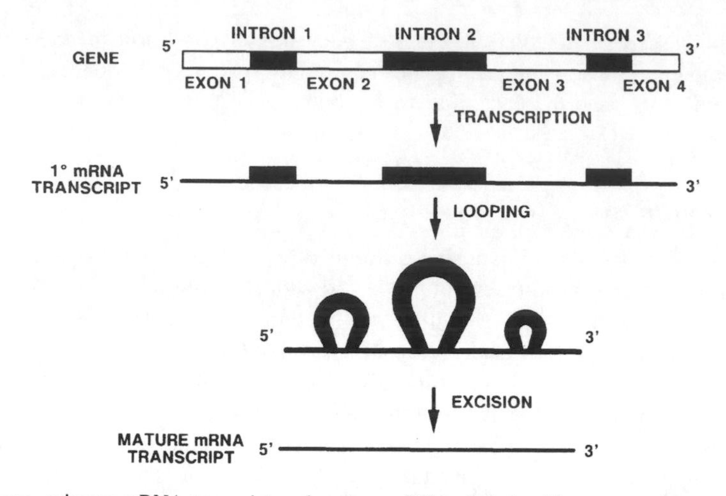

Figure 5. Gene, primary mRNA transcript and mature mRNA. Relationships among the gene, its primary mRNA transcript, and the mature mRNA transcript. The primary mRNA transcript is a faithful copy of the gene, which consists of four exons and three introns. Excision of the introns occurs at consensus donor-splice sites by looping out of the intron regions, and rejoining to form a mature transcript consisting only of exon sequences.

other had to be the chemical basis of genes. Initially, scientists were skeptical that DNA was the primary genetic material, since it was not at all clear how complex genetic information could be encoded in a macromolecule made up of only four individual bases. However, it was also clear that each DNA strand could act as a template for its own replication, whereas there was no known physicochemical mechanism for proteins to replicate themselves. Finally, pioneering experiments by Avery, MacLeod and others demonstrated unequivocally that the genetic material was DNA.

By the 1960s, it had already been shown that the final products of gene expression were specific proteins. Furthermore, the concept that each gene encoded a single protein, or more correctly, a single polypeptide, was well established. It was also known that protein synthesis, while directed by the information in DNA, took place through a mRNA intermediate. It was thus presumed that the bases in DNA and RNA were somehow organized so that they could be translated by the cell into the amino acid sequences of a protein. This presumption of a level of organization decreed the existence of a genetic code.

The breaking of the genetic code is one of the fascinating stories of molecular biology. Even before the code was broken, however, some of its characteristics had been deduced. It had been suspected, for example, that a minimum sequence of three bases was needed to specify each amino acid. This number was predicted from the existence of 20 amino acids commonly found in proteins. Since there are only four bases in DNA, some combination of bases was obviously necessary to specify a particular amino acid. Two bases could only specify 16 amino acids (4^2); this was clearly insufficient. By contrast, three bases would specify 64 (4^3).

It was subsequently shown that three bases, do, in fact, specify one amino acid, and that there is degeneracy, or redundancy in the code. Thus, one amino acid can be specified by more than one base triplet, or codon. Using synthetic polynucleotides, groups led by Nirenberg and by Khorana worked out the specification of codons for amino acids, a task which was essentially completed by 1969. The number of codons for each amino acid corresponds to the frequency with which that amino acid is used in proteins. Thus, leucine, which is abundant in proteins, can be specified by six codons (UUA, UUG, CUU, CUC, CUA, CUG), while methionine (AUG) and tryptophan (UGG) have a single codon, and occur relatively infrequently.

3.8. Protein Synthesis

After processing of the primary transcript, the mature mRNA moves from the cell nucleus into the cytoplasm, where it attaches to ribosomes, which use the mRNA as a template to direct polypeptide synthesis (Fig. 6). Protein synthesis initiates at a specific AUG site in the region of the 5′ terminus of the mRNA. The base sequences of mRNA corresponding to codons are read as nonoverlapping triplets. Thus, the initating AUG codon dictates a reading frame, i.e., how the sequential bases will be read as triplets and direct the incorporation of the correct amino acids into protein. Assembly of the polypeptide chain requires the participation of tRNA. tRNA is a stable form of single-stranded RNA, complementary portions of which undergo internal hybridization to form a double-stranded, clover-leaf structure containing looped regions (Fig. 6). One of the loops contains a base triplet, which is the complement of the codon sequences in mRNA. This triplet is called an anticodon. Every tRNA has an anticodon; it follows that a unique tRNA exists for every functional codon. At the opposite end of the tRNA molecule there is an amino acid binding site. When an amino acid is attached to this site, the tRNA is considered "charged" or activated.

During protein synthesis, the charged tRNA binds to mRNA through codon–anticodon pairing. Enzymes specialized to carry out protein synthesis, aminoacyl tRNA transferases, link the amino acids sequentially in the growing polypeptide chain, which remains attached to the ribosome until its synthesis is complete. Like nucleic acids, proteins also possess polarity: The first amino acid in the polypeptide chain has a free amino group, while the last has a free carboxyl group. Protein synthesis takes place directionally; the first amino acid specified by the initiating codon constitutes the amino terminus, with chain growth proceeding through the entire molecule to the carboxyl end. During this process, the ribosome traverses the mRNA template, reading it and directing the synthesis of the polypeptide. The same mRNA can be read by numerous ribosomes simultaneously, so that one polypeptide may be reaching completion at the 3′ end of the assembly line while a new one is being started at the 5′ end.

3.9. Protein Processing

After translation of mRNA, the finished polypeptide chain may undergo additional processing. Many proteins are made up of polypeptide subunits. Such proteins are termed multimeric or oligomeric proteins. Their subunits may be identical or different. Different subunits can arise as products of unique genes or as alternate splice products of a primary transcript from a single gene. Regardless of the complexities involved, multimeric proteins are formed by assembly of their constituent subunits into the larger protein product.

Many proteins in higher organisms also undergo additional modification before they are fully functional. One common form of posttranslational modification is the addition of extensively branched polysaccharides to the polypeptide backbone. This process, glycosylation, is frequently observed in enzymes and in neurotransmitter receptors. Addition of sugar residues takes place at specific amino acid sequences in the polypeptide chain. One common consensus sequence is formed by the amino acids asparagine-X-serine/threonine (*asn-X-ser/thr*), where X can be any amino acid and either serine or threonine can be found in the third position. Addition of the polysaccharide chain takes place on the asparagine residue.

Addition of a phosphate group (phosphorylation) is another important form of posttranslational protein modification. Phosphorylation takes place on serine or tyrosine residues, and is

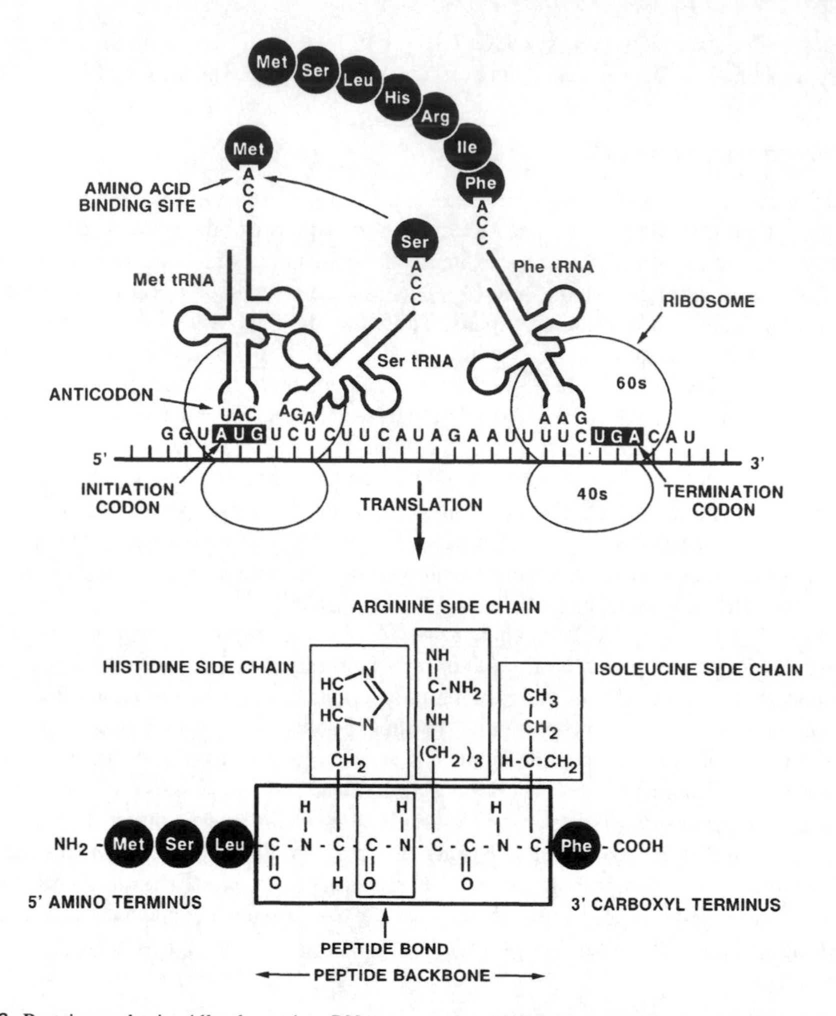

Figure 6. Protein synthesis. All eukaryotic mRNAs possess an AUG initiation codon (methionine) that designates the start site for protein synthesis. Thus, every protein should begin with a methionine at its amino terminus. However, most proteins undergo a processing step which removes this terminal *met* residue. (Top) How amino acid-charged tRNAs bind to mRNA through codon—anticodon pairing. A new peptide chain is being synthesized at the 5′ end of the mRNA. The tRNA bearing *met* is bound to the mRNA, and another tRNA bearing the second amino acid (*ser*) is about to bind. An aminoacyl tRNA transferase will catalyze the formation of a peptide bond between *met* and *ser*. The ser–met dipeptide will remain bound to the ser–tRNA, and a third tRNA will bind to the ribosome at the next codon. This process will be repeated until the translation termination codon is reached. Near the 3′ end of the message, a *phe*-tRNA is shown binding to the mRNA, bearing the assembled polpeptide chain. (Bottom) The critical components to protein structure. The peptide bond joins each of the amino acids to form the polypeptide backbone of the protein: The protruding sidechains distinguish the individual amino acids and are important in the folding of the molecule. Side chains are shown for histidine, arginine, and isoleucine.

essential to the activation of many enzymes. In addition, many proteins, including receptors, enzymes, and ion channel proteins, undergo precisely regulated cyclic changes in their activity by the addition and removal of phosphate groups. Phosphorylation of intracellular proteins by receptor-activated protein kinases is one of the most actively investigated areas of contemporary neurobiology and is covered elsewhere in this volume.

4. DNA CLONING

DNA cloning has emerged as one of the most powerful tools in modern biology. Not only has it revealed a vast body of protein sequence information, which would have taken many years to obtain by traditional protein sequencing techniques, but DNA cloning has also permitted us to analyze gene structure and to transfer genes from one host to another. The principles of DNA cloning are relatively straightforward, although the actual processes can be complicated. For purposes of this discussion, however, simplified cases serve quite nicely.

There are two principal types of DNA cloning. The first is to clone DNA from mRNA transcripts; this is particularly useful if one is interested in isolating DNAs that represent expressed messages in a specific tissue. The second method is to clone genes themselves from genomic DNA. We will first focus on cloning DNA from mRNA transcripts (cDNA cloning).

4.1. cDNA Cloning

The process of DNA cloning begins by constructing DNA copies of mature mRNA transcripts. Since these forms of DNA are complementary copies of RNA, they are termed cDNA. The formation of cDNA is made possible by an enzyme, reverse transcriptase, which is present in RNA viruses. Such viruses, of which the human immunodeficiency (HIV) virus is one well-known example, do not possess DNA. Instead, their genetic information is encoded in RNA. Since single-stranded RNA does not replicate itself, it was a mystery how RNA viruses could infect cells and reproduce themselves. The problem was solved with the discovery that RNA viruses do indeed utilize DNA in the viral replication cycle; reverse transcriptase permits them to synthesize DNA from a RNA template.

The discovery of reverse transcriptase was one critical step that led to DNA cloning. The other was the discovery of restriction endonucleases, bacterial enzymes that cut DNA at specific sequences, termed restriction sites. This was an immensely important discovery, because it paved the way not only for DNA cloning, but for the use of DNA markers in genetic linkage analysis as well. The use of DNA markers in analysis of restriction fragment length polymorphisms (RFLPs) is discussed elsewhere in this volume.

The process of DNA cloning is summarized in Figure 7. mRNA is first prepared from the tissue of interest. The mRNA serves as a template for reverse transcriptase, which synthesizes a cDNA complement of the mRNA beginning from the 3′ end, creating a DNA–RNA hybrid. When it reaches the 5′ end of the mRNA, reverse transcriptase does not stop; instead, it turns back on itself for a short distance, creating a loop in the cDNA by inserting several additional bases using the newly synthesized DNA strand as a template.

At this stage, the cDNA is a single-stranded structure hybridized to mRNA, with a loop at one end. The mRNA is no longer needed, so it is removed chemically. Next, the end of the cDNA loop is used as a primer for continued DNA synthesis by DNA polymerase, which uses the first strand as a template to create a complementary strand. A hairpin-like structure thus arises, consisting of two complementary strands of cDNA joined by the loop.

In the next step, the loop is excised enzymatically, creating a true double-stranded DNA. The first strand synthesized is the complement of mRNA, so it is homologous to the negative strand of DNA in the gene. The second cDNA strand is the complement to the first, so it is a duplicate of the mRNA and represents the coding region of the mature transcript.

The cDNA must undergo additional processing steps before it is ready to be inserted into a cloning vector; these include methylation, cutting with restriction endonucleases, and adding tails or linker arms. Numerous vectors are available for cloning, including bacteriophages and plasmids. The choice of cloning vector is determined by the strategy to be used in identifying

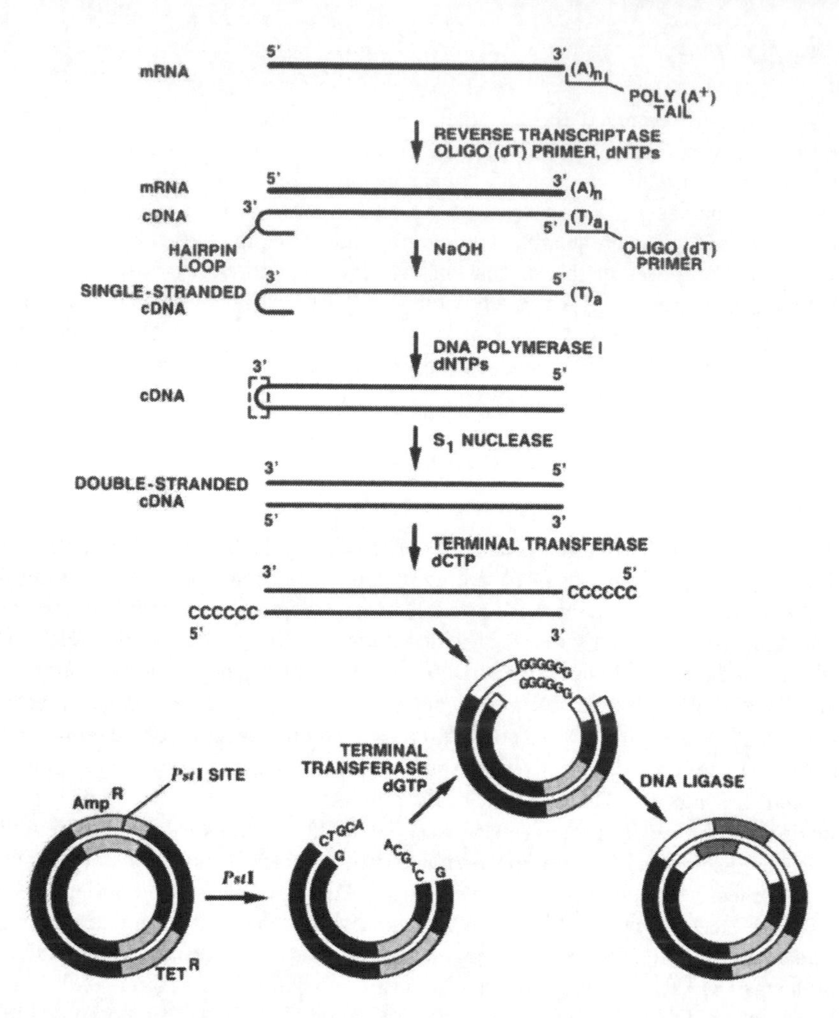

Figure 7. cDNA cloning in plasmid vectors. Poly(A+)RNA is prepared and a copy of cDNA is made using the enzyme reverse transcriptase and the four deoxynucleotide triphosphates (dNTPs). An oligo(dT) primer annealed to the poly (A+) tail serves as initiation site for first-strand synthesis. The mRNA is then chemically digested with NaOH, leaving a single-stranded cDNA with a hairpin loop. This hairpin loop will serve as a template for second-strand cDNA synthesis using DNA polymerase I and the four dNTPs. A hairpin-shaped cDNA is constructed. The hairpin loop is cleaved with S1 nuclease, creating a double-stranded cDNA. The cDNA is then inserted into a plasmid by cutting the vector with a suitable restriction enzyme, adding the appropriate annealing tails to the ends of both the cut vector and cDNA, followed by ligation of the insert into the vector. Here the plasmid contains two genes for antibiotic resistance: *amp* confers resistance to ampicillin, and *tet* confers resistance to tetracycline. The *amp* gene contains a site for the restriction enzyme PstI. We have chosen this site for cloning of our cDNA insert. After ligation, the plasmid can be infected into a bacterial host for amplification. This host will then be ampicillin sensitive because of the interruption in the *amp* gene.

the desired cDNA. For the sake of simplicity, this discussion focuses on cloning, using plasmid vectors.

From a map of the restriction sites in the vector, the experimenter selects one or more sites at which the vector is cut to insert the cDNA. The vector is then treated with the appropriate restriction enzyme. The vector is opened and the cDNA ligated into the free ends. Linker arms are usually used to attach the cDNA to the vector cut site.

At this stage, the experimenter has a mixture of cDNAs inserted into the vector. Since the cDNAs were synthesized from existing cellular mRNA, the frequency with which any individual cDNA is found corresponds to the abundance of individual mRNAs. The number of copies of any particular message is usually small (sometimes on the order of 1 in 100,000) so the experimenter generally will next amplify all the messages. This is done by inoculating bacteria with the cDNA-charged vector and allowing the "transformed" bacteria to replicate. In this way, the experimenter creates an amplified "library" of cDNAs. If the number of individual messages before amplification was 100,000, after amplification it may reach one million or more, so that 10 to 100 copies of each message are now present in the library.

The library is then screened for the cDNA of interest, using the specific tools available to the experimenter; these, in turn, depend on what is known about the protein of interest before the cloning strategy was developed. If amino acid sequences in the protein are known, the experimenter may construct short synthetic nucleic acids (oligonucleotide probes) to bind all cDNAs that contain these sequences. Reference back to the genetic code and the principles of complementarity and polarity permit synthesis of an oligonucleotide probe that will bind specifically to the cDNA of interest. Alternatively, if the protein has been purified but not sequenced, the experimenter may use antibodies directed against the protein to identify the cDNAs encoding it. If the experimenter chooses this strategy, he will probably have used a bacteriophage such as λgtll as a cloning vector. Success with this approach requires that the bacterial host can synthesize the protein of interest. Some mammalian proteins appear to be toxic to bacteria, so this method may not be uniformly applicable.

Regardless of the methods chosen to identify cDNAs, some procedure for identifying positive bacterial colonies or phage plaques is essential. Once a bacterial colony expressing the desired cDNA is identified, usually as a spot on a paper filter, that colony is isolated and grown in larger quantities. The process of bacterial growth and selection is repeated until a pure colony (clone) expressing an individual cDNA is isolated. The transformed bacteria are then grown in sufficient quantities to permit isolation of the clone on a large scale. After preparation of sufficient quantities of the clone, the experimenter then excises the cDNA insert from the vector using appropriate restriction enzymes. He now has abundant quantities of purified cDNA for further experimentation.

To cite a few examples, once isolated, a cDNA can be chemically sequenced to reveal the primary structure of the protein it encodes; this has led to amino acid sequence information on a vast number of proteins within a very short time. The cDNA may be introduced into a suitable host cell (transfection) to produce the encoded protein, so that its properties can be studied in a controlled environment; this is a very important method for studying novel proteins or proteins that have not been previously purified, such as receptors. The cDNA may be hybridized to genomic DNA to isolate the genes encoding the target protein and examine its structure.

4.2. Genomic DNA Cloning

Analysis of gene structure requires construction of a DNA library, which is made by cutting chromosomal DNA with a restriction enzyme and then introducing the mixture of DNA fragments into a cloning vector, using a form of λ bacteriophage. Isolation of the gene of interest can be done using many methods, but the most common involves screening the library with a radioactive DNA probe containing sequences complementary to those found in the gene. cDNA probes are commonly used for this purpose. Once identified, the gene can be isolated, purified, and propagated in sufficient amounts for analysis of its structure. The specific techniques are somewhat different than those just described for DNA cloning, but the principles are the same.

4.3. Mutations

The consequences of genetic mutations are well known to the readers of this volume. Several types of mutation have been described. The simplest are substitution mutations that arise during DNA replication when an incorrect base is inserted into the daughter strand (e.g., A for G). If base substitution occurs within an exon, it usually causes a single amino acid change; this may or may not have important functional consequences. As described later, base substitutions at exon—intron junctions can eliminate them and have devastating consequences.

Another group of mutational events, termed "frame-shift" mutations, arises from addition or deletion of a single base during DNA replication. This process is illustrated in Fig. 8. With the advent of recombinant DNA technology and subsequent analysis of gene structure, other important types of mutations are now known as well. These include exon deletion, whereby some or all of an exon is omitted from the gene; the resultant protein is severely truncated and usually nonfunctional. Another type of mutation is exon duplication, in which a segment, or even an entire exon, is repeated in tandem. This results in repetition of a segment of the protein, and usually obliterates its function as well. Another type of mutation results in the loss of an exon–intron boundary, so that the mRNA encodes amino acid sequences not normally found in the protein. These are usually associated with attenuation or total loss of function. As the structure of more genes is revealed, and particularly as previously unknown disease genes are identified and mapped, mutations causing deletions of large segments of genes are becoming more commonly observed.

4.4. Molecular Biologic Analyses of Central Nervous System Disorder

Molecular biologic techniques have been extremely helpful in contributing to our understanding of genetic diseases. We have chosen three diseases that have been well studied to illustrate the contributions of recombinant DNA technology: phenylketonuria, Lesch–Nyhan syndrome, and the fragile X-syndrome (reviewed by Wong and Ciaranello[3]). All are associated with mild to severe mental retardation, as are many other genetic diseases affecting the CNS.

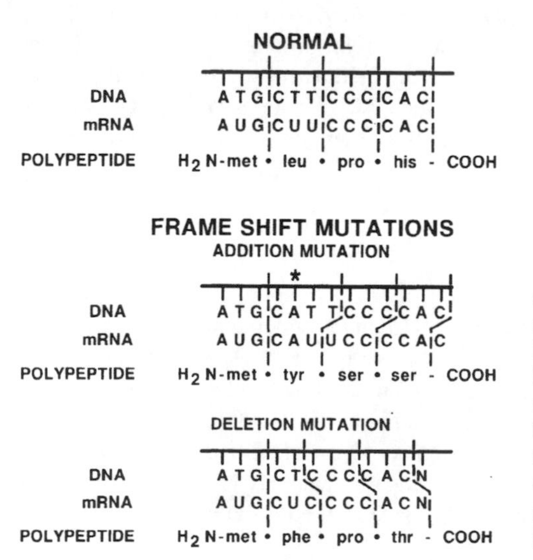

Figure 8. Frame-shift mutations. A common form of mutation is a substitution mutation, in which one base is substituted for another. Such mutations often result in amino acid substitutions, but they do not disturb the reading frame. Frame-shift mutations arise when a base is either inserted or deleted during DNA replication. These mutations can have profound effects on the course of protein synthesis, usually resulting in the generation of a translation termination signal and the production of a truncated protein chain.

4.5. Phenylketonuria

Phenylketonuria (PKU) was the first of many inherited diseases in which mental retardation was linked to a metabolic disorder. The hallmark of PKU is elevated serum phenylalanine, caused by a deficiency in liver phenylalanine hydroxylase. Because phenylalanine cannot be converted to the amino acid tyrosine, its levels and those of its metabolites, phenylacetic acid and phenylpyruvic acid, accumulate. Although the pathogenesis of PKU still has not been entirely worked out, phenylalanine metabolites are thought to be toxic to the developing brain and responsible for the occurrence of mental retardation. PKU is inherited as an autosomal-recessive trait. Thus, the maternal heterozygote is almost always a clinically unafflicted carrier of the disease. During gestation, the maternal liver metabolizes fetal phenylalanine, so its levels do not accumulate in the fetus. However, after birth, the infant's liver cannot carry out normal phenylalanine metabolism. Since the end stages of neuronal differentiation in brain development continue in children through the first 7 or 8 years after birth, the toxic effects of PKU can be manifest at any time during this period. However, PKU can be effectively treated by restricting dietary phenylalanine such that levels of 5.5 and 10 mg/dl are maintained during the critical childhood years.

In classic PKU, phenylalanine hydroxylase (PH) enzymatic activity is completely lacking. The enzyme is either not synthesized at all or its structure is drastically altered, since antibodies to rat liver phenylalanine hydroxylase, which cross-react with the human enzyme, do not detect any enzyme protein. Human phenylalanine hydroxylase consists of two nonidentical subunits of molecular weight 50,000. Recent gene-cloning studies indicate the existence of a single gene for phenylalanine hydroxylase,[4] so the difference in enzyme subunits may represent alternate forms of mRNA splicing or differential posttranslational protein processing.

The structure of the phenylalanine hydroxylase gene has been determined through the cloning of four overlapping clones spanning >125 kb of the PH gene locus. Contained within the PH gene are 13 exons and 12 introns. The transcribed region of the PH gene comprises 90 kb, with the introns ranging in size from 1 to 23 kb. Most of the larger introns are localized to the 5′ regions of the gene. Since the mature mRNA for phenylalanine hydroxylase is only 2.4 kb, phenylalanine hydroxylase represents an example of an enzyme having a gene that consists almost entirely of intron and regulatory regions (> 97%).

Identification of the PH gene structure was facilitated by the cloning of a human cDNA for phenylalanine hydroxylase. Preparation of the human cDNA was achieved by first preparing a full-length clone from a rat liver cDNA library and then using this to screen a human cDNA library.[5] Several partial-length cDNAs were generated, as well as one full-length clone that encoded the entire enzyme. The partial and full-length clones have been used as probes in restriction fragment length polymorphism (RFLP) analysis in classic PKU. Screening of DNA isolated from normal persons and patients with PKU showed identical hybridization patterns, suggesting that classic PKU did not result from a deletion or rearrangement in the gene region specified by the partial probes. However, when genomic DNA from randomly selected persons was analyzed with three restriction endonucleases, polymorphic patterns were observed at the phenylalanine hydroxylase locus.[5] Several allelic variants were identified, permitting analysis of classical PKU patients and families by RFLP analysis. In all, the existence of 12 haplotypes was suggested from this work. Chromosomal mapping of the PKU locus showed linkage to chromosome 12 in the region 12q22.2–12q24.1.[6]

The full-length cDNA probe has been used in clinical studies to screen families with one or two affected and unaffected children.[7] Classic PKU was verified in affected children by testing for phenylalanine tolerance. Lymphocytes were collected from all family members, and isolated DNA was analyzed for RFLPs. In all cases, a restriction fragment pattern could be linked to the mutant allele; segregation of the mutant allele and the disease state was concordant in families with two affected siblings, and segregation was discordant between affected and unaffected siblings.

The full-length phenylalanine hydroxylase cDNA has also provided additional information about PKU-associated RFLPs that may be valuable in the diagnosis of the disease state.[6] Eight restriction enzymes generated polymorphic gene patterns. When these RFLP haplotypes were used to study the phenylalanine hydroxylase gene in PKU and normal families, two RFLP haplotypes were more common to 75% of the control genes, whereas two different RFLP haplotypes predominated in 75% of the PKU genes.

These two haplotypes represent distinct mutations. They have been further characterized through the cloning of the mutant DNA. The more common mutation consists of a GT → AT substitution at the 5′ donor splice site of intron 12. This change leads to the exclusion of exon 12 from the mature mRNA transcript. Transfection of the mutant cDNA into cells showed that although mRNA is synthesized in appropriate amounts, a truncated protein that is enzymatically and immunologically inactive is formed. The second mutation also represents a base substitution. In this case, a C → T change occurs in exon 12 of the gene. The first base in codon 408 is thus altered. In this case, the enzyme protein is of normal length, but bears the amino acid substitution *arg → trp*. Transfection experiments suggest that normal amounts of the mutant mRNA are synthesized, but the resultant protein is inactive both enzymatically and immunologically.

Phenylalanine hydroxylase cDNA has also been used in prenatal diagnosis for two PKU families at risk for the disease.[7] Southern analysis, using the full-length human cDNA, was performed on genomic clones constructed from the DNA from lymphocytes of normal and phenylketonuric individuals. No rearrangements or deletions in the gene structure were observed. However, several RFLPs were identified that distinguished normal phenylalanine hydroxylase genes from mutant genes. DNA purified from parent and proband lymphocytes was analyzed to determine the restriction fragment with which the PKU allele segregated for each family. At 16 weeks gestation, amniocentesis was performed, amniotic cells propagated, and the DNA isolated and analyzed for RFLPs. In the first family, the fetus was determined to be homozygous for PKU. It had the same genotype as the PKU proband and had inherited the restriction fragment segregating with the PKU allele from mother and father. In the second family, the fetus was heterozygous. In this case, however, the parents were of the same genotype, and the mutant genes segregated with alternative restriction fragments. Thus, a prenatal diagnosis would not have been possible if the fetus were affected or free of the PKU trait. These two examples illustrate both the power of RFLP analysis in prenatal diagnosis and its limitations.

Finally, recombinant techniques appear to offer particular promise for the identification of maternal PKU, a new variant of PKU. Maternal PKU reflects the unanticipated, but metabolically predictable, consequence of successful treatment, which enables PKU individuals to lead normal lives and attain reproductive fitness as adults. In maternal PKU, afflicted (homozygous) mothers whose dietary phenylalanine levels are not controlled during pregnancy give birth to infants who show signs of severe retardation and other symptoms of classic PKU. Interestingly, most of these children are heterozygotes, so their phenylalanine hydroxylase should be functional, and should have undergone normal developmental maturation. The basis of fetal pathology seems to be elevated phenylalanine arising from the maternal circulation, which cannot be cleared enzymatically by the mother or the fetus, ultimately leading to fetal brain damage, congenital heart disease, low birthweight, microcephaly, and hyperphenylalaninemia.[8,9] It is possible that prolonged fetal exposure to elevated levels of phenylalanine leads to particularly severe brain damage that cannot be corrected by the subsequent maturation of fetal liver phenylalanine hydroxylase. The emergence of maternal PKU as a clinical entity indicates that it is essential that female phenylketonurics be placed back on dietary phenylalanine restriction during pregnancy.

4.6. Lesch–Nyhan Syndrome

Lesch–Nyhan syndrome was first described as a familial disorder of uric acid metabolism and CNS function.[10] The syndrome is inherited as an X-linked recessive disorder in which hypoxanthine phosphoribosyltransferase (HPRT) is virtually absent. Characteristics of the syndrome include excess uric acid production, mental retardation, and neurologic defects such as self-mutilation, muscle spasticity, and choreoathetosis. Most infants with Lesch–Nyhan syndrome are normal at birth, but begin showing evidence of developmental retardation by 3–4 months. Compulsive self-mutilation begins around 2 years and is characterized by involuntary biting of the fingers, lips and buccal mucosa. Self-mutilation can be extremely severe and can only be prevented by restraining the limbs or extracting the teeth.

Hypoxanthine phosphoribosyltransferase is a critical enzyme in the salvage pathway of the brain for purine biosynthesis. The native enzyme is thought to exist as a tetramer, with the molecular weight of each monomeric subunit ranging from 24 to 26 kDa.[11] HPRT displays considerable heterogeneity in normal humans, and several electrophoretic variants have been described.[12] These variants do not differ much in immunogenicity; thus, major alterations in primary structure are probably absent. In these cases, heterogeneity may be due to posttranslational modification of the enzyme protein. However, family studies have revealed some structural variation in HPRT, since changes in thermostability, electrophoretic mobility, and kinetic parameters are observed. This latter evidence clearly points to the presence of heterogeneity at the HPRT locus.

The cDNA encoding human HPRT has been isolated by two different groups, each producing a full-length clone for this enzyme.[13,14] These have been used to characterize the HPRT gene, HPRT-specific mRNA, and the steps involved in processing of mRNA. Although the structural gene for HPRT spans 34kb, the mature mRNA is only 1.6 kb in length. The difference in size between gene and messenger RNA arises from eight introns ranging in length from 0.1 to 10.8 kb. The coding sequences are represented by nine exons, ranging in length from 77 to 593 bp.[15]

The HPRT gene has been mapped to the distal portion of the X chromosome between the genes for PRPP synthetase and glucose-6-phosphate dehydrogenase.[16,17] Assorted evidence suggests that HPRT deficiency may be genetically heterogeneous. In contrast to many genetic disorders, there is no known selective advantage or ethnic association. Many mutant forms of the enzyme exist. A number of these represent single amino acid changes; the functional consequences of these substitutions can be very diverse, however, ranging from changes in enzyme concentration to changes in kinetic constants.

The search for RFLPs in cultured cells from Lesch–Nyhan patients has been one approach to studying the question of genetic diversity in this disorder. Subfragments of HPRT cDNA, representing specific coding regions of the genome, have been used to probe DNA restriction fragments from 28 patients. In five subjects, major deletions or rearrangements in the HPRT locus were observed.[18] Three subjects exhibited deletion mutations, in which portions of exons were missing. Interestingly, they also showed cytogenetically normal X chromosomes. A fourth individual demonstrated a mutation caused by exon duplication. A fifth mutation occurred in a subject with several exons exhibiting structural alterations of an undetermined nature.

When the mRNA from the various subjects was examined,[18] those with deletion mutations produced no mRNA, several normal individuals showed appropriate-sized mRNA, one patient with normal DNA produced no mRNA, and the exon duplication mutation produced a larger mRNA, as expected with partial gene duplication.

4.7. Fragile X Chromosome

Clinical and genetic studies clearly show that a form of X-linked mental retardation exists. Nonspecific mental retardation became the name designated for these forms of retardation, in which mothers displayed normal intelligence and sons had a 50% probability of being affected. A puzzling aspect, however, was the extreme heterogeneity associated with nonspecific retardation, with some patients showing only retardation and others showing physical abnormalities as well. It is now generally accepted that at least four forms of mental retardation with X-linked inheritance exist.[19,20] One form, the fragile X syndrome is distinguished from other forms in that a fragile site is present at the distal end of the long arm of the X chromosome.

In fragile X syndrome, mental retardation is accompanied by macroorchidism and other distinct physical and neurological manifestations. These include increased birthweight; prominent forehead; relatively hypoplastic midface; high, arched palate; and large, often malformed ears.

The fragile X chromosome site, fra(X)(q27), was first described in 1969.[21] The fragile X site shares several morphological features with other fragile chromosomal sites, the most notable being the triradial figure generated by breakage and nondisjunction at the fragile site.

A characteristic specifically associated with the fragile X site is its folate sensitivity. In media lacking folic acid and thymidine, the triradial chromosome is elicited in lymphocytes and skin fibroblasts of affected males. In the heterozygous female, the fragile X site also appears if those carriers show some intellectual retardation. However, expression appears age dependent, since it is often absent in older heterozygotes. The fragile X site has also been demonstrated in cultured amniotic fluid cells and cultured lymphocytes from fetal blood, suggesting a potential prenatal test for this disorder.[22,23]

Unfortunately, both genetic counseling and prenatal diagnosis by testing for the fragile marker X in cultured cells are fraught with ambiguity. First, fragile X is difficult to demonstrate in some female carriers. Second, males may have the fragile X and not be retarded, or vice versa. Third, there are difficulties associated with expression of fragile X in tissue culture; a lack of expression may not ensure absence of the chromosomal abnormality.

Since fragile X is not a treatable disorder, genetic screening and prenatal diagnosis are critical components to its detection. Molecular genetic approaches may provide us with viable alternatives to the screening methods described above. The linkage studies described below have already revealed critical information about the mapping of the gene locus associated with the fragile X site and other closely segregating gene loci. Since the defective gene in the fragile X syndrome has not been identified, RFLPs of closely segregating genes may provide helpful diagnostic information.

The heritable fragile site associated with fragile X syndrome has been localized to Xq27.3.[24,25] Closely segregating with this site is the gene for glucose 6-phosphate dehydrogenase (G6PD). Segregation analysis has mapped the gene encoding G6PD to Xq28.[26] Recombinant DNA technology has provided additional linkage information, showing close association between the fragile X and coagulation factor IX genes; further analysis shows that these genes flank the fragile X site, lying between it and the gene encoding HPRT. A tentative mapping of the genes around the fragile X site on the long arm of the X chromosome is therefore:

xg/RC8/L1.28(centromere)DXYS1/S21/HPRT/52A/Fact. IX/*frag X* /G6PD/DX13 (telomere)

The apparent linkage for 52A, factor IX, and fragile X has been investigated further; it now appears that, since recombinations among these sites have been found they are not as closely linked as previously suspected. Thus, the close linkage of the fragile X site with factor IX may not be preserved in all families, and factor IX linkage may be an unreliable probe for genetic counseling and prenatal diagnosis.

5. APPLICATIONS OF MOLECULAR BIOLOGY FOR CHILD PSYCHIATRY

5.1. The Near Future

What can we expect to be the contributions of these new molecular technologies to child psychiatry for the immediate future. As this chapter is being written, groups of child psychiatry researchers have already begun using RFLPs in linkage studies for familial infantile autism, Tourette's syndrome, and affective disorders. RFLPs have important limitations, however. Most are linkage markers, not the target genes themselves, and they may be located many thousands or even millions of bases away from the actual gene. This limits their absolute accuracy as diagnostic tools. Furthermore, as illustrated with one of the PKU haplotypes, even if a mutant gene is present, it may not always be detected by RFLP analysis. Nonetheless, despite these and other limitations, linkage analysis using cDNA markers has become the standard first step in identifying candidate loci for genetic diseases. Clearly, the limitations of the method diminish or disappear altogether as the genes responsible for a disease are identified, localized, and cloned.

In the short term, the immediate value of recombinant methodologies will be in establishing the boundaries of a clinical phenotype and identifying maternal carriers and afflicted offspring within a particular disease category. Establishing the major characteristics of a clinical phenotype, as well as determining which other disorders might be related to it (e.g., autism and communication disorders) is essential but has heretofore not been possible in psychiatric diagnoses. The presence of a reliable biologic marker will be an immense aid in refining our nosology and extending our understanding of the disorders with which we are concerned.

Identification of carriers is necessary for reliable genetic counseling. It provides parents with an elective choice to initiate a pregnancy, or to carry an ongoing pregnancy to term. While this raises difficult and sometimes morally agonizing decisions, the physician's responsibility is to provide the best possible information so that parents may make the most informed choice. Molecular markers and gene-specific diagnostic techniques will provide data of a quality far superior to what we now have.

Early identification of afflicted children is also important because it permits development of intervention strategies which may be initiated before the developmental repertoire becomes immutable. Early intervention in developmental disorders is mandatory when corrective therapy is available, particularly when the treatment can prevent the normal adverse course of the disorder. PKU provides a classic example of the urgency of early intervention. With most developmental disorders, however, we must currently rely on the appearance of symptom markers before a definitive diagnosis is made. While necessary to avoid mislabeling and stigmatizing children, years of valuable time may be lost before treatment is instituted, time when CNS plasticity might permit unaffected brain areas to acquire functions that could compensate for impaired development of other centers.

5.2. A Look into the More Distant Future

In addition to early diagnosis and intervention, what are the practical benefits that molecular biology will offer child psychiatry in the longer term. Although our crystal ball is no clearer than anyone else's, it seems to us that the benefits will be in (1) an increased understanding of the cellular mechanisms gone awry in developmental disorders, and (2) various modes of gene replacement or gene attenuation therapy.

5.2.1. Understanding Cellular Mechanisms

Biochemical research in psychiatry has proceeded on the assumption that psychiatric diseases are diseases of metabolism, models of which are phenylketonuria or Lesch–Nyhan syndrome. All of our major hypotheses—the transmethylation hypothesis of schizophrenia, the biogenic amine hypotheses of affective illness, and most recently the dopamine hypothesis of schizophrenia—have been constructed around the assumption of a metabolic defect. However, there is not much evidence, despite an intensive research effort, to support a metabolic disease model for any psychiatric illness. Yet there is clear evidence that several major psychiatric disorders are genetic diseases, including some of the developmental disorders of interest to child psychiatry researchers.

Accordingly, it may be time to turn to other conceptual frameworks. Fortunately, alternative models abound. In mice, in which many neurologic mutants have been carefully mapped and described, most genetic disorders arise from mutations in neurohistogenesis, cell migration, or differentiation. For the most part, these mutations do not have metabolic consequences of the sort discussed earlier. In some cases, the mutations are expressed structurally; cells do not migrate, they fail to develop, or they develop and die. In other examples, the assembly of ion channel proteins is defective. In still others, arborization of dendrites is impaired. In theory, every step in CNS development can be the target of mutation; in practice, however, some proportion of these is likely to cause embryonic death. What we now appreciate from research in mouse neurogenetics is the diversity of processes which can be affected and how the consequences are expressed functionally. In our view, this research offers a rich source of information for researchers in search of models. We suggest that mutations in some of the developmental processes cited as examples above may well be implicated in the pathology of developmental disorders. Examples of the syndromes we have in mind include familial infantile autism, childhood schizophrenia, most retardation syndromes, unipolar and bipolar disorders, Tourette syndrome, and some attention-deficit and communication disorders.

There are other reasons that recombinant DNA techniques offer the best, and perhaps the only, way to understand the molecular genetics of developmental disorders. We are not likely to understand the biology of these disorders without some appreciation of the protein defects. These will be extremely difficult to decipher for psychiatric diseases, since in most cases we have no clue as to where to begin looking. Thus, the only way to make headway with the developmental disorders is by RFLP analysis, gene mapping, and ultimately understanding the structure of the candidate genes to identify their protein products and establish their role in cellular dysfunction. We need only cite the work in Duchenne's muscular dystrophy or in familial Alzheimer's disease in support of this contention.

5.2.2. Gene-Replacement Therapies

Perhaps one day in the not too distant future, molecular biology will also provide us with therapeutic strategies through the use of gene replacement therapy. Although this field is still in its infancy, progress is being made at an astonishing rate. Viral vectors are used to insert cloned genes into cells that lack the candidate gene altogether or that express a defective gene product. However, the frequency with which the transferred genes are expressed is relatively low, and many of the viral vectors are oncogenic; thus, this strategy is limited for the present to inserting genes into cultured cells or laboratory animals. In another strategy, foreign genes have also been stably incorporated and expressed in rodents through bone marrow transplantation using virally transformed marrow cells. Finally, it has been possible to create transgenic mice by inserting foreign genes into fertilized ova. A small number of such ova undergo normal maturation and development, and the mice exhibit incorporation of the gene into the host genome with expression of functional protein. Technical obstacles currently limit the practicality of gene replacement therapies. However, these limitations are rapidly being solved. As this chapter is being written, researchers are already seeking approval for limited experimentation on gene transfer in humans through transplantation of genetically altered bone marrow cells.

5.2.3. Gene Attenuation Therapies

Other forms of gene therapy may also be available in the future. Some therapies may take advantage of chemical modification of the gene. We know, for example, that the amount of methylation at specific regions of a gene are responsible for its expression. Usually, hypermethylation inactivates transcription while hypomethylation leads to gene activation. By regulating the extent of gene methylation, it may be possible to turn genes on or off selectively. This would be useful in dominant disorders, in which a single gene copy suffices for disease expression. Individuals with dominant diseases are usually heterozygous, so turning off the disease gene might permit expression of the normal gene.

Another form of gene attenuation has recently emerged. This new and exciting area of research involves the inhibition of gene expression with antisense RNA. The strategy takes advantage of the fact that when mRNA forms a double-stranded molecule it cannot be translated. In effect, this results in inhibition of gene expression. The object of this form of gene attenuation is to introduce into cells a single-stranded nucleic acid that is complementary to a specific mRNA.

Since mRNA encodes the information to direct protein synthesis, it is defined as a *sense* RNA molecule. By definition, an RNA molecule that is complementary to sense mRNA, and that usually does not encode information for directing protein synthesis, is termed *antisense* RNA. In theory, one can specifically inhibit expression of a given gene by introducing into the cell a single-stranded nucleic acid that is complementary to that gene's mRNA.

Suggestions that this method would be effective to inhibit eukaryotic gene expression come from several sources. First, it has been shown that several bacterial genes use antisense RNA as a natural means to modulate gene expression.[27,28] Second, antisense nucleic acids have been used to inhibit in vitro translation.[29,30] Finally, antisense nucleic acids have been successfully used to inhibit viral replication in animal cells.[30]

In 1984 two groups demonstrated that antisense RNA could inhibit gene expression in eukaryotic cells.[31,32] Several groups have now been able to inhibit gene expression in a number of species by injecting antisense RNA into either oocytes or embryos.[33–35] The field has expanded and now includes such exciting results as synthetic antisense oligonucleotides inhibiting oncogene expression[36] and antisense inhibition of gene expression in transgenic plants[37] and animals.[38] Furthermore, antisense RNA has the potential to inhibit expression of deleterious genes such as oncogenes or pathogenic viral genes and thereby serve as a therapeutic modality. Finally, it is possible that antisense RNA plays an important role in the regulation of gene expression in eukaryotic cells.

6. CONCLUSIONS

For most genetic diseases, and for all the psychiatric diseases for which a genetic etiology is likely, the protein or gene defect underlying the clinical disorder is unknown. Recombinant DNA technologies provide us with promising avenues for identification of gene defects and understanding cellular dysfunction in these disorders. The use of RFLP analysis has permitted the identification of the gene locus or closely associated loci for a number of diseases. RFLPs can now be used as proximate markers to identify carriers and affected individuals. Precisely locating the candidate gene after a chromosomal marker has been found has proven a laborious but achievable goal and remains the objective of all linkage work. Once the candidate gene is isolated, its structure can be determined and its product identified. These can be enzymes, structural proteins, or regulatory proteins. We then have the information and tools for further investigation of abnormal expression and developmental events leading to clinical dysfunction.

The cloned gene or cDNA will also provide a direct diagnostic tool for the disease carriers and disease state. We can expect that identification of RFLP markers or molecular cloning of cDNAs and genes will provide us with tools for pre- and perinatal diagnosis and genetic counseling. This will enable clinicians to identify maternal carriers and afflicted offspring with certainty. While initially such methods would have greatest application in the severe developmental disorders, in the longer term one would expect them to be used whenever there is a likelihood they would provide useful information.

The potential merits of gene-replacement therapies or treatment strategies that use antisense RNA to inhibit gene expression are clear, despite their technical obstacles, but ethical and emotional issues must be factored into decisions to use these techniques when the day arrives that they are available. Use of recombinant techniques for prenatal diagnosis, gene therapy or gene suppression raise important moral, ethical, and legal considerations. Obviously, the intent is that modern technologies cure genetic defects or provide symptom relief and peace of mind, but for some, this new knowledge and its implications may cause anguish and uncertainty, while for a few, they will be perceived as interference with divine or natural order. For the present, perhaps we should adjust to this uncertain future by letting molecular biology and recombinant DNA technology help us demonstrate the abnormalities associated with the devastating developmental disorders. At the same time, however, it is not too early to begin the critical societal discussions about where these paths will take us.

The past decade has witnessed an extraordinary period of growth in our understanding of the molecular basis of cellular function. This has been due to a powerful convergence of forces: the advent of recombinant DNA technology, the parallel emergence of neuroscience as a full-fledged field, the union of molecular biology and neuroscience into a new discipline of molecular neurobiology, and the impact of all these events on developmental neurobiology, changing it from an anatomic to a molecular science.

As a field, child psychiatry has not been as quick to embrace the tools of molecular biology as have other clinical specialties. However, we hope that after reading this chapter, the reader will appreciate that recombinant techniques offer enormous potential to researchers interested in genetic disorders of CNS development.

REFERENCES

1. Maniatis T, Goodbourn S, Fischer JA: Regulation of inducible and tissue-specific gene expression. Science 236:1237–1252, 1987
2. Ptashne M: How eukaryotic transcriptional activators work. Nature (Lond) 335:683–689, 1988
3. Wong DL, Ciaranello RD: Molecular Biological Approaches to Mental Retardation, in Meltzer HY (ed), Psychopharmacology, The Third Generation of Progress. New York, Raven, 1987, p 861–872

4. Ledley FD, Grenett HE, DiLella AG, et al: Gene transfer and expression of human phenylalanine hydroxylase. Science 228:77–79, 1985
5. Woo SL, Lidsky AS, Guttler R, et al: Cloned human phenylalanine hydroxylase gene allows prenatal diagnosis and carrier detection of classical phenylketonuria. Nature (Lond) 306:151–155, 1983
6. Woo SL, Guttler F, Ledley FD, et al: The human phenylalanine hydroxylase gene, in Berg K (ed): Medical Genetics: Past, Present, Future. New York, Alan R. Liss, 1985, pp 123–135
7. Lidsky AS, Guttler F, Woo SL: Prenatal diagnosis of classic phenylketonuria. Lancet 1:549–551, 1985
8. Tourian A, Sidbury JB: in Stanbury JB, Wyngarrden JB, Fredrickson DR (eds): The Metabolic Basis of Inherited Disease. New York, McGraw-Hill, 1983, pp 270–286
9. Levy HL: Maternal PKU, in Berg K (ed): Medical Genetics: Past, Present, Future. New York, Alan R. Liss, 1985, pp 109–122
10. Lesch M, Nyhan WL: A familial disorder of uric acid metabolism and central nervous system function. Am J Med 36:561–570, 1964
11. Holden JA, Kelly WN: Human hypoxanthine-guanine phosphoribosyltransferase: Purification and subunit structure. J Biol Chem 253:4459–4463, 1978
12. Kelly WN, Arnold WJ: Human hypoxanthine–guanine phosphoribosyltransferase: Studies on the normal and mutant forms of the enzyme. Fed Proc 32:1656–1659, 1973
13. Michelson AM, Markham AF, Orkin SH: Isolation and DNA sequence of a full length cDNA clone for human X chromosome-encoded phosphoglycerate kinase. Proc Natl Acad Sci USA 80:472–476, 1983
14. Jolly DJ, Esty A, Bernard HO, Friedmann T: Isolation of a genomic clone partially encoding human hypoxanthine phosphoribosyltransferase. Proc Natl Acad Sci USA 79:5038–5041, 1982
15. Jolly DJ, Okayama H, Berg P, et al: Isolation and characterization of a full-length expressible cDNA for human hypoxanthine phosphoribosyltransferase. Proc Natl Acad Sci USA 80:477–481, 1983
16. Becker MA, Yen RC, Itkin P, et al: Regional localization of the gene for human phosphoribosylpyrophosphate synthetase on the X chromosome. Science 203:1016–1019, 1979
17. McKusick VA: The human genome through the eyes of a clinical geneticist. Cytogenet Cell Genet 32:7–23, 1982
18. Yang TP, Patel PI, Chinault AC, et al: Molecular evidence for new mutations at the HPRT locus in Lesch–Nyhan patients. Nature (Lond) 310:412–414, 1984
19. Tariverdian G, Weck B: Nonspecific X-linked mental retardation—A review. Hum Genet 62:95–109, 1982
20. Turner G, Opitz JM: Editorial Comment: X-linked mental retardation. Am J Med Genet 7:407–415, 1980
21. Lubs HA: A marker X chromosome. Am J Hum Genet 21:231–244, 1969.
22. Jenkins EC, Brown WT, Duncan CJ, et al: Feasibility of fragile X chromosome prenatal diagnosis demonstrated. Lancet 2:1291, 1981
23. Rhoades FA, Oglesby AC, Mayer M, Jacobs PA: Marker X syndrome in an oriental family with probable transmission by a normal male. Am J Med Genet 12:205–217, 1982
24. DeArci MA, Kearns A: The fragile X syndrome: The patients and their chromosomes. J Med Genet 21:84–91, 1984
25. Brookwell R, Turner G: High resolution banding and the locus of the Xq fragile site. Genet 63:77, 1983
26. Szabo P, Purrello M, Rocchi M, et al: Proc Natl Acad Sci USA 81:7855–7859, 1984
27. Simons RW, Kleckner N: Translational control of IS10 transposition. Cell 34:683–691, 1983
28. Coleman M, Green PJ, Inouye M: The use of RNAs complementary to specific mRNAs to regulate the expression of individual bacterial genes. Cell 37:429–436, 1984
29. Paterson BM, Roberts BE, Kuff EL: Structural gene identification and mapping by DNA–mRNA hybrid-arrested cell-free translation. Proc Natl Acad Sci USA 74:4370–4374, 1977
30. Stephenson ML, Zamecnik PC: Inhibition of Rous sarcoma viral RNA translation by a specific oligodeoxynucleotide. Proc Natl Acad Sci USA 75:285–288, 1978
31. Izant JG, Weintraub H: Inhibition of thymidine kinase gene expression by anti-sense RNA: A molecular approach to genetic analysis. Cell 36:1007–1015, 1984
32. Rubenstein JLR, Nicolas J-F, Jacob F: L'ARN non sens (nsARN): Un outil pour inactiver specifiquement l'expression d'un gene donne *in vivo*. CR Acad Sci Paris 299:271–274, 1984
33. Melton DA: Injected anti-sense RNAs specifically block messenger RNA translation *in vivo*. Proc Natl Acad Sci USA 82:144–148, 1985
34. Bevilacqua A, Erickson RP, Hieber V: Antisense RNA inhibits endogenous gene expression in mouse preimplantation embryos: Lack of double-stranded RNA "melting" activity. Proc Natl Acad Sci USA 85:831–835, 1988
35. Rosenberg UB, Preiss A, Seifert E, et al: Production of phenocopies of *Kruppel* antisense RNA injection into *Drosophila* embryos. Nature (Lond) 313:703–706, 1985

36. Heikkila R, Schwab G, Wickstrom E, et al: A *c-myc* antisense oligoecoxynucleotide inhibits entry into S phase but not progress from G_0 to G_1. Nature (London) 328:445–446, 1987
37. van der Krol AR, Lenting RE, Veenstra J, et al: An anti-sense *chalcone* synthase gene in transgenic plants inhibits flower pigmentation. Nature (Lond) 333:866–869, 1988
38. Katsuki M, Sato M, Kimura M, et al: Conversion of normal behavior to shiverer by myelin basic protein antisense cDNA in transgenic mice. Science 241:593–595, 1988

BIBLIOGRAPHY

Readers interested in obtaining a more in-depth understanding of molecular biology will find the following texts useful:

Lewin B. Genes, ed 3. New York, Wiley, 1987

Watson JD, Hopkins NH, Roberts JW, et al: Molecular Biology of the Gene, ed 4. Menlo Park, California, Benjamin/Cummings, 1987

2

Neurotransmitter Assessment in Neuropsychiatric Disorders of Childhood

Ann M. Rasmusson, Mark A. Riddle, James F. Leckman, George M. Anderson, and Donald J. Cohen

1. INTRODUCTION

An appreciation for the potential role of anomalous neurobiologic processes in the etiology of childhood psychiatric illness did not emerge as a consistent theme of clinical research until the early 1960s. Before that time, etiologic constructs were premised on an interaction between psychodynamic factors and poorly defined constitutional factors. However, observation of the powerful effects of neuropharmacologic agents on behavior and growing knowledge of the functional neuroanatomy of neurotransmitter systems provided a strong impetus to biologically oriented research in the area of child and adolescent psychiatry. Since then, a pattern has emerged whereby clinical investigators incorporate newly identified and measurable components of the neural substrate into heuristic schemes of causation, in an attempt to advance our understanding of the pathophysiology of child psychiatric disorders.

The focus of early neurobiologic studies in child psychiatry on the possible role of serotonin and dopamine in the pathophysiology of autism is illustrative. Serotonin was discovered in the brain by Twarog and Page[1] and by Amin et al.[2] in 1953 and 1954, respectively. In 1954, lysergic acid diethylamide (LSD) was found to block peripheral serotonin receptors, leading Gaddum[3] and Woolley and Shaw[4] to suggest that serotonin not only might be a neurotransmitter but might also have a role in maintaining "sanity." Then, in 1961, Shain and Freedman[5] measured plasma serotonin (5-HT) and urinary 5-hydroxyindoleacetic acid (5-HIAA), the major metabolite of serotonin, in groups of children with autism and mental retardation. These investigators found high blood serotonin in the autistic subjects and raised the question of central nervous system (CNS) serotonergic dysfunction in this disorder. However, at that time, there were no methods to explore this hypothesis more directly.

In 1963, the technique of fluorescence histochemistry, developed by Falck et al.[6] in 1962 for measurement of catecholamines in brain tissue, was adapted for use in cerebrospinal fluid

Ann R. Rasmusson, Mark A. Riddle, James F. Leckman, George M. Anderson, and Donald J. Cohen • Yale Child Study Center, Yale University School of Medicine, New Haven, Connecticut 06510.

(CSF).[7,8] This technique was modified in 1968 to permit measurement of 5-HIAA and homovanillic acid (HVA), the major metabolite of dopamine, in small enough volumes of CSF to permit clinical studies in children.[9] In the meantime, the importance of CNS dopamine in mediating stereotypic behavior in animals had raised the question of dopaminergic involvement in autism.[10–12] Subsequently, the first study of CSF serotonin and dopamine metabolites in children with autism was reported;[13] results were inconclusive, however.

Investigation of the role of other neural entities in psychiatric disease has followed a similar pattern. During the 1970s, CNS peptide neuromodulators were first identified and measured using sensitive immunohistochemical assays.[14,15] Since then, they have come under clinical scrutiny in a variety of childhood disorders.[16–18] Currently, at the basic science level, the functions of second messenger systems[19–22] and neuronal ion channels[23] are being delineated. Consequently, clinical theories of pathogenesis have begun to invoke them.[24–26] Clinical probes of these molecular level systems in the CNS may be more difficult to develop, however.

Nevertheless, despite the imperatives of advancing neurobiologic technology, this chapter focuses on the clinical use of classic neurotransmitter and metabolite measurement as a means of enhancing our understanding of the biologic underpinnings of childhood psychopathology. This approach has been pursued by a small cadre of child psychiatric investigators in collaboration with clinically oriented neuroscientists for nearly three decades. A critical look at the results of this approach, and at the methods used to generate them, should illuminate areas of progress, as will as help in the design of future clinical neurobiologic studies.

2. REVIEW OF PAST FINDINGS

Unfortunately, past studies of neurotransmitter and metabolite levels in children with psychiatric disorders have often yielded variable or contradictory results. Some of this confusion may be due to the nature of the conditions studied, since many may be biologically heterogeneous or have complex and unstable phenotypes. In addition, a number of methodological difficulties limit the interpretive power of these studies. Therefore, in the interest of clarity and brevity, only results from selected disorders that constitute robust and reproducible departures from normative neurotransmitter and metabolite levels are presented. In some instances, these results are interpreted within the context of observations made using other clinical probes and in view of neurophysiologic mechanisms delineated at more basic levels of scientific inquiry.

2.1. Attention-Deficit Hyperactivity Disorder

Within each child psychiatric diagnostic category, invocation of the *neurotransmitter hypothesis* has followed a unique line of logic, often provoked by some well-documented effect of a psychoactive medication. In the case of attention-deficit hyperactivity disorder (ADHD), it resulted from the observation of the positive therapeutic effects of stimulant medications, such as methylphenidate and amphetamine.[27,28] The clinical neuropsychopharmacologic studies that followed then yielded concrete evidence for neurotransmitter system dysfunction in ADHD.[29]

The dopaminergic system was first implicated because of abnormalities in CSF HVA. Although baseline CSF HVA levels were found to be normal, administration of amphetamine caused a decline in HVA in the ADHD subjects.[30] Although it was suggested that the drug should have increased CSF HVA by increasing dopamine release, controls were not treated with amphetamine, and the decline in HVA may have been due to the monoamine oxidase

(MAO) inhibitory activity of amphetamine.[31] In another study, CSF HVA was measured after probenecid loading in a group of 6 drug-free boys with ADHD and in 26 controls undergoing evaluation for other disorders.[32] In this study, the CSF HVA-to-probenecid ratio was found to be significantly lower in the boys with ADHD, suggesting that dopamine turnover may be low in this disorder. However, information regarding dopamine receptor status was lacking; therefore, no judgment can be made regarding overall dopaminergic neurotransmission. Support for dopamine involvement in ADHD has also been provided by an animal model in which rat pups, given relatively selective dopamine neuron lesions by the intracisternal administration of 6-OH-dopamine,[33–35] develop hyperactivity. Breese et al.[36] observed functional dopamine receptor changes to occur in this model, and now, with the development of new scientific techniques such as intracerebral microdialysis,[37–38] an even more rigorous evaluation of its potential explanatory value may be permitted.

A variety of indirect evidence, however, argues for salient involvement of the noradrenergic system in ADHD (see Zametkin and Rapoport[39] for a thoughtful review). For instance, children with ADHD have greater vascular pressor responses on standing than controls, despite equal increases in plasma norepinephrine.[40] This is consistent with, but not conclusively indicative of, postsynaptic α-adrenergic receptor supersensitivity. In support of this interpretation are several studies that have shown ADHD subjects to have lower baseline urinary 3-methoxy-4-hydroxy-phenylglycol (MHPG) excretion than that of normals.[41–44] However, normal urinary MHPG[45] and increased urinary norepinephrine[46] have been observed in other studies.

Pharmacologic profiles of response in ADHD also support the noradrenergic hypothesis.[39] Symptoms have been found to be responsive to drugs that augment the neurotransmitter content of monoaminergic synapses nonselectively: dextroamphetamine,[47] methylphenidate,[27,48] and the MAO-A inhibitors, clorgyline and tranylcypromine,[49] as well as to the relatively specific norepinephrine reuptake blocker, the tricyclic antidepressant (TCA), desipramine.[48,50–53] Such agents as levodopa, mianserin, piribedil, and tryptophan, which affect dopamine and serotonin systems more selectively, are ineffective.[39] In addition, dextroamphetamine, the MAO inhibitors, and desipramine, which are among the most effective agents in the treatment of ADHD, reproducibly suppress MHPG levels in plasma and urine.[41,42,52,54–56]

Initially, then, it might seem that the clinical mechanism of action of these drugs is to decrease CNS noradrenergic activity. The increases in synaptic norepinephrine induced by these agents may exert inhibitory feedback effects on presynaptic neurons and, in fact, MAO inhibitors do acutely decrease the firing rate of locus ceruleus neurons.[57] However, increases in synaptic norepinephrine could also increase noradrenergic neurotransmission via postsynaptic effects. Linnoila et al.[58] suggest, for instance, that although desipramine decreases CNS norepinephrine turnover, it increases the efficiency of noradrenergic neurotransmission.

An important issue of interpretation is at stake but, unfortunately, there are not sufficient relevant clinical data from which to draw firm conclusions. This is underscored by the fact that while the noradrenergic system is implicated in ADHD,[39] it is not consistently cited for either overactivity or underactivity. Indeed, the lack of direct information regarding drug-induced changes in noradrenergic receptor sensitivity cautions against overinterpretation of the data.

It may be useful, then, to consider relevant observations made at the basic science level. Conventionally, changes in receptor sensitivity have been thought to counter changes in neurotransmitter availability in synapses homeostatically; thus, increases in the synaptic content of a neurotransmitter or its pharmacologic agonist are associated with decreases in receptor number or sensitivity, while decreases in the synaptic content of a neurotransmitter or its agonist are associated with increases in receptor number or sensitivity. However, such processes tend to confine neurotransmitter systems to operation not only within fixed parameters, which in some cases might be pathologic, but they could even thwart therapeutic intervention, since receptor alterations might ultimately counter intended drug effects on neurotransmitter release.

However, it was recently demonstrated that such homeostatic processes do not universally operate. In a rigorous electrophysiologic study of the acute and long-term effects of antidepressants on serotonergic neurotransmission, Blier et al.[57] detailed coordinated changes in pre- and postsynaptic receptor sensitivity, neuronal firing rate, and neurotransmitter release. Interestingly, while the various antidepressants had different effects on pre- and postsynaptic receptors, the overall effect was to enhance the efficiency of serotonergic neurotransmission.

For example, selective serotonergic reuptake blockers acutely decreased presynaptic neuron firing rate, while chronic administration led to desensitization of presynaptic somatodendritic receptors, so that serotonergic neuron firing rate eventually returned to normal. Chronic treatment also caused downregulation of the terminal autoreceptors of presynaptic neurons, enhancing the amount of serotonin released per impulse. In addition, the functional sensitivity of postsynaptic receptors was not altered. Thus, the net effect was an increase in serotonergic tone.

Similar strategies could be employed to study the timing and directionality of the impact of ADHD-effective medications on noradrenergic neurotransmission. The results might then permit an assessment of the noradrenergic hypothesis in this disorder. Perhaps the animal model of ADHD developed by Shaywitz et al.[33] could be adapted for this purpose. It is intriguing, for instance, that hyperactivity and cognitive deficits induced in rats by lesioning the mesolimbic dopamine system resolve with simultaneous lesioning of the ventral noradrenergic bundle.[59] In addition, although functional comparison of the noradrenergic system in this model[33,34] has not been made to that in human ADHD subjects, monkeys with methylphenyltetrahydropyridine (MPTP) lesions of the dopaminergic nigrostriatal system have increased norepinephrine turnover.[60,61] Positron-emission tomography (PET) studies of noradrenergic receptor status in children with ADHD may also prove useful; however, radiolabeled noradrenergic receptor ligands have not yet been developed,[62] and systematic application of this technique to children is still not possible.[63]

2.2. Autism

In autism, by far the most robust and well-replicated neurobiologic derangement found thus far, is the presence of hyperserotonemia, affecting 30–40% of patients. There is not, however, clear-cut evidence to suggest that a similar aberration exists in the CNS of these patients. Studies of CSF 5-HIAA with and without probenecid have revealed levels to be normal or only slightly low. Nevertheless, the serotonin hypothesis of autism continues to be explored.[64]

Derangements suggested to account for the hyperserotonemia in autism have included (1) abnormalities in the rate of serotonin degradation by MAO, (2) abnormalities in platelet transport and storage of serotonin, and (3) abnormalities in tryptophan metabolism.[64] Any of these potential peripheral abnormalities could similarly affect central serotonergic activity. Alternatively, an abnormality in tryptophan metabolism restricted to the periphery could secondarily affect CNS serotonergic activity, since peripheral free tryptophan levels influence tryptophan transport into the CNS and subsequent central 5-HT production.[65] In addition, high peripheral tryptophan can augment CNS production of toxic metabolites, such as quinolinic acid.[66]

Thus far, however, peripheral catabolism of 5-HT has been found to be normal in autism,[67,68] while tryptophan-loading studies have produced conflicting and confusing results.[67] Thus, a defect in platelet transport and storage of 5-HT now appears more likely to be the pathophysiologic mechanism, since it could account for hyperserotonemia in the face of normal to low 5-HIAA production. However, since studies of platelet uptake of 5-HT have been inconclusive,[64] there is now speculation that either intraneuronal vesicular incorporation or release of 5-HT, or both, might be problematic in this disorder.

Todd and Ciaranello[69] used an immunologic approach in the search for the cause of hyperserotonemia in autism. Interestingly, these investigators found anti-5-HT, la antibodies in the blood of a drug-naive autistic child, while general antibrain antibodies were not elevated in a group of 20 autistic children as compared with depressed, retarded, and normal controls.[70] Todd has therefore proposed an immunologic injury hypothesis to account for these laboratory findings.[71] Although the replicability of the finding of anti-5-HT, la antibodies in autism is yet to be established by a more thorough study now under way (R. Todd, personal communication), it is interesting to speculate about how such an aberration might be related to hyperserotonemia: perhaps vesicular incorporation or release of 5-HT, or both, are normally coupled to 5-HT, la receptor binding by intracellular second messenger processes.

Other neurotransmitter systems have also drawn scrutiny in autism. CSF HVA has been reported to be low, high, and, most recently, normal in studies with and without probenecid.[64] Nevertheless, involvement of the dopaminergic system in autism is supported by the response of autistic subjects to haloperidol and possibly fenfluramine[72] and by animal studies in which pharmacologic manipulation of dopaminergic neurotransmission induces hyperactivity and stereotypies,[10–12,33,34,73] symptoms commonly seen in this disorder.

Involvement of the opiate system in autism has also been considered.[74] Plasma humoral endorphin levels were found to be low in a group of 10 autistic subjects compared with normal controls, while levels in a co-morbid group of subjects with childhood schizophrenia fell between those in the autistic and normal control groups.[16] By contrast, 55% of a group of 20 autistic children had higher CSF fraction II endorphin levels than any of eight normal controls.[17]

The proposed variety of neurotransmitter and metabolite abnormalities in autism and the disparate findings of various clinical studies suggest that either multiple biologic problems underlie a relatively homogeneous behavioral profile or that several neurotransmitter systems are affected by a single underlying anomaly. The first possibility is supported by the association of autism with several different genetic neurobiologic disorders and with metabolic or infectious diseases that have prenatal or perinatal periods of activity.[75] The second possibility is supported by the same evidence viewed in light of recent findings in basic developmental neurobiology.

During early critical periods of brain development, an isolated neurotransmitter aberration might effect changes in the function of multiple neurotransmitter systems by altering patterns of neuronal connectivity. Haydon et al.,[76] for instance, demonstrated that serotonin can inhibit both growth cone motility and the subsequent formation of synapses in selected neurons. This effect, as well as effects of action potentials on growth cone motility and neurite elongation, appear to be mediated by local alterations in growth cone calcium levels.[77] Thus, it is conceivable that even a subtle defect in CNS serotonin storage and release might have ramifying effects on early synaptic organization of the brain and produce a devastating neuropsychiatric disorder such as infantile autism. It is also conceivable that certain subsets of autistic patients might have primary aberrations in other neurotransmitter,[78] second messenger, or ion-channel systems that lead, via similar mechanisms active during early development, to devastating states of CNS disorganization.

2.3. Tourette's Syndrome

The evidence for involvement of multiple neurotransmitter systems in mediating the symptoms of Tourette's syndrome (TS) is quite strong. It is based both on the clinical response of patients to medications that influence particular neurotransmitter systems and on findings of altered levels of neurotransmitter metabolites in the CSF, blood, and urine of these patients.

The dopamine system has been most consistently implicated. Patients' tics respond to dopamine receptor antagonists such as haloperidol, pimozide, penfluridol, and fluphenazine.[79,80] Conversely, they may be precipitated by neuroleptic withdrawal[81] and may worsen on exposure to agents that acutely enhance central dopaminergic activity.[82,83] In addition, there have been several independent reports of decreased baseline and postadministration probenecid CSF HVA levels in TS patients as compared with controls.[84] Therefore, it has been speculated that dopamine receptor hypersensitivity exists in TS. However, since tic symptoms improve, but are not eradicated by pharmacologic manipulation of the dopamine system, there may be other factors with neuromodulatory effects that act in synergy with dopamine to produce tic symptoms.

For instance, dysfunction of the noradrenergic system may also occur in TS. Urinary MHPG levels have been found to be low in these patients[84] and may indicate the presence of pre- and/or postsynaptic receptor hypersensitivity. By contrast, the presence of normal to high CSF MHPG levels[84] indicates that central receptors may be in a normal or downregulated state. Unfortunately, as CSF levels were not corrected for plasma MHPG levels,[85] they may not provide a valid indication of central norepinephrine release.

Additional but controversial evidence for noradrenergic dysfunction in TS is that clonidine, an α_2-adrenergic autoreceptor agonist, may be somewhat effective for some patients.[86] As this drug decreases norepinephrine release and turnover,[87,88] it may either decrease noradrenergic neurotransmission or possibly increase it by inducing hypersensitivity of postsynaptic receptors. In fact, abrupt withdrawal of clonidine in TS subjects is associated with increases in blood pressure, plasma MHPG and HVA, and urinary catecholamines.[89] This finding is consistent with drug-induced postsynaptic supersensitivity and/or presynaptic receptor subsensitivity. However, as pointed out,[90] clonidine also has effects on the dopaminergic and serotonergic systems.[91,92]

Until recently, serotonergic system involvement in TS seemed unlikely, since clinical responses to drugs with serotonergic effects were inconsistent.[84] However, CSF 5-HIAA levels are normal to low in TS[84] and whole blood tryptophan is low.[93] In addition, preliminary experience suggests that obsessive-compulsive symptoms in TS respond to fluoxetine, a specific serotonin reuptake blocker.[94] As this drug has been shown to enhance serotonergic neurotransmission by downregulating presynaptic 5-HT receptors,[57] underactivity of the serotonergic system may contribute to the TS symptom complex.

Clinical observations of dysfunction of multiple classic neurotransmitter systems in TS does not, however, implicate them all in the primary pathophysiologic process. Research at most basic levels has demonstrated normal interactions among these systems,[91,92,95–100] as well as between the dopaminergic and dynorphinergic systems.[101–103] Thus, a primary functional abnormality in one system could cause functional abnormalities in others. The precise mechanisms by which these neurotransmitters and neuromodulators interact, however, are far from clear. They may relate in tandem fashion or may independently exert effects on common postsynaptic neurons. There is also growing evidence for postsynaptic co-modulatory effects of co-localized and co-released classic neurotransmitter and peptides.[104]

The pattern of co-localization of these substances is not constant; a single peptide may be co-localized with a variety of classic neurotransmitters, and vice versa.[104,105] Thus, a problematic peptide system could make its presence felt in a number of neurotransmitter systems, which might then mediate TS symptoms on a secondary basis. In addition, co-release from presynaptic neurons may depend on variations in the pattern of neuronal firing,[104] which in turn may depend on environmentally related conditions, such as stress.[106–108] Thus, aberrant neuromodulation might be inapparent, except under conditions in which co-release of a neurotransmitter and a peptide should occur.

The complexity of these potential presynaptic neuronal responses can be magnified even further at the postsynaptic neuron.[109] A number of different postsynaptic receptors can modulate the same ion channel either by sharing the same second messenger systems or by exerting effects on different second messenger systems, which then converge on the same channel. Alternatively, neurotransmitters released from presynaptic neurons may interact with more than one receptor subtype on a single postsynaptic neuron and induce diverging effects. As these receptor subtypes are variably sensitive to neurotransmitter concentration, the amount of neurotransmitter released by presynaptic neurons will influence the balance in these divergent postsynaptic effects.

Owing to this complexity in synaptic processing of neurochemical information, it may be too simplistic to conceive of the pathophysiology in TS as a "neurotransmitter" problem, per se. The basic pathology could as likely lie at the level of second messenger signal transduction or in the structure of the ion channels themselves. There could also be errors in synaptic organization. For instance, one could conceive of indiscriminately formed neuronal reentry loops[110,111] that would permit reverberation of nerve impulses in thought, motor, or thought/motor pathways to precipitate obsessions, tics, or compulsions. Depending on the type of loop involved, there would appear to be different neurotransmitters mediating the symptoms.

Obviously, work in animal models and in in vitro brain preparations will have to proceed further before the relevance of such hypotheses to TS is established or rebuked. Unfortunately, progress may be slow. Within the striatal and limbic systems alone, there are myriad variations in peptide/neurotransmitter co-localization.[104,105,112,113] In addition, the relationships between neuronal firing and the patterns of peptide/neurotransmitter release, as well as the patterns of postsynaptic signal processing, are just beginning to be defined. If the basic problem lies at the levels of synaptic organization,[110,111] scientific scrutiny may never be close enough to avail us of the underlying pathology.

That is not to say, however, that investigations should proceed without great hope, as there is evidence to suggest that the activity of the peptide dynorphin is abnormal in TS.[18,116] Following this clue or others into the realm of molecular genetics could lead to satisfying answers sooner than expected (see Chapter 14, *this volume*).

3. METHODOLOGY

3.1. Research Design

3.1.1. Motivation

The neurobiologic approach to childhood psychopathology has been built on the assumption that progressively reductionistic methods will locate the primary pathologic loci of neuropsychiatric disease. However, faith in the descent to the molecular level has never guaranteed immediacy of results and, indeed, until now, our scientific intention has largely been frustrated. However, even methods that do not demonstrate primary pathology do not need to be considered fruitless or misguided. Rather than revealing ultimate clues to disease causality, these methods may reveal "biologic markers" that (1) signal a physiologically and behaviorly distinct subgroup of a single diagnostic population, e.g., low CSF 5-HIAA indicates the potential for violent behavior in depression and other psychiatric disorders[117]; (2) predict response to a particular treatment modality, e.g., normal blood serotonin and high IQ in autistic patients predict a positive response to fenfluramine[118]; (3) relate closely to a particular behavioral symptom, e.g., subnormal metabolic activity in the dorsolateral prefrontal cortex of schizophrenic subjects who perform poorly on the Wisconsin Card Sort Test[119]; or (4) identify children with neuropsychiatric disease before the manifestation of symptoms, and thus supplant less reliable, subjective diagnostic procedures.[120]

In the first three cases, the detected abnormalities are probably distant manifestations of the primary loci of disease and are therefore subject to the influence of a much greater number of biologic variables than if they arose at points more proximal to primary etiologic loci. Thus, these abnormalities tend to be complex and variable in expression. Although they may mark clinically relevant attributes of underlying disease, they often fail to robustly co-segregate with particular diagnoses. At the other extreme, and as illustrated by the fourth case, are genetic markers, which have the potential to pinpoint precisely primary etiologic loci of certain neurobiologically based disorders.[121] However, precision is gained at the cost of information about the complex series of neuropsychiatric events that result from genetic defects. Nevertheless, genetic markers promise to be extremely useful in unraveling the pathophysiology of psychiatric disease. This derives from the fact that they are stable and can therefore be used as diagnostic gold standards, to help in sorting out the complex, unstable, phenotypic expressions of the diseases that they mark.

3.1.2. Caveats

The quality of all diagnostic markers depends on the design of the studies from which they are derived. Thus, it may be useful to consider common pitfalls in this process. These have been clearly and concisely summarized in the medical literature[122] but bear reiteration. Observation of these caveats during the initial development of such diagnostic tools as the dexamethasone suppression test, for instance, might have saved investigators from inappropriate early optimism and subsequent disillusionment.

The foremost problem in the design of diagnostic marker studies is bias in recruitment, evaluation, and follow-up of study subjects, such that the spectrum of disease examined is inadequate. The potential for this bias is particularly high in pediatric studies because of hesitation in subjecting children to invasive or risky procedures. It is ethically problematic to ask children to serve as normal-matched or co-morbid controls in invasive studies of disease from which they do not suffer. It is likewise untenable to involve those with disease if the ratio of potential benefit to pain and risk is not high. Therefore such studies tend to be done only on patients manifesting the most extreme forms of a disorder. As discussed by Goldberg,[123] such selective scrutiny of the gaussian extremes of disease tends to distort the statistical significance of findings. In addition, it sabotages the value of biologic markers.

The value of a diagnostic marker depends on its predictive accuracy, which, in turn, is determined by its sensitivity and specificity for the disorder or attribute of interest. When a marker is 100% sensitive, all affected patients have it. To determine sensitivity, a large population of patients with the disorder or attribute in all its configurations must be tested. For example, a marker may appear only in the late or severe stages of a disease, or its expression may be modified by other factors, such as age, sex, or stage of development. If only subjects with these characteristics are tested, the prevalence of the marker will be high, and its sensitivity will be inflated. Subsequent diagnostic use of the marker in populations with characteristics other than those of the initial study group may then yield many false-negative results, rendering the marker useless.

When a marker is 100% specific, it is present only in those patients with the disorder or attribute of interest. Accurate establishment of specificity then depends on the use of appropriate controls, including those with common co-morbid characteristics. Otherwise, it may not be discovered that a marker actually indicates the presence of a characteristic such as age or sex instead of the disorder or attribute of interest. In other words, use of the marker may yield too many false-positive results. In neuropsychiatric disease, the psychological as well as scientific costs of such diagnostic errors may be very high and certainly are to be avoided.

In addition to problems of spectrum, however, diagnostic marker studies in psychiatry are be complicated by an additional factor. Specification of sensitivity and specificity involves comparing marker presence with disease presence. Therefore, correct diagnosis is essential. Since psychiatric disease is conventionally defined by observable behavioral attributes, "gold standard" diagnostic techniques consist of systematic clinical evaluation of patients by reliable diagnosticians, using standardized procedures. However, disease may be latent or obscured by co-morbid conditions, so that correct diagnoses, and their association with markers, may require repeated evaluation and invasive testing of subjects over time. In pediatric populations, this is again problematic because of ethical constraints, arguing for the use of noninvasive procedures.

Perhaps it will be useful to consider both the merits and the caveats of biologic marker studies in the following discussion of the clinical methods used to detect and measure neurotransmitters and metabolites. It will become evident that these methods are intrinsically limited in their ability to inform us about CNS processes, but enthusiasm for their use should also be determined by the degree to which they can be practically applied within the constraints of good diagnostic marker study design.

3.2. Measurements in Cerebrospinal Fluid

3.2.1. General Issues of Validity and Reliability

3.2.1a. Normative Data. The interpretation of baseline neurotransmitter or metabolite levels in subjects with psychopathology requires comparison of these subjects with normal controls. Unfortunately, because of ethical constraints, data from normal children are difficult to obtain. Thus, the limited number of normative CSF studies that have been done have tended to use subjects undergoing spinal taps for medical or psychiatric conditions not thought to be associated with abnormalities in the particular neurotransmitter parameters under study. While these assumptions are apparently reasonable, they cannot be tested. The findings of these studies should then be perused with this major limitation in mind.

Cephalocaudal concentration gradients. The results of a pediatric study of the cephalocaudal concentration gradients of the three classic neurotransmitter metabolites were generally consistent with those obtained in nonhuman primate and adult studies.[124–128] For instance, in 26 children and adolescents with obsessive-compulsive disorder (OCD), ADHD, or conduct disorder,[129] variable cervical to lumbar concentration gradients for HVA, 5-HIAA, and MHPG were demonstrated. These were manifest as increases in the concentration of metabolite in successive CSF aliquots collected from subjects lying in the lateral decubitus position. The slope of the HVA gradient was twice that for 5-HIAA and more than 10 times that for MHPG. The large gradient for HVA is consistent with the lack of HVA production by the spinal cord, the lack of free diffusion of this compound across the blood–brain barrier, and its active transport from the CSF into the blood.[130–132] The lack of a gradient in the case of MHPG is explained by the free diffusion of this metabolite across the blood–brain barrier,[85] while the "middling" gradient for 5-HIAA is consistent with spinal cord production of this metabolite within the context of an active CSF to plasma transport process and absence of free diffusion.[132,133] In summary, it is important to control for the specific aliquot of CSF in which metabolite measurements are made.

Developmental effects. The study just described[129] also revealed developmental criteria that bear on the interpretation of CSF metabolite levels in children. Absolute concentrations of 5-HIAA and HVA were decreased in pubertal as compared with prepubertal subjects, and the cephalocaudal concentration gradient for 5-HIAA correlated positively with Tanner stage. Developmental effects were likewise seen in a study of 50 boys and girls between the ages of 3 and 17, with acute lymphoblastic leukemia in remission.[134] CSF HVA and 5-HIAA were found to decrease with age, and to be independent of sex. In either study, these results could be explained by developmental changes in either metabolite synthesis, active transport from the CSF, or both.

Circannual variations. Seasonal fluctuations in CSF monoamine metabolites may also need to be taken into account. In a study of 20–34 normal adults extending over 3 years, Brewerton et al.[135] found that CSF 5-HIAA and HVA levels were significantly lower during spring. Unfortunately though, circannual variations in CSF metabolites have not yet been studied in children.

Diurnal variations. Normative data on diurnal variations in CSF monoamines and metabolites are limited to those derived from animal studies.[136] Norepinephrine, which does not cross the blood–brain barrier, was measured in the lateral ventricular CSF of adult male rhesus monkeys and found to be highest during the light hours and lowest during the dark hours. By contrast, MHPG and VMA were variable throughout the day. HVA seemed to reflect changes in synthesis in the caudate putamen that occurred approximately 6 hr earlier. It therefore peaked in lateral ventricular CSF toward the end of the 12-hr light period and approached its nadir toward the end of the 12-hr dark period. Although similar studies have not been done in children, these results nevertheless suggest that time of day should be controlled in investigations of CSF neurotransmitter metabolite levels.

3.2.1b. Intrinsic Meaning. Even if adequate normative data were available for CSF neurotransmitter and metabolite levels in children, what would they tell us about the function of the CNS? CSF measurements satisfy the logical desire of investigators to get as close as possible to the brain. Implicit in this desire is the belief that transmitter levels reflect the functional activity of the individual neurotransmitter systems. However, a number of factors bear on the validity of this assumption. Some are relevant to all neurotransmitter systems, while others apply uniquely to individual systems.

For instance, do CSF metabolite levels primarily reflect metabolite production or the rate of metabolite egression from the CSF into the plasma? When clinical CSF studies were first done, neurotransmitter levels were below the limits of detection, and metabolite levels were just at the limits of detection. Therefore, levels of the acidic metabolites of dopamine and serotonin, HVA and 5-HIAA, were augmented by the use of probenecid, a drug that blocks the active transport of these acidic metabolites from the CSF into the plasma.[132] Using this technique, Bowers[137] found that baseline acidic metabolite levels correlated positively with post-probenecid levels; he concluded that both baseline and post-probenecid CSF levels primarily reflect CNS metabolite production. (In the case of neutrally charged MHPG, interpretation of CSF levels is more complicated because of the free diffusion of this compound across the blood–brain barrier.)

However, since the level of probenecid attained in the CSF was subsequently found to vary from person to person and thereby variably augment acidic metabolite levels, it was later recommended that CSA HVA and 5-HIAA levels be expressed as metabolite to probenecid level ratios.[138] Recently, though, technologic progress has obviated the need for probenecid. Baseline CSF monoamine metabolite levels are now easily measured via high-pressure liquid chromatography with electrochemical detection (HPLC-ec) or gas chromatography with mass spectrometry (GC–MS).[139] Even the monoamines themselves can now be measured, though with greater difficulty.

Since CSF HVA and 5-HIAA levels do then seem generally to reflect CNS metabolite production, it has been assumed that they also reflect CNS neuronal "activity." However, as pointed out by Commissiong,[140] that may not be the case, since neurotransmitter release and metabolite production is not always coupled to neuronal impulse flow. For instance, an increase in the availability of monoamine precursors can enhance neurotransmitter synthesis and degradation without concomitantly increasing either neuronal firing rate[141] or neurotransmitter release.[142–144] By contrast, it has been shown that an increase in neuronal impulse flow and coupled neurotransmitter release does enhance the mass effect of increased precursor availability on neurotransmitter synthesis and metabolism.[145–148] In addition, it was recently demonstrated that increased precursor availability can enhance neurotransmitter release in the absence of increased neuronal firing rate.[149,150]

3.2.1c. Lack of Neuroanatomic Specificity. In addition to the fact that CSF neurotransmitter metabolite levels do not fully inform us about the general neurophysiologic activity of CNS neurons, they also provide little information about the functioning of subdivisions of the neurotransmitter systems.[139] An abnormal CSF metabolite level could represent a subtle aberration present throughout a neurotransmitter system or a substantial aberration in a subdivision of that system. On the other hand, a normal metabolite level does not rule out the presence of a subtle, but functionally significant regional abnormality in neurotransmitter turnover.

3.2.1d. Relation to Etiology. Abnormal CSF neurotransmitter or metabolite levels may also not yield much information about disease etiology. They may reflect primary pathophysiologic processes, or they may be the product of compensatory or destabilizing secondary processes. In addition, abnormal levels may not be causally related to abnormal behavior.

3.2.1e. Drug Effects. Certain drugs have been shown to exert independent influences on metabolite levels and neuronal firing. For instance, haloperidol, other neuroleptics, amphetamine, yohimbine, and various other drugs inhibit transport of dopamine metabolites from the brain.[151] Thus, a rise in CSF metabolite levels induced by these drugs does not necessarily reflect an increase in neurotransmitter release. Likewise, decreases in plasma metabolite levels associated with these drugs may be accounted for in part by decreased transport of metabolites from the CNS.[152]

3.2.2. Issues Specific to Individual Monoamine Systems

3.2.2a. Dopamine. Other factors bear on the degree to which individual CSF metabolite levels reflect CNS neurotransmission. The case for dopamine metabolites, for instance, is complicated, but possibly the most satisfying. While dihydroxyphenylacetic acid (DOPAC) is the primary CNS metabolite of dopamine in rats,[153,154] HVA is the primary metabolite of dopamine in primates, including man.[125,155] There is little diffusion of HVA across the blood–brain barrier: about 1% of peripherally administered HVA contributes to CSF levels.[130,131] In addition, brain capillaries[156] and other non-neuronal structures[157] synthesize HVA and contribute small amounts to the CSF pool. In nonprimates, diffusion of HVA from the portions of the caudate nucleus lining the lateral ventricle appears to constitute the major source of CSF HVA. Sourkes[158] estimates that this reflects about 30% of the dopamine turnover of the caudate, while the remainder of striatal HVA is presumably removed via brain capillaries.

In primates, however, HVA produced in cortical regions may also contribute in large part to both CSF and plasma HVA levels.[119,159,160] While the concentrations of dopamine and metabolites in the cortical layers are much lower than that of the subcortical structures, the total cortical mass in primates is very large.[119,158,161–163] Furthermore, Elsworth et al.[160] found that, in vervet monkeys, cisternal CSF HVA levels correlated with HVA in the dorsal frontal cortex, but not that in the basal ganglia. However, this relationship may be diminished in lumbar CSF because of the active transport of HVA from the CSF.[132] As virtually no dopamine is produced by the spinal cord, there is a large HVA concentration gradient between the ventricular, cisternal, and lumbar CSF compartments, with the ratios between these being approximately 10:4.5:1.[124,125]

Another factor also uniquely affects interpretation of CSF HVA levels. Although CSF HVA is generally thought to reflect CNS dopaminergic activity, this is not wholly true, as noradrenergic neurons also contribute HVA to the CSF; as much as one half of the dopamine in these neurons may be converted by intraneuronal MAO to HVA.[164–166] The degree to which noradrenergic neuron-derived HVA may contribute to CSF HVA has been indirectly addressed in a recent study by Kopin et al.,[60] where CSF HVA levels in three rhesus monkeys were observed to fall 15% after administration of debrisoquin, a peripheral MAO inhibitor. It was suggested that this was due to a decline in the HVA contribution by peripheral noradrenergic neurons innervating the choroid plexus. In another experiment, MPTP, a nigrostriatal dopaminergic neuron toxin,[167] was administered, after which CSF HVA levels were observed to decrease to approximately 15% of normal.[60] This residual CSF HVA therefore may have derived from the noradrenergically innervated choroid plexus, residual functional CNS dopaminergic neurons, or central noradrenergic neurons. As subsequent debrisoquin treatment did not further suppress CSF HVA levels, the latter two sources would seem to be implicated.

3.2.2b. Serotonin. Like HVA, 5-HIAA, the major metabolite of serotonin, does not readily cross from the plasma into the CSF.[132,142] Initially, it was suggested that approximately 70% of lumbar CSF 5-HIAA is derived from the brain parenchyma, while 30% is derived from the spinal cord.[168,169] However, lumbar CSF levels do not substantially change after spinal cord transection or after blockage of CSF flow in the spinal canal.[170] In fact, based on carefully executed flow studies, Bulat[133] claimed that virtually all lumbar 5-HIAA is locally produced. Therefore, as the weight of evidence seems to favor the latter position, it seems that lumbar CSF levels at best reflect central serotonergic functioning, rather than serotonergic activity of the brain per se.[171]

3.2.2c. Norepinephrine. The case for the validity of CSF MHPG, the major metabolite of norepinephrine in the CNS, as an indicator of central noradrenergic activity is less certain, because this nonpolar compound freely diffuses between the CSF and plasma.[85] Indeed, several animal and human investigations have demonstrated a substantial and significant correlation between them.[85,172] Kopin et al.,[85] for instance, found a very high correlation ($r = 0.998$, $p < 0.001$) and suggested that CSF MHPG levels reasonably reflect CNS MHPG production, as long as they are corrected for plasma MHPG:

$$[C] = ([C] - 0.9\ [P])$$

where $[C]$ is CSF MHPG concentration and $[P]$ is plasma MHPG concentration. Linnoila et al.,[58] however, suggest that this high correlation was due to inclusion of data from pheochromocytoma patients, who produce very high peripheral MHPG, which overwhelms the relatively small and constant CNS MHPG contribution to CSF levels. In their study of the relationship between CSF and plasma MHPG in a group of 21 drug-free depressed patients, there was a positive but nonsignificant correlation between them: ($r = 0.41, p > 0.05$). These investigators concluded that CSF MHPG cannot be used as an indicator of CNS MHPG production in individual patients.

Nevertheless, to try to distinguish between CNS and plasma derived MHPG in the CSF may not only be difficult, but also ill advised. As discussed by Maas and Leckman,[172] the CNS and peripheral noradrenergic systems seem to operate in a coordinated manner. Linnoila *et al.*[58] in fact advocate the integration of simultaneous peripheral and central measures in order to monitor intra-individual fluctuation in CNS noradrenergic activity in response to medication.

3.3. Measurements in Peripheral Body Fluids

The use of peripheral measurements of neurotransmitters and metabolites, as a means of scrutinizing CNS neurotransmission, has garnered the intense interest of many investigators, in good part due to the inconvenience and humanistic concerns associated with CSF studies. Clearly, experience has heightened appreciation for the many caveats that make interpretation of CSF studies difficult. The interpretation of peripheral indicators of CNS functioning is even more difficult, however. As experience has made clear, the caveats that apply to CSF studies also pertain to peripheral studies, as do additional confounds that arise because of the activity of the peripheral nervous system.

3.3.1. Dopamine

Opinion regarding the best peripheral correlate of central dopaminergic activity continues to evolve. HVA in the CNS and peripheral nervous system (PNS) is produced by the action of MAO and catechol-*O*-methyltransferase (COMT) on dopamine. The HVA produced centrally constitutes approximately one half or less of plasma homovanillic acid (pHVA),[173–176] while the remainder is derived from sympathetically innervated peripheral organs such as the heart, kidneys, and salivary glands.[177] Plasma HVA is therefore influenced both by central dopamine turnover and by the discharge of sympathetic nerves in the periphery, which in turn is influenced by stress, exercise, and diet.[178–180] Thus, it is no great surprise that, as demonstrated in vervet monkeys[160] and adult schizophrenic patients,[174] pHVA does not correlate well with CSF HVA.

However, it has been found that the peripherally derived component of pHVA can be pharmacologically suppressed by debrisoquin, a MAO inhibitor[181,182] and postganglionic sympathetic blocker[183] that does not cross the blood–brain barrier. This medication, initially developed as an antihypertension agent, prevents the peripheral conversion of dopamine to HVA and inhibits release of norepinephrine from sympathetic nerves without affecting HVA production by brain.[184] Therefore, pHVA measured after administration of debrisoquin better reflects CNS-derived HVA.

Use of debrisoquin as an enhancement agent in studies relying on pHVA as an indicator of CNS dopaminergic activity has indeed yielded favorable results.[173–176,185,186] For instance, in 1985 Maas et al.[174] administered debrisoquin to a group of adults with schizophrenia. As expected, pHVA fell to 50% of baseline. Furthermore, while the correlation between pHVA and CSF HVA prior to debrisoquin was nonsignificant, the correlation after 13 days of the drug was excellent: ($r = 0.95$).

Recent work in nonhuman primates, however, suggests further refinements in the use of debrisoquin for clinical investigations. In 1987, Kopin et al.[60] suggested that post-debrisoquin pHVA would more tightly reflect CNS dopaminergic neuron activity if peripheral noradrenergic activity could be totally blocked. These investigators extrapolated post-debrisoquin MHPG levels to zero, at which point the corresponding pHVA level was approximately 25% instead of 50% of the pre-debrisoquin level. It was then suggested that central dopaminergic activity actually accounts for only about 25% of baseline pHVA levels.

To validate this conclusion, MPTP, a selective dopaminergic nigrostriatal neuron toxin,[167] was administered to three monkeys prior to the administration of debrisoquin.[60] Under these circumstances, pHVA at the point where extrapolated MHPG levels were zero was not significantly different from zero, corroborating the assertion that extrapolated pHVA after debrisoquin is of central origin. It also suggests that any contribution by CNS noradrenergic neurons must be negligible.

If the post-debrisoquin extrapolation method does prove valid in humans, it could simplify clinical studies of the dopaminergic system in children. As Kopin et al.[61] demonstrated, it can also be applied in 24-hr urine collection studies of central dopaminergic activity.

Such progress in establishing the validity of peripheral measures of central dopaminergic activity then makes issues of reliability more pertinent. For instance, diurnal variation in pHVA has been demonstrated in adults[187] with nighttime levels being significantly higher than daytime levels. Unfortunately, a similar study of pHVA stability has not been done in normal children and adolescents. However, in children and adults with TS,[188] a drop in pHVA has been observed to occur between early morning and early afternoon. In these subjects, significant across-day variability was also demonstrated.

Yet another concern regards the rapidity with which pHVA reflects changes in central dopaminergic turnover. In a study of the effects of haloperidol on postadministration debrisoquin CSF HVA and pHVA, a significant increase in CSF HVA, but no change in pHVA was seen 6 hr after drug administration (P. Delgado, personal communication). In another study, Davidson et al.[189] demonstrated a significant increase in pHVA, but not until 24 hr after neuroleptic administration. Attribution of this increase in pHVA to changes in central HVA production seems appropriate given that debrisoquin pretreatment caused a general decline in pHVA levels but did not influence the magnitude of the increase in pHVA occurring 24 hr after haloperidol. One straightforward interpretation of these studies is that haloperidol increased CNS dopamine release by blocking negative feedback mechanisms[190] but also initially blocked transport of HVA into the periphery.[151] Therefore, this concern may apply only when certain drugs are used.

3.3.2. Serotonin

As serotonin does not cross the blood–brain barrier, peripheral levels arise primarily from the production of this neurotransmitter by the enterochromaffin cells of the gut.[67] By contrasts, CSF 5-HIAA is actively transported from the CNS into the plasma.[132] Peripheral 5-HIAA levels therefore include contributions from both the central and peripheral nervous systems. However, while CSF 5-HIAA levels have been shown to correlate well with central serotonergic activity,[191–194] the relationship between plasma or urinary 5-HIAA and CNS serotonergic activity is considered tenuous.[67] Peripheral measures of serotonin[195] and 5-HIAA are therefore generally thought to reflect CNS serotonergic activity only to the degree that common defects are present in both the central and peripheral systems. This rationale has, in fact, fueled vigorous pursuit of the etiology of the hyperserotonemia seen in autism.

3.3.3. Norepinephrine

In the CNS and peripheral nervous system, norepinephrine and its methylated metabolite, normetanephrine, are acted on by MAO to form corresponding aldehydes. These may then either be oxidized to acids or reduced to glycol derivatives.[139] The former route of metabolism prevails in the periphery; therefore, vanillylmandelic acid (VMA) levels serve as an the best indication of peripheral sympathetic nervous system activity.[196] In the CNS, however, the reductive pathway prevails, so that 3,4-dihydroxyphenethyleneglycol (DHPG) and MHPG constitute the major metabolites of norepinephrine.

Both the functional interdependence of the central and peripheral noradrenergic systems[172] and free diffusion of MHPG across the blood–brain barrier[85] contribute to the strength of the correlation between central noradrenergic activity and peripheral MHPG levels.[89,197–202] However, the validity with which any particular peripheral MHPG measurement reflects the CNS noradrenergic activity in a given individual remains unclear.[58]

Depending on resolution of this validity issue, a discussion of the reliability of peripheral MHPG measures will be more or less relevant. In adults, plasma MHPG (pMHPG) does not have a diurnal rhythm. Within-day fluctuation is attributed instead to variations in stress and general activity.[187] However, in a study of the diurnal and across-day variation of pMHPG levels in a group of TS subjects composed primarily, but not entirely, of children and adolescents,[203] diurnal, across-day, and within-subject variation was found. It has been alleged, however, that TS subjects are inordinately sensitive to stress,[90] which may have had variable influence on the subjects in this study. Unfortunately, a similar study has not yet been performed in normal children.

3.4. Challenge Studies

For the most part, measurements of neurotransmitters and their metabolites in children with psychiatric disease have been made in the baseline state. This approach has yielded a host of intriguing results but does not reveal information regarding the ability of particular neurotransmitter systems to respond to perturbations. In order to unmask the dynamics of neurotransmission, challenge studies have been developed. Although differing in their means of perturbation, they commonly consist of measurement of the behavioral and/or metabolic baseline state, the introduction of a perturbing maneuver, and measurement of the outcome state. The behavioral and metabolic changes in the target population are then compared with changes seen in the control group.

This strategy has been used extensively and fruitfully in adult psychiatry. Lactate infusion, tryptophan infusion, sleep derivation, and the administration of various drugs such as the 5-HT2 agonist *m*-chlorophenylpiperazine (MCPP) have helped unmask abnormalities in neuroregulation in subjects with panic disorder,[204] depression,[205,206] and obsessive-compulsive disorder (OCD),[207,208] respectively. In addition, so-called "activation" studies, in which regional cerebral blood flow is measured before and during selected cognitive tasks, have revealed deficits in prefrontal cortical function in schizophrenia.[119]

In children, use of the challenge strategy has been less prevalent. In the earliest biologic studies of autism, tryptophan and 5-hydroxytryptophan loading were used to probe the serotonin system.[67] This strategy was subsequently advocated as a promising approach to the study of children with ADHD by Hunt et al.[209] in 1982. In children with major depressive disorder and in controls, studies of neuroendocrine function have likewise followed this design. For example, plasma growth hormone has been measured in response to hypoglycemia, clonidine, and desmethylimipramine.[210]

Possible problems in the design and interpretation of cause–effect research, such as challenge studies, are comprehensively addressed by Feinstein.[211] These generally concern possible disparities in subject susceptibility, application of principle and ancillary provocative maneuvers, and use of informational processes. Observation of these caveats will increase the potential of challenge studies to expand our understanding of the dysfunctional neural dynamics in child psychiatric disease.

4. FUTURE DIRECTIONS

4.1. Brain-Imaging Techniques

4.1.1. Positron-Emission Tomography

Although this review has focused on measurement of neurotransmitters and their metabolites, changes in their levels are not implied to be the sole determinants of changes in neurotransmission. In the conveyance of neurochemical information, neurotransmitters must interact with specific receptors that then interact with other intraneuronal systems. None of these systems is a static transducer of neurochemical information. Receptors, for instance, can change in conformation or number to transmit the signals carried by a barrage of neurotransmitter molecules more or less effectively. Therefore, a comprehensive understanding of neurotransmission depends on knowledge of their dynamic status.

In animal models and human autopsy studies, neurotransmitter receptor number and binding characteristics can be studied in ex vivo or in vitro brain preparations via radioisotopic tracer methods.[212] Obviously, it is not possible to use such methods in clinical studies. However, PET technology can now be used for this purpose.[62,212]

In PET, a positron-emitting isotope of a natural element (carbon-11, fluorine-18, oxygen-15, nitrogen-15) is created in a cyclotron and then incorporated into a chemical compound (ligand) known to bind to a particular type or subtype of neurotransmitter receptor. As the radiolabel decays, a positron is emitted. As it strikes an electron in the near brain surround, the positron and electron are annihilated and two γ-rays are emitted. These impinge at 180° from each other on a halolike γ-detector placed about the head of the patient. The point at which they were produced (i.e., the location of the receptor–ligand interaction) can then be calculated and fed into a computer that reconstructs transverse images of the receptor–ligand binding pattern in the brain.[212]

Currently, PET ligands are available for the study of serotonin, benzodiazepine, opiate, muscarinic cholinergic, and D1 and D2 dopamine receptors.[62] They may be used not only to determine the regional distribution of these receptors but to discern receptor-binding characteristics such as maximal binding capacity and equilibrium dissociation constants as well.[212]

In the pediatric population, the use of PET is limited by maximal radiation exposure limitations.[63] However, the recent development of a dual-detector positron-emission probe may obviate this difficulty and also reduce the cost of scanning.[213] Unfortunately, the configuration of the dual detector necessarily reduces the area of the brain scanned and the degree of spatial resolution attained. However, for the purpose of pharmacokinetic studies it has proved sufficient.[214]

4.1.2. Magnetic Resonance Imaging

Magnetic resonance imaging (MRI) spectroscopy is another technique currently under development for the study of brain metabolism.[215] When eventually applied in humans, it may reveal invaluable information about aberrations in energy production or the oxidative/reductive status of the brain in various developmental neuropsychiatric disorders. Currently, MRI is used primarily for neuroanatomic imaging. One exciting potential application of this technique is its use in combination with dynamic techniques, such as PET, for determining the regional specificity of alterations in brain metabolism and receptor status.

4.2. Postmortem Brain Studies

Standard techniques used in postmortem human brain studies have included HPLC, GC–MS, and radioimmunoassay (RIA) measurement of monoamines, monoamine metabolites, and neuropeptides. The spatial resolution of these methods is limited by the ability of investigators to dissect individual brain areas. Therefore, high-resolution in vitro techniques previously used in animal studies have recently been adapted so that regional alterations in brain function in psychopathologic disorders may be more effectively studied in postmortem human brains. These methods are currently being applied in investigations of adult and childhood neuropsychiatric disorders such as Huntington's disease, Parkinson's disease, schizophrenia, major depressive and manic-depressive disorder, Tourette's syndrome, and the pervasive developmental disorders.

4.2.1. Receptor Autoradiography

Receptor autoradiography is an in vitro technique[216] that has the capacity to provide information about receptor populations of specific layers of cortex, thereby capitalizing on the extensive information available about the neuroanatomic connectivity of these areas.[217] In addition, the sensitivity of this method is several magnitudes greater than that of receptor studies done in brain homogenates, it is quantitative, and it can be used to characterize fully the pharmacokinetics of particular receptor populations. This information can then be integrated with biochemical measures of neurotransmitters and their metabolites to obtain some sense of the state of regional neurotransmission. The major limitations of this method derive from its application at a single point in time and from difficulty in discriminating between results due to drug effects and those possibly due to underlying neuropathophysiologic processes. In addition, receptors are susceptible to degradation during the postmortem delay period,[218] although this problem can be managed by use of appropriately matched controls.

4.2.2. Immunohistochemistry

Immunohistochemical studies permit visualization of various antigens, such as neurotransmitters, receptors, and enzymes within neurons and neuronal processes, by specifically chosen antibodies. This affords single-cell resolution and constitutes the major advantage of this technique. However, while very sensitive, a number of procedures must be used to ensure that the antibody–antigen binding is specific for the neuronal constituent of interest. Another major disadvantage is that it is not quantitative. Therefore, it is best used to discern changes in the neuroanatomic distribution of neuronal constituents along with other methods that can quantitate them. For instance, this method has proved invaluable in characterizing the complex histochemistry of the striatum[112,113,219] and continues to be important in elucidating possible striatal pathology in such disorders as Tourette's syndrome.[116]

4.2.3. In Situ Hybridization

In situ hybridization[220] has most recently been added to the retinue of postmortem human brain methodology. This technique involves the identification and localization of selected types of messenger RNA (mRNA) in brain tissue sections by complementary sequences of DNA (cDNA). Again, this is a very sensitive method and possesses single-cell resolution capabilities. However, as currently applied in postmortem studies, it is not quantitative. Nevertheless, when used in conjunction with methods that quantitate and localize mRNA end products, such as biochemical measurement of neurotransmitters, autoradiographic measurement of receptors, and immunohistochemistry, it helps provide a better picture of the dynamic functioning of the brain.

5. CONCLUSION

Despite initial optimism, clinical measurements of neurotransmitters and metabolites in various body fluids have thus far failed to provide definitive answers about the etiology of childhood psychiatric disease. Indeed, as basic neuroscience develops finer powers of resolution, clinical probes continue to be dishearteningly blunt. The complex nature of underlying neuropathophysiologic processes seems clinically inscrutable, given the current repertoire of techniques available to investigators of human subjects.

However, it would be unfortunate if frustration were to provoke rejection of this approach and we would be misguided if the advent of exciting new technology made current techniques seem obsolete. Neurotransmission involves the fine integration of an almost regressionally infinite series of molecular events. Therefore, since no conceivable clinical or basic science technique could shed light on all points at which pathologic perturbations could occur, a variety of scientific methods must be used, each contributing a unique window on brain function.

For instance, the clinical measurement of neurotransmitters and their metabolites does, when interpreted within the context of relevant caveats, provide an index of CNS neurotransmitter production and release. This information should then be integrated with that obtained via methods that illuminate other aspects of neurotransmission, such as neurotransmitter receptor status and function of second messengers and ion channels.

However, aberrant neurotransmission comprises only one aspect of psychopathology. For instance, it may be useful to remember that the neural substrate serves but to carry and process information that passes between living organisms and the supportive or threatening environments in which they live; furthermore, while this capacity is in part genetically determined, it is also shaped by the nature of the information that is processed.[111] Interpretation of this information is essential to understanding the neural infrastructure as well as the content of psychopathologic states. This requires high-order analysis and is a task beyond the province of reductionistic neurobiology.

Therefore, the expert work of psychoanalysts, psychologists, behaviorists, and others will continue to be invaluable in delineating the neurobiologic underpinnings of childhood psychopathology. Furthermore, inasmuch as we can anticipate that the work of clinical investigators will continue to follow the lead of basic scientists, it will be important to continue to promote communication and cooperation between these two arenas. In addition, cooperation among institutions will be helpful if we are to capitalize on our limited resources and speed scientific progress in understanding, preventing, and optimally treating childhood psychiatric disease.

REFERENCES

1. Twarog BM, Page IH: Serotonin content of some mammalian tissues and urine and a method for its determination. Am J Physiol 175:157–61, 1953

2. Amin AH, Crawford TBB, Gaddum JH: The distribution of substance P and 5-hydroxytryptamine in the central nervous system of the dog. J Physiol (Lond) 126:596–618, 1954
3. Gaddum JH: Drugs antagonistic to 5-hydroxytryptamine, in Gaddum JH (ed): Ciba Foundation Symposium on Hypertension. Humoral and Neurogenic Factors. Boston, Little & Brown, 1954, pp 75–77
4. Wooley DW, Shaw E: A biochemical and pharmacological suggestion about certain mental disorders. Science 119:587–588, 1954
5. Schain RJ, Freedman DX: Studies on 5-hydroxyindole metabolism in autistic and other mentally retarded children. J Pediatr 58:315–320, 1961
6. Falck B, Hillarp N-A, Thieme G, et al: Florescence of catecholamines and related compounds condensed with formaldehyde. J Histochem Cytochem 10:348–354, 1962
7. Sharman DF: A fluorimetric method for the estimation of 4-hydroxy-3-methoxyphenylacetic acid (homovanillic acid) and its identification in brain tissue. Br J Pharmacol 20:204–213, 1963
8. Anden NE, Roos BE, Werdinius B: On the occurrence of homovanillic acid in brain and cerebrospinal fluid and its determination by a fluorometric method. Life Sci 7:448–458, 1963
9. Gerbode FA, Bowers MB: Measurement of acid monoamine metabolites in human and animal cerebrospinal fluid. J Neurochem 15:1053–1055, 1968
10. Levy DM: On the problem of movement restraint: Tics, stereotyped movements, hyperactivity. Am J Orthopsychiatry 14:644–671, 1944
11. Berkson G: Abnormal stereotyped motor acts, in Zubin J, Hunt HF (eds): Comparative Psychopathology—Animal and Human. New York, Grune & Stratton, 1967, pp 76–94
12. Randrup A, Munkvad I: Pharmacology and physiology of stereotyped behavior. J Psychiatr Res 11:1–10, 1974
13. Cohen DJ, Shaywitz BA, Johnson WT, et al: Biogenic amines in autistic and atypical children. Cerebrospinal fluid measures of homovanillic acid and 5-hydroxyindoleacetic acid. Arch Gen Psychiatry 31:845–853, 1974
14. Kow LM, Pfaff DW: Neuromodulatory actions of peptides. Annu Rev Pharmacol Toxicol 28:163–188, 1988
15. Kaczmarek LK, Levitan IB: Neuromodulation. The Biochemical Control of Neuronal Excitability. New York, Oxford University Press, 1987
16. Weizman R, Weizman A, Tyano S et al: Humoral-endorphin blood levels in autistic, schizophrenic, and healthy subjects. Psychopharmacology 82:368–370, 1984
17. Gillberg C, Terenius L, Lonnerholm G: Endorphin activity in childhood psychosis. Spinal fluid levels in 24 cases. Arch Gen Psychiatry 42:780–783, 1985
18. Leckman JF, Riddle MA, Berrettini WH, et al: Elevated CSF levels of dynorphin A(1–8) in Tourette's syndrome. Submitted
19. Berridge MJ: Isositol triphosphate as a second messenger in signal transduction. Ann NY Acad Sci 494:39–51, 1987
20. Nairn AC, Hemmings HC Jr, Greengard P: Protein kinases in the brain. Annu Rev Biochem 54:931–976, 1985
21. Gilman AG: G-proteins: Transducer of receptor-generated signals. Annu Rev Biochem 56:615–649, 1987
22. Worley PF, Baraban JM, Snyder SH: Beyond receptors: Multiple second messenger systems in brain. Ann Neurol 21:217–229, 1987
23. Nowycky MC, Fox AP, Tsien RW: Three types of neuronal calcium channel with different calcium agonist sensitivity. Nature (Lond) 316:440–443, 1985
24. Hoshino Y, Hisashi K, Yashima Y, et al: Plasma cyclic AMP in psychiatric diseases of childhood. Folia Psychiatr Neurol Jpn 34:9–16, 1980
215. Ebstein RP, Lerer B, Bennett ER, et al: Lithium modulation of second messenger signal amplification in man: Inhibition of phosphatidyl-inositol-specific phospholipase C and adenylate cyclase activity. Psychiatry Res 24:45–52, 1988
26. Mooney JJ, Schatzberg AF, Cole JO: Rapid antidepressant response to alprazolam in depressed patients with high catecholamineoutput and heterologous desensitization of platelet adenylate cyclase. Biol Psychiatry 23:543–559, 1988
27. Rapoport JL, Quinn PO, Bradbard G, et al: Imipramine and methylphenidate treatments of hyperactive boys. Arch Gen Psychiatry 30:789–793, 1974
28. Rapaport TL, Buchsbaum MS, Weingartner H, et al: Dextroamphetamine: Its cognitive and behavioral effects in normal and hyperactive boys and normal men. Arch Gen Psychiatry 37:933–943, 1980

29. Zametkin AJ, Rapoport JL: Neurobiology of attention deficit disorder with hyperactivity: Where have we come in 50 years. J Am Acad Child Adolesc Psychiatry 26:676–686, 1987
30. Shetty T, Chase TN: Central monoamines and hyperkinesis of childhood. Neurology (NY) 26:1000–1006, 1976
31. Braestrup C: Biochemical differentiation of amphetamine vs methylphenidate and nomifensine in rats. J Pharm Pharmacol 29:463–470, 1977
32. Shaywitz BA, Cohen DJ, Bowerw MB: CSF monoamine metabolites in children with minimal brain dysfunction: Evidence for alteration of brain dopamine. J Pediatr 90:67–71, 1977
33. Shaywitz BA, Klopper JH, Yager RD, et al: Paradoxical response to amphetamine in developing rat pups treated with 6-hydroxydopamine. Nature (Lond) 261:153–155, 1976
34. Lipton SV, McGough TP, Shaywitz BA: Effects of apomorphine on escape performance and activity in developing rat pups treated with 6-hydroxydopamine. Pharmacol Biochem Behav 13:371–377, 1980
35. Shaywitz SE and Shaywitz BA: Biological influences in attentional disorders, in Levine MD, Carey WB, Crocker AC, et al (eds): Developmental Behavioral Pediatrics. Philadelphia WB Saunders, 1982, pp 746–755
36. Breese GR, Napier TC, Mueller RA: Dopamine agonist-induced locomotor activity in rats treated with 6-hydroxydopamine at differing ages: Functional supersensitivity of D-1 dopamine receptors in neonatally lesioned rats. J Pharmacol Exp Ther 234:447–455, 1985
37. Ungerstedt U: Measurement of neurotransmitter release by intracranial dialysis, in Marsden CA (ed): Measurement of Neurotransmitter Release *In Vivo*. New York, Wiley, 1984, pp 81–105
38. Church WH, Justice JB Jr, Neill DB: Detecting behaviorally relevant changes in extracellular dopamine with microdialysis. Brain Res 412:397–399, 1987
39. Zametkin AJ, Rapaport JL: Noradrenergic hypothesis of attention deficit disorder with hyperactivity: A critical review, in Meltzer HY (ed): Psychopharmacology: The Third Generation of Progress. New York, Raven, 1987, pp 837–842
40. Mikkelsen E, Lake CR, Brown GL, et al: The hyperactive child syndrome: Peripheral sympathetic nervous system function and the effect of d-amphetamine. Psychiatr Res 4:157–169, 1981
41. Shekim WO, DeKirmenjian H, Chapel JL: Urinary catecholamine metabolites in hyperkinetic boys treated with d-amphetamine. Am J Psychiatry 134:1276–1279, 1977
42. Shekim WO, DeKirmenjian H, Chapel JL: Urinary MHPG excretion in minimal brain dysfunction and its modification by d-amphetamine. Am J Psychiatry 136:667–671, 1979
43. Shekim WO, Javaid J, Dans JM, et al: Urinary MHPG and HVA excretion in boys with attention deficit disorder and hyperactivity treated with d-amphetamine. Biol Psychiatry 18:707–714, 1983
44. Yu-cun S, Yu-Peng W: Urinary 3-methoxy-4-hydroxyphenylglycol sulfate in seventy-three schoolchildren with minimal brain dysfunction syndrome. Biol Psychiatry 19:861–870, 1984
45. Khan AU, DeKirmenjian H: Urinary excretion of catecholamine metabolites in hyperkinetic children. Am J Psychiatry 138:108–112, 1981
46. Rapoport JL, Mikkelsen EJ, Ebert MH, et al: Urinary catecholamine and amphetamine excretion in hyperactive and normal boys. J Nerv Ment Dis 166:731–737, 1978
47. Rapoport JL, Buchsbaum MS, Weingartner H, et al: Dextroamphetamine. Its cognitive and behavioral effects in normal and hyperactive boys and normal men. Arch Gen Psychiatry 37:933–943, 1980
48. Garfinkel BD, Wender PH, Sloman L, et al: Tricyclic antidepressant and methylphenidate treatment of attention deficit disorder in children. J Am Acad Child Adolesc Psychiatry 22:343–348, 1983
49. Zametkin A, Rapoport JL, Murphy DL, et al: Treatment of hyperactive children with monoamine oxidase inhibitors. Clinical efficacy. Arch Gen Psychiatry 42:962–966, 1985
50. Gastfriend DR, Biederman J, Jellinek MS: Desipramine in the treatment of adolescents with attention deficit disorder. Am J Psychiatry 141:906–908, 1984
51. Biederman J, Gastfriend DR, Jellinek MS: Desipramine in the treatment of children with attention deficit disorder. J Clin Psychopharmacol 6:359–363, 1986
52. Donnelly M, Zametkin AJ, Rapoport JL, et al: Treatment of childhood hyperactivity with desipramine: Plasma drug concentration, cardiovascular effects, plasma and urinary catecholamine levels, and clinical response. Clin Pharmacol Ther 39:72–81, 1986
53. Riddle MA, Hardin MT, Cho SC: Desipramine treatment of boys with attention-deficit hyperactivity disorder and tics: Preliminary clinical experience. J Am Acad Child Adolesc Psychiatry, In press.
54. Brown GL, Ebert MH, Hunt RD, et al: Urinary 3-methoxy-4-hydroxyphenylglycol and homovanillic acid response to d-amphetamine in hyperactive children. Biol Psychiatry 16:779–787, 1981
55. Zametkin A, Rapoport JL, Murphy DL, et al: Treatment of hyperactive children with monoamine oxidase

inhibitors. Plasma and urinary monoamine findings after treatment. Arch Gen Psychiatry 42:969–973, 1985
56. Zametkin A, Karoum F, Linnoila M, et al: Stimulants, urinary catecholamines and indoleamines in hyperactivity: A comparison of methylphenidate and dextroamphetamine. Arch Gen Psychiatry 42:251–259, 1984
57. Blier P, DeMontigny C, Chaput Y: Modifications of the serotonin system by antidepressant treatments: Implications for the therapeutic response in major depression. J Clin Psychopharmacol 7:24–35S, 1987
58. Linnoila M, Guthrie S, Lane EA, et al: Clinical studies on norepinephrine metabolism: How to interpret the numbers. Psychiatr Res 17:229–239, 1985
59. Taghzouti K, Simon H, Herve D: Behavioral deficits induced by and electrolyytic lesion of the rat ventral mesencephalic tegmentum are corrected by a superimposed lesion of the dorsal noradrenergic system. Brain Res 440:172–176, 1988
60. Kopin IJ, Bankiewicz KS, Harvey-White J: Assessment of brain dopamine metabolism from plasma HVA and MHPG during debrisoquin treatment: Validation in monkeys treated with MPTP. Neuropsychopharmacology 1:119–125, 1988
61. Kopin IJ, Bankiewicz K, Harvey-White J: Effect of MPTP-induced parkinsonism in monkeys on the urinary excretion of HVA and MHPG during debrisoquin administration. Life Sci 43:133–141, 1988
62. Andreasen NC: Brain imaging: Applications in Psychiatry. Science 239:1381–1388, 1988
63. Gainey MA, Capitanio MA: Recent advances in pediatric nuclear medicine. Radiol Clin North Am 26:409–418, 1988
64. Anderson GM: Monoamines in autism: An update of neurochemical research on a pervasive developmental disorder. Med Biol 65:67–74, 1987
65. Wurtman RJ: Dietary treatments that affect brain neurotransmitters. Effects on calorie and nutrient intake. Ann NY Acad Sci 499:179–190, 1987
66. During MJ, Freese A, Heyes MP: Extracellular concentration of serotonin and quinolinic acid, neurotransmitter and neurotoxin, in rat corpus striatum following systemic administration of L-tryptophan. Submitted.
67. Anderson GM, Hoshino Y: Neurochemical studies of autism, in Cohen DJ, Donnellan AM (eds): Handbook of Autism and Pervasive Developmental Disorders. New York, Wiley, 1987, pp 166–191
68. Young SN, Kavanagh ME, Anderson GM, et al: Clinical neurochemistry of autism and associated disorders. J Aut Dev Disord 12:147–165, 1982
69. Todd RD, Ciaranello RD: Demonstration of inter- and intraspecies differences in serotonin binding sites by antibodies from an autistic childhood. Proc Natl Acad Sci USA 82:612–616, 1985
70. Todd RD, Hickok JM, Anderson GM, et al: Antibrain antibodies in infantile autism. Biol Psychiatry 23:644–647, 1988
71. Todd RD: Pervasive developmental disorders and immunological tolerance. Psychiatr Dev 2:147–165, 1986
72. Campbell M: Annotation. Fenfluramine treatment of autism. J Child Psychol Psychiatry 29:1–10, 1988
73. Antelman SM, Szechtman H, Chin P, et al: Tail-pinch induced eating, gnawing, and licking behavior in rats: Dependence on nigrostriatal dopamine system. Brain Res 99:319–337, 1975
74. Deutsch SI: Rationale for the administration of opiate antagonists in treating infantile autism. Am J Ment Defic 90:631–635, 1986
75. Gillberg C: The neurobiology of infantile autism. J Child Psychol Psychiatry 29:257–266, 1988
76. Haydon PG, McCobb DP, Kater SB: Serotonin selectively inhibits growth cone motility and synaptogenesis of specific identified neurons. Science 226:561–564, 1984
77. Cohan CS, Connor JA, Kater SB: Electrically and chemically mediated increases in intracellular calcium in neuronal growth cones. J Neurosci 7:3588–3599, 1987
78. McCobb DP, kater SB: Serotonin inhibition of growth cone motility is blocked by acetylcholine. Soc Neurosci Abs 12:1117, 1986
79. Shapiro AK, Shapiro E: Clinical efficacy of haloperidol, pimozide, penfluridol, and clonidine in the treatment of Tourette syndrome. in Friedhoff AJ, Chase TN (eds): Gilles de la Tourette Syndrome. New York, Raven, 1982, pp 383–386
80. Shapiro AK, Shapiro E: Controlled study of pimozide vs. placebo in Tourette's syndrome. J Am Acad Child Adolesc Psychiatry 23:161–173, 1984
81. Klawans HD, Nausieda PA, Goetz CL, et al: Tourette-like symptoms following chronic neuroleptic therapy, in Friedhoff AJ, Chase TN (eds): Gilles de la Tourette syndrome, Adv Neurol Vol 35, 1982, pp 415–418

82. Feinberg M, Carroll BJ: Effects of dopamine agonists and antagonists in Tourette's disease. Arch Gen Psychiatry 36:979–985, 1979
83. Lowe TL, Cohen DJ, Detlor J: Stimulant medications precipitate Tourette's syndrome. JAMA 247:1729–1731, 1982
84. Leckman JF, Walkup JT, Riddle MA, et al: Tic disorders, in Meltzer HY (ed): Psychopharmacology: The Third Generation of Progress. New York, Raven, 1987, pp 1239–1246
85. Kopin IJ, Gordon EK, Jimerson DC, et al: Relation between plasma and cerebrospinal fluid levels of 3-methoxy-4-hydroxyphenethyleneglycol. Science 219:73–75, 1983
86. Leckman JF, Detlor J, Harcherik DF, et al: Short- and long-term treatment of Tourette's disorder with clonidine: A clinical perspective. Neurology (NY) 35:343–351, 1985
87. Anden NE, Corrodi H, Fuxe K, et al: Evidence for a central noradrenaline receptor stimulation by clonidine. Life Sci 9:513–523, 1970
88. Svensson TH, Bunney BS, Aghajanian GK: Inhibition of both noradrenergic and serotonergic neurons by the alpha-adrenergic agonist clonidine. Brain Res 92:291–306, 1975
89. Leckman JF, Ort S, Caruso KA, et al: Rebound phenomena in Tourette's syndrome after abrupt withdrawal of clonidine. Behavioral, cardiovascular, and neurochemical effects. Arch Gen Psychiatry 43:1168–1176, 1986
90. Leckman JF, Cohen DJ, Price RA, et al: The pathogenesis of Tourette syndrome. A review of data and hypotheses, in Shah NS, Donald AG (eds): Movement Disorders. New York, Plenum, 1986, pp 257–272
91. Geyer MA, Lee EHY: Effects of clonidine, piperoxane, and locus coeruleus lesion on the serotonergic and dopaminergic systems in raphe and caudate nucleus. Biochem Pharmacol 33:3399–3404, 1984
92. Bunney BS, DeRiemer SA: Effects of clonidine on dopaminergic neuron activity in the substantia nigra: Possible indirect mediation by noradrenergic regulation of the serotonergic raphe system. Adv Neurol 35:99–104, 1982
93. Leckman JF, Anderson GM, Cohen DJ, et al: Whole blood serotonin and tryptophan levels in Tourette's disorder: Effects of acute and chronic clonidine treatment. Life Sci 35:2497–2503, 1984
94. Riddle MA, Leckman JF, Hardin MT, et al: Fluoxetine treatment of obsessions and compulsions in patients with Tourette's syndrome. Am J Psychiatry, in press.
95. Collingridge GL, James TA, MacLeod NK: Neurochemical and electrophysiological evidence for a projection from the locus coeruleus to the substantia nigra. J Physiol (Lond) 290:44P, 1979
96. Dray A, Gonye TJ, Oakley NR, et al: Evidence for the existence of a raphe projection to the substantia nigra in rat. Brain Res 113:45–57, 1976
97. Waldmeier PC: Serotonergic modulation of mesolimbic and frontal cortical dopamine neurons. Experientia 36:1092–1094, 1980
98. Fuenmayor LD, Bermudez M: Effect of the cerebral tryptaminergic system on the turnover of dopamine in the striatum of the rat. J Neurochem 44:670–674, 1985
99. Benkirane S, Arbilla S, Langer SZ: A functional response to D1 dopamine receptor stimulation in the central nervous system: Inhibition of the relase of [^{3}H]serotonin from the rat substantia nigra. Naunyn-Schmiedebergs Arch Pharmacol 335:502–507, 1987
100. Manier DH, Gillespie DD, Saunders-Bush E, et al: The serotonin/noradrenaline-link in brain. The role of noradrenaline and serotonin in the regulation of density and function of beta adrenoceptors and its alteration by desipramine. Naunyn-Schmiedebergs Arch Pharmacol 335:109–114, 1987
101. Quirion R, Gaudreau P, Marter J-C, et al: Possible interactions between dynorphin and dopaminergic systems in rat basal ganglia and substantia nigra. Brain Res 331:358–362, 1985
102. Nylander I, Terenium L: Chronic haloperidol and clozapine differentially affect dynorphin peptides and substance P in basal ganglia of the rat. Brain Res 380:34–41, 1986
103. Li S, Sivan SP, Hong JS: Regulation of the concentration of dynorphin A(1–8) in the striatonigral pathway by the dopaminergic system. Brain Res 398, 390–392, 1986
104. Bartfai T, Iverfeldt K, Fisone G: Regulation of the release of coexisting neurotransmitters. Annu Rev Pharmacol Toxicol 28:285–310, 1988
105. Lundberg JM, Hokfelt T: Coexistence of peptides and classical neurotransmitters. Trends Neurosci 6:325–633, 1983
106. Antelman SM, Chiodo LA: Stress: Its effect on interactions among biogenic amines and role in the induction and treatment of disease, in Iversen LL, Iversen SD, Snyder SH (eds): Handbook of Psychopharmacology, Vol. 18. New York, Plenum, 1984, pp. 279–341
107. Chiodo LA, Antelman SM, Caggiula AR, et al: Reciprocal influences of activating and immobilizing stimuli on the activity of nigrostriatal dopamine neurons. Brain Res 176:385–390, 1979

108. Chiodo LA, Antelman SM, Caggiula AR, et al: Sensory stimuli alter the discharge rate of dopamine (DA) neurons: Evidence for two functional types of DA cells in the substantia nigra. Brain Res 189:544–549, 1980
109. Nicoll, RA: The coupling of neurotransmitter receptors to ion channels in the brain. Science 241:545–551, 1988
110. Shepherd GM: The Synaptic Organization of the Brain. New York, Oxford University Press, 1979
111. Edelman GR: Neural Darwinism. The Theory of Neuronal Group Selection. New York, Basic Books, 1987
112. Haber SN, Watson SJ: The comparative distribution of enkephalin, dynorphin and substance P in the human globus pallidus and basal forebrain. Neuroscience 14:1011–1024, 1985
113. Lindefors N, Brodin E, Theodorsson-Norheim E, et al: Regional distribution and in vivo release of tachykinin-like immunoreactivities in rat brain: Evidence for regional differences in relative proportions of tachykinins. Regul Pept 10:217–230, 1985
114. Crawley JN, Stivers JA, Blumstein LK, et al: Cholecystokinin potentiates dopamine-mediated behavior: Evidence for modulation specific to a site of co-existence. J Neurosci 5:1972–83, 1985
115. Smith Y, Parent A: Neuropeptide Y-immunoreactive neurons in the striatum of cat and monkey: Morphological characteristics, intrinsic organization and co-localization with somatostatin. Brain Res 372:241–252, 1986
116. Haber SN, Kowall NW, Vonsattel JP et al: Gilles de la Tourette's syndrome. A postmortem neuropathological and immunohistochemical study. J Neurol Sci 75:225–41, 1986
117. Asberg M, Nordstrom P, Traskman-Bendz L: Cerebrospinal fluid studies in suicide. Ann NY Acad Sci 487:243–255, 1986
118. Ritvo ER, Freeman BJ, Yuwiler A, et al: Fenfluramine therapy for autism: Promise and precaution. Psychopharmacol Bull 22:133–159, 1986
119. Weinberger DR, Berman KF, Illowsky BP: Physiologic dysfunction of dorsolateral prefrontal cortex in schizophrenia. A new cohort and evidence for a monoaminergic mechanism. Arch Gen Psychiatry 45:609–615, 1988
120. Meissen GJ, Myers RH, Mastromauro CA, et al: Predictive testing for Huntington's disease with use of a linked DNA marker. N Engl J Med 318:535–542, 1988
121. Kidd, KK: Searching for major genes for psychiatric disorders, in Ciba Foundation Symposium 130 on Molecular Approaches to Human Polygenic Disease. Chichester, Wiley, 1987, pp 184–196
122. Ransohoff DF, Feinstein AR: Problems of spectrum and bias in evaluating the efficacy of diagnostic tests. N Engl J Med 299:926–930, 1978
123. Goldberg, SC: Persistent flaws in the design and analysis of psychopharmacology research, in Meltzer HY (ed): Psychopharmacology: The Third Generation of Progress. New York, Raven, 1987, pp 1005–1012
124. Garelis E, Sourkes TL: Sites of origin in the central nervous system of monoamine metabolites measured in human cerebrospinal fluid. J Neurol Neurosurg Psychiatry 36:625–629, 1973
125. Wiesel F-A: Mass fragmentographic determination of acidic dopamine metabolites in human cerebrospinal fluid. Neurosci Lett 1:219–224, 1975
126. Ziegler MG, Wood JH, Lake CR, et al: Norepinephrine and 3-methoxy-4-hydroxyphenylglycol gradients in human cerebrospinal fluid. Am J Psychiatry 134:565–568, 1977
127. Moir ATB, Ashcroft GW, Crawford TBB, et al: Cerebral metabolites in cerebrospinal fluid as a biochemical approach to the brain. Brain 93:357–368, 1970
128. Sjostrom R, Ekstedt J, Anggard E: Concentration gradients of monoamine metabolites in human cerebrospinal fluid. J Neurol Neurosurg Psychiatry 38:666–668, 1975
129. Kruesi MJP, Swedo SE, Hamburger SD, et al: Concentration gradient of CSF monoamine metabolites in children and adolescents. Biol Psychiatry 24:507–514, 1988
130. Pletscher A, Bartholini G, Rissot R: Metabolic fate of 1-[^{14}C]-DOPA in cerebrospinal fluid and blood plasma of humans. Brain Res 4:106–109, 1967
131. Elchisak MA, Polinsky RJ, Ebert MH, et al: Contribution of plasma homovanillic acid (HVA) to urine and cerebrospinal fluid HVA in the monkey and its pharmacokinetic disposition. Life Sci 23:2339–2348, 1978
132. Ashcroft GW, Dow RC, Moir ATB: The active transport of 5-hydroxyindol-3-ylacetic acid and 3-methoxy-4-hydroxyphenylacetic acid from a recirculatory perfusion system of the cerebral ventricles of the unanesthetized dog. J Physiol (Lond) 199;397–425, 1968
133. Bulat M: On the cerebral origin of 5-hydroxyindoleacetic acid in the lumbar cerebrospinal fluid. Brain Res 122:388–391, 1977
134. Riddle MA, Anderson GM, McIntosh S, et al: Cerebrospinal fluid monoamine precursor and metabolite

levels in children treated for leukemia: Age and sex effects and individual variability. Biol Psychiatry 21:69–83, 1986
135. Brewerton TD, Berrettini WH, Nurnberger JI Jr, et al: An analysis of seasonal fluctuations of CSF monoamine metabolites and neuropeptides in normal controls: Findings with 5-HIAA and HVA. Psychiatr Res 23:257–265, 1988
136. Perlow MJ, Lake CR: Daily fluctuations in catecholamines, monamine metabolites, cyclic AMP, and gamma-aminobutyric acid, in Wood JH (ed): Neurobiology of Cerebrospinal Fluid. New York, Plenum, 1982, pp 63–69
137. Bowers MB Jr: Clinical measurements of central dopamine and 5-hydroxytryptamine metabolism: Reliability and interpretation of cerebrospinal fluid acid monoamine metabolite measures. Neuropharmacology 11:101–111, 1972
138. Korf J, Van Prag HM: Amine metabolism in the human brain: Further evaluation of the probenecid test. Brain Res 221–230, 1971
139. Cooper JR, Bloom FE, Roth RH: The Biochemical Basis of Neuropharmacology, ed 5. New York, Oxford University Press, 1986
140. Commissiong JW: Monoamine metabolites: Their relationship and lack of relationship to monoaminergic neuronal activity. Biochem Pharmacol 34:1127–1131, 1985
141. Trulson ME: Dietary tryptophan does not alter the function of brain serotonin neurons. Life Sci 37:1067–1072, 1985
142. Moir ATB, Eccleston D: The effects of precursor loading in the cerebral metabolism of 5-hydroxyindoles. J Neurochem 15:1093–1108, 1968
143. Edwards DJ, Rizk M, Spiker DG: Effects of L-DOPA on the excretion of alcoholic metabolites of catecholamines and trace amines in rat and human urine. Biochem Med 25:135–148, 1981
144. Gibson CJ, Wurtman RJ: Physiological control of brain norepinephrine synthesis by brain tyrosine concentration. Life Sci 22:1399–1406, 1978
145. Lehnert H, Reinstein DK, Strowbridge BW, et al: Neurochemical and behavioral consequences of acute, uncontrollable stress: Effects of dietary tyrosine. Brain Res 303:215–223, 1984
146. Conlay LA, Maher TJ, Wurtman RJ: Tyrosine accelerates catecholamine synthesis in hemorrhaged hypotensive rats. Brain Res 333:81–84, 1985
147. Hery F, Simonnet G, Bourgoin S, et al: Effect of nerve activity on the in vivo release of [^{3}H]serotonin continuously formed from L-[^{3}H]tryptophan in the caudate nucleus of the cat. Brain Res 169:317–334, 1979
148. Ulus I, Wurtman RJ: Choline increases acetylcholine release. Lancet 1:624, 1987
149. Kennett GA, Joseph MH: Does in vivo voltammetry in the hippocampus measure 5-HT release? Brain Res 236:305–316, 1982
150. During MJ, Acworth IN, Wurtman RJ: Effects of systemic L-tyrosine on dopamine release from rat corpus striatum and nucleus accumbens. Brain Res 452:378–380, 1988
151. Westerink BHC, Kikkert RJ: Effect of various centrally acting drugs on the efflux of dopamine metabolites from the rat brain. J Neurochem 46:1145–1152, 1986
152. Davila R, Manero E, Zumarraga M, et al: Plasma homovanillic acid as a predictor of response to neuroleptics. Arch Gen Psychiatry 45:564–567, 1988
153. Wilk S, Watson E, Travis B: Evaluation of dopamine metabolism in rat striatum by a gas chromatographic technique. Eur J Pharmacol 30:238–243, 1975
154. Gordon EK, Markey SP, Sherman RL, et al: Conjugated 3,4-dihydroxy phenyl acetic acid (DOPAC) in human and monkey cerebrospinal fluid and rat brain and the effects of probenecid treatment. Life Sci 18:1285–1292, 1976
155. Wood JH: Sites of origin and cerebrospinal fluid concentration gradients: Neurotransmitters, their precursors and metabolites, and cyclic nucleotides, in Wood JH (ed): Neurobiology of Cerebrospinal Fluid. New York, Plenum, 1982, pp 53–62
156. Bartholini G, Tissot R, Pletscher A: Brain capillaries as a source of homovanillic acid in cerebrospinal fluid. Brain Res 27:163–68, 1971
157. Hefti F, Melamed E, Wurtman RJ: The site of dopamine formation in rat striatum after L-DOPA administration. J Pharmacol Exp Ther 217:189–197, 1981
158. Sourkes TL: On the origin of homovanillic acid (HVA) in the cerebrospinal fluid. J Neural Trans 34:153–57, 1973
159. Stanley M, Traskman-Bendz L, Dorovini-Zis K: Correlations between aminergic metabolites simultaneously obtained from human CSF and brain. Life Sci 37:1279–1286, 1985

160. Elsworth JD, Leahy DJ, Roth RH, et al: Homovanillic acid concentrations in brain, CSF and plasma as indicators of central dopamine function in primates. J Neural Trans 68:51–62, 1987
161. Bacopoulos NG, Maas JW, Hattox SE, et al: Regional distribution of dopamine metabolites in human and primate brain. Commun Psychopharmacol 2:281–286, 1978
162. Blinkov SM, Glezer II: The Human Brain in Figures and Tables. New York, Plenum, 1968
163. Maas JW, Contreras SA, Seleshi E, et al: Dopamine metabolism and disposition in schizophrenic patients. Arch Gen Psychiatry 45:553–559, 1988
164. Anden N-E, Grabowska-Anden M: Formation of deaminated metabolites of dopamine in noradrenaline neurons. Naunyn-Schmiedebergs Arch Pharmacol 324:1–6, 1983
165. Scatton JB, Dennis T, Curet O: Increase in dopamine and DOPAC levels in noradrenergic terminals after electrical stimulation of the acending noradrenergic pathways. Brain Res 298:193–196, 1984
166. Anden N-E, Brabowska-Anden M, Lindgren S: Very rapid turnover of dopamine in noradrenaline cell body regions. Naunyn-Schmiedebergs Arch Pharmacol 329:258–263, 1985
167. Burns RS, Chiueh CC, Markey SP, et al: A primate model of parkinsonism: Selective destruction of dopaminergic neurons in the pars compacta of the substantia nigra by *N*-methyl-4-phenyl-1,2,3,6-tetrahydropyridine. Proc Natl Acad Sci USA 80:4546–4550, 1983
168. Garelis E, Sourkes TL: Sites of origin in the central nervous system of monoamine metabolites measured in human cerebrospinal fluid. J Neurol Neurosurg Psychiatry 36:625–629, 1973
169. Weir RL, Chase TN, Ng LKY, et al: 5-hydroxyindoleacetic acid in spinal fluid: relative contribution from brain and spinal cord. Brain Res 52:409–412, 1973
170. Post RM, Goodwin FK, Gordon E: Amine metabolites in human cerebrospinal fluid: Effects of cord transection and spinal fluid block. Science 179:897–899, 1973
171. Garelis E, Young SN, Lal S, et al: Monoamine metabolites in lumbar CSF: The question of their origin in relation to clinical studies. Brain Res 79:1–8, 1974
172. Maas JW, Leckman JF: Relationship between central nervous system function and plasma and urinary MHPG and other norepinephrine metabolites, in Maas JW (ed): MHPG: Basic Mechanisms and Psychopathology. Orlando, Florida, Academic, 1983, pp 33–43
173. Sternberg DE, Heninger GR, Roth RH: Plasma homovanillic acid as an index of brain dopamine metabolism: Enhancement with deprisoquin. Life Sci 32:2447–2452, 1983
174. Maas JW, Contreras SA, Bowden CL, et al: Effects of debrisoquin on CSF and plasma HVA concentrations in man. Life Sci 36:2163–2170, 1985
175. Riddle MA, Leckman JF, Cohen DJ, et al: Assessment of central dopaminergic function using plasma-free homovanillic acid after debrisoquin administration. J Neural Transm 67:31–43, 1986
176. Davidson M, Losonczy MF, Mohs RC: Effects of debrisoquin and haloperidol on plasma homovanillic acid concentration in schizophrenic patients. Neuropharmacology 1:17–23, 1987
177. Snider SR, Kuchel O: Dopamine: An important neurohormone of the sympathoadrenal system. Significance of increased peripheral dopamine release for the human stress response and hypertension. Endocr Rev 4:291–306, 1983
178. Van Loon, GR, Schwartz L, Sole MJ: Plasma dopamine responses to standing and exercise in man. Life Sci 24:2273–2278, 1979
179. Kendler KS, Mohs RC, Davis KL: The effects of diet and physical activity on plasma homovanillic acid in normal human subjects. Psychiatr Res 8:215–23, 1983
180. Davidson M, Giordani A, Mohs RC, et al: Control of extraneous factors affecting plasma homovanillic acid concentrations. Psychiatr Res 20:307–312, 1987
181. Pettinger WA, Korn A, Spieger H: Debrisoquin, a selective inhibitor of intraneuronal monoamine oxidase in man. Clin Pharmacol Ther 10:667–674, 1969
182. Medina MA, Giachetti A, Shore PA: On the physiologic disposition and possible mechanism of the anti-hypertensive action of debrisoquin. Biochem Pharmacol 18:891–901, 1969
183. Moe RA, Bates HM, Palokski ZM: Cardiovascular effects of 3,4-dihydro-2(1H)isoquinoline carboxamidine (Declinax). Curr Ther Res 6:299–318, 1964
184. Maas JW, Hattox SE, Landis DH: Differential effects on brain catecholamines by debrisoquin. Biochem Pharmacol 28:3153–3156, 1979
185. Swann AC, Maas JW, Hattox SE, et al: Catecholamine metabolites in human plasma as indices of brain function: Effects of debrisoquin. Life Sci 27:1857–1862, 1980
186. Riddle MA, Leckman JF, Anderson GM, et al: Assessment of dopaminergic function in children and adults: Long and brief debrisoquin administration combined with plasma homovanillic acid. Psychopharmacol Bull 23:411–414, 1987

187. Sack DA, James SP, Doran AR: The diurnal variation in plasma homovanillic acid level persists but the variation in 3-methoxy-4-hydroxyphenylglycol level is abolished under constant conditions. Arch Gen Psychiatry 45:162–166, 1988
188. Riddle MA, Leckman JF, Anderson GM: Plasma-free homovanillic acid: Within- and across-day stability in children and adults with Tourette's syndrome. Life Sci 40:2145–2151, 1987
189. Davidson M, Giordani AB, Mohs RC, et al: Short term haloperidol administration acutely elevates human plasma homovanillic acid concentration. Arch Gen Psychiatry 44:189–190, 1987
190. Bunney BS, Walters JR, Roth RH, et al: Dopaminergic neurons: Effect of antipsychotic drugs and amphetamine on single cell activity. J Pharmacol Exp Ther 185:560–571, 1973
191. Bowers MB: 5-hydroxyindoleacetic acid in the brain and cerebrospinal fluid of the rabbit following administration of drugs affecting 5-hydroxytryptamine. J Neurochem 17:827–828, 1970
192. Eccleston D, Ashcroft GW, Crawford TB, et al: Effect of tryptophan administration on 5-HIAA in cerebrospinal fluid in man. J Neurol Neurosurg Psychiatry 33:269–272, 1970
193. Eccleston D, Ashcroft GW, Moir ATB, et al: A comparison of 5-hydroxyindoles in various regions of dog brain and cerebrospinal fluid. J Neurochem 15:947–957, 1968
194. Modigh K: The relationship between the concentration of tryptophan and 5-hydroxy-indoleacetic acid in rat brain and cerebrospinal fluid. J Neurochem 25:351–352, 1975
195. Anderson GM, Feibel FC, Cohen DJ: Determination of serotonin in whole blood, platelet-rich plasma, platelet-poor plasma and plasma ultrafiltrate. Life Sci 40:1063–1070, 1987
196. Leckman JF, Maas JW: Preliminary characterization of plasma MHPG in man, in Maas JW (ed): MHPG: Basic Mechanisms and Psychopathology. Orlando, Florida, Academic, 1983, pp 107–128
197. Crawley JN, Hattox SE, Maas JW, et al: 3-Methoxy-4-hydroxyphenethyleneglycol increase in plasma after stimulation of the nucleus locus coeruleus. Brain Res 141:380–384, 1978
198. Elsworth JD, Redmond DE, Roth RH: Plasma and cerebrospinal fluid 3-methoxy-4-hydroxyphenylethylene glycol (MHPG) as indices of brain norepinephrine metabolism in primates. Brain Res 235:115–124, 1982
199. Jimmerson DC, Ballenger JC, Lake RM, et al: Plasma and CSF MHPG in normals. Psychopharmacol Bull 17:86–87, 1981
200. Davis KL, Hollister LE, Mathe AA, et al: Neuroendocrine and neurochemical measurements in depression. Am J Psychiatry 138:1555–1562, 1981
201. Maas JW, Kocsis, JH, Bowden CL, et al: Pre-treatment neurotransmitter metabolites and response to imipramine or amitriptyline treatment. Psychol Med 12:37–43, 1982
202. Leckman JF, Maas JW, Redmond DE Jr, et al: Effects of oral clonidine on plasma 3-methoxy-4-hydroxyphenethyleneglycol (MHPG) in man: preliminary report. Life Sci 26:2179–2185, 1980
203. Riddle MA, Leckman JF, Anderson GM, et al: Plasma MHPG: Within- and across-day stability in children and adults with Tourette's syndrome. Biol Psychiatry 24:391–398, 1988.
204. Pitts FN, McClure JN: Lactate metabolism in anxiety neurosis. N Engl J Med 227:1329–1336, 1967
205. Banki CM, Arato M: Amine metabolites and neuroendocrine responses related to depression and suicide. J Affective Disord 5:223–232, 1983
206. Heninger GR, Charney DS, Sternberg DE: Serotonergic dysfunction in depression. Prolactin response to intravenous tryptophan in depressed patients and healthy subjects. Arch Gen Psychiatry 41:398–402, 1984
207. Zohar J, Mueller EA, Insel TR, et al: Serotonergic responsivity in obsessive-compulsive disorder. Comparison of patients and healthy controls. Arch Gen Psychiatry 44:946–951, 1987
208. Charney DS, Goodman WK, Price LH, et al: Serotonin function in obsessive-compulsive disorder. Arch Gen Psychiatry 45:177–185, 1988
209. Hunt RD, Cohen DJ, Shaywitz SE, et al: Strategies for study of the neurochemistry of attention deficit disorder in children. Schizophr Bull 8:236–52, 1982
210. Puig-Antich J: Affective disorders in children and adolescents: Diagnostic validity and psychobiology, in Meltzer HY (ed): Psychopharmacology: The Third Generation of Progress. New York, Raven, 1987, pp 843–859
211. Feinstein AR: Clinical Epidemiology. The architecture of clinical research. Philadelphia, WB Saunders, 1985, pp 191–311
212. Sedvall G, Farde L, Persson A, et al: Imaging of neurotransmitter receptors in the living human brain. Arch Gen Psychiatry 43:995–1005, 1986
213. Bice AN, Wagner HN, Frost JJ, et al: Simplified detection system for neuroreceptor studies in the human brain. J Nucl Med 27:184–191, 1986

214. Jeffries KJ, Tamminga CA, Wong DF, et al: Validation of a positron emission probe for neuroreceptor studies in human brain. Soc Neurosci Abs 14:105, 1988
215. Gadian DG: Nuclear Magnetic Resonance and Its Application to Living Systems. New York, Oxford University Press, 1982
216. Kuhar MJ, DeSouza EF, Unnerstall JR: Neurotransmitter receptor mapping by autoradiography and other methods. Ann Rev Neurosci 9:27–59, 1986
217. Goldman-Rakic PS: Modular organization of prefrontal cortex. Trends Neurosci 7:419–424, 1984
218. Whitehouse PJ, Lynch D, Kuhar MJ: Effects of postmortem delay and temperature on neurotransmitter receptor binding in a rat model of the human autopsy process. J Neurochem 43:553–559, 1984
219. Graybiel AM, Hirsch EC, Agid YA: Differences in tyrosine hydroxylase-like immunoreactivity characterize the mesostriatal innervation of striosomes and extrastriosomal matrix at maturity. Proc Natl Acad Sci USA 84:303–307, 1987
220. Uhl GR (ed): In Situ Hybridization in Brain. New York, Plenum, 1986

3

Role of the GABA–Benzodiazepine Receptor Complex in Stress

In Vivo Approaches and Potential Relevance to Childhood Psychopathology

Stephen I. Deutsch, Abraham Weizman, Ronit Weizman, Frank J. Vocci, Jr., and Karin A. Kook

1. INTRODUCTION

An appreciation of the structure and function of the $GABA_A$–benzodiazepine receptor–chloride ionophore complex is necessary in order to understand the therapeutic mechanism of action of several major classes of anxiolytic and sedative-hypnotic drugs. The chapter provides the background necessary to appreciate these structural and functional considerations. Studies reporting that genetic differences in the density of central benzodiazepine receptors exist between strains of animals differing in the traits of *emotionality* and *fearfulness* are presented. The demonstration, isolation, and synthesis of several inverse agonists have spurred theoretical speculation about the existence of endogenous ligands for the benzodiazepine receptor, as well as a role for this receptor in normal and pathological responses to stress. This chapter selectively reviews studies describing the plasticity of the benzodiazepine–GABA receptor complex in response to environmental stress. Several of these stress paradigms were naturalistic ones showing that the application of specific stressors during the early stages of an animal's development results in enduring changes in benzodiazepine receptor sensitivity and behavior in the adult animal. There are also data showing that the benzodiazepine receptor is involved in the mediation and modulation of aggressive behavior. The potential relevance of these observa-

Stephen I. Deutsch • Psychiatry Service, Veterans Administration Medical Center, and Department of Psychiatry, Georgetown University School of Medicine, Washington, D.C. 20007. *Abraham Weizman* • Geha Psychiatric Hospital, Beilinson Medical Center, Petah Tiqva 49100; and Sackler Faculty of Medicine, Tel Aviv University, Tel Aviv 69978, Israel. *Ronit Weizman* • Pediatric Department, Hasharon Hospital, Petah Tiqva 49372; and Sackler Faculty of Medicine, Tel Aviv University, Tel Aviv 69978, Israel. *Frank J. Vocci, Jr.* • Medications Development Program, Division of Preclinical Research, National Institute on Drug Abuse, Rockville, Maryland 20857. *Karin A. Kook* • Division of Neuropharmacological Drug Products, Food and Drug Administration, Rockville, Maryland 20857.

tions to child psychiatry is obvious. Evidence implicating peripheral hormones in the regulation of central benzodiazepine receptor sensitivity in response to stress is presented. A stress-induced modification of the complex would suggest that adaptive responses may be accompanied by changes in γ-aminobutyric acid (GABA)ergic transmission that are mediated postsynaptically; some of these changes appear to occur rapidly (i.e., within 1 min of exposure to the stress) and may reflect post-translational modification of the complex.[1] Most of the data on the benzodiazepine receptor and its modification by environmental stress were obtained with classical in vitro techniques, especially filtration-binding assays. These techniques are performed under conditions that are not physiological with respect to temperature and salt concentrations; they disrupt the local neuronal circuitry involved in the regulation of benzodiazepine receptor sensitivity in the intact animal. Therefore, an in vivo approach to the measurement of benzodiazepine receptors in intact animals that avoids the artifacts associated with in vitro and ex vivo techniques has been developed.[2] The method involves the intravenous injection of a tracer quantity of the radiolabeled antagonist Ro15-1788. Using this in vivo approach, a relationship was shown between benzodiazepine receptor occupancy and the pharmacological potencies of several benzodiazepines.[3] The application of this method to studying the effects of environmental stress and the mechanism of stress-induced modifications of binding are also reviewed.

2. STRUCTURAL AND FUNCTIONAL CONSIDERATIONS OF THE BENZODIAZEPINE–GABA RECEPTOR COMPLEX

The major pharmacological actions of the benzodiazepine agonists, inverse agonists, and antagonists can be explained by their ability to modulate GABA-mediated chloride ion flux within the central nervous system (CNS).[4] Benzodiazepine agonists, such as diazepam, are anxiolytic agents that are effective only in the presence of GABA, acting to adjust the "gain" or fine tuning of GABAergic transmission. They act postsynaptically by binding to receptors that are structurally and functionally associated with a recognition site for GABA and a chloride channel in an oligomeric protein complex.[5] The potencies of benzodiazepine agonists in increasing the punished responding of animals in a conflict paradigm correspond positively with their abilities to inhibit [^{3}H]diazepam binding in vitro. Inverse agonists are anxiogenic and proconvulsant compounds that bind with high affinity to central benzodiazepine receptors; β-carboline carboxylate ethyl ester (β-CCE) was the first such anxiogenic compound to be discovered.[6] Inverse agonists lead to behavioral agitation; increase heart rate, blood pressure, and plasma levels of "stress" hormones (i.e., cortisol, epinephrine, and norepinephrine); and induce the subjective experience of extreme anxiety in healthy human volunteers.[7,8] Antagonists also bind to the central benzodiazepine receptor inhibiting the major central actions of agonists and inverse agonists at doses that are devoid of any significant intrinsic central actions.[9]

In 1977, two independent groups of investigators reported the existence of specific binding sites for [^{3}H]diazepam in rat brain.[10,11] The specific binding of [^{3}H]diazepam to a crude synaptosomal fraction of rat forebrain and cerebral cortex was measured over a range of concentrations. A Scatchard analysis of this data revealed a single population of saturable high-affinity binding sites; the dissociation constant for diazepam was in the low nanomolar range. Nonspecific binding was measured with an excess of unlabeled diazepam and was about 5–10% of the total bound at a concentration of 1.5 nM [^{3}H]diazepam. Benzodiazepines with an asymmetric carbon atom (C-3) showed stereospecificity in terms of their ability to inhibit specific [^{3}H]diazepam binding competitively; dextrorotatory enantiomers were more potent than the levorotatory ones. The specific [^{3}H]diazepam binding site was enriched in the synaptosomal fraction, accounting for about 60% of the specific binding observed in whole homogenate of cerebral cortex. The sites were also unevenly distributed across the various brain regions; binding was highest in the cerebral cortex and lowest in the pons-medulla. The specific [^{3}H]diazepam binding site identified in vitro appeared to be relevant to the pharmacological actions of benzodiazepines. Significant correlations were observed between the ability of a series of benzodiazepines to inhibit binding of [^{3}H]diazepam to this site and their potencies in tests predictive of anxiolytic, anticonvulsant, and sedative effects.

The benzodiazepine receptor is functionally coupled to the $GABA_A$ recognition site; this has been shown in binding studies performed in vitro. GABA and muscimol, a $GABA_A$ agonist, increased the specific binding of [^{3}H]diazepam to a washed membrane preparation of rat cerebral cortex in a dose-dependent manner.[12] This effect was due to an increase in the affinity of [^{3}H]diazepam for its specific binding site. GABA did not alter the maximal number of benzodiazepine receptor sites.

Electrophysiological studies have shown that benzodiazepine agonists are effective only in the presence of GABA to increase the likelihood that chloride ion flux will take place across the channel.[13] The effect of GABA on specific [^{3}H]diazepam binding was antagonized by (+)bicuculline, a competitive antagonist of GABA; in fact (+)bicuculline alone reduced specific [^{3}H]diazepam binding in control preparations of washed membrane, suggesting that binding was influenced by the endogenous GABA in the preparation. By contrast, the binding affinity of the methyl ester of β-carboline-3-carboxylic acid (β-CCM), an inverse agonist or proconvulsant, was reduced by GABA in a bicuculline-sensitive manner.[14] The affinity of Ro15-1788, a benzodiazepine antagonist, for the benzodiazepine receptor was virtually unchanged by GABA agonists. Thus, the sensitivity of the benzodiazepine receptor can be changed by an allosteric effect of GABA on the affinity of this binding site for benzodiazepine agonists and inverse agonists. The functional coupling of the recognition sites for GABA and the benzodiazepines can be understood on a structural basis.

The supramolecular complex also contains a domain for *channel agents* whose actions occur on or near the actual chloride ionophore.[5] The existence of this domain can explain the action of several sedative-hypnotic and convulsant agents. Picrotoxin is a convulsant drug of plant origin that blocks GABAergic transmission by binding to a site related to the chloride ion channel; this site is distinct from the $GABA_A$ recognition site and benzodiazepine receptor. In vitro studies suggest that barbiturates enhance GABA and benzodiazepine agonist binding in an anion-dependent manner that is reversed by picrotoxin.[5,15] Electrophysiologically, the potentiation of GABAergic transmission by pentobarbital can be explained by a stabilization of the chloride ion channel in the open position with a prolongation in the duration of chloride ion conductance.[16] The in vitro binding data and electrophysiology suggest a complex functional coupling between the picrotoxin–barbiturate receptor site and the recognition sites for $GABA_A$ and benzodiazepines on the supramolecular protein complex that includes allosteric interactions. For example, whereas pentobarbital causes a "shift" in the benzodiazepine receptor toward an increased affinity for agonists, the affinity for inverse agonists (e.g., β-CCM) and antagonists (e.g., Ro15-1788) is decreased and unchanged, respectively.[17,18] Pentobarbital has also been shown to potentiate the positive effect of GABA on [^{3}H]diazepam binding.[19]

The specific benzodiazepine binding site is a structural part of a tetrameric complex that contains a recognition site for a subclass of GABA agonists (designated $GABA_A$) and a chloride ionophore; the subunit structure of this benzodiazepine–GABA receptor complex is α_2–β_2.[20] The α-subunit contains the benzodiazepine binding site and the β-subunit the $GABA_A$ receptor. The separate subunits combine to form a chloride channel in a neurotransmitter-gated receptor complex. The complex has been purified to homogeneity from bovine cerebral cortex with affinity chromatography and the peptide sequences of each subunit used to synthesize oligodeoxyribonucleotide probes. The probes were used to screen complementary

DNA (cDNA) libraries derived from bovine brain and calf cerebral cortex constructed in the phage λ gt10. The protein sequences and hydropathy profiles of the two $GABA_A$ subunits suggest a common evolutionary origin. In fact, the two $GABA_A$ receptor subunits share a limited homology with the subunits of the nicotinic acetylcholine receptor (AChR), to which they display an identical number and distribution of hydrophobic transmembrane regions. These data suggest the existence of a *superfamily* of ligand-gated receptor subunits derived from a common ancestral protein. In the most current model, shared features of this superfamily of ligand-gated subunits include an extracellular β-structural loop formed by the disulfide bonding of two conserved cysteine residues, and four transverse hydrophobic membrane domains. The β-subunits of the $GABA_A$ receptor contain a unique intracellular site for the cAMP-dependent phosphorylation of a serine residue. This site does not seem to be shared with all the other ligand-gated receptor subunits. Phosphorylation of an intracellular serine residue may be the basis of rapid posttranslational modification of GABA-mediated chloride ion flux, especially in response to stress.[1,21–23] More recently performed cloning experiments suggest the existence of genetic variants of the alpha and beta subunits, and an additional class of subunit.[20] Expression of the alpha subunit variants appears to be influenced by anatomic brain region and may explain earlier pharmacological distinctions between benzodiazepine receptors (i.e., Types I and II). For example, the subunit designated $alpha_1$ is enriched in cerebellum which contains mainly "Type I" receptors, whereas the $alpha_3$ subunit is a component of "Type II" receptors and shows greater sensitivity to the allosteric effects of $GABA_A$ agonists (i.e., a higher potentiation of benzodiazepine binding in the presence of GABA than the other alpha subunit variants). Clearly, the application of molecular biological approaches will contribute to our understanding of differences among brain regions in the pharmacological properties of benzodiazepine/$GABA_A$ receptors.[20a]

3. GENETIC STRAIN DIFFERENCES AND ENVIRONMENTAL STRESS-INDUCED ALTERATIONS IN THE BENZODIAZEPINE–GABA RECEPTOR COMPLEX

Strains of mice and rats differ in terms of their maximal density of benzodiazepine receptors; these genetic differences in the density of receptors may be correlated with strain differences in response to stress and novelty. Moreover, alterations in the maximal density of benzodiazepine receptors within a strain can be produced by differences between animals in their handling, and exposure to conflict and other stressful situations.

Specific [^{3}H]diazepam binding in vitro was higher in the hippocampus and hypothalamus of a strain of rat bred selectively for low "fearfulness" (Maudsley nonreactive), as compared with a strain bred for high fearfulness (Maudsley reactive).[24] The less fearful rats also showed significantly increased binding in the midbrain, medulla-pons, and thoracic spinal cord. This apparent inverse relationship between fearfulness and benzodiazepine receptor binding was due to an increase in the maximal density of binding sites; the affinity of the benzodiazepine receptor for diazepam was unchanged. However, a reduction in [^{3}H]flunitrazepam binding was not detected in an autoradiographic study of Maudsley reactive and nonreactive rats.[25] An inverse relationship between fearfulness or emotionality and specific [^{3}H]diazepam binding in vitro was also seen in experiments with inbred strains of mice differing in their emotionality in an open field test.[26] Benzodiazepine receptor binding was decreased in whole-brain homogenate of the more emotional BALB/cJ strain of mouse, as compared with the binding in whole-brain homogenates of three nonemotional strains (AKR/J, C57BL/10J, and C57BL/6J). The decreased binding in the more emotional strain was due to a reduction in the maximal density of benzodiazepine receptor sites, rather than to an alteration in the apparent affinity constant.

The ability of a compound to increase the punished responding of food- and water-deprived rats in a conflict situation is predictive of its anxiolytic efficacy. A typical conflict paradigm involves the application of shock through an electrified drinking spout to water-deprived animals. In this paradigm, an active anxiolytic compound would be identified by its ability to increase the amount of shock accepted or tolerated by a treated group of animals over that amount accepted by animals treated with vehicle alone. The functional relevance of the benzodiazepine receptor was supported by the positive correlation observed between the ability of a series of benzodiazepines to inhibit specific [^{3}H]diazepam binding in vitro competitively and to increase punished responding in a conflict paradigm.[27]

While the central benzodiazepine receptor is relevant to the understanding of the pharmacological effects of benzodiazepines, the physiological role of this receptor is less certain. A reduction of specific [^{3}H]diazepam binding in vitro was reported in the frontal cortex of food- and water-deprived rats exposed to a standard conflict paradigm for 5 min, as compared with animals that were similarly deprived but not exposed to the conflict stress.[27] In this paradigm, animals received an electric shock (200 μA) on a 5-sec on–off schedule when they drank a 10% dextrose solution through an electrified spout. The magnitude of the reduction in specific [^{3}H]diazepam binding was about 25% and was significant ($p < 0.01$). In a related experiment, deprived rats receiving inescapable food shock (100 msec of 300 μA every 15 sec for 5 min) showed smaller (approximately 14%), but still significant reductions of specific [^{3}H]diazepam binding in frontal cortex, as compared with control animals.[27] The data suggested that alterations in either the number of available sites or the binding parameters of benzodiazepine receptors could mediate or reflect state-dependent changes associated with anxiety. Also, the reduction of specific binding sites could be due to occupation of benzodiazepine receptors by an unidentified endogenous ligand whose release was increased in response to anxiety-provoking stimuli.

Food-deprived rats trained to respond for food reinforcement on a variable interval schedule show a conditioned suppression of this behavior in response to a warning signal (conditioned stimulus (CS), i.e., paired with footshock.[28] This classically conditioned suppression of food-reinforced responding is referred to as a conditioned emotional response (CER). Low-dose diazepam administered intraperitoneally 30 min prior to the testing situation (e.g., 5 mg/kg) restored the rate of response to normal. Thus, the anxiolytic actions of benzodiazepines can be tested in this CER paradigm. The development of CER in rats was associated with about a 25% reduction ($p < 0.05$) in the maximal density of benzodiazepine receptors in the cerebral cortex.[28] In this study, the application of footshock alone in the absence of the development of CER was not associated with a significant change in the density of benzodiazepine receptors in the cerebral cortex. Thus, a learned emotional component in the development of CER seemed to be responsible for the reduction in binding, as opposed to the application of the aversive stimulus alone.

Several different and prolonged or repetitive stressful paradigms produced effects on specific benzodiazepine receptor binding that were not unidirectional.[29] The direction of change differed according to the nature of the specific stressor. Stress did not always result in a change in specific binding; moreover, in those paradigms in which an effect was observed, the magnitude of change was modest. The small magnitude of change could reflect adaptation to the repetitive or chronic nature of the stressful paradigms that obscured rapid and early changes. In one paradigm, a group of food-deprived rats were kept separately 18 hr/day in a compartment with an electrified grill floor. During this 18-hr period, they received nine randomized foot shocks (0.5 sec; 1 mA) interspersed with nine signal combinations of a tone and light flash. The experimental paradigm lasted for 21 days. The animals subjected to this chronic stressful paradigm showed a modest (13%) but significant reduction ($p < 0.01$) in the maximal density of [^{3}H]flunitrazepam binding sites in frontal cortex as compared with non-deprived and nonstressed controls. This chronic electric footshock paradigm did not change the binding in the striatum or occipital cortex. Rats subjected to 2 hr/day of forced immobilization for 4 days showed a small (9%) but significant increase ($p < 0.05$) in specific [^{3}H]flunitrazepam binding in frontal cortex. Isolation of rat pups less than 24 hr after birth from their mothers for daily periods of 15 or 30 min for 10 days resulted in small but consistent decreases in the maximal density of binding sites in whole cortex up to 36 days after birth. In the rat, there is a rapid proliferation of benzodiazepine receptors during the first 10 days of postnatal life.

Two other chronic or repetitive stress paradigms did not affect specific binding to benzodiazepine receptors measured in vitro.[29] Rats subjected to daily cold water swim stress consisting of three 5-min swimming periods at 15°C, interspersed with 5 min of rest, for 4 days did not show an alteration of specific [^{3}H]flunitrazepam binding in frontal cortex, occipital cortex, hippocampus, and striatum. Rats implanted subcutaneously with a slow-release Silastic pellet containing amphetamine base showed the expected behavioral manifestations but did not show any change in specific [^{3}H] diazepam binding in striatum and frontal cortex 5 days after implantation. Finally, specific [^{3}H]diazepam binding in forebrain did not change in male NMRI mice made aggressive by isolation for 3 weeks.[29] In view of the ability of benzodiazepine agonists to inhibit aggression in this paradigm, the absence of an effect of isolation-induced aggression on specific binding to benzodiazepine receptors was surprising. Also, the stress of isolation did result in aggression and an alteration in benzodiazepine receptor binding in other mouse strains. These data suggest that changes in specific benzodiazepine receptor binding do not occur as a general phenomenon in response to stress, but the direction and magnitude of change, and the brain region(s) involved may relate to specific qualities of the stressor. For example, immobilization stress, a continuous form of stress, resulted in an increase in specific binding in the frontal cortex, whereas intermittent footshock produced a reduction. Genetic factors can also influence the degree and direction of change in benzodiazepine receptor binding in response to stress. Moreover, artifacts associated with the in vitro binding assay, especially homogenization and filtration of tissue, may limit the appreciation of the true status of environmental stress-induced modifications of benzodiazepine receptor binding in intact animals.

In another study, housing of individual male mice in isolation for 4 weeks resulted in increased aggression and was associated with significant reductions of specific [^{3}H]diazepam binding in cerebral cortex, diencephalon, and cerebellar cortex.[30] The mouse strain was inbred (CF-1) and differed from the inbred strain (NMRI) used in the study that failed to show an effect of isolation-induced aggression on benzodiazepine receptors.[29] Essman and Valzelli distinguished between two types of benzodiazepine receptors based on the thermal stability (type II) or lability (type I) of specific [^{3}H]diazepam binding to heating at 60°C for 10 min.[30] The isolated and aggressive mice showed about a 40% reduction ($p < 0.02$) in the maximal density of so-called type I receptors in the cerebral cortex with a small increase in the binding affinity as compared with group-housed controls. There were also significant reductions in the maximal number of type II receptors in the diencephalon (37%; $p < 0.02$) and cerebellar cortex (28%; $p < 0.02$) of the isolated and aggressive mice. Different functional properties were ascribed to these two types of benzodiazepine receptors; type I receptors were considered to be more closely associated with the anticonflict effects of benzodiazepine agonists and type II receptors to be associated with sedative effects. The data support hypotheses implicating benzodiazepine receptors or unidentified endogenous ligands for these receptors in an animal's response to stress. The absence of an effect of isolation-induced aggression on specific [^{3}H]diazepam binding in the forebrain of male NMRI mice suggests that genetic factors are important in determining the role of benzodiazepine receptors in an animal's response to stress. Methodological differences in measuring specific binding to benzodiazepine receptors in vitro could have contributed, however, to these discrepant results.[29,30]

A subgroup of male Wistar rats (21%) housed individually in cages with nontransparent walls for 3 months showed aggressive (muricidal) behavior; benzodiazepines were not effective in reducing the aggressive behavior of these animals.[31] Isolated and aggressive rats showed a reduction in the specific binding of [^{3}H]flunitrazepam in the cerebral cortex, hippocampus, midbrain, and cerebellum; rats that were isolated but nonaggressive showed a reduction only in the hippocampus and cerebellum. These data suggest that induction of muricidal activity may be related to alterations in benzodiazepine receptor binding in the cerebral cortex and midbrain. Genetic factors could account for the isolation-induced reduction in binding in the cerebral cortex and midbrain of some animals and for the failure of benzodiazepine agonists to attenuate aggressivity. Benzodiazepines exert antiaggressive effects in several strains of mice maintained in isolation. As compared with group-housed animals, the B_{max} was reduced in the cerebral cortex, midbrain, and cerebellum of isolated-aggressive rats; a decreased affinity was observed in the cerebral cortex and hippocampus.

Prenatally stressed female rats showed a deficit in a well-defined maternal behavior as adults.[32] While in utero, the mothers of these animals were exposed to three sessions of stress per week consisting of 5 min of bell noise alternating with 5 min of flashing light throughout the pregnancy; the duration of each session was 4 hr. The adult female offspring of this prenatally stressed condition were mated with normal males at 2 months; maternal behavior was tested with a pup retrieval paradigm. The mothers were required to retrieve their pups under control and stressed conditions. The control condition involved traversing an alley in order to retrieve the pups and the stressed condition involved traversing the same alley with an airstream directed into it. Prenatally stressed mothers and normal mothers were identical in terms of their pup retrieval under the control condition. However, in the conflict situation (i.e., airstream), the prenatally stressed mothers performed significantly worse on a variety of measures. Fewer prenatally stressed mothers retrieved all their pups. Moreover, they showed an increased latency in making contact with the first pup and took longer to retrieve the first pup and all the subsequent pups. A replicate group of female rats stressed prenatally in an identical fashion showed a 36% reduction in the maximal density of specific [^{3}H]flunitrazepam binding sites in the hippocampus. There was no change in the affinity of the receptor for the ligand. Thus, the deficit in maternal behavior was associated with a reduction in the number of hippocampal benzodiazepine receptors.

Rat pups handled postnatally for the first 21 days of life and tested at 100 days of age showed a reduction in their latency to start eating when placed in a novel environment, as compared with nonhandled controls.[33] Handling consisted of removing the pups from their mother and the nest daily for 15 min during the first 3 weeks of life. Appetitive behaviors are known to be suppressed after exposure to a novel environment or a punishing stimulus. The maximal density of specific [^{3}H]flunitrazepam binding sites were increased in whole-brain homogenate of a replicate group of postnatally handled rats; there was no difference in the apparent affinity for [^{3}H]flunitrazepam. These data suggest that alterations in benzodiazepine receptor sensitivity are involved in an animal's response to the stress of novelty. Benzodiazepine agonists are able to reduce the suppression of appetitive behaviors upon exposure to novelty.

The ability of benzodiazepine agonists to facilitate GABAergic transmission and to suppress the effects of stressful manipulations in animals prompted a study of the effects of several different acute stress paradigms on GABAergic transmission in rats.[34] Four different acute stress paradigms increased the threshold convulsant dosage of picrotoxin significantly suggesting that these stressors augmented GABAergic transmission. In addition to showing an increase of the threshold convulsant dosage of picrotoxin, rats subjected to a single cold water swim stress (6°C, 3 min) showed an increase in the threshold convulsant dosage of pentetrazol, delayed onset of seizures following a standard intraperitoneal dose of thiosemicarbazide and isoniazid, and an increase in the maximal density of specific [^{3}H]flunitrazepam binding sites in cortex. These animals were compared with rats subjected to 3 min of forced swimming at 35°C. However, the anticonvulsant potency of diazepam and its stimulant effect on short-term food intake were not changed following cold water swim stress. These data are consistent with the lack of significant change in the affinity of the benzodiazepine receptor and suggest that the anticonvulsant efficacy and ability of diazepam to stimulate food intake occur at doses that do not saturate the available receptors. The results also suggest that rapid stress-induced alterations in the number of central benzodiazepine receptors can occur.

Swim stress is a profound continuous stressor that can cause significant elevations of the plasma corticosteroid concentration in rats.[35] Inconsistent findings on the ability of this stressor to alter the density of central benzodiazepine receptors have been noted (see for example Refs. 35 and 36). Rapid reversibility and a unique temporal course of changes in benzodiazepine receptor binding for specific brain regions may account for some of the failure to observe consistent patterns of stress-induced alterations.[37] For example, rats exposed to 15 min of swim stress at 18°C showed initial reductions in specific [^{3}H]flunitrazepam binding in vitro in the cerebral cortex (30%; $p < 0.001$) and hippocampus (27%; $p < 0.01$) immediately following the stress. However, at 45 min after the termination of stress, binding increased significantly in the hippocampus (52%; $p < 0.001$) and returned to baseline in the cerebral cortex. Immediately after stress, there were no changes detected in any of seven other brain regions that were examined. Although most of the temporal changes in binding in the hippocampus could be attributed to changes in the maximal number of receptors, a significant decrease in affinity was reported 24 hr after termination of the stressful event. The authors speculated that the high density of steroid receptors in the hippocampus and the stress-induced elevation of adrenal steroids could be related to the persistent deviation of benzodiazepine receptor binding from baseline in this tissue.[37]

Rats that are unable to exert any control over the presentation of an aversive stimulus (e.g., inescapable tailshock) show a deficit in the acquisition of escape learning in a shuttlebox task 24 hr later.[38] The behavioral syndrome displayed by these rats is termed *learned helplessness;* it has been proposed as an animal model of depression. The anxiogenic inverse agonist *N*-methyl-β-carboline-3-carboxamide (FG-7142; 10 mg/kg, i.p.) administered 24 hr before escape training in a shuttlebox resulted in a deficit in the acquisition of escape learning that was equal to the deficit seen with administration of inescapable tail shock (80 shocks, 1 mA, 5 sec).[38] The induction of at least some features of *learned helplessness* by FG-7142 was mediated by the benzodiazepine receptor, since the development of the deficit in escape learning was prevented by pretreatment with Ro15-1788, a benzodiazepine receptor antagonist. Furthermore, chlordiazepoxide (5 mg/kg, i.p.) administered 30 min before a session of inescapable shock antagonized the delayed development of an escape learning deficit.[39] These data suggest that the benzodiazepine receptor is involved in the delayed behavioral effects of exposure to inescapable stress, and benzodiazepine agonists may be protective in antagonizing the delayed emergence of learned helplessness.

The ability of halide ions to enhance the binding of [^{3}H]flunitrazepam has been interpreted to reflect the coupling of the chloride ionophore with the benzodiazepine receptor.[22,23] The inconsistent and modest effects of acute stress on benzodiazepine receptor binding prompted examination of the effects of acute swim stress on the chloride ionophore, or the "effector" component of the supramolecular complex. Stress-induced changes in the effector component were assessed by measuring changes in the ability of halide ions to enhance [^{3}H]flunitrazepam binding and alterations in the binding characteristics of a "cage" convulsant. Cage convulsants are thought to act by binding to a site in or near to the channel domain of the supramolecular complex obstructing chloride ion flux. The potency and efficacy of halide ions to increase [^{3}H]flunitrazepam binding were increased in homogenates of cerebral cortex and hippocampus prepared from rats exposed to a session of ambient temperature swim stress. In the absence of chloride ions, there was no difference in the basal binding of [^{3}H]flunitrazepam between stressed and unstressed animals. In stressed animals, chloride ions appeared to increase the affinity of the benzodiazepine receptor for [^{3}H]flunitrazepam. Acute swim stress was not only reflected in an alteration in the coupling of the chloride ionophore to the benzodiazepine receptor but also seemed to be associated with a direct modification of the chloride channel itself.[21,22] Ambient temperature swim stress (10 min) increased the affinity and maximal density of specific binding sites for [^{35}S]-*t*-butylbicyclophosphorothionate ([^{35}S]-TBPS), a radiolabeled cage convulsant thought to bind at or near the chloride ionophore. Thus, alterations in the coupling of components of the supramolecular complex to each other and a modification of the effector component itself could be the mechanism of rapid stress-induced modulation of GABAergic transmission.

In another study, rapid alterations in the functional activity of the $GABA_A$-benzodiazepine-receptor–chloride ionophore complex were shown in the cerebral cortex and hippocampus of cold water, swim-stressed rats.[23] The functional activity of the complex was assessed by the ability of muscimol, a $GABA_A$ agonist, to stimulate the uptake of chloride ions into synaptoneurosomes, a novel subcellular preparation that includes both pre- and postsynaptic elements. The potency and efficacy of muscimol to stimulate chloride ion uptake were increased in synaptoneurosomes prepared from the cerebral cortex and hippocampus of rats swim-stressed in cold water (15–17°C). Adrenalectomy performed 3 weeks before sacrifice prevented the enhancement of muscimol-stimulated chloride ion flux; adrenalectomy did not influence muscimol-stimulated flux significantly in unstressed control rats. Thus, peripheral mechanisms are involved in the regulation of the functional activity of the complex.

Rats adrenalectomized 2 weeks before sacrifice showed a significant increase in the maximal density of specific [^{3}H]flunitrazepam binding sites in the hippocampus, striatum, and hypothalamus.[40] Adrenalectomy did not alter the apparent affinity of the benzodiazepine receptor for this radiolabeled ligand in these regions. There was no change in specific [^{3}H]flunitrazepam binding in the cerebral cortex, cerebellum, and olfactory bulb. A 1-week course of replacement therapy of the adrenalectomized animals with dexamethasone phosphate (0.25 mg/kg per day, s.c.) prevented the adrenalectomy-induced increases in the maximal density of binding sites. Dexamethasone administration to sham-operated animals did not alter the affinity or maximal density of benzodiazepine receptor sites. Adrenalectomy-induced alterations in binding appeared to be confined to brain regions enriched in glucocorticoid receptors. The data suggest that glucocorticoids in the blood could act via their central receptors to influence benzodiazepine receptor sensitivity in specific regions. These glucocorticoid-mediated effects are not likely to be involved in the rapid regulation of benzodiazepine receptor sensitivity. Moreover, the effects of dexamethasone did not seem to be direct ones, since increasing concentrations of dexamethasone did not alter the binding characteristics of [^{3}H]flunitrazepam to striatal homogenates of control nonoperated rats. Adrenalectomy was also reported to influence the ability of muscimol to alter the K_D for [^{3}H]flunitrazepam binding in specific regions of rat brain.[41] Ordinarily, muscimol increases the affinity of the benzodiazepine receptor for [^{3}H]flunitrazepam through an allosteric effect on the supramolecular complex. This effect of adrenalectomy was reversed or prevented by replacement therapy with dexamethasone. Thus, peripheral corticosteroid levels may be involved in the tonic maintenance of the coupling or allosteric interaction between components of the supramolecular complex. The effects of adrenalectomy on the enhancement of [^{3}H]flunitrazepam binding by muscimol were observed in brain regions enriched with steroid receptors.

A recent study showed that the status of the coupling of components of the supramolecular complex to each other, and the maximal density of [^{35}S]-TBPS binding sites are tonically maintained and exquisitely sensitive to environmental conditions.[1] Rats housed in a soundproofed environment protected from the usual activities of a conventional animal room facility showed a significant reduction in the density of [^{35}S]-TBPS binding sites and efficacy of chloride ions to enhance [^{3}H]flunitrazepam binding, as compared with conventionally housed animals. The protected animals also showed smaller, but significant, increases in these binding parameters after 10 min of ambient temperature swim stress as compared with rats taken from a conventional facility. The protected animals were extremely sensitive to the removal of a cohort from a common cage. The second rat showed increases in the density of [^{35}S]-TBPS binding sites and efficacy of chloride ions to enhance [^{3}H]flunitrazepam binding in as little as 15 sec after removal of the first cohort. The alteration in the chloride ion enhancement of [^{3}H]flunitrazepam binding in the second cohort appeared to be region specific; it was observed in membranes prepared from the cerebral cortex and hippocampus but was not seen with cerebellar membranes. These data suggest that the swim-stress-induced increases in [^{35}S]-TBPS binding and chloride ion enhancement of [^{3}H]flunitrazepam binding are not likely to be nonspecific effects resulting solely from increased motor activity. Also, the rapid alterations in these binding parameters seen in the second cohort from the protected environment preceded changes in the plasma levels of corticosterone, adrenocorticotrophic hormone (ACTH), β-endorphin, and α-MSH. Thus, stress-induced changes in the coupling of components of the complex to each other, and modifications of the ionophore may occur independently of hypothalamic–pituitary–adrenal activation.[1] Environmental conditions may influence the setpoint of the supramolecular complex in terms of the coupling of its components, and the complex is itself exquisitely sensitive to rapid changes in the environment. The data also demonstrate that the changes in the environment to which the supramolecular complex is sensitive include social ones. Specifically, the second rat housed in the protected environment was sensitive to, and stressed by, the removal of the first cohort from the common cage.

Acute alterations in the binding parameters and function of the supramolecular complex may be mediated by naturally occurring ring A-reduced metabolites of progesterone and deoxycorticosterone, specifically, 3α-hydroxy-5α-dihydroprogesterone (3α-OH-DHP) and 3α,5α-tetra-hydrodeoxycorticosterone (3α-THDOC).[43] These naturally occurring metabolites may function as endogenous barbiturate-like compounds. They inhibit the binding of [^{35}S]-TBPS in an uncompetitive fashion and increase the apparent affinity of [^{3}H]flunitrazepam binding; these metabolites are about 1000-fold more potent that pentobarbital in altering these binding parameters. The steroid metabolites alone stimulated chloride ion uptake into a synaptoneurosomal preparation, as well as potentiating muscimol-stimulated chloride ion uptake into these subcellular vesicles.[42,43] They also enhance GABA-activated chloride ion conductance in cultured rat hippocampal and spinal cord neurons. Levels of these metabolites fluctuate naturally during the estrous cycle and pregnancy, and in response to ACTH secretion.[43,44] These naturally occurring peripheral hormones may play a profound role in regulating the setpoint of GABAergic transmission within the brain by their direct interaction with the channel domain of the supramolecular complex.

The mechanism(s) of stress-induced alterations in benzodiazepine receptor sensitivity (as reflected in behavioral measures and benzodiazepine receptor binding) are not completely known, but are likely to involve structural alterations of the $GABA_A$ receptor complex, release of endogenous ligands, or a combination of the two. The predominant mechanism may be determined by qualities of the stress (including its chronicity), and genetic factors. The literature on "endogenous" ligands is extensive and inconclusive; however, some relevant recent contributions consistent with their existence are reviewed. The possible existence of endogenous "benzodiazepine-like" compounds was supported by work showing that pretreatment of mice with benzodiazepine antagonists (i.e., Ro15-1788 and CGS 8216) increased the rate of acquisition and retention of a T-maze discrimination task (45, 46). This task requires young mice to discriminate between the two arms of the T-maze in order to avoid mild electric shock. The data are consistent with the natural occurrence of diazepam-like ligands. Recently, an endogenous *B*-carboline, *n*-butyl beta carboline-3-carboxylate (*B*-CCB), was detected in the cerebral cortex of acutely stressed rats and shown to possess properties consistent with those of an inverse agonist (i.e., proconvulsant and anxiogenic ones) (47). An endogenous benzodiazepine(s) that shares structural properties with N-desmethyldiazepam (nordiazepam) has been reported to occur in mammalian brain, including human brain stored in paraffin fifteen years before the chemical synthesis of benzodiazepines was achieved (48, 49). Monoclonal antibodies that bind benzodiazepine agonists with high-affinity were used in the identification and isolation of this compound(s). These antibodies had much lower afinities for inverse agonists (beta-carbolines) and antagonists (i.e. Ro15-1788 and CGS-8216). Moreover, lorazepam in concentrations as high as 1 ng/ml was detected in the plasma of drug-naive rats using HPLC methodology with gas chromatographic/mass spectrometric detection (50). The natural occurrence of a potent dichlorinated benzodiazepine would have profound implications for understanding the functional role of the $GABA_A$/benzodiazepine receptor complex. Of course, these results could be artifactual due to a laboratory contaminant or the unrecognized ingestion of benzodiazepines. For example, HPLC was used to resolve aqueous acid extracts of wheat grains and potatoes into fractions that inhibited the binding of benzodiazepines to rat brain membranes (51). Chemical analyses of these extracts showed they include diazepam and lormetazepam. Thus, there is the possibility that benzodiazepines may be naturally occurring products contained in plants.

4. IN VIVO APPROACH TO THE MEASUREMENT OF BENZODIAZEPINE RECEPTOR BINDING AND STRESS-INDUCED ALTERATIONS WITH RADIOLABELED Ro15-1788

The data reviewed suggest that benzodiazepine receptor binding is exquisitely regulated, and rapid changes in binding can occur in response to environmental and pharmacological perturbations. Classic in vitro and ex vivo approaches employed in studying benzodiazepine receptor binding disrupt the local neuronal circuitry that may be involved in regulating receptor sensitivity and introduce artifacts associated with filtration and homogenization of tissue. For example, an erroneous increase in the estimate of the maximal density of functionally available receptors could occur due to tissue homogenization and the unmasking of cryptic receptors. Alternatively, homogenization may decrease the estimate of receptor density due to alterations in labile receptors. Therefore, in vivo approaches to the measurement of benzodiazepine receptor binding have been sought that would obviate these difficulties and permit an accurate assessment of the relationship between the potencies of various benzodiazepines and the number of receptors occupied in intact animals.[3,52]

Recently, the specific binding of [^{3}H]-Ro15-1788 in vivo has been reported in mice using a method that avoids the possible introduction of artifacts associated with the ex vivo procedures of homogenization and filtration.[2] The original report by Goeders and Kuhar[2] was replicated and extended, and the method was applied to studying the regulation of benzodiazepine receptors in response to stressful situations.[3,52] Ro15-1788 is an imidazodiazepine that antagonizes the major central actions of benzodiazepines in animals and man at doses devoid of any intrinsic pharmacological action.[9] The ligand interacts with the central type of benzodiazepine receptor in a competitive high-affinity manner. The distribution of the maximal number of [^{3}H]-Ro15-1788 binding sites parallels closely the distribution described for [^{3}H]clonazepam, a potent agonist, in in vitro and autoradiographic studies. Although [^{3}H]-Ro15-1788 and [^{3}H]clonazepam bind to the same type of central benzodiazepine receptor, their mode of interaction with the receptor differs. The dissociation constant of [^{3}H]-Ro15-1788 binding does not alter in response to GABA or pentobarbital. Moreover, at temperatures above 21°C, the phase-transition temperature for membrane lipids, thermodynamic parameters of the binding reaction distinguish [^{3}H]-Ro15-1788 from [^{3}H]clonazepam.

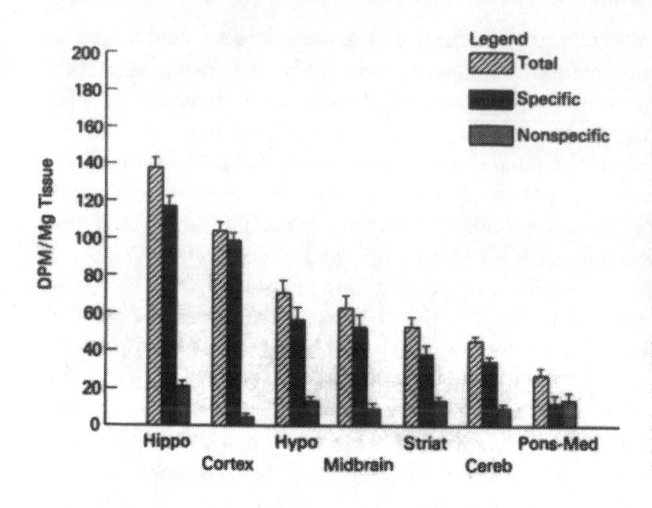

Figure 1. Regional localization of [^{3}H]-Ro15-1788 bound in vivo. Total binding was determined 20 min after intravenous injection of 5 μCi [^{3}H]-Ro15-1788. Nonspecific binding was determined by the intraperitoneal administration of clonazepam (5 mg/kg) 30 min prior to the intravenous injection of the radioligand. Total binding was determined with 17–26 mice and nonspecific binding with 7–9 animals for each brain region. Cereb, cerebellum; Hippo, hippocampus; Hypo, hypothalamus; Pons-Med, pons-medulla; Striat, striatum. Values are the mean and SEM.

The results of our characterization of specific [^{3}H]-Ro15-1788 binding in vivo and examples of the application of the method to the study of environmental stress-induced modification of binding are presented. At 20 min after intravenous injection of [^{3}H]-Ro15-1788 (2.5 and 5.0 μCi/mouse, i.v.), specific [^{3}H]-Ro15-1788 binding was unevenly distributed in various brain regions, with the highest binding observed in the hippocampus and cerebral cortex and the lowest in the pons-medulla. In the hippocampus and cerebral cortex, 80% or more of total binding was specific as defined by displacement with clonazepam (5 mg/kg, i.p.) administered 30 min prior to the administration of the radioligand (Fig. 1).

At 20 min after intravenous administration of [^{3}H]-Ro15-1788, good correlation was observed between the distribution of specific binding in vivo and the maximal density of [^{3}H]-Ro15-1788 binding sites in rats for the seven brain regions measured in vitro.[9] Significant correlations were observed with both the 2.5 and 5.0 μCi doses of injected [^{3}H]-Ro15-1788 (Fig. 2). By contrast, the binding of [^{3}H]-Ro15-1788 measured 1 min after injection showed little variation among brain regions, and there was no correlation with the density of [^{3}H]-Ro15-1788

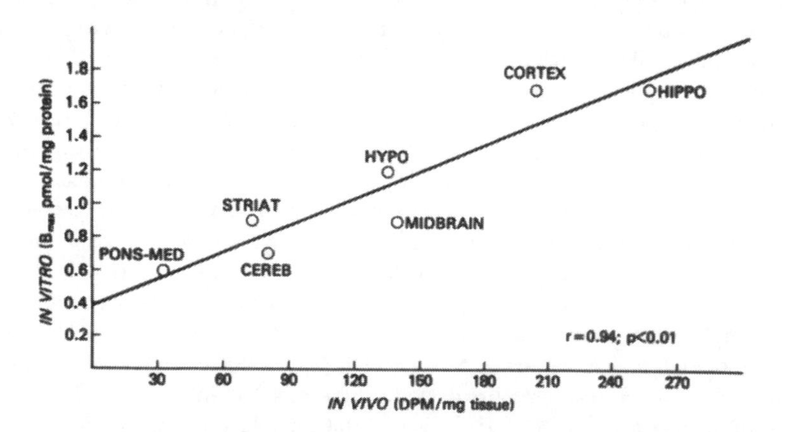

Figure 2. Regional localization of [^{3}H]-Ro15-1788 binding: correlation of in vitro and in vivo data. CEREB, cerebellum; HIPPO, hippocampus; HYPO, hypothalamus; PONS-MED, pons-medulla; STRIAT, striatum.

binding sites as determined in vitro (r = 0.116, NS). These data suggest that the tissue concentration of radioactivity at 1 min postinjection may be reflective of perfusion, whereas the amount of radioactivity contained in specific brain regions 20 min after the injection of [^{3}H]-Ro15-1788 reflects a pharmacologically meaningful drug–receptor interaction.

The binding of [^{3}H]-Ro15-1788 in brain was stereoselective, as shown by its sensitivity to displacement by the active enantiomer B10(+)-Ro11-6896, a benzodiazepine agonist, and its insensitivity to displacement by the inactive enantiomer B10(−)-Ro11-6893.

The technique was used to examine the effects of three acute stressors on benzodiazepine receptor binding in intact mice. Single sessions of ambient temperature swim stress of 2 and 10 minutes in duration resulted in significant increases in the occupancy of benzodiazepine receptors by [^{3}H]-Ro15-1788 in vivo. The apparent affinity of [^{3}H]-Ro15-1788 for the benzodiazepine receptor measured in vitro is not shifted by changes in the concentration of GABA; thus, stress-induced increases in the binding observed in vivo are not likely to be due to elevations in brain GABA levels. In contrast to the acute swim–stress paradigm, 10 min of intermittent electric footshock (10 sec of 2.0 mA scrambled gridshock every 30 sec) caused a significant reduction in specific [^{3}H]-Ro15-1788 binding in vivo in the cerebellum and hippocampus. Finally, a single trial of electroconvulsive shock (ECS) (120 V; 0.6 sec) resulted in no change in the binding as measured in vivo. The data suggest that the alterations are specific and may relate to some unique quality of the stressor. Moreover, the fact that bidirectional changes were observed with acute swim stress and intermittent electric footshock, and no change was seen after a single electroconvulsive episode, argue against these changes in binding being merely artifacts of changes in cerebral blood flow, permeability properties of the blood–brain barrier, or pharmacokinetic distribution of [^{3}H]-Ro15-1788.

A chronic swim–stress paradigm was adopted to examine long-term adaptive changes in benzodiazepine receptor binding in animals exposed to repeated stress. For these studies, mice were swim stressed (2 and 10 min) in ambient temperature water for 7 consecutive days. Specific [^{3}H]-Ro15-1788 binding in vivo was assessed 24 hr after the last session of swim stress in order to minimize potential artifacts due to stress-induced changes in blood flow, blood–brain barrier, and/or ligand distribution. This chronic stress paradigm caused significant reductions in binding in the hypothalamus, hippocampus, striatum, and cerebral cortex.

The effect of defeat stress, a naturalistic stress paradigm of defeat in a social confrontation, on specific [^{3}H]-Ro15-1788 binding in vivo was examined.[53] In this paradigm, male mice are removed from their home cages (intruder mice) and placed in the homes cages of other male mice (resident mice). The intruders are attacked and bitten and display a syndrome that resembles learned helplessness. This social stressor resulted in rapid and short-lived increases in the specific binding of [^{3}H]-Ro15-1788 in vivo in the hypothalamus, cerebral cortex, and cerebellum of defeated mice. Interestingly, intruder mice and unstressed controls were not distinguished from each other on the basis of specific [^{3}H]flunitrazepam binding measured in vitro in cerebral cortical membranes. These data suggest that classic in vitro techniques can obscure stress-induced changes that are discernible with this in vivo approach. Peripheral mechanisms were implicated in the regulation of benzodiazepine receptor sensitivity in the defeated mice. Adrenalectomy prevented the emergence of the stress-induced distinctions in in vivo binding between defeated mice and unstressed controls. Corticosterone replacement was able to restore the defeat stress-induced increases in specific [^{3}H]-Ro15-1788 binding in vivo in adrenalectomized mice. The central benzodiazepine receptor has been implicated in the regulation of plasma glucocorticoid levels.[35] Therefore, it is possible that stress-induced increases in benzodiazepine receptor binding could represent a central mechanism contributing to the regulation of the adrenocortical response to acute stress.[53] In any event, these data support a relationship between a naturalistic social stressor and rapid changes in the characteristics of benzodiazepine receptor binding in intact animals.

5. CONCLUSION

The benzodiazepine receptor appears to play a prominent functional role in mediating an animal's response to stressful changes in the environment, including social changes. Rapid and short-lived stress-induced changes in binding have been reported, as well as persistent changes in adult animals exposed to stressful and other stimuli during early development. Classic in vitro methods for measuring benzodiazepine receptors are limited by the nonphysiological conditions used in the binding assay and by the artifacts associated with filtration and homogenization of tissue. The availability of an in vivo approach to the measurement of benzodiazepine receptors with [^{3}H]-Ro15-1788 has stimulated a variety of studies designed to examine the functional role of this receptor in intact animals. The application of this in vivo technique could lead to an appreciation of the pathogenetic role of this receptor in psychiatric disorders and to the development of novel intervention strategies.

REFERENCES

1. Trullas R, Havoundjian H, Zamir N, et al: Environmentally-induced modification of the benzodiazepine/GABA receptor coupled chloride ionophore. Psychopharmacology 91:384–390, 1987
2. Goeders NE, Kuhar MJ: Benzodiazepine receptor binding *in vivo* with [^{3}H]Ro15-1788. Life Sci 37:345–355, 1985
3. Miller LG, Greenblatt DJ, Paul SM, et al: Benzodiazepine receptor occupancy *in vivo:* Correlation with brain concentrations and pharmacodynamic actions. J Pharmacol Exp Ther 240:516–522, 1987
4. Haefely W, Polc P: Physiology of GABA enhancement by benzodiazepines and barbiturates, in Olsen RW, Venter JC (eds): Benzodiazepine/GABA Receptors and Chloride Channels: Structural and Functional Properties. New York, Alan R. Liss, 1986, p 97
5. Olsen RW: GABA–benzodiazepine barbiturate receptor interactions. J Neurochem 37:1–13, 1981
6. Braestrup C, Nielsen M, Olsen CE: Urinary and brain beta-carboline-3-carboxylates as potent inhibitors of brain benzodiazepine receptors. Proc Natl Acad Sci USA 77:2288–2292, 1980
7. Ninan PT, Insel TR, Cohen RM, et al: Benzodiazepine receptor mediated experimental anxiety in primates. Science 218:1332–1334, 1982
8. Dorow R, Horowski R, Paschelke G, et al: Severe anxiety induced by FG 7142, a β-carboline ligand for benzodiazepine receptors. Lancet 9:98–99, 1983
9. Mohler H, Richards JG: Agonist and antagonist benzodiazepine receptor interaction *in vitro*. Nature (Lond) 294:763–765, 1981
10. Mohler H, Okada T: Benzodiazepine receptor: Demonstration in the central nervous system. Science 198:849–851, 1977
11. Squires RF, Braestrup C: Benzodiazepine receptors in rat brain. Nature (Lond) 266:732–734, 1977
12. Tallman JF, Thomas JW, Gallager DW: GABAergic modulation of benzodiazepine binding site sensitivity. Nature (Lond) 274:383–385, 1978
13. Study RE, Barker JL: Diazepam and (−) pentobarbital: Fluctuation analysis reveals different mechanisms for potentiation of GABA responses in cultured central neurons. Brain Res 268:171–176, 1981
14. Braestrup C, Nielsen M: GABA reduces binding of [^{3}H]-methyl-β-carboline-3-carboxylate to brain benzodiazepine receptors. Nature (Lond) 294:472–474, 1981
15. Borea PA, Supavilai P, Karobath M: Differential modulation of etazolate or pentobarbital enhanced [^{3}H]-muscimol binding by benzodiazepine agonists and inverse agonists. Brain Res 280:383–386, 1983
16. Barker JL, Owen DG: Electrophysiological pharmacology of GABA and diazepam in cultured CNS neurons, in Olsen RW, Venter JC (eds): Benzodiazepine/GABA Receptors and Chloride Channels: Structural and Functional Properties. New York, Alan R. Liss, 1986, p 135
17. Honore T, Nielsen M, Braestrup C: Barbiturate shift as a tool for determination of efficacy of benzodiazepine-receptor ligands. Eur J Pharmacol 100:103–107, 1984
18. Wong EHF, Snowman AM, Leeb-Lundberg LMF: Bariturates allosterically inhibit GABA antagonist and benzodiazepine inverse agonist binding. Eur J Pharmacol 102:205–212, 1984
19. Skolnick P, Paul SM, Barker JL: Pentobarbital potentiates GABA-enhanced [^{3}H]-diazepam binding to benzodiazepine receptors. Eur J Pharmacol 65:125–127, 1980

20. Schofield PR, Darlison MG, Fujita N, et al: Sequence and functional expression of the of $GABA_A$ receptor shows a ligand-gated receptor super-family. Nature (Lond) 328:221–227, 1987
20a. Pritchett DB, Luddens H, Seeburg PH: Type I and type II $GABA_A$-benzodiazepine receptors produced in transfected cells. Science 245:1389–1392, 1989
21. Havoundjian H, Paul SM, Skolnick P: Rapid, stress-induced modification of the benzodiazepine receptor-coupled chloride ionophore. Brain Res 375:401–406, 1986
22. Havoundjian H, Paul SM, Skolnick P: Acute, stress-induced changes in the benzodiazepine/GABA receptor complex are confined to the chloride ionophore. J Pharmacol Exp Ther 237:787–793, 1986
23. Schwartz RD, Wess MJ, Labarca R, et al: Acute stress enhances the activity of the GABA-gated chloride ion channel in brain. Brain Res 411:151–155, 1987
24. Robertson HA, Martin IL, Candy JM: Differences in benzodiazepine receptor binding in Maudsley reactive and Maudsley non-reactive rats. Eur J Pharmacol 50:455–457, 1978
25. Tamborska E, Insel T, Marangos PH: "Peripheral" and "central" type benzodiazepine receptors in Maudsley rats. Eur J Pharmacol 126:281–287, 1986
26. Robertson HA: Benzodiazepine receptors in "emotional" and "non-emotional" mice: Comparison of four strains. Eur J Pharmacol 56:163–166, 1979
27. Lippa AS, Klepner CA, Yunger L, et al: Relationship between benzodiazepine receptors and experimental anxiety in rats. Pharmacol Biochem Behav 9:853–856, 1978
28. Lane JD, Crenshaw CM, Guerin GF, et al: Changes in biogenic amine and benzodiazepine receptors correlated with conditioned emotional response and its reversal by diazepam. Eur J Pharmacol 83:183–190, 1982
29. Braestrup C, Nielsen M, Nielsen E, et al: Benzodiazepine receptors in brain as affected by different experimental stresses: The changes are small and not unidirectional. Psychopharmacology 65:273–277, 1979
30. Essman M, Valzelli L: Brain benzodiazepine receptor changes in the isolated aggressive mouse. Pharmacol Res Commun 13:665–671, 1981
31. Petkov VV, Yanev S: Brain benzodiazepine receptor changes in rats with isolation syndrome. Pharmacol Res Commun 14:739–744, 1982
32. Fride E, Dan Y, Gavish M, et al: Prenatal stress impairs maternal behavior in a conflict situation and reduces hippocampal benzodiazepine receptors. Life Sci 36:2103–2109, 1985
33. Bodnoff SR, Suranyi-Cadotte B, Quirion R: Postnatal handling reduces novelty-induced fear and increases [^{3}H]flunitrazepam binding in rat brain. Eur J Pharmacol 144:105–107, 1987
34. Soubrie P, Thiebot M, Jobert A, et al: Decreased convulsant potency of picrotoxin and pentetrazol and enhanced [^{3}H]flunitrazepam cortical binding following stressful manipulations in rats. Brain Res 189:505–517, 1980
35. LeFur A, Guilloux F, Mitrani N, et al: Relationship between plasma corticosteroids and benzodiazepines in stress. J Pharmacol Exp Ther 211:305–308, 1979
36. Skerritt JH, Trisdikoon P, Johnston GAR: Increased GABA binding in mouse brain following acute swim stress. Brain Res 215:398–403, 1981
37. Medina, J, Novas M, Wolfman C, et al: Benzodiazepine receptors in rat cerebral cortex and hippocampus undergo rapid and reversible changes after acute stress. Neuroscience 9:331–335, 1983
38. Drugan RC, Maier SF, Skolnick P, et al: An anxiogenic benzodiazepine receptor ligand induces learned helplessness. Eur J Pharmacol 113:453–457, 1985
39. Drugan RC, Ryan SM, Minor TR, et al: Librium prevents the analgesia and shuttlebox escape deficit typically observed following inescapable shock. Pharmacol Biochem Behav 21:749–754, 1984
40. DeSouza E, Goeders NE, Kuhar MJ: Benzodiazepine receptors in rat brain are altered by adrenalectomy. Brain Res 381:176–181, 1986
41. Goeders NE, DeSouza EB, Kuhar MJ: Benzodiazepine receptor GABA ratios: Regional differences in rat brain and modulation by adrenalectomy. Eur J Pharmacol 129:363–366, 1986
42. Majewska MD, Harrison NL, Schwartz RD, et al: Steroid hormone metabolites are barbiturate-like modulators of the GABA receptor. Science 232:1004–1007, 1986
43. Morrow AL, Suzdak PD, Paul SM: Steroid hormone metabolites potentiate GABA receptor-mediated chloride ion flux with nanomolar potency. Eur J Pharmacol 142:483–485, 1987
44. Schambelan M, Biglieri EG: Deoxycorticosterone production and regulation in man. J Clin Endocrinol Metab 34:695–703, 1972
45. Lal H, Kumar B, Forster MJ: Enhancement of learning and memory in mice by a benzodiazepine antagonist. FASEB J 2(11):2707–2711, 1988

46. Kumar BA, Forster MJ, Lal H: CGS 8216, a benzodiazepine receptor antagonist, enhances learning and memory in mice. Brain Res 460(1):195–198, 1988
47. Novas ML, Wolfman C, Medina JH, et al: Proconvulsant and "anxiogenic" effects of n-butyl beta carboline-3-carboxylate, an endogenous benzodiazepine binding inhibitor from brain. Pharmacol Biochem Behav 30(2):331–336, 1988
48. De Blas AL, Sangameswaran L: Demonstration and purification of an endogenous benzodiazepine from the mammalian brain with a monoclonal antibody to benzodiazepines. Life Sci 39(21):1927–1936, 1986
49. Sangameswaran L, Fales HM, Friedrich P, et al: Purification of a benzodiazepine from bovine brain and detection of benzodiazepine-like immunoreactivity in human brain. Proc Natl Acad Sci (USA) 83(23): 9236–9240, 1986
50. Wildmann J, Ranalder U: Presence of lorazepam in the blood plasma of drug free rats. Life Sci 43 (15):1257–1260, 1988
51. Wildman J: Increase of natural benzodiazepines in wheat and potato during germination. Biochem Biophys Res Commun 157(3):1436–1443, 1988
52. Deutsch SI, Miller LG, Weizman R, et al: Characterization of specific [^{3}H]Ro15-1788 binding *in vivo*. Psychopharm Bull 23:469–472, 1987
53. Miller LG, Thompson ML, Greenblatt DJ, et al: Rapid increase in brain benzodiazepine receptor binding following defeat stress in mice. Brain Res 414:395–400, 1987

4

Brain Recognition Sites for Methylphenidate and the Amphetamines

Their Relationship to the Dopamine Transport Complex, Glucoreceptors, and Serotonergic Neurotransmission in the Central Nervous System

Richard L. Hauger, Itzchak Angel, Aaron Janowsky, Paul Berger, and Bridget Hulihan-Giblin

1. NEUROPHARMACOLOGY OF STIMULANTS

The stimulants methylphenidate, cocaine, and amphetamine remain among the most widely abused psychotropic drugs. However, amphetamine has long been considered a prototypic anorectic agent and, in the past, was used in the treatment of obesity.[1–6] Although amphetamine has been used as a therapeutic agent, amphetamine can be a potent psychotomimetic, and tolerance can develop to its anorectic effects.[1–9] Consequently, the medical use of amphetamine has now been restricted to narcolepsy and childhood hyperkinesis. Methylphenidate is currently used in the treatment of attention-deficit disorder and minimal brain dysfunction in children to increase attention span and reduce hyperactivity.[10] Methylphenidate also has been proposed as an antidepressant and can potentiate conventional antidepressant medications in patients suffering from atypical depression.[10] Nevertheless, there is widespread

---

Richard L. Hauger • San Diego Veterans Administration Medical Center, and Department of Psychiatry, University of California–San Diego, School of Medicine, La Jolla, California 92093. *Itzchak Angel* • Department of Biology, Laboratoires d'Etudes de Recherches Synthelabo, 75013 Paris, France. *Aaron Janowsky* • Research Service, Department of Psychiatry, Veterans Administration Medical Center, and Oregon Health Sciences University, Portland, Oregon 97207. *Paul Berger and Bridget Hulihan-Giblin* • Clinical Neuroscience Branch, National Institute of Mental Health, National Institutes of Health, Bethesda, Maryland 20892.

illicit use of psychostimulants, in particular, because of the euphoriant and fatigue-reducing effects of methylphenidate, cocaine, and various amphetamine compounds.

Amphetamine, cocaine, and other phenylethylamine stimulant drugs cause many central and systemic effects which include euphoria, psychomotor hyperactivity, anorexia, vasoconstriction, and hyperthermia, while methylphenidate mainly acts as a psychostimulant.[1,2,11–13] Their pharmacologic properties relate to their central as well as peripheral action. The presynaptic neurochemistry of these psychostimulants has been studied intensively and, in general, these drugs can alter biogenic amine metabolism and influence the release, uptake, and turnover of dopamine, norepinephrine, and serotonin in the central nervous system (CNS).[1,2,6,7,11,14–16]

Although both stimulants inhibit the presynaptic uptake of monoamines, especially dopamine, they differ in their biochemical mechanisms for effecting dopamine release. Methylphenidate and related drugs stimulate dopamine release from a vesicular pool constituting a storage system that can be depleted by reserpine.[1,5,10,14,15,17] Amphetamine and its congeners release dopamine from a reserpine-insensitive pool that is replenished by newly synthesized amine.[1,2,5,14–17] In recent years, receptor-binding studies using radiolabeled drugs have proved invaluable in delineating the membrane sites that mediate the actions of many of these agents.[18,19] This chapter describes recent studies which have used radiolabeled stimulants to determine neuronal mechanisms which contribute to psychostimulant drug action.

2. STIMULANT RECOGNITION SITES ASSOCIATED WITH THE BRAIN DOPAMINE UPTAKE COMPLEX: POTENTIAL PROBES OF PRESYNAPTIC DOPAMINERGIC NEUROTRANSMISSION IN HUMAN PSYCHOPATHOLOGY

2.1. [^{3}H]Threo-(±)-Methylphenidate-Binding Sites and Stimulant-Induced Motoric Behavior

2.1.1. Characteristics

Threo-(±)Methylphenidate (methyl ritalinate) is a phenylethylamine derivative containing a piperidine ring that can be considered a cyclized derivative of amphetamine.[1] Although methylphenidate and related esters of ritalinic acid possess motor stimulant properties in several species, including man, the anorectic and cardiovascular actions of these compounds are minimal as compared with amphetamine and related phenylethylamines.[20,21] Current hypotheses concerning the mechanism(s) of action of methylphenidate in neuropsychiatric disorders have focused on the well-described effects of this compound in either releasing neuronal catecholamines or blocking their presynaptic reuptake, or both.[22,23]

Binding studies with [^{3}H]threo-(±)-methylphenidate have been conducted to investigate the mechanisms of action of methylphenidate and other nonamphetamine psychomotor stimulants. Specific and saturable binding sites for [^{3}H]methylphenidate have been characterized in rat brain.[24] The density of [^{3}H]methylphenidate binding in rat brain has been found to be highest in the striatum, while the specific binding in the cerebellum and hippocampus is less than 10% of that observed in the striatum. When crude synaptosomal membranes from rat striatum are prepared, this fraction has a greater than eightfold enrichment in specific binding compared with fractions containing myelin or microsomes. In the crude synaptosomal fraction (P_2) of striatum, Scatchard analysis has demonstrated the presence of a single class of [^{3}H]methylphenidate-binding sites with an apparent dissociation constant (K_d) of 235 ± 44 nM and a maximum binding capacity (B_{max}) of 13.4 ± 1.9 pmoles/mg protein.[24]

2.1.2. Association with the Dopamine Transport Complex

The binding of [^{3}H]methylphenidate to striatal membranes is strongly dependent on the presence of sodium ions and optimal binding required the presence of 120 nM sodium.[24] The effect of sodium on [^{3}H]methylphenidate binding indicates that these sites may be associated with a neurotransmitter uptake or transport system, since the previously reported high-affinity [^{3}H]imipramine- or [^{3}H]desipramine-binding sites are also sodium dependent and have been shown to be associated with the neuronal uptake mechanism for serotonin and norepinephrine, respectively.[19,25,26] Moreover, the high density of [^{3}H]methylphenidate-binding sites in the synaptosomal membrane fraction of striatum, coupled with previous reports,[22,23] implicating dopaminergic mechanisms in mediating the stimulant actions of methylphenidate, suggests that [^{3}H]methylphenidate binding may be associated with the dopamine uptake site. In subsequent studies, the most potent inhibitors of both [^{3}H]methylphenidate binding and [^{3}H]dopamine uptake have been a series of aryl-1,4-dialkylpipedazine derivatives previously reported to be selective inhibitors of dopamine uptake.[27] In examining a series of dopamine uptake inhibitors (n = 14) a highly significant correlation ($r = 0.88$, $p < 0.001$) has been obtained between the potencies of these compounds as inhibitors of [^{3}H]methylphenidate binding and [^{3}H]dopamine uptake in rat striatum.[24]

2.1.3. Correlation with Psychomotor Actions of Ritalinic Acid Esters

Ritalinic acid esters exhibit a range of psychomotor potencies in mice.[28] When the relative potencies of these compounds as locomotor stimulants are compared to their relative potencies in displacing [^{3}H]methylphenidate from striatal membranes, a highly significant correlation ($r = 0.85$, $p < 0.001$) has also been observed.[24] These observations suggest that specific binding sites for [^{3}H]methylphenidate, which appear to be associated with a dopamine uptake site or carrier, may mediate the motor stimulant properties of methylphenidate and other ritalinic acid esters.

2.1.4. Effects of Chemical and Surgical Denervation

To examine further the association between the [^{3}H]methylphenidate binding site and the dopamine transport complex, the effects of chemical and surgical denervation of dopaminergic afferents to the striatum on specific [^{3}H]methylphenidate binding have been investigated. Injection of the dopaminergic neurotoxin 6-hydroxydopamine (6-OHDA) into the right lateral ventricle produces a significant decrease in striatal [^{3}H]methylphenidate binding.[29] Scatchard analysis of binding data from lesioned animals has demonstrated a 50% reduction in the apparent B_{max} with no significant change in the apparent K_d compared with sham-lesioned controls. Furthermore, surgical lesions of the medial forebrain bundle, which result in a dramatic reduction in striatal dopamine uptake, reduce striatal [^{3}H]methylphenidate binding on the ipsilateral (lesioned) side to only 10% of the contralateral (nonlesioned side).[29] By contrast, intracerebroventricular administration of the serotonergic neurotoxin 5,7-dihydroxytryptamine (5,7-DHT) or the cholinergic neurotoxin AF64A do not alter striatal [^{3}H]methylphenidate binding.[29] The exhaustion of intraneuronal catecholamine storage vesicles caused by chronic administration (21 days) of reserpine (0.75 mg/kg per day) does not alter striatal [^{3}H]methylphenidate binding, although reserpine-induced monoamine depletion abolishes the psychostimulant effects of methylphenidate and related drugs. These findings provide additional support for the hypothesis that [^{3}H]methylphenidate binding sites in the striatum are localized to dopaminergic nerve terminals and may be associated with the dopamine transport complex.

2.1.5. Studies in Human Brain

The binding of [3H]methylphenidate has been demonstrated in crude striatal membranes prepared from postmortem human brain tissue (A. Janowsky, unpublished data). The binding to human striatal membranes is quite similar to rat striatum, except that the estimated B_{max} is somewhat lower than that found in fresh rat striatum. Since the apparent B_{max} for [3H]methylphenidate binding in recently frozen rat striatal tissue is similar to human striatum, this difference may be due to the subsequent freezing and thawing of brain tissue. Structure activity studies of the human striatal sites have revealed a rank order of potency similar to rat striatum for the displacement of specific [3H]methylphenidate binding by various ritalinic acid esters (A. Janowsky, unpublished data). Consequently, the specific [3H]methylphenidate binding site in human striatum also appears to be related to the dopamine transport complex and, on the basis of competition experiments, appears to mediate the psychomotor stimulant actions of the methylphenidate class of stimulants. It is possible that the analysis of [3H]methylphenidate binding in postmortem human brain tissue may reveal information concerning the pathophysiology of various disorders such as schizophrenia, Parkinson's disease, and hyperkinetic syndromes of childhood.

2.2. [3H]-GBR-12935 and [3H]-GBR-12783 as Selective Ligands for the Dopamine Uptake Site

2.2.1. Characteristics of [3H]-GBR-12935 and [3H]GBR-12783 Binding Sites

The diphenyl-substituted piperazine derivatives {1-[2-(diphenylmethoxy)ethyl]-4-(3-phenyl-propyl)piperazine} GBR-12935 and GBR-12783 are at present the most selective and potent inhibitors of dopamine (DA) uptake by presynaptic neurons.[27,30] For example, these GBR compounds are at least 10–50 times more potent as an inhibitor of neuronal DA uptake than mazindol (an appetite suppressant that also strongly inhibits norepinephrine uptake) and are approximately 100–800 times more potent than cocaine.[27,31,32] It is interesting to note that GBR compounds do not appear to have anorectic actions, suggesting that mazindol's anorectic effects may not be mediated through an interaction with the dopamine reuptake carrier protein. Although other neurotransmitter mechanisms may be important in psychostimulant action, GBR-12935 and GBR-12783 are 25- to 2000-fold less potent as inhibitors of norepinephrine (NE) or serotonin (5-HT) uptake in brain slices or crude synaptosomes prepared from hypothalamus or cerebral cortex.[27,30,33] Consequently, [3H]-GBR-12935 or [3H]-GBR-12783 are the most selective ligands for the DA uptake system and should prove very useful in further characterizing and eventually purifying the DA transport complex. Other ligands, such as [3H]cocaine, [3H]mazindol, [3H]nomifensine, and [3H]methylphenidate, have a low affinity and/or poor selectivity for the DA transport complex.

Since the inhibition of dopamine uptake by these piperazine derivatives occurs at nanomolar drug concentrations, the binding of radiolabeled GBR-12935 or GBR-12783 to brain tissue has been recently characterized. Specific [3H]-GBR-12935 binding in rat brain is saturable, stereospecific, of high affinity (K_d 10^{-9} M), and sodium dependent.[31,34] With respect to the effects of sodium, the specific binding of [3H]-GBR-12935 to striatal membranes is maximal at a sodium concentration of 240 mM. Scatchard analysis of equilibrium binding data from the striatum has revealed a twofold higher affinity for [3H]-GBR-12935 binding at 240 mM compared with 30 mM NaCl (0.82 ± 0.08 nM versus 1.53 ± 0.39 mM) without any change in the maximal number of specific GBR-binding sites. Specific binding in the presence of sodium is greatest in the striatum, nucleus accumbens, and olfactory tubercle, where the DA-containing terminals are the greatest.

[3H]-GBR-12935 binding is low in other brain regions, such as frontal cortex, cerebellum, brainstem, and hippocampus, where sodium inhibits binding.[31] The enhancing effect of the sodium ion on [^{3}H]-GBR-12935 binding to the corpus striatum is specific, since the addition of $MgCl_2$, $CaCl_2$, LiCl, or Mg acetate inhibits binding compared with binding in the absence of cation or presence of 120 mM NaCl. Only KCl (50 mM) has no significant effect on [^{3}H]-GBR-12935 binding. The sodium dependency of GBR binding to the striatum and nucleus accumbens, and the fact that a classic biogenic amine uptake mechanism requires sodium, suggests that the regional distribution of sodium-dependent [^{3}H]-GBR-12935 binding sites closely approximates the localization of dopaminergic nerve terminals in the central nervous system (CNS). Sodium-independent [^{3}H]-GBR-12935 binding is of considerably lower affinity, is not saturable in some brain regions and most likely represents the cis-flupenthixol-sensitive "piperazine acceptor" site.[123]

A recent autoradiographic study of [^{3}H]-GBR-12935 binding has demonstrated a marked correlation between the distribution of GBR binding sites and DA cell bodies and nerve terminals.[36] Furthermore, GBR binding is highly correlated with the localization of DA type-1 and type-2 receptors in mesostriatal and mesolimbocortical brain systems.[37]

The potencies of a series of CNS stimulants and drugs in inhibiting specific [^{3}H]-GBR-12935 binding is highly correlated ($r = 0.96$, $p<0.01$) with their potencies in inhibiting DA uptake in crude striatal synaptosomes.[31,34] Specific [^{3}H]-GBR-12935 binding is not inhibited by agonists or antagonists of α-adrenergic, β-adrenergic, dopaminergic, muscarinic, or nicotonic-cholinergic receptors. In addition, the biogenic amines serotonin and norepinephrine at concentrations ≥50 mM have no effect on specific [^{3}H]-GBR-12935 binding, whereas DA inhibits specific binding with an IC_{50} value of approximately 15 mM. Drugs that block the uptake of 5-HT and NE are also ineffective in displacing [^{3}H]-GBR-12935 binding from striatal membranes at concentrations up to 10 mM. This finding is consistent with the fact that GBR binding sites are almost undetectable in brain areas associated with NE or 5-HT uptake sites where there is little DA innervation such as the hypothalamus.

[^{3}H]-GBR-12783 also labels a homogeneous population of high-affinity sites associated with the neuronal DA uptake system.[32] In particular, [^{3}H]-GBR-12783 binding is thermal sensitive and sodium dependent in the striatum, nucleus accumbens, and tuberculum olfactorium. The inhibition of GBR-12783 binding to striatum by various drugs is highly correlated with their rank order of potency for inhibiting striatal [^{3}H]-DA uptake.[32] Furthermore, GBR-12783 possesses an approximately 150-fold lower affinity for the NE uptake system labeled with [^{3}H]desipramine in the cerebral cortex than for the DA transport complex labeled with [^{3}H]-GBR-12783.

GBR-12783 binding sites in the striatum are associated with ascending DA fibers in the medial forebrain bundle (MFB). For example, the unilateral section of the MFB produces a marked reduction in [^{3}H]-GBR-12783 binding to the ipsilateral striatum, which correlates with the decrease in [^{3}H]-DA uptake in the lesioned striatal tissue.[32] Earlier work with the [^{3}H]-GBR-12935 ligand has demonstrated that 6-hydroxydopamine lesions of dopaminergic neurons in the striatum produces a 45% decrease in [^{3}H]-DA uptake corresponding to the loss of GBR-12935 binding sites in this brain area.[31] The complete depletion of striatal DA content by pretreating rats with 5 mg/kg reserpine and 250 mg/kg α-methyl-para-tyrosine (α-MT), however, does not change either the binding affinity (K_d) or capacity (B_{max}) of [^{3}H]-GBR-12783.[32] Dopamine in the 10^{-5} M concentration range does compete with [^{3}H]-GBR-12783 for binding sites similar to results with [^{3}H]-GBR-12935 binding.[31,32]

(+)-Amphetamine strongly interacts at the DA uptake system, as indicated by its inhibitory potency on striatal [^{3}H]-DA uptake being similar in magnitude to methylphenidate but greater than the effect of cocaine on uptake (Table I). However, (+)-amphetamine is remarkably less effective than these stimulants in inhibiting [^{3}H]-GBR-12783 binding (e.g., 25-fold less potent than cocaine, more than 70-fold less potent than methylphenidate, and more than 12,000-fold less potent than GBR-12783).[32] Similar discrepancies between relative binding affinities (K_i) and inhibition potencies at the DA uptake system have been noted in studies on

Table I. Interactions of Psychostimulants at Brain Recognition Sites for Stimulant Drugs[a]

		Inhibitory potencies at stimulant binding sites					
Psychostimulants	Inhibition of DA uptake	Hypothalamic amphetamine binding site	Hypothalamic mazindol binding site	Striatal mazindol binding site (Na^+-dependent)	Striatal cocaine binding site (Na^+-dependent)	Striatal GBR binding sites (Na^+-dependent)	Brain MDMA binding sites
Amphetamine	1^+–2^+	+++	+	+	+	0–+	0
Mazindol	4^+	+++	++++	++++		+++	
Cocaine	3^+		++–+++	+++	++++	++	
GBR	>4^+					Greater than ++++	
MDMA	0						
PCA	1^+–2^+	++++					++++ ++++

[a]+ = lowest binding affinity, ++++ = highest binding affinity.

[^{3}H]mazindol and [^{3}H]cocaine binding in the brain (Table I). These findings suggest that presynaptic DA probes such as GBR-12935 and GBR-12783 may bind to a component of the DA uptake complex that is different from the site involved in DA transport. A similar hypothesis has already been made for [^{3}H]cocaine and [^{3}H]nomifensine binding sites.[35,38,39]

2.2.2. Studies of [^{3}H]-GBR-12935 Binding in Human Brain

Specific binding of [^{3}H]-GBR-12935 has been demonstrated in frozen human caudate membranes, which is saturable, of high affinity (K_d = 3.4 nM) and low capacity (2.7 pmoles/mg protein), and inhibited by a variety of stimulants.[40,41] The K_i for stimulant drugs at the [^{3}H]-GBR-12935 binding site correlates with the rank order of potency of these drugs in inhibiting DA uptake in the striatum. Consequently, [^{3}H]-GBR-12935 binding sites in the human striatum appear to have similar characteristics to rat striatum. Recently, a linear decline in the density of human caudate [^{3}H]-GBR-12935 binding sites has been observed with age.[41] This gradual loss of presynaptic [^{3}H]-GBR-12935 binding sites in the caudate nucleus with age is associated with a decline in DA cell bodies and terminals, dopamine-synthesizing enzymes in neurons, DA content, and postsynaptic concentrations of DA (D_2) receptor sites occurring during senescence.[42]

Since Parkinson's disease is the best described degenerative disease of DA-containing afferents to the striatum, the binding of [^{3}H]-GBR-12935 in human caudate membranes prepared from patients with Parkinson's disease has been compared with binding in matched controls. The specific binding of [^{3}H]-GBR-12935 is reduced by approximately 45% in the caudate from patients with Parkinson's disease, but no significant differences in binding has been observed in the nucleus accumbens from patients and controls.[40] These findings suggest that the severe degeneration of the nigrostriatal (but not mesolimbic) dopamine pathway in Parkinson's disease is paralleled by a highly significant and specific decrease in [^{3}H]-GBR-12935 binding. Previous reports, however, have described a decrease in the concentrations of DA and its metabolites in the nucleus accumbens.[43] Thus, it appears that the decrease in DA and its metabolites may not always coincide with changes in the presynaptic DA transport complex.

2.3. [^{3}H]Cocaine Binding Sites

2.3.1. Characteristics

The alkaloid cocaine is a powerful psychomotor stimulant that produces euphoria and provokes changes in cardiovascular function.[44] A secondary property of cocaine is its local anesthetic effect.[45] Initial studies on [^{3}H]cocaine binding in mouse brain cortical membranes revealed sodium-independent binding sites with relatively low affinity.[46] The potencies of various cocaine analogues in displacing [^{3}H]cocaine from these binding sites were correlated with their central stimulatory potency, but not with their local anesthetic effects.[47] Consequently, these binding sites were postulated to mediate the increased centrally-mediated locomotion and behavioral stimulation produced by cocaine. Since sodium facilitates the specific binding of various ligands to brain binding sites associated with the neuronal uptake of several types of neurotransmitters, subsequent studies have examined the effect of sodium ions on [^{3}H]cocaine binding to rat striatal membranes.[35,48,115,116] Sodium chloride increases the density of high-affinity [^{3}H]cocaine binding sites in the corpus striatum, and this facilitation of [^{3}H]cocaine binding by sodium ions is clearly limited to the corpus striatum, an area with a high dopamine content. The sodium-induced increase in striatal cocaine binding is abolished by 6-OHDA denervation of DA neurons in the corpus striatum, while the destruction of glutamatergic terminals by cortical ablation increases the caudate concentration of sodium-sensitive cocaine recognition sites.[35] Kainic acid, a neurotoxin that destroys neuronal cell bodies but spares axons and nerve terminals, does not alter striatal [^{3}H]cocaine binding measured in the presence of sodium.[35]

Furthermore, the capacity of a series of drugs to inhibit sodium-dependent cocaine binding has been significantly correlated with their potency to inhibit [^{3}H]-DA uptake into rat striatal synaptosomes.[35] Taken together, these findings suggest that sodium-dependent cocaine binding sites are localized presynaptically on dopaminergic nerve terminals in corpus striatum and are related to dopamine uptake sites similar to the striatal [^{3}H]methylphenidate and [^{3}H]-GBR-12935 binding sites. Preliminary work has recently demonstrated that positron emission tomography using cocaine labelled on carbon-11 can identify cocaine binding in the human striatum.[117] Cocaine binding sites have also been identified in the fetal rat brain suggesting that *in utero* exposure to cocaine may cause postnatal withdrawal effects and long-term behavioral changes by altering cocaine binding in the fetal central nervous system.

Sodium-insensitive cocaine binding sites in the caudate nucleus may be located related to the serotonin reuptake transporter protein on 5-HT terminals, while [^{3}H]cocaine binding sites in the brain areas other than corpus striatum may be associated with separate neurotransmitter uptake systems, e.g., serotonin, and may mediate other actions of cocaine.[116] Cocaine has also been shown to interact at neurotransmitter receptor sites in addition to the DA uptake system. While cocaine possesses significant binding affinities at muscarinic-acetylcholine receptors (AChR) and at sigma receptors,[48] the latter binding site being associated with the psychotomimetic properties of certain opiates,[49] cocaine generally has a very low affinity at most brain neurotransmitter receptors. A recent study has indicated that the active stimulant isomers: (−)-cocaine and (+)-amphetamine display micromolar affinities for the brain sigma receptor site.[50] Although (−)cocaine acting as a competitive inhibitor at sigma sites may mediate its dysphoric and/or psychotomimetic properties, the relative potencies of cocaine analogues at sigma receptors in cerebellar membranes do not correlate with their potencies in producing behavioral reinforcement in animals.[50,51] The reinforcing properties of cocaine and related compounds which lead to cocaine self-administration appear to be specifically mediated by the DA transporter in brain.[48]

2.3.2. Studies of [^{3}H]Cocaine Binding in Human Brain

Sodium-dependent [^{3}H]cocaine binding has also been demonstrated in postmortem human putamen tissue, although the affinity for [^{3}H]-GBR-12935 binding to human brain is more than 200-fold higher than the affinities for the two reported saturable [^{3}H]cocaine sites (K_d = 0.21 μM and 26.4 μM, respectively).[52] When the higher-affinity [^{3}H]cocaine sites associated with neuronal dopamine uptake are measured in putamen samples from patients with Parkinson's disease, a highly significant 60% decrease in [^{3}H]cocaine binding has been found compared with nondiseased tissue.[52] However, [^{3}H]-GBR-12935 has a much higher affinity for the dopamine transporter and should be a more sensitive probe to study the neuropathology of Parkinson's disease and other related disorders of brain dopaminergic neurons.

2.4. [^{3}H]Nomifensine Binding Sites

Nomifensine is more potent than tricyclic antidepressants in inhibiting the neuronal uptake of dopamine in the striatum and has also been reported to be effective in the treatment of depression.[53] [^{3}H]Nomifensine has been shown to label a site primarily associated with the neuronal uptake of dopamine in striatal membranes.[39,54] The distribution of [^{3}H]nomifensine binding sites in rat brain has also been studied by quantitative autoradiography. Specific sodium-dependent [^{3}H]nomifensine binding sites (benztropine-displaceable) are most abundant in the caudate-putamen, olfactory tubercle, and nucleus accumbens, where intraventricular 6-OHDA administration dramatically decreases this binding.[54] [^{3}H]Nomifensine also labels (desipramine-displaceable) sites in several nuclei of the thalamus and hypothalamus and in the bed nucleus of the stria terminalis,[54] regions with an extremely dense noradrenergic innervation[55] and a high density of [^{3}H]desipramine binding sites.[26] Since [^{3}H]nomifensine binds to presynaptic uptake sites for both dopamine and norepinephrine, this ligand has less selectivity for dopamine uptake sites than does [^{3}H]-GBR-12935.

3. STIMULANT RECOGNITION SITES ASSOCIATED WITH BRAIN GLUCORECEPTORS AND NEURONAL (Na^+,K^+)-ATPASE

3.1. Relationship of Brain [^{3}H](+)-Amphetamine Binding Sites to Stimulant Anorexia, Feeding Regulation, and Central Glucostats

3.1.1. Characteristics of [^{3}H](+)-Amphetamine Binding Sites

Amphetamine stimulants produce large changes in the metabolism of biogenic amines by stimulating neuronal release and inhibiting reuptake of monoamines.[1,2,11,14–16] In addition to their potent psychostimulant effects, amphetamine and its halogenated derivatives (e.g., fenfluramine) strongly suppress food intake in humans and animals.[3–6,11] However, the anorectic effects of amphetamine are distinct from its general psychostimulant properties.[4,6,12,56] Amphetamine's anorectic actions most likely involve an increased turnover of central catecholamines, such as norepinephrine and dopamine, or the indoleamine, serotonin, although contradictory data concerning the involvement of these neurochemical events in mediating amphetamine-induced anorexia have been reported. For example, the anorectic effects of fenfluramine may be mediated by its action on serotonergic neurotransmission.[6] Numerous other neurotransmitters have also been proposed to be central regulators of appetite and feeding behavior; these neuroregulators may also mediate the anorexigenic actions of amphetamine-like drugs.[57–64] However, no unified theory of appetite has been firmly established. Since the regulation of appetite represents a complex homeostatic process, the elucidation of the mechanisms involved in the anorectic action of amphetamine may enhance our understanding of the mechanisms underlying appetite and satiety. In this regard, the radioligands [^{3}H](+)-amphetamine and [^{3}H]mazindol have been used to delineate specific binding sites in the CNS that appear to mediate the anorectic actions of amphetamine and its analogues.

Several studies have demonstrated the presence of saturable and stereospecific binding sites for [^{3}H](+)-amphetamine in rat hypothalamic membranes.[65–67] Scatchard analysis of the complete saturation isotherm using filtration and centrifugation methods has revealed a single population of binding sites with an apparent binding affinity (K_d) of approximately 5–10 μM and maximum binding capacity (B_{max}) of approximately 300–400 pmoles/mg protein.[67] Since the brain levels of amphetamine following pharmacologically active doses are in the range of 2–50 μM[14] and thus correspond closely to the apparent K_d for the [^{3}H](+)-amphetamine binding sites in rat brain,[67] such low-affinity sites may be pharmacologically relevant. In addition, the subcellular distribution of [^{3}H](+)-amphetamine binding sites has shown that the synaptosomal fraction possesses the highest density of binding sites relative to the mitochondrial, nuclear, and microsomal fractions.[67]

Physiological concentrations of sodium markedly inhibit specific [^{3}H](+)-amphetamine binding. This finding suggests that [^{3}H](+)-amphetamine does not label a monoaminergic uptake site as do various radiolabeled antidepressants and specific uptake blockers, since the binding of psychotropic drugs to such uptake sites are generally dependent on the presence of sodium. Unilateral surgical interruption of the ascending medial forebrain bundle or 6-hydroxydopamine denervation of dopaminergic fibers to the corpus striatum results in significant increases in the B_{max} of striatal [³H](+)-amphetamine binding sites.[114] Since similar 6-hydroxydopamine or surgical lesions of dopamine-containing afferents to the corpus striatum reduces the density of striatal methylphenidate and cocaine binding sites,[29,35] [³H](+)-amphetamine may bind to postsynaptic sites in the rat striatum. Although many studies suggest that amphetamine produces its pharmacologic actions by promoting the release of extravesicular dopamine from presynaptic nerve fibers,[1,2,14–17] this finding suggests a possible postsynaptic site of action for amphetamine.

Inhibition of lysosomal uptake by chloroquine does not influence the binding of [³H](+)-amphetamine.[69] In addition, recent work has demonstrated that [³H](+)-amphetamine does not bind to monoamine oxidase-A (MAO-A),[69] despite the finding that the MAO-A inhibitor harmaline displaces [³H](+)-amphetamine from its binding sites.[70] These findings suggest that MAO-A is not the acceptor site labeled by [³H](+)-amphetamine. Consequently, [³H](+)-amphetamine binding sites in synaptosomal membranes most likely are associated with a membrane-bound enzyme complex or other large membranous constituent such as an acceptor site rather than a classic neurotransmitter receptor site.

Measurement of the regional distribution of [³H](+)-amphetamine binding sites in the brain has shown that the brainstem has the highest density of specific binding sites followed by the hypothalamus and striatum (Figure 1). The cerebellum has a binding site concentration similar to that of the pituitary. Specific [³H](+)-amphetamine binding appears to be highly localized to the CNS, since peripheral tissues such as liver, kidney, and heart have very low to undetectable levels of specific binding.[67]

Initial structure–activity experiments demonstrated that the relative affinities of a series of phenylethylamine derivatives for displacing [³H](+)-amphetamine binding from hypothalamic membranes were highly correlated with their potencies as anorectic agents in rats.[66] Furthermore, there was no significant correlation between displacement of [³H](+)-amphetamine binding and the motor stimulant potencies of a series of amphetamine derivatives.[66] A recent study has also demonstrated a positive correlation ($r = 0.78$, $p < 0.0001$) between the potencies of 19 phenylethylamines and related compounds to inhibit food intake and their relative

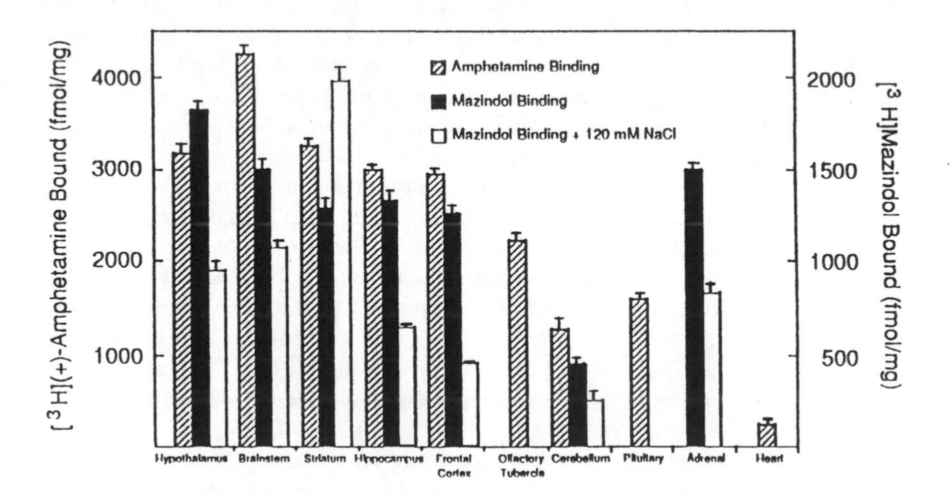

Figure 1. Regional distribution of specific [³H](+)-amphetamine (100 nM) and [³H]mazindol (10 nM) binding sites in the rat central nervous system.

binding affinities (IC_{50}) at the hypothalamic [^{3}H](+)-amphetamine binding site,[71] although their observed IC_{50} values for (+)-amphetamine (60 μM) were approximately one order of magnitude lower than the previously reported K_d for the hypothalamic [^{3}H](+)-amphetamine binding site.[67,72,73] Therefore, future studies will be required to examine the functional relationship between hypothalamic amphetamine binding sites and appetite regulation.

3.1.2. Glucostatic Regulation of Brain [^{3}H](+)-Amphetamine Binding Sites

3.1.2a. Effect of Food Deprivation on [^{3}H](+)-Amphetamine Binding Sites. The hypothalamus serves as the monitor and integrator of multiple sensory inputs associated with feeding behavior.[57,74–76] Since serum glucose concentrations are known to decrease during fasting in rats, the relationship between changes in [^{3}H](+)-amphetamine binding to hypothalamic membranes and the resultant hypoglycemia has been examined. Fasting in rats typically reduces serum glucose levels to 55% and 64% of control values after 24 and 72 hr, respectively. Hypothalamic and brainstem [^{3}H](+)-amphetamine binding begin to decline following the development of significant hypoglycemia during food deprivation.[73] Scatchard analysis of saturation binding data indicates that animals fasted for 72 hr develop an approximately 50% reduction in the B_{max} but no change in the apparent K_d.[73] A similar reduction in [^{3}H](+)-amphetamine binding by food deprivation has also been observed in the brainstem. No reduction in [^{3}H](+)-amphetamine binding site density has been found in other brain areas, such as the frontal cortex, striatum, or cerebellum, or in peripheral tissues, such as the liver, where marked adaptational changes during starvation are known to occur.[77] Furthermore, other neurotransmitter/drug recognition sites in hypothalamus or other brain area, (e.g., benzodiazepine, 5-HT_2 serotonin and β-adrenergic receptors, are not altered by food deprivation.[73]

When rats that had been food deprived for 72 hr are allowed to consume food pellets (4 hr), hypothalamic [^{3}H](+)-amphetamine binding is actually increased above control levels, while the binding in brainstem is not significantly different from control (i.e., fed) rats.[73,78] Moreover, a highly significant correlation ($r = 0.86$, $p < 0.001$) has been observed between the number of hypothalamic [^{3}H](+)-amphetamine binding sites and blood glucose in control, starved, and refed animals.[73,78] This rapid reversal of the food deprivation-induced decrease in [^{3}H](+)-amphetamine binding has also been observed in rats given access to a 10% glucose solution for 4 hr.[73] Therefore, food deprivation is associated with an acute, specific, and time-dependent decrease in hypothalamic [^{3}H](+)-amphetamine binding site density that is rapidly reversible.

In acute starvation, the body undergoes a series of hormonal and metabolic changes in order to maintain an ample glucose supply to the CNS.[77] Hypothalamic and brainstem [^{3}H](+)-amphetamine binding sites appear to be directly influenced by changes in circulating and/or brain glucose concentration. The fact that the food deprivation-induced decrease of hypothalamic [^{3}H](+)-amphetamine binding sites is rapidly reversed by a brief period of refeeding, which dramatically elevates serum glucose concentrations from hypoglycemic to hyperglycemic levels, supports the hypothesis that the [^{3}H](+)-amphetamine binding site in the hypothalamus and brainstem may be directly coupled to the mechanisms responsible for sensing physiological changes in blood glucose.

The glucostatic regulation of [^{3}H](+)-amphetamine binding sites in the hypothalamus has been further examined by directly increasing serum glucose concentration in normal rats. The parenteral administration of a 10% D-glucose solution to normal rats over 4 hr significantly increases hypothalamic [^{3}H](+)-amphetamine binding compared to saline-injected controls.[73,78] Injection of L-glucose, the physiologically inactive glucose stereoisomer, in an identical manner fails to alter [^{3}H](+)-amphetamine binding. The induction of hypothalamic

[^{3}H](+)-amphetamine binding sites by parenteral glucose administration is associated with a small but significant elevation in circulating glucose levels. That such changes in blood glucose may serve as a satiety signal under certain conditions has already been postulated.[79–90] These data suggest that at least one of the anorectic effects of amphetamine (i.e., regulation of carbohydrate intake) may be modulated by glucose via membrane-bound glucoreceptors in the CNS.

3.1.2b. Effect of Glucose on Hypothalamic [^{3}H](+)-Amphetamine Binding Sites. The effect(s) of D-glucose on [^{3}H](+)-amphetamine binding has also been examined in vitro using hypothalamic slices. The addition of D-glucose (1–100 mM) to hypothalamic slices incubated in Krebs–Ringer bicarbonate buffer (KRBB) at 37°C results in a significant dose- and time-dependent stimulation of [^{3}H](+)-amphetamine binding that is maximal within 10 min and at a D-glucose concentration of approximately 30 mM.[72] Scatchard analysis indicates that incubation with D-glucose results in an almost fourfold increase in the number of hypothalamic [^{3}H](+)-amphetamine binding sites compared with slices preincubated in the absence of glucose. Incubation of hypothalamic slices in KRBB in the absence of glucose results in a time-dependent decrease in the number of [^{3}H](+)-amphetamine binding sites.[72]

L-Glucose, 3-*O*-methoxy-D-glucose, and 2-deoxy-D-glucose alone fail to alter hypothalamic [^{3}H](+)-amphetamine binding sites, whereas 2-deoxy-D-glucose inhibits the D-glucose-stimulated increase in [^{3}H](+)-amphetamine binding in vitro.[72] The addition of 10 mM D-glucose following a 10-min preincubation in the absence of glucose fully restores the number of hypothalamic [^{3}H](+)-amphetamine binding sites to that observed in slices incubated with D-glucose throughout the entire 20-min incubation.[72]

3.1.2c. Coupling of (Na^{+},K^{+})-ATPase and Hypothalamic [^{3}H](+)-Amphetamine Binding Sites. The cardiac glycoside ouabain antagonizes the stimulatory effect of D-glucose on [^{3}H](+)-amphetamine binding in a dose-dependent manner, the EC_{50} being approximately 10 μM.[72] Since ouabain specifically inhibits (Na^{+},K^{+})-ATPase, activation of the neuronal (Na^{+},K^{+})-ATPase may contribute the stimulatory effect of D-glucose on hypothalamic [^{3}H](+)-amphetamine binding may be dependent (Na^{+},K^{+})-ATPase since the glucose-induced changes in specific [^{3}H](+)-amphetamine and [^{3}H]ouabain binding in hypothalamic slices are significantly correlated ($r = 0.89$, $p < 0.01$),[72] the glucostatic regulation of hypothalamic [^{3}H](+)-amphetamine binding involves functional changes in the activity of neuronal (Na^{+},K^{+})-ATPase.

3.1.2d. Glucoprivic Feeding and Hypothalamic [^{3}H](+)-Amphetamine Binding Sites. The peripheral or central administration of 2-deoxy-D-glucose (2-DG) initiates glucoprivic feeding and produces hyperglycemia.[82,83,86,89,90–92] 2-DG is believed to initiate hunger and feeding by blocking central glucoreceptors, which are responsible for satiety. Since hypothalamic [^{3}H](+)-amphetamine binding sites appear to be involved in food deprivation-induced hunger and satiety, the effect of 2-DG administration on [^{3}H](+)-amphetamine binding in the hypothalamus has been studied. The parenteral administration of 2-DG (10–200 mg/kg) significantly increases [^{3}H](+)-amphetamine binding site density in the hypothalamus when compared to saline-injected controls.[78] The increase in [^{3}H](+)-amphetamine binding is observed as early as 15 min after 2-DG administration and lasts approximately 4–6 hr. The effect of 2-DG in increasing [^{3}H](+)-amphetamine binding in the hypothalamus is only observed if the animals are not given access to food.[78]

Previous studies have documented that alloxan produces experimental diabetes in animals by irreversibly altering membrane-bound glucoreceptors on β islet cells of the pancreas.[93] Alloxan administered intraventricularly can also alter brain glucoreceptors and block the effect of 2-DG on feeding behavior.[90–92] When rats are treated with alloxan intraventricularly 2–3 weeks prior to 2-DG administration, the latter fails to increase both food intake and [^{3}H](+)-amphetamine binding site density in the hypothalamus.[78] 2-DG administration, however, does produce a hyperglycemic response in alloxan-pretreated animals. These findings further support a role for hypothalamic [^{3}H](+)-amphetamine binding sites in processing signals of hunger and satiety produced by changes in peripheral glucose levels. More specifically, it appears that [^{3}H](+)-amphetamine binding site may be associated with membrane-bound glucoreceptors in the hypothalamus. Thus, central glucoreceptors appear to respond to changes in circulating glucose and then influence hypothalamic [^{3}H](+)-amphetamine binding. The data suggest that glucose and amphetamine may act at a common locus in the hypothalamus and brainstem.

3.1.2e. Genetic Obesity and Hypothalamic [^{3}H](+)-Amphetamine Binding Sites. The genetically obese (*ob/ob*) mouse displays hyperphagia and develops marked obesity associated with hyperglycemia. In 40-day-old ob/ob mice, which had obtained a body weight >150% of their lean littermates, [^{3}H](+)-amphetamine binding in the hypothalamus is increased by approximately 60% above the levels of their lean littermates.[94] Since the weight gain in ob/ob mice is most prominent during the first 1–3 months of life,[95] an increase in hypothalamic [^{3}H](+)-amphetamine binding may be involved in the evolution of obesity. Since ob/ob mice are markedly hyperglycemic,[95] it is also possible that the increase in [^{3}H](+)-amphetamine binding site density in the hypothalamus is secondary to elevated blood glucose. Conversely, it is possible that central glucoreceptors are altered in genetically obese mice, so that an inadequate or deficient satiety signal is relayed by brain glucoreceptors.

3.2. Brain [^{3}H]Mazindol Binding Sites

3.2.1. Characteristics of [^{3}H]Mazindol Binding Sites

3.2.1a. Characteristics of Sodium-Dependent [^{3}H]Mazindol Binding Sites. Although the chemical structure of mazindol is very different from amphetamine, both share common pharmacological properties, including their ability to increase rotational behavior in rats with unilateral 6-hydroxydopamine lesions of the nigrostriatal pathway and to produce psychomotor activation.[76] Since the above effects are inhibited by dopamine receptor antagonists,[97] these actions of mazindol may be mediated by DA neurotransmission. In fact, a significant correlation has been observed between the inhibitory potencies of several mazindol derivatives on in vitro [^{3}H]-DA uptake and their potencies in stimulating locomotor activity or rotational behavior in 6-OHDA-lesioned rats.[96,98] Recent work has also demonstrated that mazindol can reduce stimulant-induced euphoria and drug craving caused by cocaine and other psychostimulants, presumably by inhibiting the DA transport complex in presynaptic neurons.[119] In addition to its interaction at dopamine uptake sites, mazindol is a potent inhibitor of norepinephrine uptake in neurons.[99] The ability of mazindol to act as an appetite suppressant may be mediated by dopaminergic mechanisms,[97] although one report found no significant correlation between the anorectic effects of mazindol and its analogs and their potencies as DA uptake blockers.[100]

Mazindol binding sites in the CNS can be differentiated on the basis of whether physiological concentrations of sodium chloride enhance or inhibit [^{3}H]mazindol binding to brain membranes. In the presence of sodium, high-affinity binding of [^{3}H]mazindol has been demonstrated in various brain regions. The sodium-dependent [^{3}H]mazindol binding is predominantly localized to the corpus striatum, since the number of sodium-dependent sites in other brain areas is less than 15% of striatal binding site concentrations (Fig. 1), and autoradiographic localization of mazindol binding sites has confirmed that sodium-dependent mazindol binding is mainly to nigrostriatal DA reuptake sites.[120]

The inhibition of specific sodium-dependent binding of [^{3}H]mazindol to striatal membranes by various drugs such as mazindol, nomifensine, benzatropine, cocaine, and amphetamine closely correlates with the inhibition of synaptosomal DA uptake in the striatum by these agents. (+)-Amphetamine is 200 times less potent in displacing high-affinity sodium-dependent [^{3}H]mazindol binding from striatal membranes and twofold less potent in inhibiting hypothalamic [^{3}H](+)-amphetamine binding than mazindol. With respect to hypothalamic [^{3}H]mazindol binding in the absence of sodium, (+)-amphetamine is somewhat less potent in inhibiting specific binding than is mazindol. These findings demonstrate that mazindol and (+)-amphetamine have a similar ability to inhibit food intake and to bind at anorectic drug sites in the hypothalamus. In contrast to mazindol, (+)-amphetamine, is considerably less potent at striatal DA uptake sites and has a significantly lower affinity at sodium-dependent [^{3}H]mazindol binding sites in the corpus striatum that possibly mediate stimulant-induced motoric behavior.

Although sodium-dependent [^{3}H]mazindol binding sites appear to be associated with striatal DA uptake sites, dopamine is a more potent inhibitor of DA uptake than of [^{3}H]mazindol binding in the corpus striatum.[99] Since the apparent K_d for sodium-dependent [^{3}H]mazindol binding to rat striatum (18.2 nM) is lower than the apparent K_d for [^{3}H]-GBR-12935 binding (1 nM), [^{3}H]-GBR-12935 is a more selective ligand for labeling the neuronal dopamine uptake site in the corpus striatum. Moreover, [^{3}H]mazindol in the presence of sodium appears to label norepinephrine uptake sites in the cerebral cortex and submaxillary/sublingual gland as well as striatal dopamine uptake sites in the striatum.[99] [^{3}H]-Mazindol binding sites are reduced in the corpus striatum following destruction of dopaminergic neurons by 6-OHDA and in the cerebral cortex following destruction of noradrenergic neurons by *N* -(2-chloroethyl)-*N*-ethyl-2-bromobenzylamine.[99] Consequently, GBR ligands appear to be more specific labels of the brain dopamine transport than mazindol.

3.2.1b. Characteristics of [^{3}H]Mazindol Binding Sites Inhibited by Sodium Chloride. In the absence of sodium ions, a high concentration of saturable low-affinity binding sites for [^{3}H]mazindol has been measured in the rat hypothalamus.[101,102] Studies on the subcellular distribution of these binding sites have shown that the synaptosomal fraction has the highest density of sites, while the lowest density is found in the crude mitochondrial fraction.[102] Kinetic analysis of the binding of [^{3}H]mazindol to hypothalamic membranes has demonstrated a single class of noninteracting binding sites with an apparent affinity constant (K_d) of 9.3 ± 1.4 μM and maximum number of binding sites (B_{max}) of 433 ± 88 pmoles/mg protein.

The binding of [^{3}H]mazindol is unevenly distributed in various brain regions with the highest densities of sites being observed in the hypothalamus, followed by the striatum and brainstem (Fig. 1).[102] No marked differences in binding site concentrations have been observed in the striatum, hippocampus, and cerebral cortex. Using microdissection methods in hypothalamic tissue, the highest level of [^{3}H]mazindol binding has been localized to the paraventricular nucleus (PVN) with a significantly lower binding in the lateral hypothalmus.[122] Since injection of norepinephrine, opioid peptides, and neuropeptide Y directly into the PVN stimulates feeding,[57,58,61,62,64] the PVN is an important site for the regulation of food intake.[103] Furthermore, lesions or knife cuts around the PVN disrupt the satiety action of intraperitoneally administered cholecystokinin.[104] Consequently, the high concentration of [^{3}H]mazindol binding sites in the PVN suggests that these sites may be involved in the inhibition of food intake. Preliminary autoradiography studies have also revealed a high concentration of [^{3}H]mazindol binding sites in the median eminence and hypothalamic nuclei such as the PVN and arcuate and in other brain areas including the dorsal raphe, locus coeruleus, olfactory tubercle, nucleus accumbens, amygdala, and bed nucleus of the stria terminalis.[120] Since many of these brain areas contain CRF cell bodies and terminals, the anorectic mazindol binding site may be coupled to CRF neurons.

Specific [^{3}H]mazindol binding is rapidly reversible, temperature sensitive, labile to pretreatment with proteolytic enzymes and inhibited by physiological concentrations of sodium. In all brain regions studied except the striatum, binding is markedly inhibited by NaCl (120 mM).[102] However, specific [^{3}H]mazindol binding to striatal membranes is enhanced by sodium, presumably because of the presence of sodium-dependent [^{3}H]mazindol binding to the dopamine transporter. In peripheral tissues, very low levels of specific binding has been observed in the liver and kidney, whereas relatively high levels were found in the adrenal gland.[102] When presynaptic noradrenergic and dopaminergic nerve terminals are destroyed by the intracerebroventricular administration of 6-hydroxydopamine, sodium-stimulated binding of [^{3}H]mazindol to the corpus striatum was inhibited by 50–60% in the absence of any changes in hypothalamic[^{3}H]mazindol binding in the presence or absence of 120 mM NaCl.[122] By contrast, when serotonergic afferents to the hypothalamus are lesioned with the specific neurotoxin 5,7-DHT, the binding of [^{3}H]mazindol to hypothalamic membranes is increased two- to threefold.[122] These findings suggest that hypothalamic [^{3}H]mazindol binding sites have a postsynaptic location, where these sites are possibly influenced by serotonergic neurotransmission.

Since mazindol is a potent anorectic drug and inhibitor of [^{3}H](+)amphetamine binding,[71] the potencies of a series of anorectic phenylethylamine derivates in inhibiting [^{3}H]mazindol binding in hypothalamic membranes and in reducing food intake has been examined.[101,102] Parachloroamphetamine was the most potent drug tested (K_i = 1.8 uM) and phendimetrazine the least potent (K_i = 1.4 mM). Furthermore, a highly significant correlation (r = 0.84, p <0.01) has been observed between this binding displacement data and the anorectic potencies of these drugs. By contrast, no correlation was observed between the potencies of these drugs in inhibiting [^{3}H]mazindol binding and their motor stimulant properties (r = 0.45, p = NS), nor with their potencies in inhibiting drinking behavior (r = 0.56).[102] These findings suggest that the sodium-independent [^{3}H]mazindol binding site in the hypothalamus (like [^{3}H](+)-amphetamine binding) is not associated with dopamine uptake sites but may mediate the anorectic actions of mazindol, amphetamine, and related phenylethylamines.

3.2.2. Glucostatic Regulation of [^{3}H]Mazindol Binding Sites

Similar to previous findings for [^{3}H](+)-amphetamine binding sites,[73] food deprivation for 72 hr results in a significant reduction in [^{3}H]mazindol binding in whole hypothalamus.[122] Analysis of subhypothalamic sites has shown that the largest food deprivation-induced decrease in [^{3}H]mazindol binding occurs in the paraventricular nucleus.[122] The loss of [^{3}H]mazindol binding sites in the hypothalamus during food deprivation can be reversed by refeeding food-deprived rats.[122]

The levels of [^{3}H]mazindol binding in the hypothalamus and brainstem have also been found to be significantly higher in genetically obese (ob/ob) mice compared to their lean littermates (?/+),[105] similar to previous findings for brain [^{3}H](+)-amphetamine and [^{3}H]ouabain binding sites in obese mice.[78,94] Furthermore, these changes in binding were highly correlated with the circulating levels of glucose in obese mice.[105]

4. BRAIN RECOGNITION SITES FOR STIMULANTS THAT INFLUENCE SEROTONERGIC NEUROTRANSMISSION

4.1. Brain [^{3}H]-p-chloro-Amphetamine Binding Sites

The halogenated amphetamine analogue, para-chloroamphetamine (PCA), has been shown to have selective effects on serotonergic neurotransmission.[106] PCA also has prominent anorexogenic activity, demonstrated by the observation that its inhibitory potency in suppressing food intake is the highest among phenylethylamine anorectics.[1–3,12] Saturable and low-affinity (K_d ~4 μM) binding sites for [^{3}H]-PCA have been demonstrated in rat hypothalamic membranes.[107] These PCA binding sites have structure–activity characteristics similar to hypothalamic [^{3}H]mazindol binding sites.[101,102] Consequently, PCA recognition sites in the hypothalamus may also be involved in the anorectic actions of mazindol and various phenylethylamine derivates.

4.2. Methylenedioxymethamphetamine (Ectasy) Binding Sites in the Central Nervous System

The psychotropic agent 3,4-methylenedioxymethamphetamine (MDMA) is a ring-substituted derivative of methamphetamine and possesses both stimulant and hallucinogenic actions.[108] MDMA has been shown to inhibit in vitro [^{3}H]serotonin uptake and release, to reduce the brain content of serotonin (5-HT) and its metabolite 5-hydroxyindoleacetic acid, and to decrease in vivo the maximal uptake of 5-HT and the activity of tryptophan hydroxylase.[108] Consequently, MDMA may have selective effects on serotonergic neurotransmission.

The relative binding potencies of MDMA at various brain recognition sites has been recently examined. MDMA exhibited the highest affinity at serotonin recognition sites in the rat cerebral cortex possessing a nearly equal affinity for 5-HT$_1$ and 5-HT$_2$ sites.[109,110] The interactions of MDMA at pre- and postsynaptic serotonin recognition sites may mediate its mood-enhancing effects. However, MDMA also has significant binding affinities for α_2-adrenoceptors and M-1 muscarinic receptors in the frontal cortex[110] that correlate with its hypotensive and salivation effects, respectively.[108] Although the dopamine agonist-like activity of MDMA may be indirectly mediated by changes in presynaptic DA release,[111] MDMA exhibits low binding affinities at postsynaptic D-1 and D-2 dopamine receptors in the corpus striatum.

Recently, specific binding sites for MDMA have been identified in rat brain.[112] Scatchard analysis of [^{3}H](+)-MDMA binding has revealed an apparent dissociation constant (K_d) of 99 nM and a B_{max} of 30 fmoles/mg protein. Both (+)-*p*-chloroamphetamine and (+)-methamphetamine significantly inhibit MDMA binding; however, (+)-amphetamine exhibits little affinity at brain MDMA sites. Even though MDMA can produce in vitro a large release of serotonin by brain tissue and in vivo a prolonged reduction of striatal serotonin levels—neurochemical effects that can be locked by coadministration of the serotonergic reuptake inhibitor citalopram—citalopram does not appear to interact at brain [^{3}H](+)-MDMA binding sites.[112] Consequently, MDMA sites do not appear to be located in presynaptic nerve terminals.[112]

5. CONCLUSION

A prevailing mechanism of psychostimulants is the inhibition of the presynaptic uptake of dopamine. Behavioral stimulants also strongly promote the synaptic release of dopamine. Amphetamine psychostimulants are potent releasers of dopamine from the reserpine-insensitive cytoplasmic pool of DA, which is dependent upon replenishment by newly formed amine.[1,2,5,14–17] In contrast, the non-amphetamine class of stimulants such as methylphenidate, cocaine, and nomifensine can be distinguished from amphetamine in that these latter stimulants promote the release of dopamine from the granular storage pool of vesicular DA which is reserpine-sensitive.

Recent radioligand binding studies have demonstrated the direct action of psychostimulants at the DA transporter located on presynaptic DA neurons (Table II). [^{3}H]Threo-(+)-Methylphenidate (ritalin), the diphenyl-substituted piperazine derivatives [^{3}H]-GBR-12935 and [^{3}H]-GBR-12783, [^{3}H]cocaine, and [^{3}H]nomifensine is preferentially bound to presynaptic sites on DA nerve terminals.[24,31,32,34–36,39] Similar to the characteristics for the binding of radiolabeled antidepressants and specific uptake blockers, the interaction of the above psychostimulants with the presynaptic DA uptake site is highly dependent on sodium. The interaction of motor stimulant drugs at ritalin binding sites in the corpus striatum appears to mediate psychostimulant action on the dopamine transport complex in striatal nerve terminals. All of the other psychostimulants (e.g., GBR, cocaine, nomifensine) appear to define a similar site related to the striatal dopamine transport system which, at least in part, mediates the central action of non-amphetamine psychostimulants.

Although the amphetamine binding site in the striatum does not appear to have a presynaptic location, binding of amphetamine stimulants to the presynaptic DA carrier site may be an important initial event in the psychostimulant effects of amphetamine (Table I). This hypothesis is supported by the observations that methylphenidate inhibits *in vitro* amphetamine-induced DA release and *in vivo* amphetamine-stimulated behaviors. Other stimulants such as phencyclidine may also interact at presynaptic DA transport sites.

Therefore, this recent work on the labeling of the DA transport complex with tritiated ritalin, GBR, cocaine, mazindol, and nomifensine, and the classical studies of psychostimulant mechanisms have unequivocably demonstrated that stimulants act on presynaptic monoamine neurons in the central nervous system (Tables I–II). Although psychostimulants do not directly interact with postsynaptic receptors, stimulant drugs can mimic catecholamines at their postsynaptic receptor sites via their stimulation of biogenic amine release and inhibition of neuronal reuptake inactivation. These presynaptic effects can subsequently produce regulatory changes at postsynaptic monoamine receptor sites.

In addition to these indirect effects of stimulants on postsynaptic catecholamine neurons, recent studies have identified postsynaptic binding sites in the central nervous system for [^{3}H](+)-amphetamine and [^{3}H]mazindol (Table II).[65–67,71,101,102] These putative anorectic sites decrease in number during changes in glucose metabolism secondary to food deprivation. Glucose can directly regulate amphetamine and mazindol binding sites in the hypothalamus and the glucose-stimulated increases in these sites can be blocked by glucose transport inhibitors (Table II). 2-Deoxy-glucose can also increase hypothalamic amphetamine and mazindol binding sites in association with its induction of glucoprivic feeding and hyperglycemia. The inactivation of central glucoreceptors by intraventricular alloxan abolishes these glucoregulatory effects on hypothalamic anorectic binding sites.

Central glucoreceptors most likely mediate the inhibition of food intake produced by 2-DG acting at brain glucostats which initiate feeding behavior.[55,113] The glucostatic theory of food intake hypothesizes that central glucoreceptors respond to changes in circulating glucose levels and glucose utilization to regulate appetite.[79,80] Hypothalamic amphetamine and mazindol binding sites may be coupled to a central glucostat where the anorectic effects of stimulants are expressed. Neurons in the lateral and ventromedial hypothalamus become hyperpolarized in response to glucose. [^{3}H](+)-Amphetamine and [^{3}H]mazindol may bind to glucose-sensitive neurons in regulating appetite. Furthermore, since prominent glucoregulatory effects have been demonstrated for mazindol binding sites in the PVN and other hypothalamic nuclei which contain abundant CRF neurons, stimulants may act a central locus composed of glucoreceptors and CRF neurons to express their anorectic action.

Serotonergic anorectics such as fenfluramine and para-chloro-amphetamine (PCA) also bind to the [^{3}H](+)-amphetamine binding site.[66] Although different neurochemical mechanisms have been proposed for fenfluramine and amphetamine, the anorectic mechanism of action for these stimulant drugs may involve the amphetamine/mazindol binding site.[2–4,6,63] However, PCA and the novel stimulant MDMA (Ectasy) both preferentially influence serotonergic neurotransmission and have their own separate binding sites in the central nervous system (Table II).[107,112] Brain PCA and MDMA binding sites may mediate the central effects of

Table II. Characteristics of Brain Recognition Sites for Psychostimulants[a]

Stimulant binding site	Binding affinity (K_d)	Anatomic localization	Sodium effects	Effect of selective destruction of monoaminergic terminals		Effect of catecholamine depletion by reserpine	Effect of glucose	Possible function
				DA	5-HT			
Methylphenidate	~200 nM	Highest in presynaptic DA terminals of striatum	Na^+-dependent	↓	0	0		Mediation of psychomotor activation via inhibition of presynaptic DA uptake
GBR	<1 nM	Highest in presynaptic DA terminals in striatum and nucleus accumbens	Na^+-dependent	↓		0		Component of DA uptake complex separate from the site involved in DA transport. GBR is most selective and potent inhibitor of DA uptake. Parkinson's disease associated with ↓ GBR sites in caudate nucleus
Cocaine	200–300 nM	Highest in presynaptic DA terminals in striatum	Na^+-dependent	↓	0			Cocaine binding site on DA transporter of DA terminals mediates inhibition of DA uptake and potentiation of DA transmission. Possible site mediating cocaine reinforcement. Parkinson's disease associated with ↓ cocaine sites in caudate putamen
Nomifensine	100 nM	Highest in presynaptic DA terminals in caudate–putamen, nucleus accumbens, olfactory tubercle	Na^+-dependent	↓				Probe for presynaptic uptake sites for DA and norepinephrine. Possesses antidepressant action in depressed patients.

Amphetamine	5–10 μM	Highest in hypothalamus, brainstem, striatum	Na^+-inhibited	↑ in striatum		↑	Membrane-bound complex coupled to brain glucoreceptors that may be related to the anorectic action of amphetamine
Mazindol							
Na^+-dependent	~20 nM	Highest in presynaptic DA terminals in striatum	Na^+-inhibited	↓			Presynaptic DA uptake sites in striatum and NE uptake sites in cerebral cortex
Na^+-independent	5–10 μM	Highest in hypothalamus (especially PVN), brainstem, striatum	Na^+-inhibited	0 in hypothalamus	↑ in hypothalamus	↑	Membrane-bound glucostatic complex acting as an anorectic site for appetite regulation
PCA	~1–5 μM	Hypothalamus	Na^+-inhibited				Anorectic binding site labelled by 5-HT neurotoxin
MDMA	~100 nM		Na^+-independent				Binding site for 5-HT neurotoxins (not localized to presynaptic 5-HT terminals) which may mediate psychostimulant and hallucinogenic actions of MDMA analogs

[a] ↑ = increased binding, ↓ = decreased binding, 0 = no effect.

stimulants which influence serotonergic neurotransmission. Future studies will be required to assess whether some of the neuropharmacologic effects of serotonergic-active stimulants such as PCA and MDMA are also mediated by [^{3}H](+)-amphetamine and [^{3}H]mazindol binding sites in the central nervous system which undergo glucostatic regulation or by their own unique brain binding sites.

ACKNOWLEDGMENTS. The work described in this article was done as part of our employment with the federal government and is therefore in the public domain. The studies on the binding sites for methyphenidate, GBR, amphetamine, and mazindol performed by the authors have been published in a series of research articles.[24,29,31,34,40,65–67,72,73,94,101,102,102a,105,107,114,119,122] Parts of this chapter (particularly passages in Sections 2.1, 2.4, and 3.1) were first published in previous reviews[78,121] and are included in this revised version of that material. We also wish to acknowledge Ms. Maria Bongiovanni's excellent preparation of this chapter for which we are grateful.

REFERENCES

1. Biel JH, Bopp BA: Amphetamines: Structure–activity relationships, in Iverson LL, Iverson SD, Snyder SH (eds): Handbook of Psychopharmacology, Vol. 11. New York, Plenum, 1979, pp 1–39
2. Moore KE: Amphetamines: Biochemical and behavioral actions in animals,in Iverson LL, Iverson SD, Snyder SH (eds): Handbook of Psychopharmacology, Vol. 11. New York, Plenum, pp 41–98
3. Hoebel BC: The psycholopharmacology of feeding, in Iverson LL, Iverson SS, Snyder SH (eds): Handbook of Psychopharmacology, Vol. 8. New York, Plenum, 1977, pp 94–112
4. Samanin R, Garattini S: Neuropharmacology of feeding, in Silverstone T (ed): Drugs and Appetite, Orlando, Florida, Academic, 1982, pp 23–39.
5. Ridley RM: Psychostimulants, in Grahame-Smith DG (ed): Psychopharmacology 2. Part 1: Preclinical Psychopharmacology. New York, Elsevier, 1985, p 183–205
6. Rowland NE, Carlton J: Neurobiology of an anorectic drug: Fenfluramine. Prog Neurobiol 27;13–62, 1986
7. Segal DS, Schuckit MA: Animal models of stimulant-induced psychosis, in Creese I (ed): Stimulants: Neurochemical, Behavioral, and Clinical Perspectives, New York, Raven, 1983, p 131–167
8. Robbins TW, Sahakian BJ: Behavioral effects of psychomotor stimulant drugs: Clinical and neuropsychological implications, in Creese I (ed): Stimulants: Neurochemical, Behavioral, and Clinical Perspectives, New York, Raven, 1983, p 301–331
9. Robinson TE, Becker JB: Enduring changes in brain and behavior produced by chronic amphetamine administration: A review and evaluation of animal models of amphetamine psychosis. Brain Res Rev 11:157–198, 1986
10. Klein DF, Gittelman R, Quitkin F, et al (eds): Diagnosis and Drug Treatment of Psychiatric Disorders: Adults and Children, ed 2. Baltimore, Williams & Wilkins, 1980
11. Bizzi AB, Bonaccorsi S, Jespersen A, et al: Pharmacological studies on amphetamine and fenfluramine, in Costa E, Garattini S (eds): Amphetamines and Related Compounds, New York, Raven, 1970, pp 577–595
12. Cox RH, Maickel RP: Comparison of anorexigenic and behavioral potency of phenylethylamines. J Pharmacol Exp Ther 181:1–9, 1972
13. Segal DS: Behavioral characterization of d- and 1-amphetamine. Science 190:475–477, 1975
14. Axelrod J: Amphetamine: Metabolism, physiological disposition and its effect on catecholamine storage, in Costa E, Garattini S (eds): Amphetamines and Related Compounds. New York, Raven, 1970, pp 207–216
15. Carlsson A: Amphetamine and brain catecholamines, in Costa E, Garattini (eds): Amphetamines and Related Compounds. New York, Raven, 1970, p 289–300
16. Glowinski J, Axelrod J: Effect of drugs on the uptake, release and metabolism of H^3-norepinephrine in the rat brain. J Pharmacol Exp Ther 149:43–49, 1965
17. McMillen BA: CNS stimulants: Two distinct mechanisms of action for amphetamine-like drugs. Trends Pharmacol Sci 19:429–432, 1983
18. Skolnick P, Paul SM: in Smythies JR, Bradley R (eds): Int Rev Neurobiol 23;103–115, 1982

19. Hauger RL, Rehavi M, Angel I, et al: Brain recognition sites for heterocyclic antidepressants: Their role in the therapeutic action of antidepressants in mood disorders. Adv Human Psychopharmacol 4:291–317, 1987
20. Perel J, Dayton P: The neuropharmacology of psychostimulants, in Usdin E, Forest I (eds): Psychotherapeutic Drugs, Part II. New York, Marcel Dekker, 1976, p 1287–1305
21. Krueger G, McGrath W: Behavioral and cardiovascular effects of stimulants, in Gordon M (ed): Psychopharmacological Agents, Vol. 1. Orlando, Florida, Academic, 1984, p 225–240
22. Chieneb C, Moore K: Blockade by reserpine of methylphenidate-induced release of lesion dopamine, J Pharmacol Exp Ther, 193:559–570, 1975
23. Breese A, Cooper B, Hollister A: Psychopharmacology (Berl) 44:5–10, 1975
24. Schweri MM, Skolnick P, Rafferty M, et al: [^{3}H]-threo(+)-methylphenidate binding to 3,4-dihydroxyphenylethylamine uptake sites in corpus striatum: Correlation with the stimulant properties of ritalnic acid esters. J Neurochem 45:1062–1070, 1985
25. Paul SM, Rehavi M, Rice KC, et al: Does high affinity [^{3}H]-imipramine binding label serotonin reuptake sites in brain and platelet? Life Sci 28:2753–2760, 1981
26. Rehavi M, Skolnick P, Brownstein MJ, et al: High-affinity binding of [^{3}H]-desipramine to rat brain: A presynaptic marker for noradrenergic uptake sites. J Neurochem 38:889–895, 1982
27. van der Zee P, Koger HS, Goojes J, et al: Aryl 1,4-dialk(en)ylpiperazines as selective and very potent inhibitors of dopamine uptake. Eur J Med Chem 15:363–370, 1980
28. Portoghese P, Malspeis LJ: Relative hydrolysis rates of certain alkyl (d1)-α-(2-piperidyl)-phenylacetates, J Pharmacol Sci, 50:494–501, 1961
29. Janowsky A, Schweri MM, Berger P, et al: The effects of surgical and chemical lesions on striatal [^{3}H]threo-(+)-methylphenidate binding: Correlation with [^{3}H]dopamine uptake. Eur J Pharmacol 108:187–191, 1985
30. Heikkila RE, Manzino L: Behavioral properties of GBR 12909, GBR 13069, and GBR 13098. Specific inhibitors of dopamine uptake. Eur J Pharmacol 103:241–248, 1984
31. Janowsky A, Berger P, Vocci F, et al: Characterization of sodium-dependent [^{3}H]GBR-12935 binding in brain: A radioligand for selective labelling of the dopamine transport complex. J Neurochem 46:1272–1276, 1986
32. Bonnet J-J, Protais P, Chagraoni A, et al: High-affinity [^{3}H]GBR-12783 binding to a specific site associated with the neuronal dopamine uptake complex in the central nervous system. Eur J Pharmacol 126:211–222, 1986
33. Bonnet J-J, Costentin J: GBR 12783, a potent and selective inhibitor of dopamine uptake: Biochemical studies in vivo and ex vivo. Eur J Pharmacol 121:199–206, 1986
34. Berger P, Janowsky A, Vocci F, et al: [^{3}H]GBR-12935: A specific high affinity ligand for labelling the dopamine transport complex. Eur J Pharmacol 107:289–290, 1985
35. Kennedy LT, Hanbauer I: Sodium-sensitive cocaine binding to rat striatal membranes: Possible relationship to dopamine uptake sites. J Neurochem 41:172–178, 1983
36. Dawson TM, Gehlert DR, Wamsley JK: Quantitative autoradiography localization of the dopamine transport complex in the rat brain: Use of a highly selective radioligand: [^{3}H]GBR-12935. Eur J Pharmacol 126:171–173, 1986
37. Dawson TM, Gehlert DR, Wamsley JK: Quantitative autoradiographic localization of central dopamine D-1 and D-2 receptors, in Creese I, Breese G (eds): Dopamine Receptor Function and Biochemistry. New York, Plenum, 1987, pp 78–97
38. Javitch JA, Blaustein RO, Snyder SH: [^{3}H]mazindol binding associated with neuronal dopamine and norepinephrine uptake sites. Mol Pharmacol 26:35–44, 1987
39. Dubocovich ML, Zahniser NR: Binding characteristics of the dopamine uptake inhibitor [^{3}H]nomifensine to striatal membranes. Biochem Pharmacol 34:1137–1148, 1985
40. Janowsky A, Vocci F, Berger P, et al: [^{3}H]GBR-12935 binding to the dopamine transporter is decreased in the caudate nucleus in Parkinsons disease. J Neurochem 49:617–621, 1982
41. Zelnik N, Angel I, Paul SM, et al: Decreased density of human striatal dopamine uptake sites with age. Eur J Pharmacol 126:175–176, 1986
42. Wong DF, Wagner HN, Dannels RF, et al: Effects of age on dopamine and serotonin receptors measured by positron tomography in the living human brain. Science 226:1393–1396, 1984
43. Price K, Farley I, Hornykiewicz O: Neurochemistry of Parkinsons disease: Relation between striatal and limbic dopamine. Adv Biochem Psychopharmacol 19:293–300, 1978
44. Byck R, Van Dyke C: Plasma concentrations and central effects following cocaine administration, in

Petersen RC, Stillman RC (eds): Cocaine. 1977. Washington, DC, U.S. Government Printing Office, 1977, p 97–108
45. Barash PG: Cocaine and related stimulants, in Petersen RC, Stillman RC (eds): Cocaine, 1977. Washington, DC, U.S. Government Printing Office, 1977, p 193–202
46. Reith MEA, Meisler BE, Sershen H, Lajtha A: Structural requirements for cocaine congeners to interact with dopamine and serotonin uptake sites in mouse brain and to induce stereotyped behavior. Biochem Pharmacol 35:1123–1129, 1986
47. Sershen H, Reith MEA, Lajtha A: Neuropharmacology 19:1145, 1980
48. Ritz MC, Lamb RJ, Goldberg SR, et al: Cocaine receptors on dopamine transporters are related to self-administration of cocaine. Science 237:1219–1222, 1987
49. Martin WR, Eades CG, Thompson JA, et al: The effects of morphine and morphine-like drugs in the nondependent and morphine-dependent chronic spinal dog. J Pharmacol Exp Ther 197:517–528, 1976
50. Sharkey J, Glen KA, Wolfe S, et al: Cocaine binding at sigma receptors. Eur J Pharmacol 149:171–174, 1988
51. Slifer BL, Balster RL: Reinforcing properties of stereoisomers of the putative sigma agonists N-allylnormetazoocine and cyclazocine in rhesus monkeys. J Pharmacol Exp Ther 225:522–531, 1983
52. Pimoule C, Schoemaker H, Javoy-Agid F, et al: Decreased [^{3}H]cocaine bindings in the human putamen of patients with Parkinson's disease. Eur J Pharmacol 95:145–152, 1983
53. Willner P: The role of dopamine in depression. Brain Res Rev 6:211–241, 1983
54. Scatton B, Dubois A, Dubocovich ML, et al: [^{3}H]Nomifensine binding to the striatal dopamine uptake site. Life Sci 36:815–822, 1985
55. Moore RY, Bloom FE: Localization of monoaminergic neurons in the central nervous system. Annu Rev Neurosci 2:113–143, 1979
56. Kuczenski R: Biochemical actions of amphetamine and other stimulants, in Creese I (ed): Stimulants: Neurochemical, Behavioral, and Clinical Perspectives. New York, Raven, 1983, p 31–61
57. Leibowitz SF: Neurochemical systems of the hypothalamus, in Morgane PJ, Panksepp J (eds): Handbook of the Hypothalamus, Vol. 3. New York, Marcel Dekker, 1980, pp 299–437
58. Leibowitz SF: Hypothalamus catecholamine systems in relation to control of eating behavior and mechanisms of reward, in Hoebel BC, Novin D (eds): The Neural Basis of Feeding and Rewards. Maine, Haer Institute, 1982, pp 241–257
59. Grossman SP: Contemporary problems concerning our understanding of brain mechanisms that regulate food intake and body weight, in Stunkard AJ, Stellar E (eds): Eating and Its Disorders, New York, Raven, 1984, pp 5–13
60. Hoebel BG: Neurotransmitters in the control of feeding and its rewards: Monoamines, opiates and brain-gut peptides, in Stunkard AJ, Stellar E (eds): Eating and Its Disorders, New York, Raven, 1984, pp 15–38
61. Leibowitz SF: Brain neurotransmitters and appetite regulation. Psychopharmacol Bull 21:412–418, 1985
62. Morley JE, Levine AS: Pharmacology of eating behavior. Annu Rev Pharmacol Toxicol 25:127–146, 1985
63. Blundell JE: Psychopharmacology of centrally acting anorectic agents, in Sandler M, Silverstone T (eds): Psychopharmacology and Food. Oxford, Oxford University Press, 1986, p 71–89
64. Morley JE, Blundell JE: The neurobiological basis of eating disorders: Some formulations. Biol Psychiatry 23:53–78, 1988
65. Paul SM, Hulihan B, Hauger R, et al: High affinity and stereospecific binding of [^{3}H]d-amphetamine to rat brain. Eur J Pharmacol 78:145–147, 1982
66. Paul SM, Hulihan-Giblin B, Skolnick P: (+)-Amphetamine binding to rat hypothalamus: Relation to anorexic potency of phenylethylamines. Science 218:487–490, 1982
67. Hauger RL, Hulihan-Giblin B, Skolnick P, et al: Characteristics of [^{3}H](+)-amphetamine binding sites in the rat central nervous system. Life Sci 34:771–782, 1984
68. Bonisch H: Biochemical mechanisms for psychostimulant effects. Arch Pharmacol 327:267, 1984
69. Lesage A, Strolin Benedetti M, Rumigny JF: Evidence that (+)[^{3}H]-amphetamine binds to acceptor sites which are not MAO-A. Biochem Pharmacol 34:3002–3005, 1985
70. Lesage A, Strolin Benedetti M, Rumigny JF: High affinity binding site for (+)amphetamine in rat hypothalamus: Fact or artifact? Neurochem Int 6:283–286, 1984
71. Blosser JC, Barrantes M, Parker RB: Correlation between anorectic potency and affinity for hypothalamic (+)-amphetamine binding sites of phenylethylamines. Eur J Pharmacol 134:97–103, 1987
72. Angel I, Hauger RL, Luu MD, et al: Glucostatic regulation of (+)-[^{3}H]amphetamine binding in the hypothalamus: Correlation with Na^+, K^+-ATPase activity. Proc Natl Acad Sci USA 82:6320–6324, 1985
73. Hauger RL, Hulihan-Giblin B, Skolnick P, et al: Glucostatic regulation of hypothalamic and brainstem

[^{3}H](+)-amphetamine binding during food deprivation and refeeding. Eur J Pharmacol 124:267–275, 1986

74. Grossman SP: The biology of motivation. Annu Rev Psychol 30:209–242, 1979
75. Morley JE, Levine AS: The central control of appetite. Lancet 1:398–401, 1983
76. Stellar E: The physiology of motivation. Psychol Rev 61:5–22, 1954
77. Cahill GF: Starvation. Trans Am Clin Climatol Assoc 94:1–21, 1982
78. Hauger R, Hulihan-Giblin B, Angel I, et al: Glucose regulates [^{3}H](+)-amphetamine binding and Na^+K^+ ATPase activity in the hypothalamus: A proposed mechanism for the glucostatic control of feeding and satiety. Brain Res Bull 16:281–288, 1986
79. Mayer J, Bates MW: Blood glucose and food intake in normal and hypophysectomized, alloxan-treated rats. Am J Physiol 168:812–821, 1952
80. Mayer J, Thomas DW: Regulation of food intake and obesity. Science 156:328–337, 1967
81. Le Magnen J: Interactions of glucostatic and lipostatic mechanisms in the regulatory control of feeding, in Novin D, Wyrwicka W, Bray G (eds): Hunger: Basic Mechanisms and Clinical Implications, New York, Raven, 1976, pp 89–101
82. Smith GP, Epstein AN: Increased feeding in response to decreased glucose utilization in the rat and monkey. Am J Physiol 217:1083–1091, 1969
83. Balaguia S, Kanner M: Hypothlamic sensitivity to 2-deoxy-D-glucose and glucose: Effects on feeding behavior. Physiol Behav 7:251–262, 1971
84. Louis-Sylvestre J, Le Magnen J: A fall in blood glucose level precedes meal onset in free-feeding rats. Neurosci Biobehav Rev 4(suppl 1):13–21, 1980
85. Le Magnen J: The body energy regulation: The role of three brain responses to glucopenia. Neurosci Biobehav Rev 4(suppl 1):65–74, 1980
86. Smith GP, Gibbs J, Strohmayer AJ, et al: Threshold doses of 2-deoxy-D-glucose for hyperglycemia and feeding in rats and monkeys. Am J Physiol 222:77–86, 1972
87. Oomura Y, Ooyama H, Sujimori M, et al: Glucose inhibition of the glucose-sensitive neurone on the rat lateral hypothalamus. Nature (Lond) 247:284–286, 1974
88. Oomura Y: Glucose as a regulator of neuronal activity, in Szabo AJ (ed): Advances in Metabolic Disorders. Orlando, Florida, Academic, 1983, p 31–65
89. Ritter RC, Slusser PG, Stone S: Glucoreceptors controlling feeding and blood glucose: Location in the hindbrain. Science 213:451–453, 1981
90. Ritter SJ, Murnane M, Ladenheim EE: Glucoprivic feeding is impaired by lateral or fourth ventricle alloxan injection. Am J Physiol 243:R312–317, 1982
91. Miselis RR, Epstein AN: Feeding induced by intracerebroventricular 2-deoxy-D-glucose in the rat. Am J Physiol 220:1438–1447, 1975
92. Woods SC, McKay LD: Intraventricular alloxan eliminates feeding elicited by 2-deoxyglucose. Science 202:1209–1211, 1978
93. Dunn JS, Sheehan JL, McLetchie NGB: Necrosis of Islets of Langerhans produced experimentally. Lancet 1:484–487, 1943
94. Hauger R, Hulihan-Giblin B, Paul SM: Increased number of hypothalamic [^{3}H](+)-amphetamine binding sites in genetically obese (ob/ob) mice. Neuropharmacology 25:327–330, 1986
95. Bray GA, York DA: Hypothalamic and genetic obesity in experimental animals: An autonomic and endocrine hypothesis. Physiol Rev 59:719–809, 1979
96. Heikkila RE, Babington RG, Houlihan HJ: Pharmacological studies with several analogs of mazindol: Correlation between effects on dopamine uptake and various in vivo responses. Eur J Pharmacol 71:277–286, 1981
97. Carruba MO, Zambotti F, Vincentini L, et al: Pharmacology and biochemical profile of a new anorectic drug: Mazindol, in Garattini S, Samanin R (eds): Central Mechanisms of Anorectic Drugs, New York, Raven, 1978, p 145–164
98. Ross SB: The central stimulatory action of inhibitors of dopamine uptake. Life Sci 24:159–167, 1979
99. Javitch JA, Blaustein RO, Snyder SH: [^{3}H]Mazindol binding associated with neuronal dopamine and norepinephrine uptake sites. Mol Pharmacol 26:35–44, 1984
100. Heikkila RE, Cabat FC, Manzinoi LM, et al: Unexpected differences between mazindol and its homologs on biochemical and behavioral responses. J Pharmacol Exp Ther 217:745–749, 1981
101. Angel I, Paul SM: Demonstration of specific binding sites for [^{3}H]mazindol in rat hypothalamus: Correlation with the anorectic properties of phenylethylamines. Eur J Pharmacol 113:133–134, 1985

102. Angel I, Luu M-D, Paul SM: Characterization of [^{3}H]mazindol binding in rat brain: Sodium-sensitive binding correlates with the anorectic potencies of phenylethylamines. J Neurochem 48:491–497, 1987
103. Gold RM, Jones AP, Sawchenko PE: Paraventricular area: Critical focus of a longitudinal neurocircuitry mediating food intake. Physiol Behav 18:1111–1119, 1977
104. Crawley JN, Kiss JZ: Paraventricular nucleus lesions abolish the inhibition of feeding induced by systemic cholecystokinin. Peptides 6:927–935, 1985
105. Angel I, Goldman ME, Paul SM: Defective glucostatic regulation of anorectic drug recognition sites and Na^+K^+ATPase in genetically obese mice. Soc Neurosci Abs 12(pt 2):795, 1986
106. Sanders-Bush E, Massari VJ: Actions of drugs that deplete serotonin. Fed Proc 36:2149–2153, 1977
107. Angel I, Luu M-D, Hauger R, et al: Specific [^{3}H]mazindol and [^{3}H]p-chloroamphetamine binding sites in the hypothalamus: Correlation with anorectic properties of phenylethylamines. Soc Neurosci Abs 11:670, 1985
108. Shulgin AT: The background and chemistry of MDMA. J Psychoactive Drugs 18:291–299, 1986
109. Lyon RA, Glennon RA, Titeler M: 3,4-Methylenedioxymethamphetamine (MDMA): Stereoselective interactions at brain 5-HT, and 5-HT_2 receptors. Psychopharmacology 88:525–526, 1986
110. Battaglia G, Brooks BP, Kulsakdinun C, et al: Pharmacologic profile of MDMA (3,4-methylenedioxymethamphetamine) at various brain recognition sites. Eur J Pharmacol 149:159–163, 1988
111. Johnson MP, Hoffman AJ, Nichols DE: Effects of the enantiomers of MDA, MDMA and related analogues on [^{3}H]serotonin and [^{3}H]dopamine release from superfused rat brain slices. Eur J Pharmacol 132:269, 1986
112. Gehlert DR, Schmidt CJ, Wu L, et al: Evidence for specific methylenedioxymethamphetamine (ectasy) binding sites in the rat brain. Eur J Pharmacol 119:135–136, 1985
113. Herberg LJ: Hunger reduction produced by injecting glucose into the lateral ventricle of the rat. Nature (Lond) 245–246, 1960
114. Hulihan-Giblin B, Hauger RL, Janowsky A, et al: Dopaminergic denervation increases [^{3}H](+)-amphetamine binding in the rat striatum. Eur J Pharmacol 113:141–142, 1985
115. Kuhar MJ, Ritz MC, Sharkey J: Cocaine receptors on dopamine transporters mediate cocaine-reinforced behavior. NIDA Research Monograph 88:14–21, 1988.
116. Hanbauer I: Modulation of cocaine receptors. NIDA Research Monograph 88:44–54, 1988
117. Volkow ND, Fowler JS, Wolf AP, et al: Cocaine binding in the human brain with positron emission tomography. Soc Neurosci Abstr 15:802, 1989
118. Meyer JS, Collins L: Cocaine binding sites in feta rat brain. Soc Neurosci Abstr 15:255, 1989
119. Berger P, Gawn F, Kosten TR: Treatment of cocaine abuse with mazindol. Lancet ii:283, 1989
120. Vincent GP, Levin BE: *In vitro* autoradiographic mapping of a putative "anorectic" binding site with [^{3}H]mazindol in rat brain. Soc Neurosci Abstr 15:1131, 1989
121. Hauger RL, Hulihan-Giblin B, Janowsky A, et al: Central recognition sites for psychomotor stimulants: methyl phenadate and amphetamine, in O'Brien RA (ed.): Receptor Binding in Drug Research, New York, Marcel Dekker Inc., 1986, pp. 167–182
122. Angel I, Janowsky J, Paul SM: The effects of serotonergic and dopaminergic lesions and sodium ions on [^{3}H]mazindol binding in rat hypothalamus and corpus striatum. Brain Research, 1990, in press.

5

Neuropeptide Modulation of Development and Behavior

Implications for Psychopathology

Curt A. Sandman and Abba J. Kastin

1. INTRODUCTION

Peptides are very simple molecules composed of amino acids linked by bonds resulting from the elimination of water between an amino group in one molecule and a carboxyl group in an adjacent molecule. Neuropeptides are peptides that either are found in, or influence, the nervous system. The number of neuropeptides identified changes constantly and currently exceeds 60 (cf. Krieger[1] and Miller[2]). Slight changes in the sequence of amino acids in neuropeptides can profoundly alter their influences on the brain and behavior.[3,4] The discovery that neuropeptides were located in the nervous system and participated in the communication among cells initially prompted studies of their modulatory influence on classic neurotransmitters. Because neuropeptides did not satisfy some of the rigid criteria of a neurotransmitter and were frequently co-localized with classic transmitters, it was assumed their primary function was to assist in neural communication. Subsequently, peptides were anointed as a new class of neurotransmitter[1] and in an insightful analysis, Hokfelt et al.[5] described critical differences between classic neurotransmission and peptidergic transduction. These investigators noted that, compared with classic transmitters, peptidergic transmission was slow and that peptides were released in very small quantities (three- to sixfold less than classic transmitters). A speculative conclusion from this analysis was that peptides produced long-lasting effects (they need not act fast) that were very potent (very little is needed to initiate and sustain neuronal communication). Currently, neuropeptides are considered both as important modulators of aminergic transmission and as primary transmitters. Thus, the addition of a new class of

Curt A. Sandman • State Developmental Research Institute, Fairview, Costa Mesa, California 92626; and Department of Psychiatry and Human Behavior, University of California–Irvine Medical Center, Orange, California 92668. *Abba J. Kastin* • Veterans Administration Medical Center, Tulane University School of Medicine, New Orleans, Louisiana 70146.

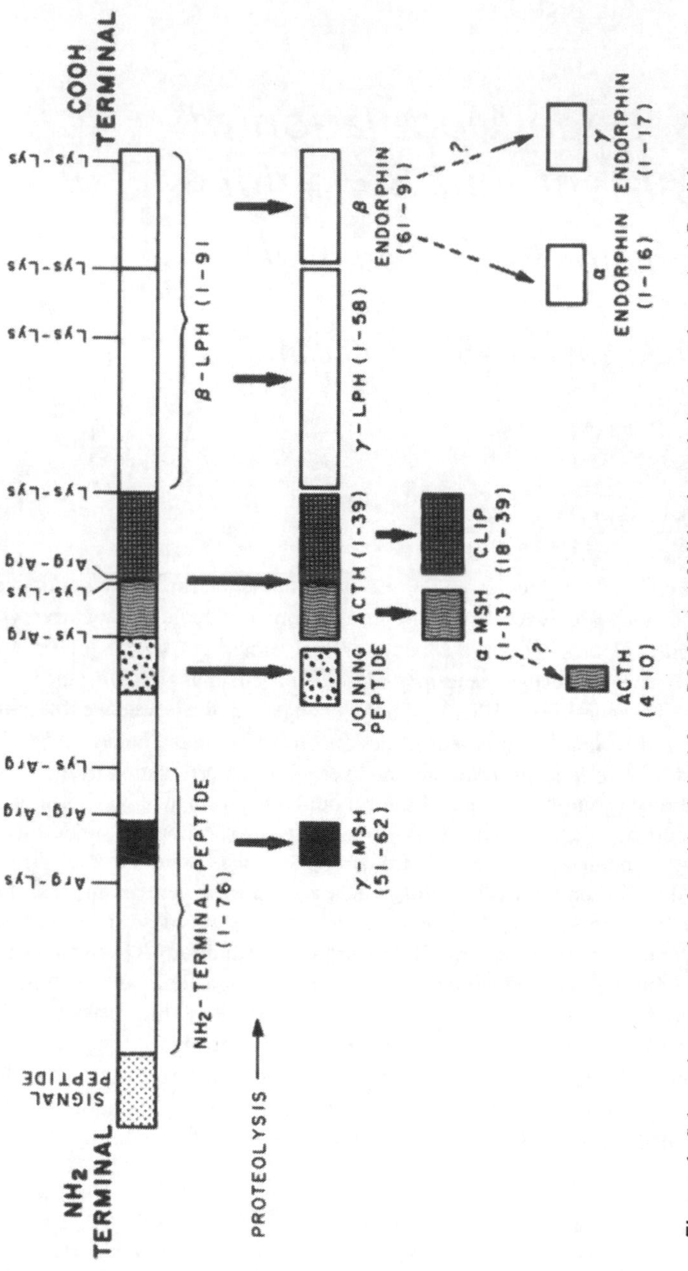

Figure 1. Schematic representation of peptides derived from POMC. (—) Validated translational products. (---) Possible products.

transmitters and their co-localization with classic transmitters encouraged the development of models for complex rather than simple (see, e.g., Dale[6]) signal transmission.

The discovery by O'Donohue et al.[7] and by Akil et al.[8] of co-localization of two peptides (α-MSH and β-endorphin) in the same cells of the brain extended the complex model of neural communication. The possibility that multiple peptides coexisted in single brain cells suggested that the information contained and transmitted in cells was far greater than previously conceived. These two peptides are packaged in the 31-KDa biologically inactive precursor protein molecule pro-opiomelanocortin (POMC), and are intracellularly, post-translationally processed into α-MSH and β-endorphin (and several other neuropeptides) (Fig. 1). The ultimate extent of peptide participation in neural transmission will undoubtedly exceed predictions from current models. One factor contributing to the continuing development of neuropeptide participation in neural transmission is the role of enzymes in liberating bioactive molecules. As described recently,[9] the translation of POMC by enzymes is controlled by both genetic and environmental factors. The cleaved products do not result from tissue specific POMC (which is located in a variety of cell types), but from differential regulation of enzymes. Somehow, through physiological or environmental triggers, specific enzymes are available and POMC is transformed into bioactive peptides. The implication that neuromodulators or neurotransmitters were available on demand, especially environmental demand, confers special relevance for psychopathology and, as this chapter explores, brain/behavior development.

2. ORGANIZATIONAL ENDOCRINE/PEPTIDE INFLUENCES

Considerable evidence has accumulated over the past 30 years indicating that neurochemical balance influences the organization of the brain. It is generally believed that there are critical periods during which profound changes in brain organization can occur. These critical periods often relate to developmental periods when the brain is undergoing transformations such as cell migration or receptor development. Since many of these events occur in utero, fetal changes in neurochemical state may have greater and longer-lasting influences than alterations in neonates or adults because immature, especially fetal, organisms are more susceptible to "outside" changes.[10] It is proposed that some environmental influences (i.e., stress) experienced by a pregnant woman that result in increased levels of endogenous molecules can be transduced to the fetus. In this way, environmental events could contribute to the organization or disorganization of the fetal brain and behavior. The theme will be developed that psychopathology might result from fetal exposure to excessive levels of these "endogenous teratogens."

The early observations of Phoenix et al.[11] provided important clues that the prenatal endocrine environment could influence the organization of the brain and behavior. These investigators discovered that prenatal treatment of guinea pigs with testosterone masculinized the adult females. These findings suggested that the developmental genetic "program" of femaleness could be greatly altered. Later studies[12] showed that early exposure to sexually dependent steroids sensitized adult animals to subsequent treatment. Male rats exposed in utero to high concentrations of estrogens responded as adults to low doses of estrogen with stereotypic female behavior. Other brain/behavior systems have been associated with early endocrine changes, including responses to stress,[13] physiological reflexes,[14] and learning and memory.[15]

As reviewed elsewhere,[16,17] extensive literature supports a reciprocal relationship between the POMC neuropeptides MSH/ACTH and β-endorphin. In general, the MSH/ACTH peptides enhance and β-endorphin depresses neural and perhaps behavioral efficiency. Improvement in learning and memory in rats, in normal volunteers, and in mentally retarded patients has been reported after administration of MSH/ACTH.[18–27] Early exposure to MSH/ACTH produced a "permanent" advantage for rats to learn active avoidance, passive avoidance, and discrimination problems.[2,8,29] Prenatally, α-MSH, and not ACTH, is the most neuroactive peptide. It has growth-promoting properties in utero, and normal intrauterine growth and labor are dependent on normal pituitary/MSH function.[29–31] MSH may be the primary tropic hormone for the adrenals,[29] stimulates secretion of the sebaceous vernix,[30] and is present in the brain during early stages of development before it declines.[31] Thus, the MSH-like peptides may have important functions that ensure the normal growth and development of the organism. However, the focus of this chapter is on the disorganizing influence of β-endorphin on the brain and behavior.

3. β-ENDORPHIN AND BIRTH

Over the course of gestation, concentrations of β-endorphin in maternal and fetal circulation change (Table I). Genazzani et al.[32] reported the highest concentrations of amniotic β-endorphin in the first trimester (173 fmoles/ml), decreasing to 75 fmoles/ml in the second trimester and to 14 fmoles/ml by the third trimester. Maternal plasma values increased over the same period of time. From the tenth week to term, plasma β-endorphin, but not Met-enkephalin, levels rose about 140%.[33,34] Similar studies confirmed the increase in maternal plasma β-endorphin to term[35,36] and the decrease in amniotic levels.[37] This effect appears to be specific for β-endorphin because *N*-POMC (a fragment of the parent POMC containing γ- and β-MSH) in amniotic fluid at gestational age 16–20 weeks was 10 times higher than in maternal plasma.[38] Thus, the changes may not be due to a greater abundance of POMC but to specific enzymatic activity responsible for cleaving β-endorphin.

Maternal plasma levels of β-endorphin are significantly elevated during labor with the highest concentrations measured after delivery.[39] Concentrations of this peptide in cord blood also were elevated but not as high as in the maternal samples. These maternal changes were considered to be a reflection of either stress or of an endogenous analgesic for the mother during birth. The relationship between maternal and fetal concentrations was unclear. One study compared 15 mothers and their infants with normal and uneventful vaginal delivery (full labor), 13 mothers and their infants delivered by cesarean section after labor onset (early labor), and 12 mother–infant pairs for whom elective cesarean section was chosen (no labor).[40] All mothers

Table I. β-Endorphin in Plasma and Amniotic Fluid during the Prenatal Period in Normal Pregnancy[a]

Studies	Findings
Gennazzani et al.,[32] Divers et al.[37]	Amniotic BE decreased to term
Evans et al.[33]	10th week to term, BE elevated 140% in plasma
Newnham et al.,[34] Browning et al.,[35] Wilkes et al.,[36] Divers et al.,[37] Csontos et al.[39]	Plasma BE increased to term
Kofinas et al.[40]	Plasma BE increased only in vaginal delivery compared with cesarean (neonatal BE higher than mother only for C section)
Goland et al.[43]	No correlation between maternal and cord plasma BE

[a]BE, β-endorphin.

and infants were relatively healthy. β-Endorphin concentration was measured in plasma collected 1 min before induction of anesthesia or at vaginal delivery. Blood (mixed) from the umbilical cord was collected within 1 min after the cord was clamped and cut.

Maternal plasma β-endorphin concentrations were elevated only in the vaginally delivered group (40.3 fmoles/ml versus 8.5 and 8.2 fmoles/ml in the other two groups), suggesting that the stress and pain of the birth process in these healthy subjects were the trigger for a fivefold increase. Neonatal (cord) β-endorphin concentrations were not different in the three groups. However, concentrations of neonatal β-endorphin was higher than maternal concentrations in both cesarean section groups. Only in the vaginally delivered group was the maternal concentration higher than the infant concentration. These findings indicated that maternal and cord sources for β-endorphin at birth were independent. The function of elevated maternal levels during vaginal birth may have been analgesic. As the following section suggests, changes in the fetal contribution of β-endorphin seem to be related to complications of birth.

4. β-ENDORPHIN AND BIRTH COMPLICATIONS

A growing array of evidence has implicated β-endorphin as a mediator of the central nervous system (CNS) effects of perinatal complications, especially hypoxia.[37,41–45] Experimentally induced hypoxia (10% O_2 for 30 min) did not result in increased β-endorphin either in pregnant sheep or in infants.[46] However, in a chronically catheterized fetus, circulating β-endorphin significantly increased in response to maternal hypoxia (from 125 to 503 pg/ml). Severe hypoxia (5% O_2) did produce significantly elevated β-endorphin in newborn and adult sheep.[46] Similar findings were reported in a follow-up study,[47] including parallel release of vasopressin. These studies demonstrate the vulnerability of the fetus to the effects of hypoxia and perhaps to environmental stress in general.

In the initial study of its kind, Gautray et al.[41] measured β-endorphin from amniotic fluid in women at least 37 weeks after cessation of menstruation. They reported a 2 to 20-fold increase of immunoassayable β-endorphin-like activity in cases of fetal distress. In a study of 116 women, Divers et al.[37] compared third-trimester concentrations of β-endorphin in amniotic fluid from normal pregnancies with samples from complicated courses (Table II). A two-

Table II. β-Endorphin in Amniotic Fluid and Plasma Associated with Birth Complications[a]

Studies	Birth complications	Findings
Gautray et al.[41]	Hypoxia	BE elevated in amniotic fluid at term
Divers et al.[37]	Premature labor, intrauterine growth retardation	BE elevated in amniotic fluid in third trimester
Wardlaw et al.[42]	Acidosis (hypoxia)	Correlation between pH and arterial BE ($r = -0.83$)
Shaaban et al.[44]	Fetal distress	BE elevated in cord blood
Davidson et al.[45]	Hypoxia	Elevated BE in infant plasma 2 hr after delivery
Panerai et al.[48]	Drug addicted	Up to 1000-fold increase in plasma BE in infant at 40 days of age

[a]BE, β-endorphin.

fold increase in β-endorphin was associated with premature labor and intrauterine growth retardation (IUGR).

Several studies have confirmed that elevated plasma β-endorphin is also associated with "complicated" pregnancies. Wardlaw et al.[42] measured β-endorphin in umbilical cord blood. Arterial samples were slightly higher than venous samples, suggesting a primary fetal contribution. However, their sample included cases with cesarean section before and during labor. Subsequent analysis indicated that arterial pH (acidosis, a marker of fetal hypoxia) was highly correlated with concentrations of β-endorphin. In 19 fetuses, the concentration of β-endorphin was negatively ($r = 0.83$) related to pH. In a related study,[44] umbilical cord blood was assayed for β-endorphin in infants delivered in a variety of ways with and without fetal distress (prolonged bradycardia, flattened heart rate). β-Endorphin was elevated only for infants with features of distress, regardless of means of delivery, a finding consistent with that of Kofinas et al.[40] The arteriovenous ratio was higher in normal pregnancies, suggesting a contribution from the placenta. However in the fetal distress group, the ratio was reversed. Thus, during stress, the fetal contribution to cord β-endorphin concentrations increased.

Davidson et al.[45] reported that plasma concentrations of β-endorphin were elevated in infants of low birthweight (premature and with severe to moderate asphyxia). Compared with nonasphyxiated premature infants, the oxygen-deprived groups had a threefold (12.7–40.2 pmoles/liter) increase in β-endorphin 4–6 hr after delivery. A significant increase was measured within the first 2 hr, but by 24 hr the concentrations had returned to normal.

However, a study by Panerai et al.[48] indicated that the ability of the infant pituitary to release β-endorphin is greatly enhanced in drug-addicted mothers. Infants from normal mothers had elevated concentrations of plasma β-endorphin that persisted for 5 days. The infants born to drug-addicted women had concentrations of β-endorphin that increased dramatically on days 2 and 3 (up to 1000-fold) and remained elevated for at least 40 days (as long as the investigators were able to test). Yet the effects of drug addiction on maternal concentrations of plasma β-endorphin during pregnancy were not significant.

All the above studies suggested that β-endorphin may have mediated the effects of hypoxia on the fetus. The weight of these correlational studies are impressive but more specific evidence for β-endorphin as a mediator of the CNS effects of hypoxia was illustrated by Chernick and Craig.[49] Pregnant rabbits were treated with naloxone (an opiate blocker) or with saline and then asphyxiated. The fetuses were delivered by cesarean section and evaluated by several criteria. All fetuses survived the procedure, but the pups whose mothers received naloxone had higher scores for respiration, color, muscle tone, response to stimulation, and general activity than did pups whose mothers had received saline. This study provided strong evidence that endogenous opiates mediated the effects of hypoxia and that blockade of the opiates eliminated the danger to the fetus.

The source of endorphins in fetal plasma (umbilical vessels) and amniotic fluid is not certain, although several possibilities seem obvious. The first of these is the fetal pituitary. Fachinetti et al.[50] showed that the human perinatal pituitary is capable of secreting β-endorphin, although it has only a fraction of the potency of the adult pituitary. Brubaker et al.[51] reported that the human fetal anterior pituitary contains β-endorphin as early as the thirteenth week of gestation. This is not unexpected, as the human hypothalamic–pituitary axis has been found to contain many of the biologically active peptides characteristic of the adult from the earliest stage (about weeks 6–8 of gestation[52]). The placenta is also a likely source. Nakai et al.[53] and Fraioli and Genazzani[54] found that cultured human placental cells produce β-endorphin. Odagiri et al.[55] detected large quantities of β-endorphin and β-lipotropin in placenta. In simultaneously drawn human umbilical blood samples, Shaaban et al.[44] reported that venous concentrations of β-endorphin exceeded arterial values by 42%, strongly implicating a placental contribution, at least under some circumstances.

The maternal contribution of endorphins to either circulating levels in the fetus or amniotic fluid is unknown. Several investigators who have measured fetal and maternal endorphins simultaneously have not found correlations between the two compartments. For instance, Goland et al.[43] assayed β-endorphin simultaneously in paired maternal and umbilical cord plasma samples and found no correlation. Browning et al.[35] found no correlation between maternal venous plasma concentrations of β-LPH, β-endorphin, and γ-lipotropin with umbilical levels (there was a positive correlation within maternal and fetal samples). However, the placenta has receptors for β-endorphin,[53,55] and it is conceivable that blood-borne maternal peptides (or other signals) could initiate the processes necessary for placental synthesis and release into the amniotic fluid.

Data gathered in rodents indicated that opiate receptors in the brain are topographically distinct by embryonic day 14, roughly equivalent to the end of the second trimester in humans.[56] However the bioavailability of β-endorphin in the infant pituitary (both anterior and intermediate lobe) is a fraction of that in the adult pituitary.[57] If neurobehavioral disregulation due to perinatal complications is mediated in part by β-endorphin, and if the ability of the immature pituitary is limited in its ability to release β-endorphin, perhaps the effects of hypoxia (discussed below) are on the well-developed opiate receptor system in the CNS.

5. BEHAVIOR OF ANIMALS EXPOSED PERINATALLY TO β-ENDORPHIN

The next line of argument for the relationship among perinatal complications, β-endorphin, and psychopathology is that animals exposed to β-endorphin (which is released during perinatal trauma) exhibit disregulated behavior (Table III). Pregnant rats injected peripherally with high levels (100 mg/rat) of β-endorphin during the second- and third-trimester delivered pups with a variety of neurological disorders, including delayed eye opening and environmental exploration, impaired discrimination learning, and hyposensitivity to acoustic stimulation.[16] Several rats in the treated group, but none in the controls, were born with enlarged brain ventricles and died after several days. Yet brain concentrations of β-endorphin were decreased in 40-day-old rats treated in utero with high levels of opiates.[58] In a recent study in our laboratory,[59] multiparous pregnant rats treated only during the third trimester with β-endorphin delivered pups that had many of the effects reported in the earlier study but also displayed evidence of persistent pain insensitivity (also observed after postnatal treatment).[60] Exposure to β-endorphin during the second and third trimester, but not the third alone,

Table III. Effects in Adult Rats of Pharmacologically Induced Exposure to β-Endorphin In Utero

Delayed eye opening by 2 days
Depressed activity
Significant weight gain by 4 days
Attenuation of somatomotor response to auditory stimulus by 350%
Less response to placement in open field (100%)
200% less activity in open field
Less physical contact time with paired rat
Hypersensitivity to morphine
Impaired performance of reversal learning problems.
"Permanent" subsensitivity of dopamine receptors
"Acute" supersensitivity of opiate (μ) receptors

produced apparent supersensitivity of the opiate receptor (sensitized to morphine challenge). This difference in response to pain and morphine may reflect the importance of timing or critical periods during development when chemical influences have their unique organizing (or disorganizing) effects.

In addition, at maturity, rats exposed as fetuses to β-endorphin had "permanent" decreases (subsensitivity) in the density of striatal, dopamine (D_2) receptors (B_{max}) without changes in affinity (K_d). It is not surprising that perinatal exposure to β-endorphin exerted effect on the dopamine system because there are numerous reports of co-regulation and co-localization of dopamine and endorphin systems.[61–63]

Opiate receptor density after pre- and postnatal administration of β-endorphin presents a complex pattern of effects.[64,65] Prenatal administration of β-endorphin (100 μg/mother during the third trimester) increased opiate μ-receptor density, but not affinity, measured on postnatal day 14. These effects were not apparent on postnatal day 60. However, postnatal administration of β-endorphin (50 μg/rat on days 1–7) resulted in downregulation of opiate μ-receptors on day 14. The dramatic differences observed in pre- and postnatal exposure are somewhat paralleled by the behavioral differences reported after β-endorphin treatment between second plus third trimester versus third trimester only.[16,59] Many differences in design can account for the different results,[65] including the critical period of opiate receptor development and even individual variation. For instance, we[15,59,66] and others[67] have found significant differences between male and females in response to prenatal and postnatal peptide and drug exposure.

To summarize, β-endorphin can be increased in amniotic fluid and in cord blood in response to stress, especially hypoxia. The relative contribution of fetus and mother is unknown; it appears, however, that the fetal contribution increases as a function of stress. Even if maternal β-endorphin does not pass the placental barrier, its presence in maternal plasma could be transduced by receptors on the placenta that can influence synthesis and release of β-endorphin. Exposure of the fetus to excessive concentrations of β-endorphin (as might occur due to perinatal complications) can arrest development, impair learning, and alter brain content of β-endorphin and the density of receptors in CNS.

6. ENDURING BRAIN/BEHAVIOR COMPLICATIONS RELATED TO HYPOXIA

The next connection to be made in this analysis is that hypoxia can have long-term consequences. The clinical literature is replete with examples of hypoxia-related delayed development. For example, prenatal hypoxic–ischemic brain injury is believed to be the major cause of cerebral palsy[68–70] and of the common diagnosis in mental retardation of "unknown perinatal complication."[71–73] Towbin[74] indicated that infants surviving hypoxic damage manifested diffuse cerebral dysfunction such as mental retardation, cerebral palsy, epilepsy, and moderate to severe behavioral disorders. Johnston[70] suggested that the timing and perhaps chronicity of the hypoxic episode may specify which structures and chemical systems in the brain were affected, both of which may determine the severity of outcome.

The studies of Windle et al.[75–78] provided evidence of the permanent structural damage to the brain of monkeys in all cases of perinatal asphyxia. This series of studies indicated that only a 7-min period of oxygen deprivation was sufficient to produce damage to the brainstem and thalamus. In general, the longer the period of asphyxiation, the greater the damage. Although developmental delay was almost always associated with asphyxiation, eventually the animals showed recovery. However, permanent decrements in learning and memory often, but not always,[79] are associated with milder oxygen deprivation in human patients even though evidence of structural damage may be absent.

In a study[80] of 60 children with biochemical evidence of intrapartum hypoxia (acidosis) at

Table IV. Paradoxical Responses to Sedative-Hypnotic Medication

	No. of subjects	% Paradoxical
Sib and stereotypy	22	68
Sib only	18	39
Stereotypy only	40	35
Controls	20	0

delivery, 8 (13%) had major motor deficits, 4 of those had cognitive deficits, and 1 was blind. Ten children (16%) had minor motor deficits, and one had cognitive loss. Twelve of these 18 children (9 boys and 9 girls) were mature at delivery. The apparent critical factor in determining the presence and severity of deficit was the degree and duration of fetal hypoxia. As the buffer base in the umbilical artery decreased from 30–34 mEq/liter to <22 mEq/liter, the incidence of deficits increased from 20% to 80%. Children with deficits were hypoxic for more than 1 hr.

Hypertension during pregnancy may expose the fetus to extended periods of hypoxia. The relationship between maternal hypertension and fetal hypoxia is well established.[81–85] Hypertension reduces the respiratory reserve of the placenta, thereby decreasing fetal oxygenation. In a direct study of the influence of maternal hypertension on infant outcome, Salomen and Hermonen[86] reported an increased risk of mental retardation in a sample of 136 cases and 122 controls. Maternal hypertension was a much higher risk factor than familial factors, parity, mode of birth, or mother's smoking during pregnancy. Increased prolactin in amniotic fluid of hypertensive women during weeks 14–20 of gestation is compatible with opiate involvement because opiates stimulate prolactin release.[87]

A series of studies[71–73,88] in our laboratory provides a possible, although somewhat inferential, link among hypoxia, β-endorphin, and a specific behavior. An extremely high incidence of "paradoxical" excitement, rather than sedation, to sedative-hypnotic medication was observed in a group of patients with autistic behavior [stereotypy (ST) and self-injurious behavior (SIB)[71]]. As presented in Table IV, 68% of patients with both of these autistic-like behaviors, and 35–39% of patients with one of these behaviors responded paradoxically to treatment. A follow-up study[73] confirmed this observation and discovered a relationship between paradoxical responding and a history of perinatal hypoxia. Multivariate modeling indicated that patients responding paradoxically could be separated from patients responding normally with at least 93% accuracy and that a history of perinatal hypoxia was the most influential variable in the equation. The significance of this observation will become evident in the following sections that suggest possible etiological significance of β-endorphin for SIB.

7. PATIENTS WITH SIB HAVE ELEVATED CONCENTRATIONS OF β-ENDORPHIN IN CSF AND PLASMA

Several studies have examined concentrations of β-endorphin in patients with autism and SIB. Humoral endorphin, a unique endogenous opiate that is not cross-reactive with β-, α-, γ-endorphin, dynorphin, or enkephalin, was reported to be decreased in plasma of 10 autistic patients (4 neuroleptic free) compared with 11 healthy controls.[89] A group of 12 schizophrenic children (4 medication free) also had lowered concentrations of humoral endorphin. These concentrations were lowest in the patients free from neuroleptic medication.

In a recent study,[90] blood samples were collected from 40 subjects ranging in age from 19 to 39 years, comprising 4 groups: (1) SIB and ST (N = 11); (2) SIB only (N = 6); (3) ST only (N = 12); and (4) a control group (N = 11). Plasma concentrations were determined by radioimmunoassay (RIA), with less than 0.1% cross-reactivity with related opiates (including β-LPH) at the normal serum concentration.

A comparison of morning (8 AM) and evening (8 PM) values across groups confirmed the presence of a significant diurnal rhythm with higher values in the morning than the evening. The rhythm was more evident in control patients than the SIB–ST groups. The control group (~320 pg/ml) had lower concentrations of β-endorphin than did the patient groups (~450–480 pg/ml) (Fig. 2). A threshold of 450 pg/ml was effective in distinguishing the diagnostic groups. Specificity, which is the percentage of controls who tested normal, was an acceptable 0.73. Sensitivity, which is the percentage of patients who test positive, was a very high 0.82 for the SIB + ST group but was less apparent in the other groups. Diagnostic confidence, the combined true (diagnostic) positives divided by the total number of positives (diagnostic + control) was 0.75 for the SIB–ST group. Overall, the diagnostic confidence for the use of a 450-pg/ml plasma threshold (at 8:00 AM) for the detection of ST or SIB was a significant 0.84.

A more direct study of brain concentration of β-endorphin in autism and SIB was reported by Gillberg et al.[91] Levels of fractions I and II of endogenous opiate activity (which are not correlated) were determined in the cerebrospinal fluid (CSF) of 20 autistic children and 8 control patients. Fraction I was low in patients with neurogenic pain and high in major depression. Fraction II was elevated in unipolar depression and correlated with measures of suicidal ideation and phobic symptoms.[92] It was unclear what the precursor of fractions I and II were but the authors suggested that it may not have been POMC. Slight increases in fraction I and significant increases in fraction II were detected in the autistic patients. Highest levels of both fractions were detected in autistic children who were self-injurious. An association between elevated fraction II and pain insensitivity and self-destructiveness was observed. Detailed analysis of patient characteristics indicated that the elevated β-endorphin was due only to SIB and not age, sex, or apparent responsiveness to pain.

It is not clear whether opiates are elevated in plasma or CSF in autistic or SIB patients. The three studies reviewed apparently measured different endogenous opiates. Two measured concentrations in plasma and one in CSF. Two studies indicated that opiates were elevated[90,91]

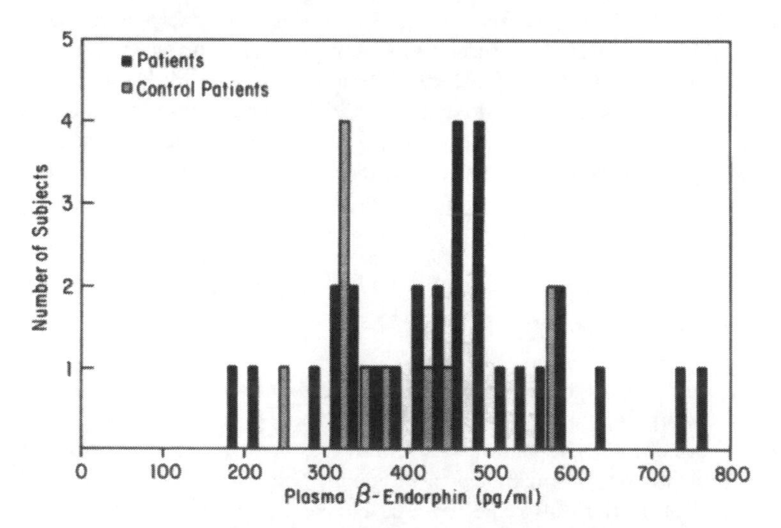

Figure 2. Concentrations of β-endorphin in patients with SIB and controls.

and one[90] found that it was lower in autistic patients. It appears that subanalysis of patients in the study by Gillberg et al.[91] supported the possibility that opiates were elevated in SIB, as we[90] have found. The study by Wiezman et al.[89] did not report data for this subgroup. If, as this chapter suggests is possible, SIB is related to perinatal complications associated with exposure to β-endorphin, whether or not it is elevated in CSF or plasma may be a sufficient but unnecessary outcome. The effects could be on receptor systems not reflected by changes in plasma or CSF. The two primary biological hypotheses proposed to account for SIB implicate opiate receptors.[71–73,88] The addiction hypothesis suggests SIB secures the "fix" of β-endorphins required by tolerant downregulated receptors. The pain hypothesis argues that SIB is associated with opiate-induced elevated sensory threshold-restricting experience, including pain. This hypothesis is consistent with supersensitive opiate receptors.

8. TREATMENT OF SIB WITH OPIATE ANTAGONISTS

Perhaps the strongest evidence linking the opiate system to SIB is found in reports of attenuation of this behavior by blockade of the opiate receptor.[93,94] There are six studies in the literature describing the effects of naloxone on this bizarre and untreatable behavior common in autistic and retarded patients. Each study differed in important ways. In our first study,[88] two profoundly retarded men in their early 20s exhibiting "intractable" SIB were given I.M. injections of either naloxone HCL (0.1, 0.2, and 0.4 mg) or placebo, in a double-blind crossover design (Table V). Each subject was observed and videotaped for 90 min. The frequency of SIB and the duration of stereotypy were recorded by independent judges viewing the videotapes for 10 time samples, each 1 min in duration. Treatment with all doses of naloxone reduced the occurrence of self-injurious episodes in both clients and attenuated self-restraining behaviors in one of the clients. These effects persisted for 60–90 min (Figs. 3 and 4).

The findings reported by Sandyk[95] closely paralleled these results. Attenuation of SIB,

Table V. Double-Blind Crossover Study of Naloxone in Patients with SIB: Protocol for the Naloxone-Self Abuse Study

Week of study	Patient 1	Patient 2
Fifth week	Videotaped behavior of both clients to accommodate to procedures	
Sixth week	Accommodation to injection	
Day 1	Placebo	Placebo
Day 3	Placebo	Placebo
Day 5	Placebo	Placebo
Seventh week		
Day 1	0.1 mg naloxone	Placebo
Day 3	0.2 mg naloxone	Placebo
Day 5	0.4 mg naloxone	Placebo
Eighth week		
Day 1	Placebo	0.1 mg naloxone
Day 3	Placebo	0.2 mg naloxone
Day 5	Placebo	0.4 mg naloxone

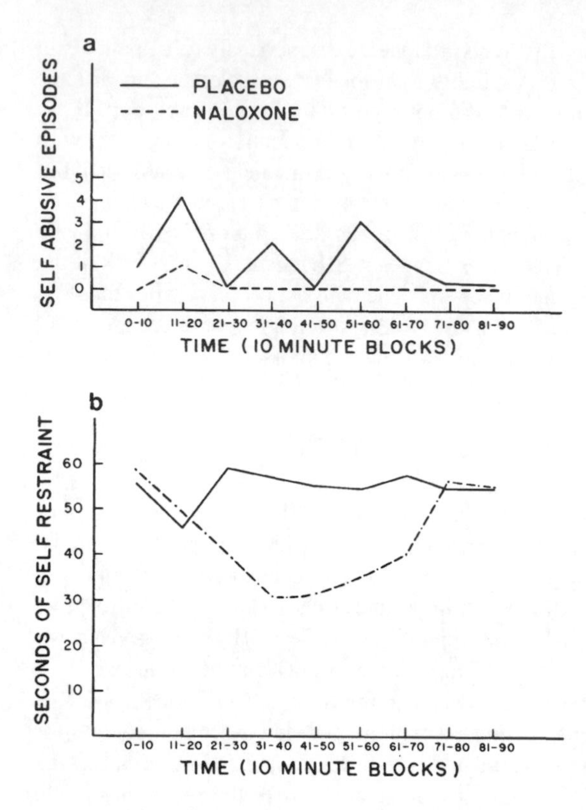

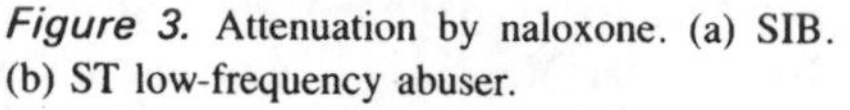
Figure 3. Attenuation by naloxone. (a) SIB. (b) ST low-frequency abuser.

hyperactivity and ST were observed 15 min after injection of 1.2 mg naloxone to a developmentally delayed 11-year-old boy; 45 min after injection, the aberrant behavior returned. Controlled, blinded testing with placebo did not reduce SIB, suggesting a central role of opiates in control of this behavior.

Two other studies also reported positive effects of naloxone. In the first, Davidson et al.[96] administered multiple im injections of 0.075 mg and 0.15 mg naloxone to a retarded 8-year-old boy. Observational periods of 5-min duration were evaluated by raters unaware of the treatment. Neither dose of naloxone reduced the frequency of SIB compared with placebo (although rates were lower in the treatment phase). However, Davidson et al. reported that SIB was less intense and that the patient "whined" after self-injury during the naloxone period. In the second study,[97] naloxone was infused (1.0 or 2.0 mg) over a 6-hr period for two consecutive days in a moderately retarded 14-year-old patient with a long history of self-mutilation. After, but not during, the infusion, SIB markedly decreased and remained attenuated for 2 days.

Recently we[98] treated with naloxone a 21-year-old woman of normal intelligence with a long history of SIB. All medications were eliminated before she was challenged with 0.4, 0.8, and 1.2 mg of naloxone and placebo in a double-blind procedure. Treatment with naloxone resulted in normalization of event-related potentials of the brain (especially at 1.2 mg), improved memory (at 0.8 and 1.2 mg), reduced anxiety (at 0.4 mg), and improved behavior measured by the Conners scale (at 0.8 mg). She did not have any SIB during the treatment phase and reported no urges to self-injure. She also reported that the voices encouraging her SIB completely disappeared during challenge with naloxone. The symptom/dose relationship is similar to the tranquilizing effects seen at low doses and stimulating effects observed at high doses reported by Campbell et al.[99]

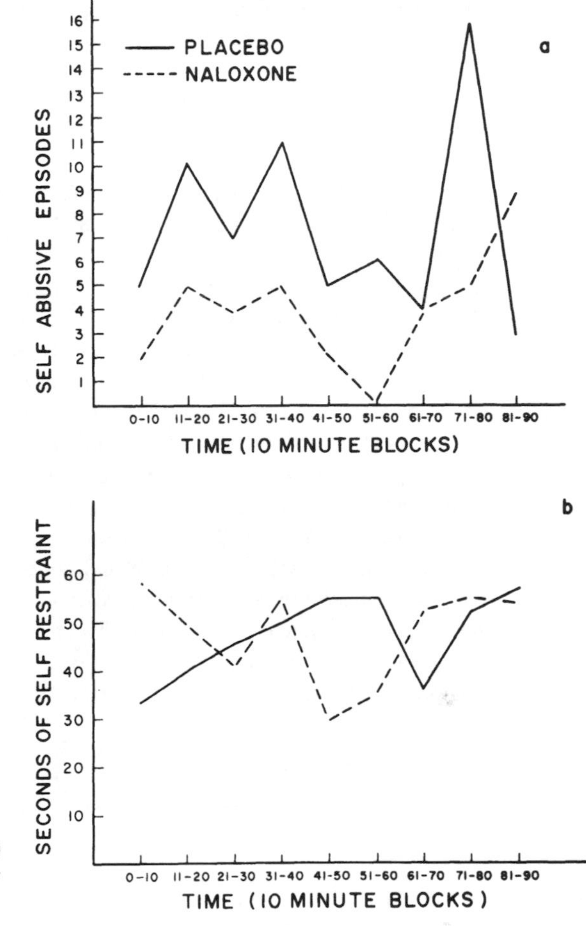

Figure 4. Attenuation by naloxone. (a) SIB. (b) Not in ST in high-frequency abusers.

The recent report[100] of a failure of naloxone to attenuate SIB was designed after our study,[88] but the characteristics of the patient were unusual (one had PKU and the second was deaf and blind), and the rater reliability was low. Furthermore, the SIB rates were low, reaching zero levels during placebo. Although sensory handicap may be prevalent among patients with SIB, especially stereotypic SIB, it is not certain how, or if, it relates to opiate dysfunction.

There are three published reports and one unpublished of the effects of naltrexone (a long-acting, orally administered opiate blocker) on autistic patients and SIB. Herman et al.[101] administered naltrexone (0.0, 0.5, 1.0, 1.5, and 2.0 mg/kg) to three drug-free boys with SIB (ages 10, 17, and 17) and observed them for five 1-min periods at 1 and 4 hr after treatment. At all the doses except the highest (2.0 mg/kg), naltrexone significantly reduced SIB. Except for this dose, the effects were dose dependent. The patient with the highest rate of SIB (most severe disorder) was the most responsive to naltrexone.

In a preliminary report of our recent study,[90] four male patients (ages 23–26) were administered fixed doses (0, 25, 50, or 100 mg) of naltrexone at 8 AM 2 days per week (Monday and Wednesday) in a double-blind design. All the patients had a long history of SIB, and interventions had proved largely unsuccessful. Three of the patients were free of drug. They were adapted to all procedures (including administration of capsules) 10 days before the study. During the treatment phase, the patients were videotaped daily for 10 continuous minutes in their normal surroundings in the morning (10:30–11:30 AM) and afternoon (1:30–2:30 PM). Each dose was evaluated for 1 week for frequency and severity of SIB, duration of ST, and level of activity. The tapes were evaluated by two raters 1 day each week, achieving interrater reliability of 0.92.

The results for SIB are presented in Fig. 5 as changes from the placebo-control condition. There are several important considerations in evaluating these data. First, the patients differed in their base rates of SIB during placebo control (MC > RM > JH > DL). Second, as noted by many investigators, this behavior is highly variable. Two patients consistently had higher levels of SIB in the morning and two in the evening. The data presented are changes from baseline when SIB was most frequent. Averaging AM and PM values did not change the dose–response effects for the highest dose. Third, consistent with the findings of Herman et al.,[100] the most severe patient (MC) was most sensitive to treatment with naltrexone. The highest dose of naltrexone eliminated the SIB of patient MC (highest base rate) for both AM and PM. Unlike the findings of Herman et al., the ≥2-mg/kg equivalent was most effective in three of four patients. The only patient for whom the highest dose was not most effective was RM who received the highest mg/kg equivalent (consistent with the findings of Herman et al.[101]). Fifth, it was clear that the attenuation of SIB was not due to decreased activity or to alterations in ST. Indeed, one patient had an apparent dose related increase in ST.

In a third study, Campbell et al.[99] administered naltrexone (0.5, 1.0, and 2.0 mg/kg per day) to seven autistic children aged 3.75–6.5. Five of the seven children were judged clinically to be responders to naltrexone, and the other two children also had positive changes. The spectrum of findings included the observation that the 0.5-mg/kg dose was tranquilizing and

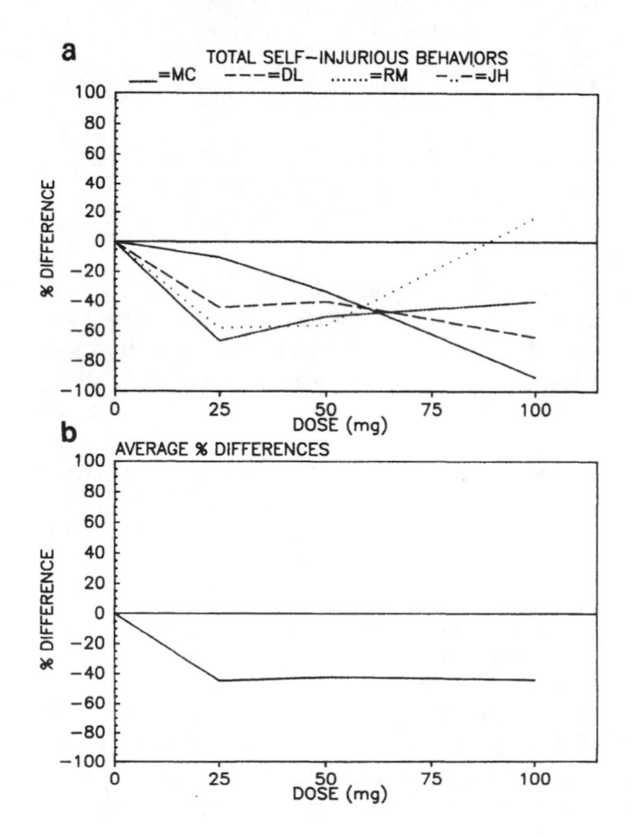

Figure 5. Dose-dependent attenuation by naloxone of SIB in four patients. (a) Individually. (b) Averaged.

the 2.0-mg/kg dose was stimulating. Positive changes in activity in a hyperactive child, improvement on the Conners scale and decreased aggressiveness (although not rated) was associated with treatment with naltrexone. A reduction in stereotypy at the 1.0- and 2.0-mg/kg doses was observed. Three of the children had SIB that was reduced by naltrexone.

Finally, in a study of two relatively low-frequency SIB patients (ages 21 and 29), Szymanski et al.[102] reported no effect of 50 and 100 mg of naltrexone. The patients were administered naltrexone daily for 3 weeks, followed by placebo periods in a multiple baseline (C-ABAB) design. The patients were rated for indeterminate periods for changes in SIB and several other behaviors. One patient evidenced declining SIB from baseline throughout the course of the study. In the second patient, the rate of SIB appeared to decline but then abruptly increased and was maintained on 100 mg naltrexone. There were two noteworthy aspects to this study. One was that daily administration of a relatively long-lasting drug was given. Thus, some cumulative effects could be possible. Given the duration of the placebo period (3 weeks), it is unlikely (but unknown) that the effects could persist. Detailed analysis of the placebo period was not discussed. The second different aspect of this study was that both patients were maintained on psychoactive medication throughout the study.

From the literature, at least nine patients with SIB have been treated with naloxone. A positive effect was reported in seven of them and a robust effect in six of those. Sixteen patients with autism (12 with SIB) have been treated with naltrexone. Of these 12 SIB patients, 10 have had positive responses. In most cases, it appears that the salutary effects are dose dependent. There appears to be sufficient evidence to implicate the opiate system in this severe neurodevelopmental disorder. Furthermore, the effectiveness of relatively small doses of the opiate blocker and the normalization of brain responses to environmental stimulation favors the supersensitive/sensory stimulation hypothesis of autism/SIB.

9. CONCLUSION

The perspective developed in this chapter has grown from a synthesis of somewhat disparate areas of basic and clinical science. Although the basic arguments in the analysis are generally supported by experimental findings, links among the arguments vary in inferential strength. The major points are as follows:

1. β-Endorphin increases in maternal plasma during pregnancy and peaks at delivery.
2. β-Endorphin increases in amniotic fluid and cord blood in individuals with perinatal complications, especially hypoxia.
3. Fetuses exposed to elevated β-endorphin have delayed patterns of behavioral and developmental changes in neural organization.
4. SIB in conjunction with paradoxical responding to sedative-hypnotics is related to perinatal complications, especially hypoxia.
5. Patients with SIB may have elevated concentrations of β-endorphin in plasma and CSF.
6. SIB can be significantly attenuated with opiate blockers.

10. MODELS EXTENDING THE IMPLICATIONS

The results reviewed have considered the impact of a stress such as hypoxia on fetal development. Hypoxia can arise from physiological complications (pre-eclampsia and eclampsia) but also from stress-related "psychological" factors. The incidence of concurrent (stress-related) hypertension with pregnancy has been considered to contribute at least 20% of problem pregnancies due to hypertension.[83] Elegant models have described the relationship between stress and physiological maladaption such as hypertension[103,104]; an extension pertaining directly to the outcome of pregnancy is presented in Fig. 6. In the model, the solid lines are relationships based on research findings, and the dotted lines are inferred or indirect relationships. The model makes three major assumptions: (1) as suggested inferentially in this chapter, opiates such as β-endorphin can serves as a final common pathway for the effects of a stress such as hypoxia on the fetus; (2) maternal hypertension, by causing fetal hypoxia, can stimulate the endorphinergic system and influence neural and behavioral outcome; and (3) it is possible that stress, in addition to its relationship with hypertension, can influence directly the outcome of pregnancy. The model suggests that in addition to the effects related to hypertension, stress may exert direct fetal actions. It is possible that stress, even without hypertension, can be transduced to the fetus by stimulating the release of β-endorphin.

The possibility of an endogenous substance such as β-endorphin as a final common pathway for the effects of stress on the fetus and on infant outcome could explain some findings in the literature. For instance, several studies have reported that the experience of stress by pregnant women is predictive of obstetric complications. Erickson[105] reported that increased stress (e.g., fear, dependency) was related to decreased Apgar scores and to complications of delivery. In an another study,[106] women with more life change events (a quantifiable index of stress) and elevated anxiety were significantly at risk for abnormal pregnancies. Negative attitude[107] and fear[108] about having children were more likely to result in perinatal death, congenital anomaly, and lowered Apgar scores.

Consistent with this model, Nuckolls et al.[109] reported that both increased life stress and unfavorable psychosocial conditions were significantly related to obstetric complications. Furthermore, taken singly, these factors were not related. In a study of 114 women, Andreoli et al.[110] found that uncontrolled life change events during the second trimester (although the third may not have been fairly assessed) were significantly greater in complicated pregnancies. Connelly and Cullen[111] suggested that stress has a timedependent relationship to pregnancy outcome. Different stresses produced different profiles of pregnancy outcome and those effects were dependent on gestational age. Brown et al.[112] reported immature motor development in infants whose mothers evidenced anxiety. These findings strongly implicate psychological stress as an important factor in the outcome of pregnancy and are summarized in Fig. 7.

Psychological factors such as life-related stress, either perceived or expected, can influence fetal development (Fig. 7). The effects of stress are not consistent. Some women may be more susceptible than others. Factors related to the vulnerability buffer (e.g., social support, ego strength) can protect the woman from the effects of stress. If this buffer is weak, maladaptive responses such as overproduction of peptides or hypertension might occur. The placenta is a barrier for many teratogenic agents, but it may pass β-endorphin or other opiates, detect the stress signal from the mother (i.e., increased β-endorphin) with receptors on the placenta, and initiate a cascade of events, including the synthesis and release of β-endorphin into the fetal environment. It is reasonable to assume that this system would exist to protect the fetus from survival in a hostile environment and the species from suboptimal life forms. However, because of advances in medicine, the fetus often survives and suffers neurological, cognitive, and behavioral deficits.

The model also suggests new areas for intervention that may influence problems related to perinatal complications such as SIB. Stress management approaches that decrease vulnerability to stress represents one possible area for intervention. A second area may be pharmacological. If β-endorphin is the major common pathway for stress and hypoxia-linked perinatal complications, blocking this system may have positive effects on outcome. The study conducted by Chernick and Craig[49] suggested that the fetal effects of asphyxia in pregnant

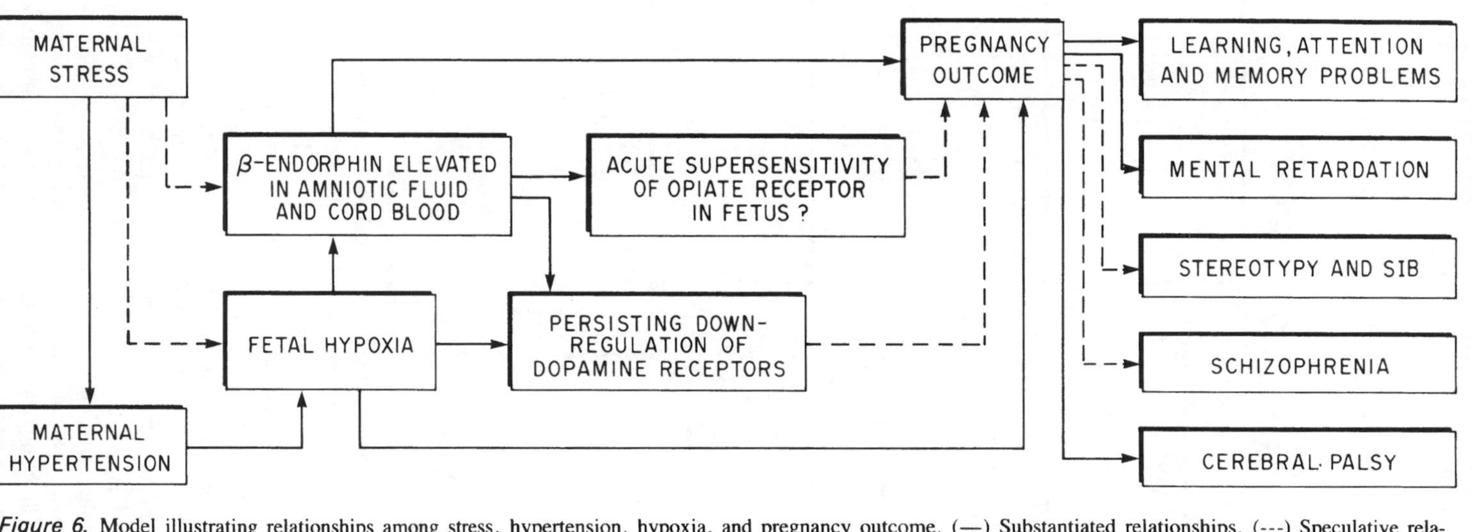

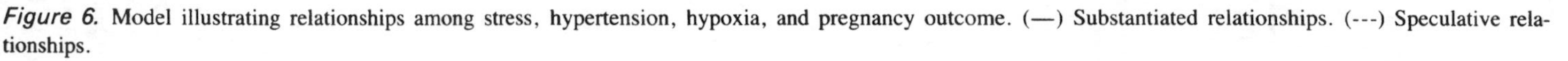

Figure 6. Model illustrating relationships among stress, hypertension, hypoxia, and pregnancy outcome. (—) Substantiated relationships. (---) Speculative relationships.

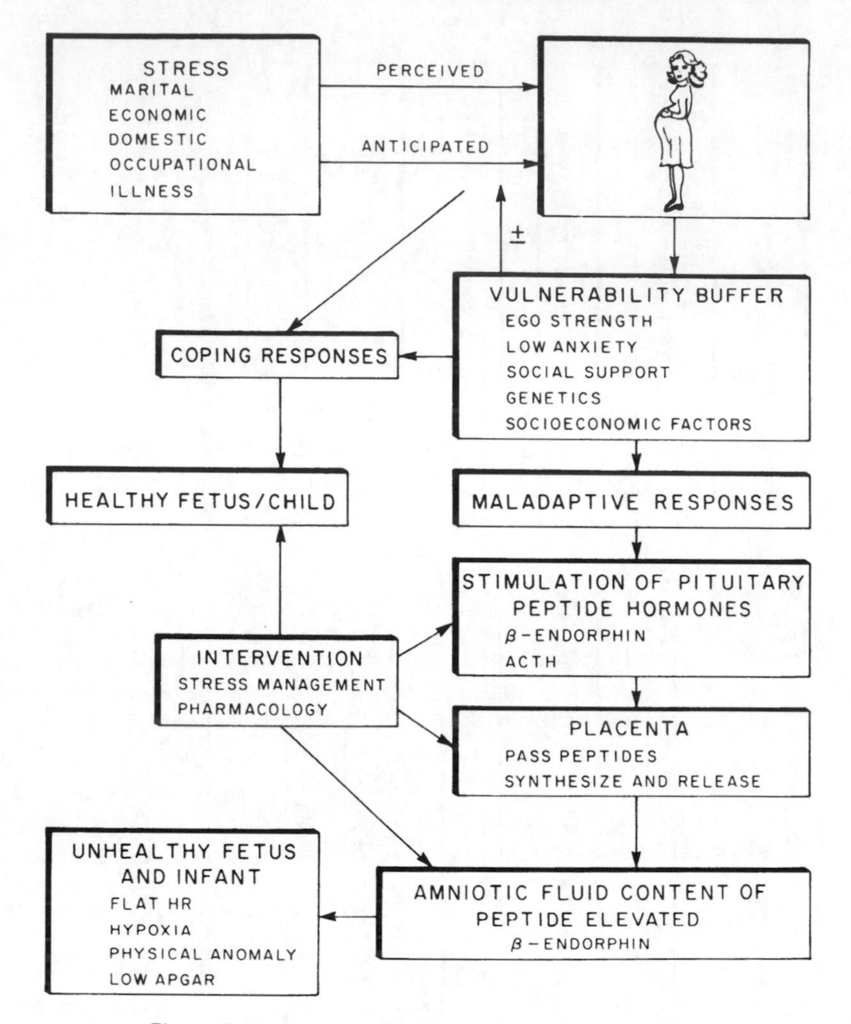

Figure 7. Factors implicating maternal stress in infant outcome.

rabbits could be blocked by naloxone. In a related clinical study, Goodlin[113] administered naloxone in eight cases of intrapartum unchanging fetal heart rate without narcotic complications (an index of hypoxia). Within 20 min, four of the eight fetuses evidenced normal beat-to-beat variability. Even though this did not prove a viable treatment because of the increased experience of pain reported by the mothers, the findings added further support for the role of opiates in mediating the effects hypoxia and perhaps stress during fetal development. Thus, the model not only describes possible etiological links between stress and outcome of pregnancy, but also intervention "windows" for mitigating the effects on the fetus of elevated endorphin.

Finally, a highly speculative neuronal model describing the relationships discussed in this chapter are presented in Fig. 8. In this model, fetal exposure to elevated β-endorphin is suggested to influence opiate and dopamine receptors. Changes in these receptors have been reported either after fetal hypoxia or exposure to β-endorphin. Although this model is incomplete, the findings to date suggest that fetal exposure to β-endorphin possibly resulted in supersensitive opiate and subsensitive dopamine postsynaptic receptors. The persisting quality of this phenomenon, especially with the dopamine system, may be expressed as a congenital

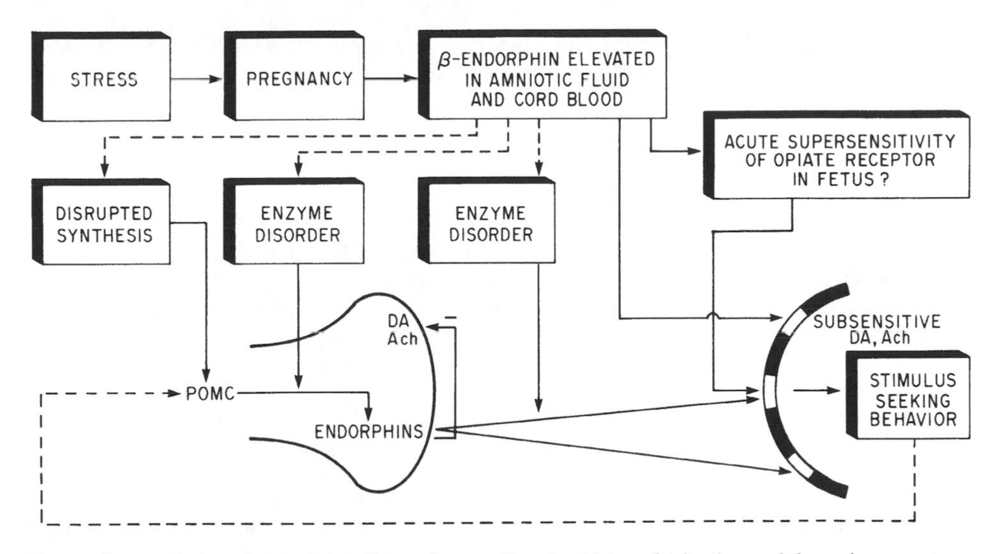

Figure 8. Speculative model of the effects of stress (β-endorphin) on fetal opiate and dopamine receptors.

defect in receptor plasticity. The mechanism for this reported effect is unknown, and several presynaptic possibilities are suggested in the model, including altered synthesis of POMC and delayed degradation of β-endorphin either intracellularly or in the synaptic cleft. Presynaptic disinhibition of dopamine by increased availability of endorphin also is possible. As over-simplified by the model, the opiate system may effect the brain and behavior as a primary transmitter and as a modulator of other transmitters such as dopamine.

For instance, some of the CNS effects of β-endorphin appear to be mediated by dopamine.[114] Even though opiates and dopamine agonists do not bind to each other's receptors (except at high doses), both pre- and postsynaptic mechanisms have been implicated in their reciprocity.[115] Dopamine antagonists have many of the same effects as morphine,[116–118] suggesting presynaptic inhibition of dopamine release possibly related to co-mapping of opiate receptors on dopamine terminals.[119–122] Inhibition of dopamine release by β-endorphin in the presence of potassium[123–125] and its disinhibition by naloxone[123] is further support for this possibility. Acute administration of morphine increases the rate of dopamine turnover (resulting in subsensitivity).[115] However, chronic use of opiates causes dopamine receptor supersensitivity.[126]

As suggested above, many of the effects of opiates in adult animals are paralleled by drugs that are dopamine antagonists. Presynaptic inhibition of dopamine release is a plausible action of β-endorphin.[115,116] Typically, in the mature nervous system, receptor antagonists or synthesis inhibition results in receptor supersensitivity.[127] Curiously, however, prenatal treatment with a variety of drugs that decreased dopamine (neuroleptics, α-methyl-*p*-tyrosine) also decreased dopamine receptors through adulthood.[128,129] Furthermore, dopamine-dependent sexual behavior can be altered permanently in rats exposed prenatally to dopamine antagonists.[130] This analysis coupled with the growing literature relating dopamine with β-endorphin suggests that the plethora of disorders ascribed to dopamine disregulation may have their roots, with varied expression, in early organization of the brain and may have as an environmental impetus, stressful episodes such as hypoxia. In summary, an endogenous substance that can act as a teratogen, β-endorphin, may be involved in the final common pathway for the many effects on the fetus due to hypoxia and other stress-related complications of pregnancy.

ACKNOWLEDGMENTS. This work was supported in part by a grant from the Stallone Foundation and by grant 1 R01 MH 41446-01A2 from the National Institute of Mental Health and by the Veterans Administration.

REFERENCES

1. Krieger CF: Brain peptides: What, where, why? Science 222:975–985, 1983
2. Miller RJ: Peptides as neurotransmitters: Focus on the enkephalins, in de Wied D, Gispen WH, van Wimersma Greidanus TJB (eds), Neuropeptides and Behavior, Vol. 1. Oxford, Pergamon, 1986, pp 95–136
3. Dewied D: Behavioral effects of intraventricular administered vasopressin and vasopressin fragments. Life Sci 19:685–690, 1976
4. Sandman CA, Beckwith BE, Kastin AJ: Are learning and attention related to the sequence of amino acids in ACTH/MSH peptides? Peptides 1:277–280, 1980
5. Hokfelt T, Johansson O, Ljungdahl A, et al: Peptidergic neurons. Nature (London) 284:515–521, 1980
6. Dale HH: Pharmacology and nerve endings. Proc Soc Med 28:319–332, 1935
7. O'Donohue TL, Miller RL, Jacobwitz DM: Identification, characterization, and stereotaxic mapping of intraneuronal α-melanocyte stimulating hormone-like immunoreactive peptide in discrete regions of the rat brain. Brain Res 175:1–23, 1979
8. Akil H, Hewlitt H, Barchas JD, et al: Binding of H-beta-endorphin to rat brain membranes: Characterization of opiate properties and interaction with ACTH. Eur J Pharmacol 64:1–8, 1980
9. Farah JM, Millington WR, O'Donohue TL: The role and regulation of post translational processing of proopiomelanocortin, Central Actions of ACTH and Related Peptides. Fidia Research Series, Symposia in Neuroscience IV-Liviana Press, Padova, 1986
10. Gluckman PD, Marti-Henneberg C, Kaplan SL, et al: Hormone ontogeny in the ovine fetus. X. The effects of β-endorphin and naloxone on circulating growth hormone, prolactin, and chronic sommatomammotropin. Endocrinology 107:76–79, 1980
11. Phoenix CH, Goy RW, Gerall AA, et al: Organizing action of prenatally administered testosterone propionate on the tissue mediating mating behavior in the female guinea pig. Endocrinology 65:369–382, 1959
12. Whalen RE: Differentiation of the neural mechanisms which control gonadotropin secretion and sexual behavior, in Diamond M (ed), Reproduction and Sexual Behavior. Bloomington, Indiana, Indiana University Press, 1968, pp 303–340
13. Levine S: Plasma-free corticosteroid response to electric shock in rats stimulated in infancy. Science 135:1585–1592, 1962
14. Eayrs JT: Age as a factor determining the severity and reversibility of the effects of thyroid deprivation in the rat. J Endocr 22:409–419, 1961
15. Beckwith BE, Sandman CA, Hothersall D, et al: Influence of neonatal injections of α-MSH on learning, memory, and attention in rats. Physiol Behav 18:63–71, 1977
16. Sandman CA, Kastin AJ: The influence of fragments of the LPH chain on learning, memory and attention in animals and man. Pharmacol Ther 13:39–60, 1981
17. Sandman CA, Kastin AJ: Behavioral actions of ACTH and related peptides, in Li CH (ed), Hormonal Proteins and Peptides. Orlando, Florida, Academic, 1987, pp 147–171
18. Sandman CA, Miller LH, Kastin AJ, et al: A neuroendocrine influence on attention and memory. J Comp Physiol Psychol 80:59–65, 1972
19. Sandman CA, Alexander WD, Kastin AJ: Neuroendocrine influences on visual discrimination and reversal learning in the albino and hooded rat. Physiol Behav 11:613–617, 1973
20. Veith JL, Sandman CA, George JM, et al: Effects of MSH/ACTH 4-10 on memory, attention and endogenous hormone levels in women. Physiol Behav 20:43–50, 1978
21. Walker BB, Sandman CA: Influences on an analog of the neuropeptide ACTH 4-9 on mentally retarded adults. Am J Ment Defic 83:346–352, 1979
22. Ward MM, Sandman CA, George JM, et al: MSH/ACTH 4-10 in men and women: Effects upon performance of an attention and memory task. Physiol Behav 22:669–673, 1979
23. Sandman CA, Walker BB, Lawton CA: An analog of MSH/ACTH 4-9 enhances interpersonal and environmental awareness in mentally retarded adults. Peptides 1:109–114, 1980

24. Sandman CA, Kastin AJ: Intraventricular administration of MSH induces hyperalgesia in rats. Peptides 2:231–233, 1981
25. Pantella K, Sandman CA, Bachman DS: Trial of an ACTH 4-9 analogue in children with intractable seizures. Neuropediatrics 13:59–62, 1982
26. Veith JL, Sandman CA, George JM, et al: The relationship of endogenous ACTH levels and enhanced visual–attentional functioning in congenital adrenal hyperplasics. Psychoneuroendocrinology 10:33–48, 1985
27. Sandman CA, Berka C, Walker BB, et al: ACTH 4-9 effects on the human visual event-related potential. Peptides 6:803–807, 1985
28. Champney TF, Shaley TC, Sandman CA: Effects of neonatal cerebral ventricular injections of ACTH 4-9 and subsequent adult injections on learning in male and female albino rats. Pharmacol Biochem Behav 5:3–10, 1976
29. Swaab DF, Boer GJ, Boer K, et al: Fetal neuroendocrine mechanisms in development and parturition, in Corner MA, Baker RE, van de Poll NE, Swaab DF, Uylings HBM (eds), Maturation of the Nervous System. Progress in Brain Research. Amsterdam, Elsevier, 1978, pp 277–290
30. Swaab DF, Boer GJ, Visser M: The fetal brain and intrauterine growth. Postgrad Med J 54 (suppl 1):63–73, 1978
31. Visser M, Swaab DF: Life span changes in the presence of α-melanocyte-stimulating hormone-containing cells in the human pituitary. J Dev Physiol 1:161–178, 1979
32. Gennazzani AR, Petraglia F, Parrini D, et al: Lack of correlation between amniotic fluid and maternal plasma contents of beta endorphin, beta lipotropin and adrenocorticotropin hormone in normal and pathologic pregnancies. Am J Obstet Gynecol 148:198–203, 1984
33. Evans MI, Fisher AM, Robichaux AG, et al: Plasma and red blood cell b-endorphin immunoreactivity in normal and complicated pregnancies. Am J Obstet Gynecol 151:433–437, 1985
34. Newnham JP, Tomlin S, RaHers SJ, et al: Endogenous opioid peptides in pregnancy. Br J Obstet Gynecol 90:535–538, 1983
35. Browning AJF, Butt WR, Lynch SS, et al: Br J Obstet Gynecol 90:1147–1151, 1983
36. Wilkes MM, Stewart RD, Bruni JF, et al: A specific homologous radioimmunoassay for human β-endorphin: Direct measurement in biological fluids. J Clin Endocrinol Metab 50:309–315, 1980
37. Divers WS, Stewart RD, Wolkes MM, et al: Amniotic fluid, β-endorphin and α-melanocyte-stimulating hormones immunoreactivity in normal and complicated pregnancies. Am J Obstet Gynecol 144:539–546, 1982
38. Lis M, Julesz J, Senicas VM, et al: N-terminal peptide of pro-opiomelanocortin in human amniotic fluid. Am J Obstet Gynecol 146:575–579, 1983
39. Csontos K, Rust M, Hollt V, et al: Elevated plasma B-endorphin levels in pregnant women and their neonates. Life Sci 25:835–844, 1979
40. Kofinas GD, Kofinas AD, Tavakoli FM: Maternal and fetal β-endorphin release in response to the stress of labor and delivery. Am J Obstet Gynecol 152(1):56–59, 1985
41. Gautray JP, Jolivet A, Vielh JP, et al: Presence of immunoassayable β-endorphin in human amniotic fluid: Elevation in cases of fetal distress. Am J Obstet Gynecol 123:211–212, 1977
42. Wardlaw SL, Stark RI, Boci L, et al: Plasma β-endorphin and β-lipotropin in the human fetus at delivery: Correlation with arterial pH and pO2. J Clin Endocrinol Metab 79:888–891, 1979
43. Goland RS, Wardlaw SL, Stark RI, et al: Human plasma β-endorphin during pregnancy, labor, and delivery. J Clin Endocrinol Metab 52:74–78, 1981
44. Shaaban MM, Hong TT, Hoffman DI, et al: Beta-endorphin and beta-lipotropin concentrations in umbilical cord blood. Am J Obstet Gynecol 144:560–568, 1982
45. Davidson S, Gil-Ad Irit, Rogovin Hana, et al: Cardiorespiratory depression and plasma β-endorphin levels in low-birth weight infants during the first day of life. American Journal of Diseases in Children 141:145–148, 1987
46. Wardlaw SL, Stark RI, Daniel S, et al: Effects of hypoxia on β-lipotropin release in fetal, newborn and maternal sheep. Endocrinology 108:1710–1715, 1981
47. Stark RI, Wardlaw SL, Daniel SS, et al: Vasopressin secretion induced by hypoxia in sheep: Developmental changes and relationship to β-endorphin release. Am J Obstet Gynecol 143:204–215, 1982
48. Panerai AE, Martini A, Abbate D, et al: β-endorphin, met-enkephalin and beta-lipotropin in chronic pain and electroacupuncture. Adv Pain Res Ther 5:543–547, 1983
49. Chernick V, Craig RJ: Nalaxone reverses neonatal depression causes by fetal asphyxia. Science 216:1252–1253, 1982

50. Fachinetti F, Bagnoli F, Bracci R, et al: Plasma opioids in the first hours of life. Pediatr Res 16:95–99, 1982
51. Brubaker PL, Baird AC, Bennet HP, et al: Corticotropic peptides in the human fetal pituitary. Endocrinology 111:115–1154, 1982
52. Chard T, Silman RE: Pituitary peptides in primary fetuses, in Novy MJ, Resko JA (eds), Fetal Endocrinology. Orlando, Florida, Academic, 1981, pp 303.
53. Nakai Y, Nakao K, Oli S, et al: Presence of immunoreactive b-lipotropin and β-endorphin in human placenta. Life Sci 23:2013–2098, 1978
54. Fraioli F, Genazzani AR: Human placental β-endorphin. Gynecol Obstet Invest. 11:37–42, 1980
55. Odagiri E, Sherell BJ, Mount CD, et al: Human placental immunoreactive corticotropin, beta-lipotropin, and beta-endorphin: Evidence for a common precursor. Proc Natl Acad Sci USA 76:2027–2031, 1979
56. Kent JL, Pert CB, Herkenham M: Ontogeny of opiate receptors in rat forebrain: Visualization by in vitro autoradiography. Dev Brain Res 2:487–504, 1982
57. Allessi NE, Khachaturian H, Watson S, et al: Postnatal ontogeny of acetylated and non-acetylated β-endorphin in rat pituitary. Life Sci 33(suppl 1):57–60, 1983
58. Moldow RL, Kastin AJ, Hollander CS, et al: Brain β-endorphin-like immunoreactivity in adult rats given β-endorphin neonatally. Peptides, Brain Research Bulletin 7:683–686, 1981
59. Sandman CA, Yessian N: Persisting subsensitivity of the striatal dopamine system after fetal exposure to β-endorphin. Life Sci 39:1755–1763, 1986
60. Sandman CA, McGivern RF, Berka C, et al: Neonatal administration of β-endorphin produces "chronic" insensitivity to thermal stimulus. Life Sci 25:1755–1760, 1979
61. Feigenbaum J, Yanai J: The role of dopaminergic mechanisms in mediating the central behavioral effects of morphine in rodents. Neuropsychobiology 11:98–105, 1971
62. MacNichol E, Kung YW, Levin S, et al: Stimulation of dopamine synthesis in caudate nucleus by intrastriatal enkephalins and antagonism by nalaxone. Science 200:522–554, 1978
63. Pollard H, Lorens C, Schwartz J: Enkephalin receptors on dopaminergic neurons in rat striatum. Nature (Lond) 268:745–747, 1977
64. Zadina JE, Kastin AJ, Coy DH, et al: Developmental, behavioral, and opiate receptor changes after prenatal or postnatal β-endorphin, CRF, or Tyr-MIF-1. Psychoneuroendocrinology 10:367–383, 1985
65. Zadina JE and Kastin AJ: Neonatal peptides affect developing rats: β-Endorphin alters nociception and opiate receptors, Corticotropin-releasing factor alters corticosterone. Dev Brain Res 29:21–29, 1986
66. Beckwith BE, O'Quin RK, Petro MS, et al: The effects of neonatal injections of α-MSH on the open field behavior of juvenile and adult rats. Physiol Psychol 5:295–299, 1977
67. Grimm FE, Frieder B: Differential vulnerability of male and female rats to the timing of various perinatal insults. Int J Neurosci 27:155–164, 1985
68. Crothers BS, Paine RS: The Natural History of Cerebral Palsy. Cambridge, Harvard University Press, 1959
69. Hill A, Volpe JJ: Seizures, hypoxic-ischemic brain injury and inteventricular hemmorhage in the newborn. Ann Neurol 10:109–121, 1981
70. Johnston MV: Neurotransmitter alteration in a model of prenatal hypoxic-ischemic brain injury. Ann Neurol 13:511–518, 1983
71. Barron J, Sandman CA: Relationship of sedative-hypnotic response to self-injurious behavior and stereotypy in mentally retarded clients. Am J Ment Defic 2:177–186, 1983
72. Barron J, Sandman CA: Self-injurious behavior and stereotypy in an institutionalized mentally retarded population. Appl Res Ment Retard 5:81–93, 1984
73. Barron J, Sandman CA: Paradoxical excitement to sedative-hypnotics in mentally retarded clients. Am J Ment Defic 2:124–129, 1985
74. Towbin A: Cerebral dysfunctions related to perinatal organic damage: Clinical–neuropathologic correlations. J Abnorm Psychol 87:617–635, 1978
75. Windle WF: An experimental approach to prevention or reduction of the brain damage of birth asphyxia. Dev Child Neurol 8:129–140, 1966
76. Windle WF: Brain damage at birth. JAMA 206:1967–1972, 1967
77. Dawes GS, Hibbard E, Windle WF: The effect of alkali and glucose infusion on permanent brain damage in rhesus monkeys asphyxiated at birth. J Pediatr 65:801–806, 1964
78. Sechzer JA, Faro MD, Barker JN, et al: Developmental behaviors: Delayed appearance in monkeys asphyxiated at birth. Science 171:1173–1175, 1971

79. Broman SH: Perinatal anoxia and cognitive development in early childhood, in Field T, Sostek AM, Goldberg S (eds), Infants Born at Risk. New York, Spectrum, 1979, pp 201–252
80. Low JA, Galbreath RS, Muir DW, et al: Factors associated with motor and cognitive deficits in children after intrapartum fetal hypoxia. Am J Obstet Gynecol 148:533–539, 1984
81. Zaspan FP: Toxemia of pregnancy in gynecology and obstetrics, Sciara F (ed), Vol. 2. Hagerstown, Maryland, Harper & Row, 1981, pp 1–20
82. Davidson JM, Lindheimar MD: Hypertension in pregnancy, Sciarra F (ed), Vol. 1. Hagerstown, Maryland, Harper & Row, 1981, pp 1–28
83. Welt SI, Crenshaw MD: Concurrent hypertension and pregnancy. Clin Obstet Gynecol 21:619–648, 1978
84. Dunlop TCH: Chronic hypertension and perinatal. Proc R Soc Med 59:838–841, 1966
85. Alvarez RA: Hypertensive disorders in pregnancy. Clin Obstet Gynecol 16:47–71, 1973
86. Salomen JJ, Hermonen OP: Mental retardation and mother's hypertension during pregnancy. J Ment Defic 28:53–56, 1984
87. McCoshen JA, Tyson JE: Altered prolactin bioactivity in amniotic fluid of hypertensive pregnancy. Obstet Gynecol 65:24–30, 1985
88. Sandman CA, Datta P, Barron J, et al: Naloxone attenuates self-abusive behavior in developmentally disabled clients. Appl Res Ment Retard 4:5–11, 1983
89. Weizman R, Weizman A, Tyano S, et al: Humoral–endorphin blood levels in autistic, schizophrenic and healthy subjects. Psychopharmacology 82:368–370, 1984
90. Sandman CA: β-Endorphin disregulation in autistic and self-injurious behavior: A neurodevelopmental hypothesis. Synapse 2:193–199, 1988
91. Gillberg C, Terenius L, Lonnerheim G: Endorphin activity in childhood psychosis. Arch Gen Psychiatry 42:780–783, 1985
92. Agren H. Terenius L, Wahlstrom A: A depressive phenomenology and levels of cerebrospinal fluid endorphins. Ann NY Acad Sci 398:388–398, 1982
93. Deutsch SI: Rationale for the administration of opiate antagonists in treating infantile autism. Am J Ment Defic 90:631–635, 1986
94. Farber JM: Psychopharmacology of self-injurious behavior in the mentally retarded. J Am Acad Adol Psychiatry 26:296–302, 1987
95. Sandyk R: Naloxone abolished self-injuring in a mentally retarded child. Ann Neurol 17:520, 1985
96. Davidson PW, Kleene BM, Carroll M, et al: Effects of naloxone on self-injurious behavior: A case study. Appl Res Ment Retard 4:1, 1983
97. Richardson JS, Zaleski WA: Naloxone and self-mutilation. Biol Psychiatry 18:99–101, 1983
98. Sandman CA, Barron JL, Crinella FM, et al: Influence of Naloxone on brain and behavior of a self-injurious woman. Biol Psychiatry 22:899–906, 1987
99. Campbell M, Small AM, Sokol MS, et al: Naltrexone in autistic children: An acute dose range tolerance trial. San Diego, Stallone Foundation, 1987
100. Beckwith BE, Couk DI, Schumacher K: Failure of naloxone to reduce self-injurious behavior in two developmentally disabled females. Appl Res Ment Retard 7:183–188, 1986
101. Herman CH, Hammock MK, Arthur-Smith A, et al: Naltrexone decreases self-injurious behavior. Ann Neurol 22:550–552, 1987
102. Szymanski L, Kedesdy J, Sulkes S, et al: Naltrexone in treatment of self injurious behavior: A clinical study. Res Dev Disabil 8:179–180, 1987
103. Selye H: The Stress of Life. New York, McGraw-Hill, 1956
104. Axelrod J, Reisine TD: Stress hormones: Their interaction and regulation. Science 224:452–459, 1984
105. Erickson MT: The relationship between variables and specific complications of pregnancy, labor, and delivery. J Psychosom Res 20:207–210, 1976
106. Gorsuch RL, Key MK: Abnormalities of pregnancies as a function of anxiety and life stress. Psychosomat Med 36:352–362, 1974
107. Luakaran VH, van de Berg BJ: The relationship of maternal attitude to pregnancy outcomes and obstetric complications. Am J Obstet Gynecol 136:374–379, 1980
108. Lederman E, Lederman RP, Work BA, et al: Maternal psychobiological and physiological correlates of fetal–newborn health status. Am J Obstet Gynecol 139;956–958, 1981
109. Nuckolls KB, Cassel J, Kaplan BH: Psychosocial assets, life crisis, and the prognosis of pregnancy. Am J Epidemiol 95:431–441, 1972
110. Andreoli C, Magni G. Rizzardo R: Stressful life events, anxiety and obstetric complications, in Pancheri

P, Zichella L, Falsachi P (eds), Endorphins, Neuroregulators and Behavior in Human Reproduction. Amsterdam, Excerpta Medica, 1984, pp 297–304

111. Connelly JA, Cullen JH: Materhnal stress and the origins of health status, in Call JD, Galenson E, Tyson RL (eds), Frontiers of Infant Psychiatry. New York, Basic Books, 1983, pp 273–281
112. Brown WA, Manning T, Grodin J: The relationship antenatal and perinatal psychological variables to the use of drugs in labor. Psychosom Med 34:119–127, 1972
113. Goodlin RC: Naloxone and its possible relationship to fetal endorphin levels and fetal distress. Am J Obstet Gynecol 139:16–19, 1981
114. Feigenbaum J, Yanai J, Moon B, et al: The effect of drugs altering striatal dopamine levels on apomorphine induced stereotypy. Pharmacol Biochem Behav 16:235–240, 1982
115. Costa E: The modulation of postsynaptic receptors by neuropeptide cotransmitters: A possible site of action for a new generation of psycho-tropic drugs. In Usdin E. Bunney WE, Davis JM (eds), Neuroreceptors—Basic and Clinical Aspects. New York, Wiley, 1981, pp 15–25.
116. Eidelberg E, Erspamer R: Dopaminergic mechanism of opiate actions in brain. J Pharmacol Exp Ther 192:50–57, 1975
117. Eidelberg E, Schwartz A: Possible mechanism of action of morphine in brain. Nature (Lond) 225:1152–1153, 1970
118. Kuschinsky K, Hornykiewicz O: Morphine catalepsy in the rat: Relation to striatal dopamine metabolism. Eur J Pharmacol 19:119–122, 1982
119. MacNichol E. Kung YW, Levin S, et al: Stimulation of dopamine synthesis in cudate nucleus by intrastriatal enkephalins and antagonism by naloxone. Science 200:552–554, 1978
120. Pollard H, Lorens J, Schwartz J: Enkephalin receptors on domaminergic neurons in rat striatum. Nature (Lond) 268:745–747, 1977
121. Pollard H, Lorens C, Schwartz et al: Localization of opinate receptors and enkephalins in the rat striatum in relationship with the nigrostriatal dopaminergic system: Lesion studies. Brain Res 151:392–398, 1978
122. Carenzi A, Guidptto R, Reveulta R, et al: Molecular mechanisms in the actions of morphine viminol (R2) on rats striatum. J Pharmacol Exp Ther 194:311–318, 1975
123. Loh H, Brase D, Sampath-Khanna S, et al: β-Endorphin in vitro inhibition of dopamine release. Nature (Lond) 264:567–568, 1976
124. Subramanian N, Mitznegg P, Sprugel W: Influence of enkephalin on K^+ evoked efflux of putative neurotransmitters in the rat brain. Arch Pharmacol 229:163–165, 1977
125. Biggio M, Casu M, Corda M, et al: Stimulation of dopamine synthesis in caudate nucleus by intrastriatal enkephalins and antagonism by naloxone. Science 200:552–554, 1978
126. Iwatsubo K, Clouet DH: Dopamine-sensitive adenylate cyclase of the caudate nuclleus of rats treated with morphine as haloperidol. Biochem Pharmacol 24: 1499–1503, 1975
127. Seeman D: Brain dopamine receptors. Pharmacol Rev 32:229–292, 1981
128. Friedhoff AJ, Miller JC: Prenatal psychotropic drug exposure and the development of central dopaminergic and cholinergic neurotransmitter systems. Monog Neural Sci 9:91–98, 1983
129. Rosengarten H, Friedhoff AJ: Enduring changes in dopamine receptor cells of pups from drug administration to pregnant and nursing rats. Science 203:1133–1135, 1979
130. Hull EM, Nishita JK, Bitran D, et al: Perinatal dopamine-related drugs demasculinize rats. Science 224:1011–1013, 1984

6

Drug Discrimination Studies in Animals

A Behavioral Approach to Understanding the Role of Neurotransmitter Receptor Complexes in Mediating Drug Effects

John Mastropaolo and Anthony L. Riley

1. ANIMAL MODELS IN BEHAVIORAL PHARMACOLOGY

Although not without controversy,[1] animal models of human pathology have been used to study a wide range of issues[2–10] (Table I). The analogy from the animal model to the human condition can take a variety of forms and can provide different types of information. Ideally, such models would accurately simulate the human syndrome, which they are purported to represent in such a way that they fulfill the requisite characteristics for animal models proposed by Robbins and Sahakian.[11] Briefly, they suggest that animal models should (1) mimic the behavioral features of the disorder, (2) have a similar etiology, and (3) show recovery in behavioral features in response to treatments that alleviate symptoms in humans. Although such criteria may represent the exemplar of an animal model, many useful models may not meet each of these requirements. For example, Kornetsky[12] described a continuum of levels at which animal models might represent a human problem of interest. In his analysis, models are seen as homologous if there is a correspondence in the etiology of the disease and the model. At another level, models may be only isomorphic. That is, while there are similarities between the model and the human state, the cause of the condition created for the model may be different from the cause in humans. Finally, the model may not have a direct resemblance to the disease but may be a nonhomologous, nonisomorphic representation that has some predictive value concerning some aspect of the disease.

John Mastropaolo • Psychiatry Service, Veterans Administration Medical Center, Washington, D.C. 20422.
Anthony L. Riley • Psychopharmacology Laboratory, Psychology Department, The American University, Washington, D.C. 20016.

Table I. Animal Models Applied to Human Conditions

Human application	Animal model	Investigators
Aggression	Isolation syndrome (mice)	Valzelli[2]
Depression	Forced swimming	Porsolt et al.[3]
	DRL 72-sec	O'Donnell and Seiden[4]
	Learned helplessness	Seligman et al.[5]
Mania	Amphetamine–barbiturate hyperactivity	Rushton and Steinberg[6]
Obesity	VMH-lesioned rat	Schacter and Rodin[7]
Alcoholism	Alcohol self-administration	Mello[8]
	Schedule-induced polydipsia	Falk et al.[9]
Minimal brain dysfunction	Lead toxicity	Silbergeld and Goldberg[10]

2. ANIMAL MODELS OF DEPRESSION

Animal models of depression may provide the best illustration of these issues. For example, employing the differential reinforcement of low rate (DRL) 72-sec schedule of reinforcement, wherein an animal is reinforced for responding only if a period of 72 sec or greater has elapsed since the last response (or reinforcement), as a tool in the study of depression is clearly an example of the final type of model, i.e., a nonhomologous, nonisomorphic model. The contribution of this model is neither based on the belief that the behavior maintained under a DRL schedule has a similar etiology to depression nor based on the notion that performance under this schedule is similar to the behavior of a depressed individual. Instead its utility is a function of the selective improvement of behavior maintained under this schedule by antidepressant drugs; i.e., a greater proportion of responses are reinforced.[4] This model, therefore, has important predictive value as a screen for drugs with potential antidepressant activity. Learned helplessness [5] is an animal model of depression that represents a different aspect of analysis available through animal models. In this procedure, a proportion of preshocked animals show an inability to acquire a simple response in order to escape from a noxious foot shock. This inability to act and to make contact with new contingencies in the animal model is clearly similar to the human condition. This model, therefore, is an isomorphic model. Although it would certainly be controversial, one could even argue that this model was homologous to the human condition. Since the typical procedure for creating this animal model employs the application of inescapable shock, it has been suggested that animals learn that outcomes are independent of behavior during this treatment.[5] It is similar, then, to the "hopelessness" that is a prominent feature in clinical depression.

3. DRUG DISCRIMINATION IN BEHAVIORAL PHARMACOLOGY

While animal models are useful tools for the assessment of therapeutic interventions into pathological conditions in a preclinical setting, the primary focus of the present chapter is an animal model of a different type. Specifically, this chapter focuses on drug discriminations, a technique for assessing the receptor activity of pharmacologically active compounds. This technique uses the ability of the animal to "report" this receptor activity by making different responses that have come under the control of the drug. Although there was a 20-year period of inactivity following the first report of a drug discrimination,[13] more recently the field has shown considerable growth, emerging as the most widely used paradigm in behavioral pharmacology.[14,15] This is not surprising, since discriminative control of behavior by drugs has both practical and theoretical significance.

4. THE DRUG DISCRIMINATION PROCEDURE

In a drug discrimination, a drug serves as a discriminative stimulus. A discriminative stimulus is any stimulus having a discriminative function in the control of behavior.[16] In other words, a discriminative stimulus is any environmental event that sets the occasion for a particular response to be reinforced. For a basic example of a discrimination, consider a rat trained to press a bar in order to obtain food. If presses on the bar produce food during portions of the session when a light is on and do not produce food during portions of the session when the light is off, the light is said to serve as a discriminative stimulus. That is, it indicates (sets the occasion) when the particular response (in this case, bar pressing) will be reinforced (in this case, produce a food pellet). The degree of discriminative control over responding by the light would be evidenced by differences in the rate of responding during those portions of the session when the light was on versus those portions of the session when the light was off. If discriminative control of responding was complete, responding would only occur when the light was on; the rate of responding when the light was off would be zero. Although typically exteroceptive stimuli such as lights and tones serve such discriminative functions, interoceptive stimuli may also serve in this way. Drugs are one such interoceptive stimulus.

The general procedure for training a drug discrimination with the popular operant technique involves training animals to select a response lever based on the substance injected prior to the operant session (Table II). For example, responses on the left lever will result in the delivery of a reinforcer (usually food) during sessions preceded by the injection of a drug, while responses on the right lever will produce a reinforcer during sessions preceded by an injection of the drug vehicle. Typically, incorrect responses will be recorded but will have no programmed consequence. Care is usually given to counterbalance the training conditions such that half of the animals are trained with the opposite lever associated with the injection stimuli. With this procedure, therefore, the lever selection should be based solely on the internal cues that follow the injection. Response selection, therefore, is thought to indicate the ability of the animal to identify some subjective aspect of drug treatment. When differential responding is established, this procedure then provides the unique situation in which the internal states produced by drugs can be subjected to an objective analysis.[17]

If following drug discrimination training, different doses of the training drug are substituted for the training dose, the amount of responding to these substitution doses is a function of their similarity to the training stimulus.[18] Thus, as with other procedures in behavioral

Table II. Operant Drug Discrimination Procedure

1. Response shaping, typically lever pressing
2. Responding under intermittent schedule, typically FR 10
3. Injections of vehicle or drug prior to sessions, typically alternating (Sessions preceded by drug injection set the occasion for one lever to produce reinforcement, typically food. Sessions preceded by vehicle set the occasion for the opposite lever to produce reinforcement.)
4. Training until performance reaches criteria, typically 90% accuracy

pharmacology, a dose–response function can be determined. This dose–response function is analogous to the generalization gradient that can be generated around an exteroceptive stimulus; where, for example, tones of different frequencies would be presented in place of the training frequency. Typically, the dose–response function (generalization gradient) that this dose-substitution procedure yields shows less drug-appropriate responding at doses lower than the training dose, while at higher doses responding is similar to that following the training dose. Substantial decreases in the animal's ability to respond, reflected in decreases in rate, often limit the highest dose that can be tested. From this generalization gradient, one can determine the ability of various doses of a drug to substitute for the training dose and thereby assess the relative potency of these doses for providing the appropriate discriminative cue.

5. IMPLICATIONS OF DRUG DISCRIMINATION

Although drug discrimination is interesting in that it extends the types of stimuli that can control operant behavior (cf. Colpaert[19]), the implications of this procedure are far more important for pharmacology. The importance of the drug discrimination paradigm for pharmacology is based on a procedure similar to that for determining control at different doses of the training drug. The important difference, however, is that a dose of a different drug, rather than a different dose of the same drug, is substituted for the training drug.[18,20] In general, after training with one drug is complete, only compounds from the same pharmacological class will produce a similar discriminative stimulus and engender drug-appropriate responding. There is some variability, however, in the specificity of the generalization to other drugs, which seems to depend on the class of the training drug (cf. Overton[20]). Typically, active drugs with different pharmacological effects will not produce responding appropriate to the drug stimulus, despite obvious behavioral effects, i.e., reductions in the rate of responding. By virtue of the specificity of the pharmacological basis for producing the drug-induced discriminative stimulus, it has been suggested that this technique can be as useful as traditional pharmacological techniques in the classification of compounds.[21,22]

In a related area, drug discrimination has also emerged as a technique to study structure–activity relationships[23]; i.e., the relationship between a particular drug activity and the nature of the molecular activities that produce that activity. For example, though only a limited number of benzodiazepine derivatives were studied, Glennon and Young[23] found a significant correlation between the ED_{50} values for compounds that produce a benzodiazepine stimulus and their previously reported "drug effect" potencies in humans.

Although there is obvious clinical utility derived from the use of drug discrimination as a tool for drug classification and structure–activity relationships, there are more direct clinical implications as well. For example, it has been suggested that drug discrimination studies are useful for predicting compounds that have potential for abuse in humans.[24] In a discussion of the opioid analgesics, Holtzman[24] notes the factors that contribute to abuse potential. These include factors unrelated to the drug's actions, per se, such as awareness and availability of the drug. He also notes that, in general, drugs that can produce morphine-like subjective effects in a dose-related manner have a high abuse potential.[25] The effects produced by different opioids vary across a broad spectrum. Among these effects are the "relaxation" and "well-being" on one end and the "dysphoria" and "psychotomimetic" effects on the other.[26] Since the "subjective effects" produced by a drug seem to be an accurate predictor of its potential for abuse, a methodology for examining these effects objectively is invaluable. Clearly the drug discrimination procedure is ideally suited to this task and should be exploited as a screen for the potential abuse liability of new compounds, particularly when a compound with demonstrated abuse potential in the drug class being examined is available as a reference for comparing new compounds. Indeed, Stolerman et al.[22] demonstrated how drug discrimination techniques can even be applied to opioids that are difficult to evaluate because they have multiple actions (e.g., cyclazocine) by training animals to discriminate drug combinations.

6. EXTEROCEPTIVE VERSUS INTEROCEPTIVE STIMULI

Despite the general similarity between drugs as interoceptive stimuli and more conventional exteroceptive stimuli, some important differences should be noted. For example, unlike exteroceptive stimuli, drug effects have a slow onset and termination. Thus, it is impossible to have an immediate transition from one stimulus condition to the next in drug discrimination training. In addition, drug stimuli present a unique situation because drug and nondrug conditions cannot be presented simultaneously.[21] These factors probably contribute to the slow acquisition of differential responding which emerges with drug stimuli. For example, Zenick and Greene[27] noted that a considerable amount of time must be devoted to the training of each subject in order to establish differential responding with drug stimuli. In fact, they indicate that it is not unusual for as many as 40 sessions to be necessary to establish relatively stable responding that is under equal control in both the drug and nondrug state.

7. DRUG DISCRIMINATION WITH CONDITIONED TASTE AVERSIONS

Recently, a paradigm has been introduced in which conditioned taste aversion learning (see Riley and Tuck[28] and references cited therein) serves as a baseline for a drug discrimination.[29,30] Conditioned taste aversions are a remarkably simple and robust classic conditioning procedure that capitalizes on the rat's tendency to avoid consumption of novel foods (tastes, which serve as the conditioned stimulus) followed by "illness" (that can be produced by the injection of a drug, which serves as the unconditioned stimulus) on a previous occasion (for review, see Riley and Tuck[31]). The basis for the use of conditioned taste aversions as a baseline for drug discrimination learning is simple. The toxicant that follows the presentation of a taste should be paired with (predicted by) the presence or absence of a drug stimulus. Thus, this drug stimulus (the "subjective effects" produced by the drug) can serve as a signal for the animal to determine whether or not the taste will be followed by an aversive stimulus. With the drug stimulus as a predictor of a future aversive stimulus, one would expect that consumption of a particular taste would vary precisely with the drug state to which the animal was subjected. For this procedure to be successful, the drug stimulus must precede the presentation of the particular taste.

A general example of the paradigm would be illustrative. In this procedure, animals are first adapted to drinking during a limited period each day; i.e., animals are allowed access to water for only 20 min at the same time each day. The water is presented in calibrated drinking tubes so that the amount consumed can be recorded. When water consumption shows no systematic upward or downward trend, consumption is considered stable and the next phase can begin. At this point, animals are assigned to groups matched on water consumption. The examination of a particular training stimulus (i.e., drug dose) requires at least two groups, although four groups provide a more comprehensive design. For a simple example, the two group design will be outlined (Table III). For these two groups, an injection of saline (or the drug vehicle) precedes the next 20-minute fluid-access period (the route of injection and the amount of time prior to consumption for the injection depends on the particular training drug to be examined). At the appropriate time following this injection of the drug vehicle, a novel tasting solution (e.g., a 0.1% saccharin solution) is presented in place of the usual water during

Table III. Taste-Aversion Drug-Discrimination Procedure[a,b]

Procedure	Days	Groups: Conditioned	Control
Water deprivation	1–12	N–W–N	N–W–N
Saccharin habituation	13–15	V–S–V	V–S–V
Conditioning	16	D–S–L	D–S–V
Recovery	17–19	V–S–V	V–S–V

[a]The 4-day cycle of conditioning and recovery is repeated until the discrimination is acquired.

[b]N, no injection; W, 20 min access to water; S, 20 min access to saccharin; V, vehicle injection; D, drug injection; L, LiCl injection.

the 20-min fluid access period. Immediately following this drinking period, animals in both groups are given an injection of distilled water. This procedure of injections before and after access to saccharin is continued for two additional days. On the day following this 3-day habituation period, during which animals become familiar with the handling inherent in the injection procedures and with the saccharin solution, animals in the two groups are treated differentially. Following the three habituation days, animals in both groups are injected with the training drug at the appropriate time prior to the presentation of the saccharin solution on this first conditioning day. Immediately after the 20-min access to saccharin, animals in the control group are injected with distilled water, while animals in the experimental group (i.e., the drug-discrimination group) are injected with an aversion-inducing agent (e.g., LiCl) at this time. On the next 3 days, both groups are treated in a manner identical to that during the first 3 days of saccharin presentation, i.e., the habituation phase. These three days, which serve as recovery days during which no aversion-inducing agents are administered, allow the animals both to recover from the "illness" induced by the LiCl and to rehydrate following any reduction in consumption that may have occurred. In this paradigm, these recovery days also serve to establish that the absence of the drug stimulus is a "safety signal" for the drug-discrimination group; i.e., the absence of the drug stimulus indicates that saccharin will not be followed by an aversive stimulus. On the day following the third recovery day, animals are again subjected to another conditioning trial. This entails both groups being injected with the drug prior to the opportunity to drink saccharin. However, as above, only the drug-discrimination group is injected with LiCl immediately after the 20-min drinking period, while the control group is injected with distilled water. This cycle of one conditioning day followed by three recovery days is repeated until the discrimination is established. As can be seen, the control group in this procedure is treated identically to the drug-discrimination group, with the exception of the LiCl (aversion-inducing agent) injections. Thus, any changes in consumption induced by the repeated administration of the training drug (on every fourth day) will not be unique to the drug-discrimination group.

It was noted earlier that the evidence for a drug discrimination in the traditional operant procedure was the selection of the appropriate lever in an operant chamber. In this paradigm, a drug discrimination is evidenced by changes in consumption. If this procedure is successful in establishing a drug discrimination, consumption should come to be a function of the injection prior to the opportunity to drink. Specifically, the drug-discrimination group should avoid drinking the saccharin solution (i.e., display an aversion) on days when the saccharin presentation is preceded by an injection of the training drug, but should continue to drink the same saccharin solution at a high level on days when the presentation of saccharin is preceded by an

Table IV. Taste-Aversion Drug-Discrimination Procedure[a,b]

Procedure	Days	Groups			
		Cond (1)	Cont (1)	Cond (2)	Cont (2)
Water deprivation	1–12	N–W–N	N–W–N	N–W–N	N–W–N
Saccharin habituation	13–15	V–S–V	V–S–V	D–S–V	D–S–V
Conditioning	16	D–S–L	D–S–V	V–S–L	V–S–V
Recovery	17–19	V–S–V	V–S–V	D–S–V	D–S–V

[a]The 4-day cycle of conditioning and recovery is repeated until the discrimination is acquired.
[b]N, no injection; W, 20 min access to water; S, 20 min access to saccharin; V, vehicle injection; D, drug injection; L, LiCl injection. Cond (1), conditioning (drug discrimination-trained) group 1; Cont (1), control group 1; Cond (2), conditioning (drug discrimination-trained) group 2; Cont (2), control group 2.

injection of the drug vehicle. Animals in the control group, however, should show no systematic change in drinking as a function of the injection given prior to the opportunity to drink. As mentioned, the inclusion of two additional groups (a second drug-discrimination group and its control) provides a more comprehensive assessment (Table IV). These groups are treated similarly to the groups already described, with the simple exception that the injections given prior to the opportunity to drink are reversed. For the drug discrimination group, a drug injection prior to the saccharin presentation serves as the stimulus which signals that saccharin will not be followed by an aversive stimulus and the injection of saline (or the appropriate drug vehicle) serves as the stimulus which signals that saccharin will be followed by an aversive stimulus (i.e., the LiCl injection). The additional control group would provide a basis for comparison to determine the effects of being injected with the training drug prior to the opportunity to drink on 3 of 4 days.

8. PCP AS A DISCRIMINATIVE STIMULUS IN TASTE AVERSIONS

The first drug examined with conditioned taste aversions as a behavioral baseline for drug discrimination learning was phencyclidine (PCP).[30] This experiment was conducted as outlined above with 1.8 mg/kg of PCP serving as the training stimulus. This experiment employed the four group design (see Table IV). For one group, the injection of 1.8 mg/kg of PCP prior to the opportunity to drink signalled that saccharin would be followed by an aversive stimulus, i.e., 1.8 mEq of 0.15 M LiCl, while for a second group this aversive stimulus was signaled by the injection of distilled water (the PCP vehicle). The relevant control groups were treated identically, except that these groups were never injected with LiCl. In this experiment, rats rapidly acquired the drug discrimination. The discrimination-trained groups were significantly different from their respective controls after only two conditioning trials, i.e., on the third injection of the stimulus that signaled saccharin would be followed by an aversive stimulus (Fig. 1).

Following this, Mastropaolo et al.[30] substituted different doses of PCP for the training stimulus. These substitutions were conducted on the second recovery day in the conditioning–recovery cycle, thus allowing for both the maintenance of the training conditions and sufficient "safety" days for the animals to recover from the possible suppressant effect of the LiCl and reduced consumption from the acquisition of the discrimination. These substitutions yielded orderly dose–response curves (Fig. 2). Specifically, for animals in group PL (animals trained to avoid drinking saccharin in the presence of the PCP stimulus), as the dose of PCP increased,

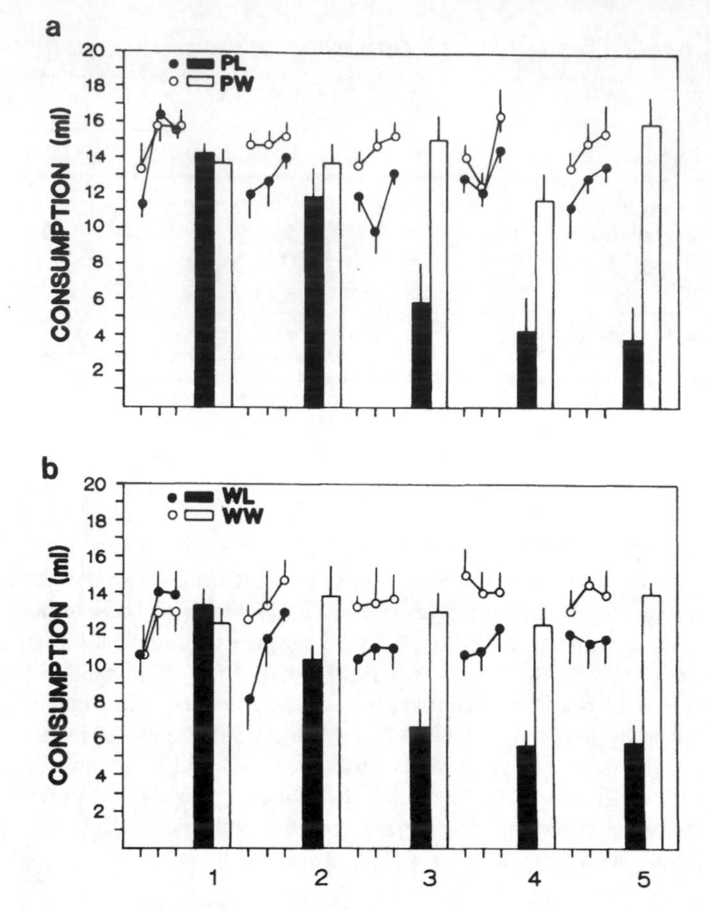

Figure 1. Acquisition of stimulus control over conditioned taste aversions with PCP. (a) Groups trained with 1.8 mg/kg of PCP signaling a saccharin–LiCl pairing (group PL) and their controls (group PW). (b) Groups trained with the absence of PCP signaling a saccharin–LiCl pairing (group WL) and their controls (group WW). Bars represent conditioning trials; connected points represent habituation (the first three points) and recovery sessions.

the amount of saccharin consumed decreased. For group WL (animals trained to avoid drinking saccharin in the absence of the PCP stimulus), as the dose of PCP increased, the amount of saccharin consumed increased, indicating that for both groups PCP was serving a discriminative function. That is, for group PL, PCP served as a signal that saccharin would be followed by LiCl, and consumption was reduced in the presence of the PCP stimulus. For group WL, PCP served the opposite function; i.e., it signaled that saccharin would not be followed by LiCl, and consumption increased in the presence of the PCP stimulus. Neither of the two control groups (groups PW and WW) showed any systematic relationship between consumption and the dose of PCP administered, although both groups showed a slight decrease in consumption at the highest dose tested (3.2 mg/kg).

To determine whether the discriminative stimulus properties were receptor specific, this experiment also examined the dose–response functions for ketamine (a drug from the same class as PCP) and *d*-amphetamine (a drug from a different drug class) in these same animals. The dose–response curves for ketamine were similar to those for PCP in that an orderly relationship between drug dose and consumption was observed (Fig. 3). Again, for group PL, as the dose of ketamine increased, the amount of saccharin consumed decreased. For group

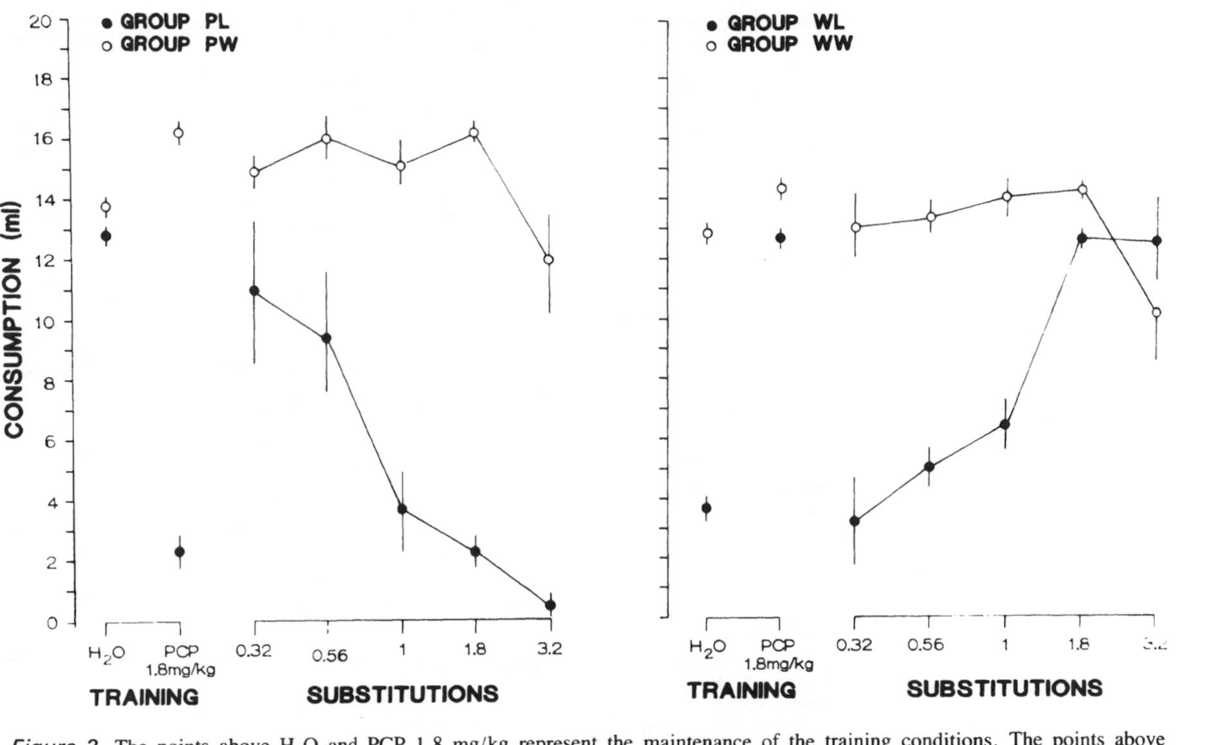

Figure 2. The points above H_2O and PCP 1.8 mg/kg represent the maintenance of the training conditions. The points above substitutions represent the dose–response relationship when various doses of PCP were injected during probe sessions. Group designations are identical to those described in Fig. 1.

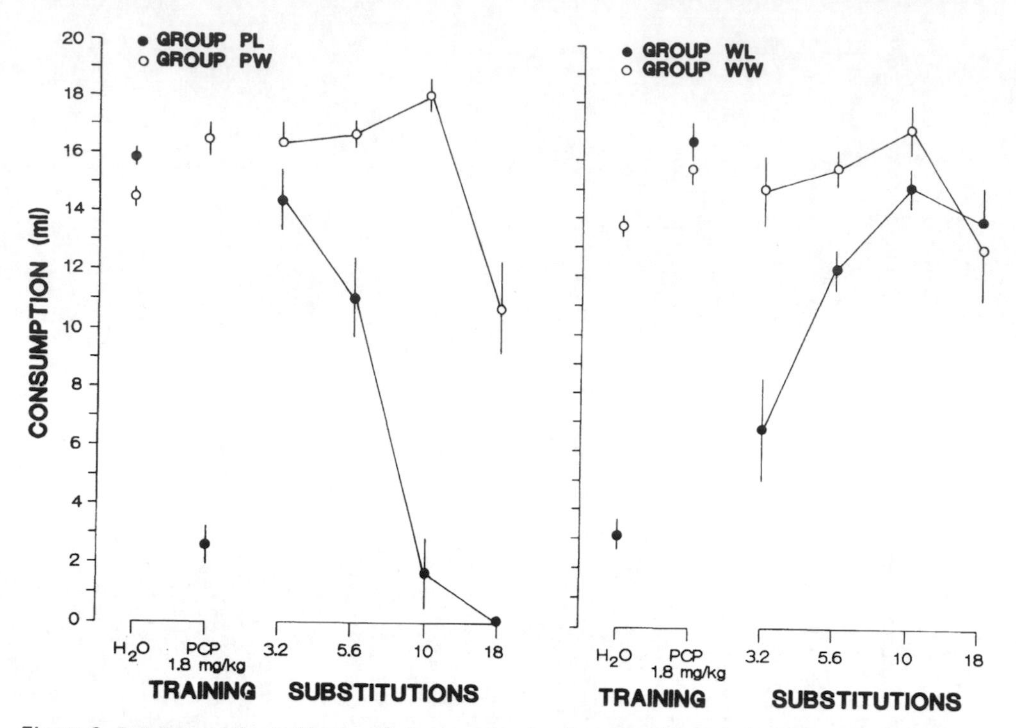

Figure 3. Data presented as in Fig. 2, with the exception that the points above substitutions represent the dose–response relationship for probe sessions with various doses of ketamine.

WL, the opposite was again observed. That is, as the dose of ketamine increased, the amount of saccharin consumed increased. The control groups again showed no systematic relationship between ketamine dose and consumption, although consumption did tend to decrease in both groups at the highest dose tested (18 mg/kg). It should be noted, however, that in this procedure ketamine was less potent than PCP on a mg/kg basis (cf. Brady and Balster[32]).

As expected, the dose–response curves for *d*-amphetamine were not similar to those for PCP (Fig. 4). In fact, with *d*-amphetamine, the dose–response curves obtained were similar for groups PL, PW, and WW. For groups PL, PW, and WW, consumption decreased as the dose of *d*-amphetamine increased (from 0.18 to 1.8 mg/kg). This decrease in consumption was a result of the adipsic effect of *d*-amphetamine and did not reflect conditioning. For group WL, consumption decreased at each of the doses of *d*-amphetamine tested, indicating that none of the doses produced a stimulus similar to PCP, which constituted the "safety" stimulus for this group.

This first demonstration of the use of conditioned taste aversions as a behavioral baseline for the assessment of drug discrimination learning was strikingly successful. As with the traditional procedures, control was established to the training drug and the control was specific to this general class. For example, ketamine, a drug from the same class as the PCP training drug, produced PCP-appropriate control over drinking, while *d*-amphetamine, a drug from a dissimilar class, did not. In addition, the substitution data reveal that this procedure produces results similar to that yielded from binding experiments in that the potency of ketamine was less than that of PCP on a mg/kg basis.

Probably the most notable result of this experiment, however, was the rate of acquisition of the discrimination. Unlike traditional procedures that can require extensive training, this procedure required only two conditioning trials before statistically significant differences

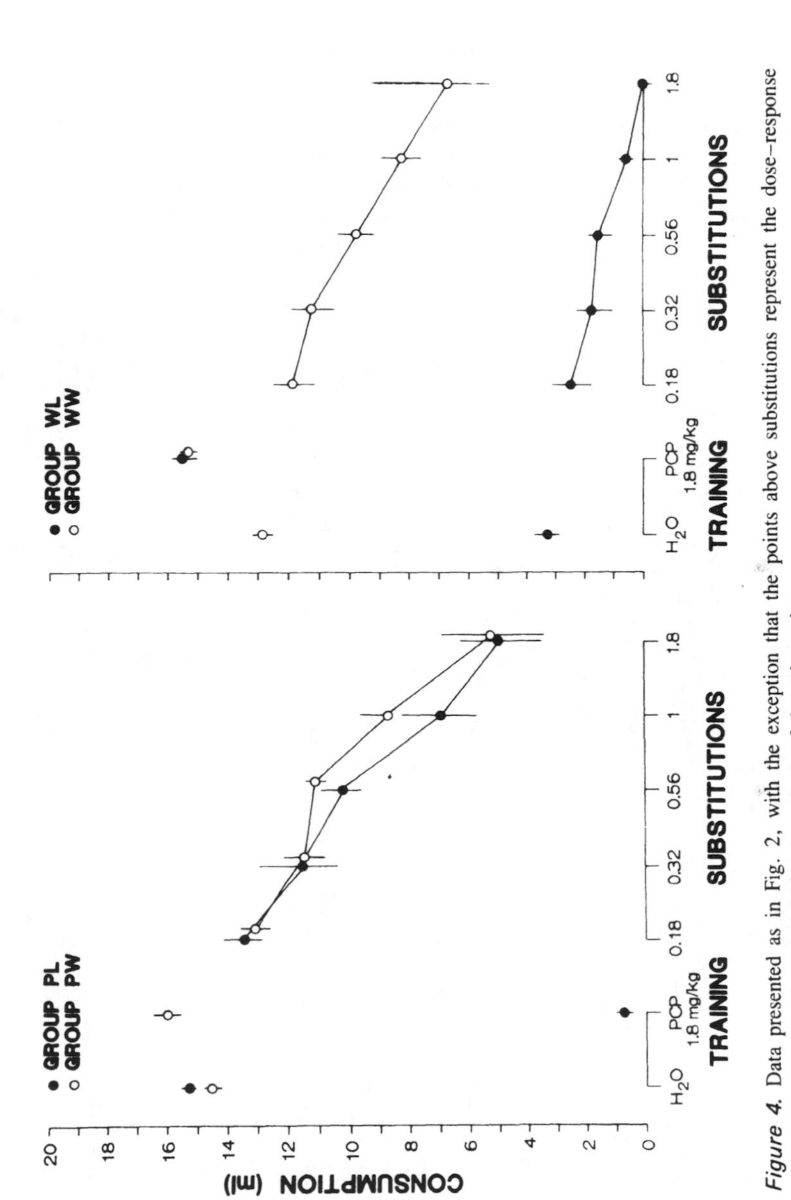

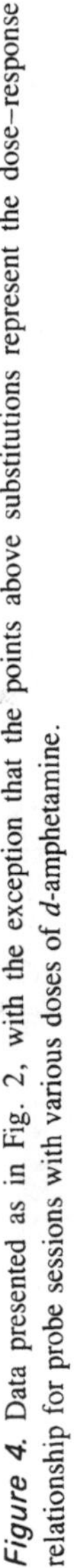
Figure 4. Data presented as in Fig. 2, with the exception that the points above substitutions represent the dose–response relationship for probe sessions with various doses of *d*-amphetamine.

emerged between the training and control groups. This rapid acquisition suggests that this procedure may be more sensitive to the discriminative properties of drugs than other procedures. If this were true, this increased sensitivity may be reflected by an ability to detect the discriminative stimulus properties of drugs that have not been observed in other paradigms.

9. DRUG DISCRIMINATION WITH NALOXONE

In a direct test of the sensitivity of this procedure in detecting discriminative stimulus properties of drugs that are difficult to demonstrate in other procedures, Kautz et al.[33] used conditioned taste aversions as a baseline to examine the discriminative stimulus properties of naloxone, a narcotic antagonist that has been shown either to fail to serve as a discriminative stimulus or to do so only at high doses and with extensive training.[34,35] Specifically, one of two doses of naloxone (1 or 3 mg/kg) served as the discriminative stimulus for different groups of animals. The two group procedure was employed for each training dose of naloxone (see Table III). As with PCP, significant differences emerged between the groups by the third conditioning trial, i.e., after only two naloxone–saccharin–LiCl presentations (Fig. 5). Fol-

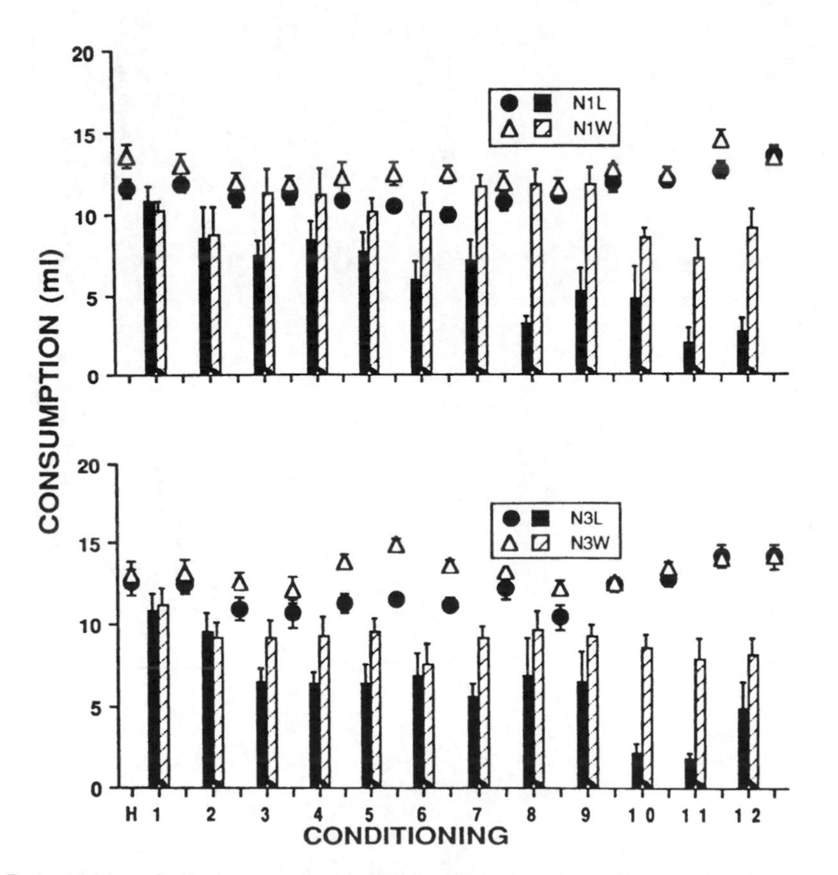

Figure 5. Acquisition of stimulus control over conditioned taste aversions with naloxone. (a) Groups trained with 1 mg/kg of naloxone signaling a saccharin–LiCl pairing (closed symbols, group N1L) and their controls (open symbols, group N1W). (b) Groups trained with 3 mg/kg of naloxone signaling a saccharin–LiCl pairing (closed symbols, group N3L) and their controls (open symbols, group N3W). Bars represent conditioning trials; points represent the mean of the three habituation (H) and recovery days.

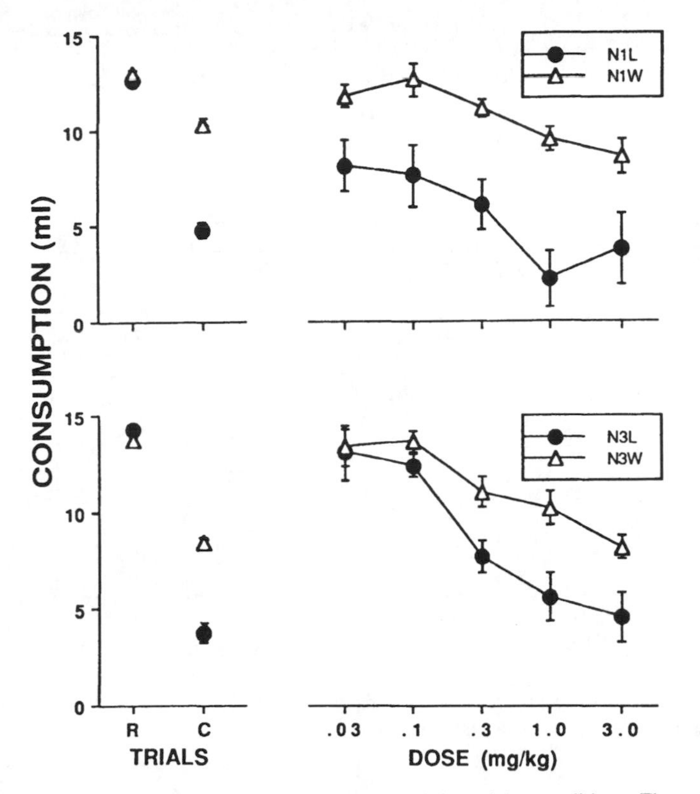

Figure 6. Points above R and C represent the maintenance of the training conditions. The points above dose represent the dose–response relationship when various doses of naloxone were injected during probe sessions. Group designations are identical to those described in Fig. 1.

lowing training, with their respective doses of naloxone, each group was used to assess the dose–response function for the training drug. Orderly dose–response functions were evident for each of the discrimination-trained groups (Fig. 6). That is, with both training doses of naloxone the drug-discrimination groups tended to decrease consumption of saccharin as the dose of naloxone increased, while for the respective control groups consumption was not systematically related to dose. These same groups were then employed to determine the dose–response function for naltrexone. Similar to naloxone, an orderly relationship between the dose of naltrexone and consumption of saccharin was evident for the drug-discrimination groups; i.e., consumption tended to decrease as the dose of naltrexone increased (Fig. 7). For the control groups, drinking was slightly suppressed at the highest doses tested. Finally, these same animals were used to determine the dose–response function for morphine. Morphine did not produce reliable decreases in consumption in any of the groups throughout the range of doses tested (3.2, 5.6, and 10 mg/kg) (Fig. 8). The results of this experiment support the notion that this behavioral baseline (conditioned taste aversion) is more sensitive than other procedures in demonstrating the discriminative stimulus properties of drugs. Subsequent work in this same laboratory (unpublished data) has demonstrated that the discriminative stimulus properties of naloxone can be mimicked by the mixed agonists/antagonists diprenorphine and nalorphine.

Although this new technique for assessing the discriminative stimulus properties of drugs has only recently been introduced, there is already a literature establishing its utility. For example, Lucki et al.[36] demonstrated that this technique was successful at rapidly establishing

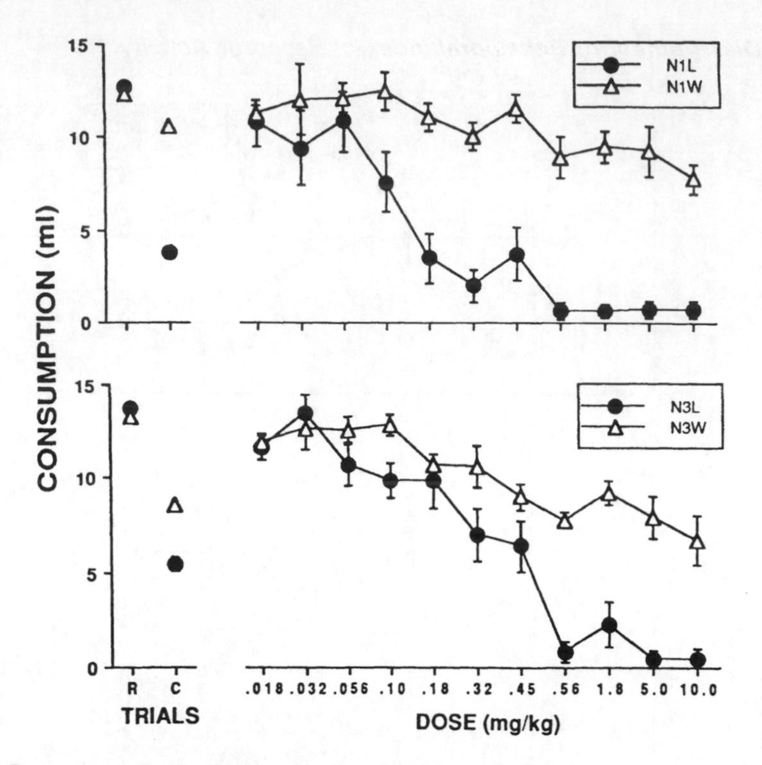

Figure 7. Data presented as in Fig. 6, with the exception that the points above dose represent the dose–response relationship for probe sessions with naltrexone.

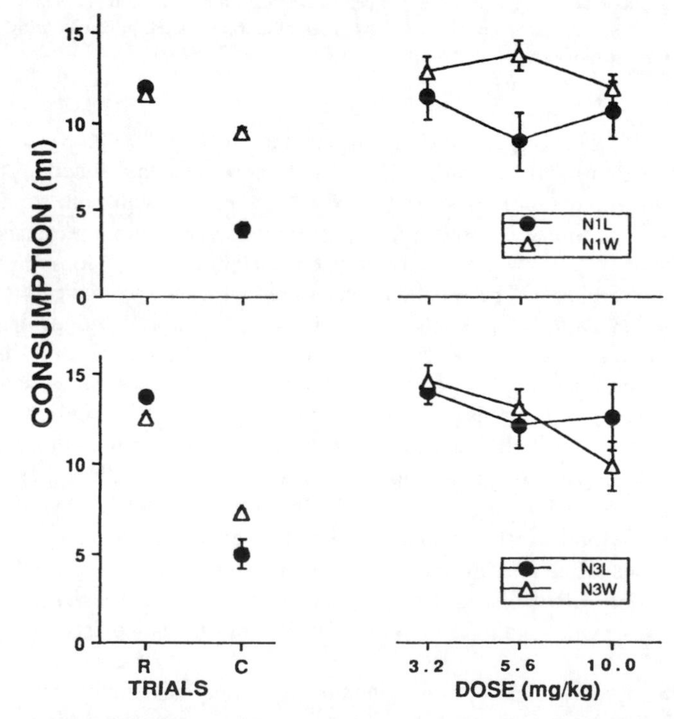

Figure 8. Data presented as in Fig. 6, with the exception that the points above dose represent the dose–response relationship for probe sessions with morphine.

the stimulus properties of selective 5-hydroxytryptamine$_{1A}$ (5-HT$_{1A}$) agonists. In this experiment, rats acquired the drug discrimination after only two to three conditioning trials. Importantly, the stimulus trained under these conditions was shown to be selective for 5-HT$_{1A}$ agonists and was not mimicked by 5-HT$_{1B}$ agonists. In addition, Martin et al.[37] employed the conditioned taste aversion baseline to assess the discriminative stimulus properties of morphine. Finally, Riley and colleagues[38] demonstrated rapid acquisition of discriminative control over taste aversions even when the dipsogenic agent, pentobarbitol, is used as the discriminative stimulus.

In summary, the use of conditioned taste aversions as a behavioral baseline to assess the discriminative stimulus properties of drugs possesses advantages over the traditional procedures typically employed. It has the practical benefit of providing a means by which many animals can be trained rapidly, with no great demand for equipment. In addition, the sensitivity of the procedure should allow for an examination of a wider variety of compounds and may reveal similarities between drugs that are not detectable with other techniques.

REFERENCES

1. Kuker-Reines B: Psychology Experiments on Animals. New England, New England Anti-Vivisection Society, 1982
2. Valzelli L: The "isolation syndrome" in mice. Psychopharmacology 31:305–320, 1973
3. Porsolt RD, Anton G, Blavet N, Jalfre M: Behavioral despair in rats: A new model sensitive to antidepressant treatment. Eur J Pharmacol 47:379–391, 1978
4. O'Donnell JM, Seiden LS: Differential-reinforcement-of-low-rate 72-second schedule: Selective effects of antidepressant drugs. J Pharmacol Exp Ther 224:80–88, 1983
5. Seligman MEP, Maier SF, Solomon RL: Unpredictable and uncontrollable aversive events, in Brush FR (ed): Aversive Conditioning and Learning. New York, Academic, 1971, pp 347–400
6. Rushton R, Steinberg H: Mutual potentiation of amphetamine and amylobarbitone measured by activity in rats. Br J Pharmacol 21:295–305, 1963
7. Schacter S, Rodin J (eds): Obese Humans and Rats. Washington D.C., Erlbaum/Wiley, 1974
8. Mello NK: A review of methods to induce alcohol addiction in animals. Pharmacol Biochem Behav 1:89–101, 1973
9. Falk JL, Samson HH, Winger G: Behavioral maintenance of high concentrations of blood ethanol and physical dependence in the rat. Science 177:811–813, 1972
10. Silbergeld EK, Goldberg AM: Lead-induced behavioral dysfunction: An animal model for hyperactivity. Exp Neurol 42:146–157, 1974
11. Robbins TW, Sahakian BJ: Animal models of mania, in Mania: An Evolving Concept. Lancaster, MTP Press, 1980, pp 143–216
12. Kornetsky C: Animal models: Promises and problems, in Hanin I, Usdin E (eds): Animal models in psychiatry and neurology. Oxford, Pergamon, 1977, pp 1–9
13. Conger JJ: The effects of alcohol on conflict behavior in the albino rat. Q J Studies Alcohol 12:1–29, 1951
14. Stolerman IP, Baldy RE, Shine PJ: Drug discrimination procedure: A bibliography, in Colpaert FC, Slangen JL (eds), Drug discrimination: Applications in CNS pharmacology. Amsterdam, Elsevier, 1982, pp 401–448
15. Stolerman IP, Shine PJ: Trends in drug discrimination research analyzed with a cross-indexed bibliography. Psychopharmacology 86:1–11, 1985
16. Catania AC (ed): Contemporary Research in Operant Behavior. Glenview, Scott, Foresman, 1968
17. Lal H, Yeiden S: Discriminative stimuli produced by clonidine in spontaneously hypertensive rats: Generalization to antihypertensive drugs with different mechanisms of action. J Pharmacol Exp Ther 232:33–39, 1985
18. Seiden LS, Dykstra LA: Psychopharmacology: A biochemical and behavioral approach. New York, Van Nostrand–Reinhold, 1977
19. Colpaert FC: Discriminative stimulus properties of narcotic analgesic drugs. Pharmacol Biochem Behav 9:863–887, 1978
20. Overton DA: Applications and limitations of the drug discrimination method for the study of drug abuse, in

Bozarth MA (ed), Methods of Assessing the Reinforcing Properties of Abused Drugs. New York, Springer-Verlag, 1987, pp 291–340
21. Barry H: Classification of drugs according to their discriminable effects in rats. Fed Proc 33:1814–1824, 1974
22. Stolerman IP, Stephensen JD, Rauch RJ: Classification of opioid and 5-hydroxytryptamine receptors by means of discriminative drug effects. Neuropharmacology 26:867–875, 1987
23. Glennon RA, Young R: The study of structure activity relationships using drug discrimination methodology, in Bozarth MA (ed), Methods of Assessing the Reinforcing Properties of Abused Drugs. New York, Springer-Verlag, 1987, pp 373–390
24. Holtzman SG: Drug discrimination studies. Drug Alcohol Depend 14:263–282, 1985
25. Jasinski DR, in Martin WR (ed): Handbook of Experimental Pharmacology. Berlin, Springer-Verlag, 1977, pp 197–225
26. Haertzen CA: Subjective effects of narcotic antagonists cyclazocine and nalorphine on the Addiction Research Center Inventory (ARCI). Psychophamacologia 18:366–377, 1970
27. Zenick H, Greene JE: Statistical and methodological considerations in drug–stimulus discrimination learning, in Ho BT, Richards DW, Chute DL (eds), Drug discrimination and state dependent learning. New York, Academic, 1978, pp 203–226
28. Riley AL, Tuck DL: Conditioned food aversions: A bibliography, in Braveman NS, Bronstein P (eds), Experimental Assessments and Clinical Applications of Conditioned Food Aversions. New York, New York Academy of Sciences, 1985, pp 381–437
29. Mastropaolo J, Moskowitz KH, Dacanay RJ, Riley AL: Conditioned taste aversions in rats as a behavioral baseline for the assessment of the stimulus properties of phencyclidine. Soc Neurosci Abs 12:912, 1986
30. Mastropaolo J, Moskowitz KH, Dacanay RJ, Riley AL: Conditioned taste aversions as a behavioral baseline for drug discrimination learning: An assessment with phencyclidine. Pharmacol Biochem Behav 32:1–8, 1989
31. Riley AL, Tuck DL: Conditioned taste aversions: A behavioral index of toxicity, in Braveman NS, Bronstein P (eds), Experimental Assessments and Clinical Applications of Conditioned Food Aversions. New York, New York Academy of Sciences, 1985, pp 272–292
32. Brady KT, Balster RL: Discriminative stimulus properties of ketamine stereoisomers in phencyclidine-trained rats. Pharmacol Biochem Behav 17:291–295, 1982
33. Kautz MA, Geter B, McBride SA, et al: Naloxone as a stimulus for drug discrimination learning. Drug Dev Res 16:317–326, 1989
34. Overton DA, Batta SK: Investigation of narcotics and antitussives using drug discrimination techniques. J Pharmacol Exp Ther 211:401–408, 1979
35. Carter RB, Leander JD: Discriminative stimulus properties of naloxone. Psychopharmacology 77:305–308, 1982
36. Lucki I, South JA, Berger R: Rapid detection of the stimulus properties of 5-hydroxytryptamine (5-HT) agonists. Soc Neurosci Abs 13:344, 1987
37. Martin GM, Gans M, van der Kooy D: Discriminative properties of morphine that modify associations between tastes and LiCl. Soc Neurosci Abs 1987
38. Riley AL, Jeffreys RJ, Pournaghash S, et al: Conditioned taste aversions as a behavioral baseline for drug discrimination learning: An assessment with the dipsogenic compound pentobarbital. Drug Dev Res 16:229–236, 1989

7

The Platelet as a Peripheral Model of Serotonergic Function in Child Psychiatry

Moshe Rehavi and Ronit Weizman

1. INTRODUCTION

Neurochemical strategies have been adopted to implicate specific neurotransmitters in psychiatric disorders and the therapeutic mechanisms of psychotropic agents. These studies have examined alterations in neurotransmitter metabolites in urine, plasma, and spinal fluid as a reflection of the release and turnover of specific neurotransmitters in the brain (a marker of presynaptic function), and binding to specific receptor sites (a marker of postsynaptic function). With the introduction of positron-emission tomography (PET), it has become feasible to evaluate the distribution and density of at least some neurotransmitter receptors in vivo in patients and healthy normal control subjects. Another strategy has involved the identification of specific neurotransmitter and psychotropic drug receptor sites on easily accessible peripheral tissues. A major limitation of these latter studies is the assumption that the binding characteristics of these "peripheral" binding sites reflects the situation in brain accurately.

Tricyclic antidepressants (TCAs) are widely used and are very effective in treating depression, although the precise mechanisms of action responsible for their beneficial therapeutic effects are still unclear. These drugs interact with many membranal sites in the brain and periphery, including pre- and postsynaptic receptors and reuptake sites. TCAs have been shown to inhibit the presynaptic active neuronal uptake of norepinephrine and serotonin.[1–3] This activity was suggested to be their major mechanisms of action. Chronic administration of TCA decreases the number and sensitivity of β-adrenergic receptors, though they have little affinity for these receptors. Long-term treatment of rodents with TCAs induces a reduction in the sensitivity of the β-adrenergic receptor coupled adenylate cyclase in the limbic forebrain, and a reduction in the maximal density of β-adrenergic and serotonergic receptor binding sites in cerebral cortex.[4–7]

Tricyclic antidepressants are able to inhibit competitively and with high affinity the

Moshe Rehavi • Department of Physiology and Pharmacology, Sackler Faculty of Medicine, Tel Aviv University, Tel Aviv 69978, Israel. *Ronit Weizman* • Pediatric Department, Hasharon Hospital, Petah Tiqva 49372; and Sackler Faculty of Medicine, Tel Aviv University, Tel Aviv 69978, Israel.

binding of radiolabeled receptor antagonists to several classes of neurotransmitter receptors in the central nervous system. These receptors include the muscarinic cholinergic,[8,9] histaminergic,[10,11] α-adrenergic,[12] and S1- and S2-serotonergic[13] receptors. The high affinity of TCAs for these receptors can explain their side effect spectrum, such as sedation, orthostatic hypotension, constipation and tachycardia. However, high-affinity binding to the above sites does not account for their therapeutic effects.[8,9]

Using tritium-labeled imipramine, specific high-affinity binding sites for TCAs were studied and characterized in rat and human brain.[14,15,25] Interestingly, binding sites with similar characteristics were also found on the membranes of human platelets.[16–18] The [^{3}H]imipramine binding was suggested to label the serotonin reuptake site in both brain and platelet membranes[18,19] and seems to be a useful and valuable tool for studying the serotonergic transport system.

The possible pharmacological relationship between the imipramine binding site in brain and platelet encouraged the use of this peripheral "marker" on platelets in clinical research designed to implicate this site in psychiatric disorders and the therapeutic action of TCAs. Several studies suggested that platelets can serve as a peripheral model for aminergic nerve endings.[20–22] Serotonin is concentrated in the platelet and presynaptic nerve terminals by an active transport mechanism which can be blocked by TCAs. The two tissues also store serotonin in specific storage granules[21] that protect it from catabolism by monoamine oxidases. Under basal conditions, serotonin is transported into platelets and synaptosomes by this saturable high-affinity low-capacity active transport system and nonsaturable passive diffusion.[22] At low serotonin concentrations, the kinetics of serotonin uptake into the platelet is explained by the active membrane transport system. Similarly, serotonin transport into presynaptic nerve terminals is explained by saturable active uptake that is inhibited by TCAs in a competitive manner.[22]

2. [^{3}H]IMIPRAMINE BINDING SITES

High-affinity (1.5 to 8-nM) stereoselective and saturable binding sites for [^{3}H]imipramine were first demonstrated in brain[14,23,25] and later in human platelets.[16,17] These binding sites are unevenly distributed in the brain: the highest densities are found in the hypothalamus and cerebral cortex, while the cerebellum has a very low density of binding sites.[23,24,25] Scatchard analysis revealed that these sites constitute a single homogeneous population of high-affinity sites[14,17,25] that are displaceable by an excess of unlabeled desipramine.[15] A low-affinity site (micromolar) was also identified,[26] but its importance is as yet unknown. High-affinity imipramine binding sites in brain membranes and human platelets appear to have similar characteristics. [^{3}H]Imipramine binding is inhibited in a competitive and stereospecific manner by other TCAs.[15,17,18,25] Atypical nontricyclic antidepressants and various neurotransmitters have weak potencies in inhibiting [^{3}H]imipramine binding. There is a good correlation between the potencies of the various drugs in inhibiting imipramine binding in human brain and platelets.[17,27] A simultaneous decrease in the density of imipramine binding sites was observed in cat brain and platelets after chronic treatment with imipramine.[28]

The rank order for the inhibition of [^{3}H]imipramine binding by a series of TCAs and other drugs was similar to that of the inhibition of serotonin uptake by these same drugs. This highly significant correlation that was found in both platelets and brain suggested that the high-affinity [^{3}H]imipramine binding site might be related to the reuptake site for serotonin in brain and platelets.[15,17,19,27]

Several studies confirmed a possible functional relationship between [^{3}H]imipramine binding sites and serotonin uptake. Fawn-Hooded rats show a hereditary deficiency in platelet serotonin content and storage[29] that was accompanied by an absence of [^{3}H]imipramine binding sites.[30] A parallel development of high-affinity serotonin uptake and imipramine binding sites was demonstrated in developing rats.[31] Electrolytic lesions of the ascending raphe nuclei caused a parallel decrease of [^{3}H]serotonin uptake and [^{3}H]imipramine binding,[19] as well as a reduction in endogenous serotonin levels.[32] Lesioning rat brain with 5,7-dihydroxytryptamine, a specific serotonergic neurotoxin, induced a simultaneous decrease in cortical [^{3}H]imipramine binding and serotonin uptake.[33,34] The "slowly dissociating" ligand 2-nitroimipramine inhibited [^{3}H]imipramine binding and [^{3}H]serotonin uptake in a parallel fashion.[35–37] Autoradiographic studies have also shown that [^{3}H]nitroimipramine binds to brain regions enriched in serotonin.[39] In addition, [^{3}H]imipramine binding sites were shown to be localized to presynaptic serotonergic nerve terminals.[38,39] [^{3}H]Imipramine binding shares a sodium dependency with serotonin uptake.[40–42] The sodium dependency permitted the differentiation of the specific high-affinity [^{3}H]imipramine binding site from the low-affinity site. Subcellular fractionation showed that the distribution of [^{3}H]imipramine binding sites and serotonin uptake resided mainly in the mitochondrial (synaptosomal) fraction with very little binding in nuclear, myelin, or microsomal fractions.[43,44] These data also support a probable functional relationship between [^{3}H]imipramine binding sites and serotonin uptake sites. However, pharmacological and thermodynamic approaches demonstrate that the [^{3}H]imipramine binding site is not identical to the transport recognition site for serotonin. The most recent conceptualization suggests that the high-affinity binding site for [^{3}H]imipramine is coupled allosterically to the serotonin carrier site.[45–47]

3. [^{3}H]IMIPRAMINE BINDING IN AFFECTIVE DISORDERS

The correlation between imipramine binding parameters and the pharmacological profile of the binding in brain and platelets prompted investigations of [^{3}H]imipramine binding in the platelets of depressed patients. A decrease in the density of [^{3}H]imipramine binding sites in platelets of untreated severely depressed patients was reported by several investigators.[48–57] However, other studies did not find differences in imipramine binding between depressed patients and controls,[58–60] and one study demonstrated an increase in these sites in depressed bipolar patients.[60] The discrepancies in the findings could probably be explained by heterogeneity within diagnostic categories of depression, severity of depression, age and drug effects, as well as methodological differences in the preparation of the platelets and binding assay.

[^{3}H]Imipramine binding to the suprachiasmatic nuclei of the anterior hypothalamus in rats was found to be highest at the end of the dark phase and lowest at the end of the light phase.[62] A similar circadian rhythm of platelet serotonin uptake was observed in normal volunteers. Platelet [^{3}H]imipramine binding was also shown to vary with circardian annual rhythm, being highest in January and lowest in September.[63,64] A later study could not confirm seasonal variation in this parameter.[65] Decreased [^{3}H]imipramine binding was observed in the frontal cortex and hypothalamus of suicide victims.[66]

Several investigators have reported a downregulation of the platelet [^{3}H]imipramine binding site after antidepressant treatment. Long-term administration of TCAs to rats induces a decrease in the maximal binding capacity (B_{max}) of [^{3}H]imipramine binding in cortex.[66] A similar TCA-induced decrease in B_{max} was observed in cat brain and platelet.[68] Such an alteration was not obtained by atypical antidepressant treatment.[66] Five weeks of lithium treatment downregulated the number of [^{3}H]imipramine binding sites.[69] Twenty-one days of electroconvulsive shock administered to rats induced a decrease in the B_{max} of [^{3}H]imipramine in rat cerebral cortex.[70] Early onset of rapid eye movement (REM) sleep is known to occur in depression[71] and REM sleep deprivation is associated with antidepressive effects.[72] Seventy-two hours of REM sleep deprivation was reported to downregulate [^{3}H]imipramine binding sites in rat cerebral cortex.[67]

Female depressed patients showed a more significant reduction in [^{3}H]imipramine binding suggesting that gender may be involved in the regulation of this site.[73,57] Bipolar patients show the most prominent reduction.[74,57] Aging is associated with an increase in the maximal density of [^{3}H]imipramine binding sites in mouse and human brain.[75–77] However, platelet [^{3}H]imipramine binding in aged depressed patients was reported to be reduced, and the magnitude of this reduction seemed greater than in younger depressed patients.[78]

Platelet [^{3}H]imipramine binding may serve as a possible predictor of response to antidepressant treatment,[79] as an indicator of the severity of depression,[80] and as a biochemical marker to distinguish among subtypes of depression.[74]

[^{3}H]Imipramine binding and serotonin uptake were studied in our laboratory in prepubertal children and adolescents with major affective disorders and other neuropsychiatric disorders not thought to relate to affective illness per se, but thought to involve serotonergic mechanisms.

The manifestations of affective disorder in children and adolescents are not always identical to those in adults. In view of the decreased platelet [^{3}H]imipramine binding reported in adult depressed patients, a similar decrease in platelet [^{3}H]imipramine binding was expected in depressed children and adolescents.[81] However, the maximal density of these sites in the depressed children did not differ significantly either from the nonaffective patients (including 13 children and adolescents suffering from conduct disorder and schizophrenia) or from age and sex matched normal controls. Furthermore, [^{3}H]imipramine binding did not distinguish between patients with unipolar and bipolar affective disorder, and could not discriminate between patients with a positive family history for depression and no family history for this disorder. It might be that the young population included in this research had not been exposed to the disease for a long enough period to obtain the modulation of platelet [^{3}H]imipramine binding sites observed in adult depressive patients. Alternatively, the decreased binding in adult depressed subjects and the unaltered [^{3}H]imipramine binding in children and adolescents could indicate that symptoms of depression in youngsters do not necessarily imply endogenous depression of the adult type. This possibility is supported by the lack of REM sleep abnormality in children suffering from acute depressive episodes[82,83] and the inability of the dexamethasone suppression test (DST) to discriminate between major depressive and other psychiatric disorders in this age group.[84,85]

4. [^{3}H]IMIPRAMINE BINDING AND SEROTONIN UPTAKE IN OBSESSIVE-COMPULSIVE DISORDER

Obsessive-compulsive disorder (OCD) is a psychiatric entity associated frequently with depressive symptoms. A serotonergic dysfunction has been implicated in the pathophysiology of this disorder. The therapeutic effect of clomipramine and L-tryptophan is consistent with hypothesized serotonergic dysfunction in OCD.[86–88] Clomipramine appears to be effective also in childhood OCD. The reported correlation between the beneficial therapeutic effect of clomipramine and reduced 5-hydroxyindoleacetic acid (5-HIAA) levels in cerebrospinal fluid[89] also supports the serotonergic hypothesis. Furthermore, a significant correlation between clinical improvement during cloimipramine treatment and reduction in platelet serotonin levels in children with OCD was recently demonstrated.[90] A positive correlation between reduction in OCD symptoms and plasma levels of clomipramine was also reported.[96]

Obsessive-compulsive disorder patients often develop depressive symptoms, and several neurobiological indicators of depression have been detected, such as abnormal DST,[91] shortened REM latency (but not density),[92,93] and blunted human growth hormone response to clonidine.[94,95]

In a study conducted in our laboratory,[97] [^{3}H]imipramine binding and serotonin uptake in platelets of eight adolescent and 10 adult OCD patients were evaluated and compared with normal age- and sex-matched controls. All the adults in this study were drug free for at least 1 year, whereas none of the adolescents had ever been treated with antidepressants. The study showed a significantly lower number of [^{3}H]imipramine binding sites in platelets of adolescent and adult OCD patients as compared with control subjects. Serotonin uptake did not differ between patients and controls. The finding of reduced platelet imipramine binding in OCD is not in accordance with a previous study[98] reporting unaltered [^{3}H]imipramine binding and serotonin uptake in adult OCD patients. The fact that in our study[97] the decrease in platelet [^{3}H]imipramine binding was not accompanied by a parallel reduction in serotonin uptake, as was shown in depression, supports the assumption that OCD is not a subgroup of depression but a separate biological and clinical entity. A psychobiological link between the two disorders is to be considered, although the nature of a link, if any, is yet unclear.

5. PLATELET IMIPRAMINE BINDING IN CONDUCT DISORDER

Low 5-HIAA in CSF of patients with behavioral problems was reported by several investigators. This parameter also correlated to violent suicide attempts and aggressiveness.[98,99] We evaluated platelet imipramine binding in 10 adolescent patients suffering from conduct disorder. The subjects were part of a control group (nonaffective patients) in a study conducted on affective disorders in children and adolescents.[81] [^{3}H]Imipramine binding in the platelets of the patients did not differ from normal controls. In contrast to our finding, a recent study[100] reports a reduction of [^{3}H]imipramine binding sites on platelets of conduct-disordered children. The controversial results could probably be explained by differences in platelet isolation and laboratory procedures. As mentioned by the investigators of this recent study, methodologic variables are known to influence platelet [^{3}H]imipramine binding parameters.

6. IMIPRAMINE BINDING AND SEROTONIN UPTAKE IN SCHIZOPHRENIA AND AUTISM

Alteration in serotonin uptake to platelets of schizophrenics was observed by some investigators,[101–103] although this result has not been replicated by others.[104,105] [^{3}H]Imipramine binding in platelets from schizophrenic patients did not differ from controls[106,107]; this parameter could not discriminate between exacerbated and remitted patients or severity of symptoms.[107] The involvement of the serotonergic system in the pathopathology of autism was demonstrated by various investigators. Hyperserotonemia was reported in a subgroup of autistic and severely retarded children,[108,109] and there is a familial resemblance between serotonin levels of autistics and their first degree relatives. Low CSF concentrations of homovanillic acid (HVA) and 5-HIAA, the main metabolites of dopamine and serotonin, were demonstrated in autistic patients.[110] Decreased CSF 5-HIAA accumulation following probenecid,[109] abnormal DST,[111] and decreased excretion of 24-hr urinary MHPG[112] were also reported in autistic patients. Moreover, suppression of blood serotonin levels by fenfluramine treatment was demonstrated to be in correlation with amelioration of autistic symptoms.[113,114] Alteration in serotonin uptake to platelets in autism was demonstrated by several investigators,[115–117] but a later study did not replicate this finding.[118]

Platelet imipramine binding in drug-free autistic subjects did not differ from normal controls.[119] We studied platelet imipramine binding in neuroleptic-treated autistic patients aged 15–24 years, as compared with treated schizophrenic patients and normal controls.[120] No difference in maximal binding capacity (B_{max}) and in equilibrium dissociation constant (K_d) was observed. The results are in accordance with previous studies on imipramine binding[103,106] or autism.[119] The lack of effect of neuroleptics on this binding site was also demonstrated previously.[103] The lack of alteration in platelet imipramine binding in autistic patients does not ultimately exclude the possibility of alteration in the activity of the central serotonergic system. For example, abnormal endocrine response to insulin stress in autistics may indicate a dysfunction of the central serotonergic system.[121] Also, abnormal cell-mediated autoimmune response toward myelin basic protein,[98] the protein that contains the serotonin binding site in its tryptophan peptide region,[122] was demonstrated in a subgroup of autistics. Furthermore, circulating antibodies directed against human brain serotonin receptors were detected in a subgroup of autistic patients.[123]

7. IMIPRAMINE BINDING AND SEROTONIN UPTAKE IN ENURESIS

Tricyclic antidepressants have a beneficial effect in treating nocturnal enuresis. The etiology of the disorder as well as the mode of action of tricyclics is as yet not clear. The antidepressive effects of these drugs requires 2–4 weeks, while the antienuretic effect occurs within days. A possible alteration of serotonergic turnover was suggested in enuresis.[124] The efficacy of imipramine treatment in functional nocturnal enuresis along with the assumption of possible serotonergic involvement in the pathogenesis of this disorder led us to investigate [^{3}H]imipramine binding and [^{3}H]serotonin uptake in platelets of enuretic children and adolescents.[125,126] A significant reduction in the density of platelet imipramine binding sites was demonstrated in enuretics as compared with controls. Neither the dissociation constant for [^{3}H]imipramine binding nor the serotonin uptake kinetic parameters showed any alterations. The reduction in platelet imipramine binding in the enuretic patients does not seem to be related to depression, since none of the patients had ever suffered from a major depressive episode and none reported depression in first-degree relatives. The lack of correlation between imipramine binding and serotonin uptake values suggests that the binding site for imipramine and the site for serotonin uptake might not be identical.

REFERENCES

1. Glowinski, J, Axelrod J: Inhibition of uptake of tritiated-noradrenaline in the intact rat brain by imipramine and structually related compounds. Nature (Lond) 204:1318–1319, 1964
2. Iversen LL: Uptake mechanism for neurotransmitter amines. Biochem Pharmacol 23:1927–1935, 1974
3. Carlsson A, Corrodi H, Fuxe K, et al: Effect of antidepressant drugs on the depletion of intraneuronal brain 5-hydroxytryptamine stores caused by 4-methyl-alpha-ethyl-meta tyramine. Eur J Pharmacol 5:357–366, 1969
4. Sulser F, Vetulani J, Hobley PL: Mode of action of antidepressant drugs. Biochem Pharmacol 27:257–261, 1978
5. Banerjee SP, Kung LS, Riggs SJ, et al: Development of β-adrenergic receptor subsensitivity by antidepressants. Nature (Lond) 268:455–456, 1977
6. Wolfe BB, Harden TK, Sporn JR, et al: Presynaptic modulation of β-adrenergic receptors in rat cerebral cortex after treatment with antidepressants. J Pharmacol Exp Ther 207:446–457, 1978
7. Peroutka SJ, Synder SH: Long-term antidepressant treatment decreases spirperidol-labeled serotonin receptor binding. Science 210:88–90, 1980
8. Rehavi M, Maayani S, Sokolovsky M: Tricyclic antidepressants as antimuscarinic drugs: In vivo and in vitro studies. Biochem Pharmacol 26:1559–1567, 1977

9. Synder SH, Yamamura HI: Antidepressants and the muscarinic acetylcholine receptor. Arch Gen Psychiatry 34:236–239, 1977
10. Green JP, Maayani S: Tricyclic antidepressants block histamine H_2 receptor in brain. Nature (Lond) 269:163–165, 1977
11. Peroutka SJ, Synder SH: [^{3}H]-Mianserin: Differential labeling of serotonin$_2$ and histamine receptors in rat brain. J Pharmacol Exp Ther 216:142–148, 1981
12. U'Prichard DC, Greenberg DA, Sheehan PP et al: Tricyclic antidepressants: Therapeutic properties and affinity for α-noradrenergic receptor binding sites in the brain. Science 199:197–198, 1978
13. Bennett JL, Aghajanian GK: D-LSD binding to brain homogenates: Possible relationship to serotonin receptors. Life Sci 15:1935–1944, 1975
14. Raisman R, Briley M, Langer SZ: Specific tricyclic antidepressant binding sites in rat brain. Nature (Lond) 281:148–150, 1979
15. Raisman R, Briley, M, Langer SZ: Specific tricyclic antidepressant binding sites in rat brain characterized by high affinity [^{3}H]imipramine binding. Eur J Pharmacol 61:373–380, 1980
16. Briley M, Raisman R, Langer SZ: Human platelets possess high affinity binding sites for [^{3}H]imipramine. Eur J Pharmacol 58:347–348, 1979
17. Paul SM, Rehavi M, Skolnick P, et al: Demonstration of specific high affinity binding sites for [^{3}H]imipramine on human platelets. Life Sci 26:953–959, 1980
18. Langer SZ, Briley M, Raisman R, et al: Specific [^{3}H]imipramine binding in human platelets. Naunyn-Schmiedebergs Arch Pharmacol 313:189–194, 1980
19. Paul SM, Rehavi M, Rice KC, et al: Does high affinity [^{3}H]imipramine binding label serotonin reuptake sites in brain and platelet? Life Sci 28:2753–2760, 1981
20. Pletscher A: Metabolism, transfer and storage of 5HT in blood platelets. Br J Pharmacol 32:1–16, 1968
21. Sneddon JM: Blood platelets as a model for monoamine-containing neurons. Prog Neurobiol 1:151–198, 1973
22. Stahl SM: The human platelet: A diagnostic and research tool for the study of biogenic amines in psychiatric and neurologic disorders. Arch Gen Psychiatry 34:509–516, 1977
23. Langer SZ, Agid FJ, Raisman R et al: Distribution of specific high-affinity binding sites for [^{3}H]imipramine in human brain. J Neurochem 37:267–271, 1981
24. Palkovitz M, Raisman R, Briley M, et al: Regional distribution of [^{3}H]imipramine binding in rat brain. Brain Res 210:493–498, 1981
25. Rehavi M, Paul SM, Skolnick P, et al: Demonstration of specific high affinity binding sites for [^{3}H]imipramine in human brain. Life Sci 26:2273–2279, 1980
26 Reith MEA, Sershen H, Allen D, et al: High and low affinity of [^{3}H]imipramine in mouse cerebral cortex. J Neurochem 40:389–392, 1983
27. Langer SZ, Moret C, Raisman R, et al: High affinity [^{3}H]imipramine binding in rat hypothalamus. Association with uptake of serotonin but not of norepinephrine. Science 210:1133–1135, 1980
28. Briley M, Raisman R, Arbilla S, et al: Concommittant decrease in [^{3}H]imipramine binding in cat brain and platelets after chronic treatment with imipramine. Eur J Pharmacol 81:309–314, 1982
29. DeParda M, Pieri L, Keller H, et al: Effects of 5,6-dihydroxytryptamine and 5,7-dihydroxytryptamine on rat central nervous system after intraventricular or intracerebral application and on blood platelets in vitro. Ann NY Acad Sci 305:595–603, 1978
30. Dumbrille-Ross A, Tang SW: Absence of high-affinity [^{3}H]imipramine binding in platlets and cerebral cortex of Fawn-Hooded rats. Eur J Pharmacol 72:137–138, 1981
31. Mocchetti I, Brunello N, Racagni G: Ontogenetic study of [^{3}H]imipramine binding sites and serotonine uptake system: Indication of possible interdependence. Eur J Pharmacol 83:151–154, 1982
32. Sette M, Raisman R, Briley M, et al: Localization of tricyclic antidepressant binding sites on serotonin nerve terminals. J Neurochem 37:40–42, 1981
33. Gross G, Gothert M, Ender HP, et al: [^{3}H]imipramine binding sites in the rat brain selective localization on serotonergic neurons. Naunyn-Schmiedebergs Arch Pharmacol 317:310–314, 1981
34. Brunello J, Chaung DM, Costa E: Different synaptic location of mianserin and imipramine binding sites. Science 215:1112–1115, 1982
35. Rehavi M, Ittah Y, Rice KC, et al: 2-Nitroimipramine: A selective irreversible inhibitor of [^{3}H]serotonin uptake and [^{3}H]imipramine binding in platelets. Biochem Biophys Res Commun 99:954–959, 1981
36. Rehavi M, Ittah Y, Skolnick P, et al: Nitroimipramines-synthesis and pharmacological effects of potent long-acting inhibitors of [^{3}H]serotonin uptake and [^{3}H]imipramine binding. Naunyn-Schmiedebergs Arch Pharmacol 320:45–49, 1982

37. Rehavi M, Tracer H, Rice KC, et al: [^{3}H]2-Nitroimipramine: A selective "slowly-dissociating" probe of the imipramine binding site ("serotonin transporter") in platelets and brain. Life Sci 32:645–653, 1983
38. Dawson TM, Wamsley JK: Autoradiographic localization of [^{3}H]imipramine binding sites: Association with serotonergic neurons. Brain Res Bull 11:325–329, 1983
39. Biegon A, Rainbow TC: Distribution of imipramine binding sites in the rat brain studied by quantitative autoradiography. Neurosci Lett 37:209–214, 1983
40. Rudnick G: Active transport of 5-hydroxytryptamine by plasma membrane vescicles isolated from human blood platelets. J Biol Chem 252:2170–2174, 1977
41. Briley M, Langer SZ: Sodium dependency of [^{3}H]imipramine binding in rat cerebral cortex. Eur J Pharmacol 72:377–380, 1981
42. Lingjaerde D: Blood platelets: As a model system for studying the biochemistry of depression, in Usdin E (ed): Frontiers in Biochemical and Pharmacological Research in Depression. New York, Raven, pp 99–111
43. Kinnier WJ, Chang DM, Gwynn G, et al: Characteristics and regulation of high affinity [^{3}H]imipramine binding to rat hippocampal membranes. Neuropharmacology 20:411–419, 1981
44. Rehavi M, Skolnick P, Paul SM: Subcellular distribution of high affinity [^{3}H]imipramine binding and [^{3}H]serotonin uptake in rat brain. Eur J Pharmacol 87:335–339, 1983
45. Sette M, Briley MS, Langer SZ: Complex inhibition of [^{3}H]imipramine binding by serotonin and non-tricyclic serotonin uptake blockers. J Neurochem 40:622–628, 1983
46. Wennogle LP, Meyerson LR: Serotonin uptake inhibitors differentially modulate high affinity imipramine dissociation in human platelet membranes. Life Sci 361:1541–1550, 1985
47. Segonzac A, Schoemaker A, Langer SZ: Temperature dependence of drug interaction with the platelet 5-hydroxytryptamine transporter. J Neurochem 48:331–339, 1987
48. Briley MS, Langer SZ, Raisman R, et al: Tritiated imipramine binding sites are decreased in platelets of untreated depressed patients. Science 209:303–305, 1980
49. Asrach KB, Shin JC, Kulcsar A: Decreased [^{3}H]imipramine binding in depressed male and females. Common Psychopharmacol 4:425–432, 1980
50. Paul SM, Rahavi M, Skolnick P, et al: Depressed patients have decreased binding of tritiated imipramine to platelet serotonin "transporter." Arch Gen Psychiatry 38:1315–1317, 1981
51. Raisman, R, Sechter, D, Briley, MS, et al: High affinity [^{3}H]imipramine binding to platelets from untreated and treated depressed patients compared to healthy volunteers. Psychopharmacology 75:368–371, 1981
52. Raisman R, Briley MS, Bouchami F, et al: [^{3}H]imipramine binding and serotonin uptake in platelets from untreated depressed patients and control volunteers. Psychopharmacology 77:332–335, 1982
53. Suranyi-Cadotte BE, Wood PL, Nair NPV, et al: Normalization of platelet [^{3}H]imipramine binding in depressed patients during remission. Eur J Pharmacol 85:357–358, 1982
54. Suranyi-Cadotte BE, Wood PL, Schwartz G, et al: Altered platelet [^{3}H]imipramine binding in schizoaffective and depressive disorders. Biol Psychiatry 18:923–927, 1983
55. Wagner A, Aberg-Wistedt A, Asberg M, et al: Lower [^{3}H]imipramine binding in platelets from untreated depressed patients compared to healthy controls. Psychiatry Res 16:131–139, 1985
56. Langer SZ, Galzin AM, Poirier MF, et al: Association of [^{3}H]imipramine and [^{3}H]paroexetine binding with the 5HT transporter in brain and platelets: relevance to studies in depression. J Receptor Res 7:499–521, 1987
57. Roy A, Everett D, Pickar D, et al: Platelet tritiated imipramine binding and serotonin uptake in depressed patients and controls. Arch Gen Psychiatry 44:320–327, 1987
58. Baron M, Barkai A, Gruen R, et al: [^{3}H]imipramine platelet binding sites in unipolar depression. Biol Psychiatry 18:1403–1409, 1983
59. Whitaker PM, Warsh JJ, Stancer HC, et al: Seasonal variation in platelet [^{3}H]imipramine bindings: Comparable values in control and depressed populations. Psychiatry Res 11:127–131, 1984
60. Muscettola G, DiLauro A, Giannini C: Platelet [^{3}H]imipramine binding in bipolar patients. Psychiatry Res 18:343–353, 1986
61. Mellrup ET, Plenge P, Rosenberg R: [^{3}H]Imipramine binding sites in platelets from psychiatric patients. Psychiatry Res 7:221–227, 1982
62. Wirz-Justice A, Krauchie K, Morimasa T, et al: Circadian rhythm of [^{3}H]imipramine binding in the rat suprachiamatic nuclei. Eur J Pharmacol 87:331–333, 1983
63. Egrise D, Desmedt D, Schoutens A, et al: Circannual variations in the density of tritiated imipramine binding sites on blood platelets in man. Neuropsychobiology 10:101–102, 1983

64. Gulzin AM, Loo H, Sechter D, et al: Lack of seasonal variation in platelet [^{3}H]imipramine binding in humans. Biol Psychiatry 21:876–882, 1986
65. Stanley M, Virgilio J, Gershon S: Tritiated imipramine binding sites are decreased in the frontal cortex of suicides. Science 216:137–139, 1982
66. Kinnier WJ, Chang DM, Gwynn G, et al: Down regulation of dihydroalprenolol and imipramine binding sites in brain of rats repeatedly treated with imipramine. Eur J Pharmacol 67:289–294, 1980
67. Mogilnicka E, Arbilla S, Depoortere H, et al: Rapid eye movement sleep deprivation decreases the density of [^{3}H]dihydroalprenolol and [^{3}H]imipramine binding sites in the rat cerebral cortex. Eur J Pharmacol 65:289–293, 1980
68. Plenge P, Mellrup ET: [^{3}H]imipramine high-affinity binding sites in rat brain. Effects of imipramine and lithium. Psychopharmacology 77:94–97, 1982
69. Langer SZ, Zarifian E, Briley M, et al: High affinity [^{3}H]imipramine binding in: a new biological marker in depression. Pharmacopsychiatry 15:4–10, 1982
70. Kupfer D: REM latency: A psychobiologic marker for primary depressive disease. Biol Psychiatry 11:159–174, 1976
71. King D: Pathologic and therapeutic consequences of sleep loss: A review. Dis Nerv Syst 38:843–879, 1977
72. Poirier M-F, Benkelfat C, Loo H, et al: Reduced B_{max} of [^{3}H]imipramine binding to platelets of depressed patients free of previous medication with 5HT uptake inhibitors. Psychopharmacology 89:456–461, 1986
73. Lewis DA, McChesney C: Tritiated imipramine binding distinguishes among subtypes of depression. Arch Gen Psychiatry 42:485–488, 1985
74. Severson JA, Marcusson JO, Osterburg HH, et al: Elevated density of [^{3}H]imipramine binding in aged human brain. J Neurochem 45:1382–1389, 1985
75. Severson JA: [^{3}H]imipramine binding in aged mouse brain: Regulation by ions and serotonin. Neurobiol Aging 7:83–87, 1986
76. Brunello N, Riva M, Volterra A, et al: Aged-related changes in 5HT uptake and [^{3}H]imipramine binding sites in rat cerebral cortex. Eur J Pharmacol 110:393–394, 1985
77. Schneider LS, Severson JA, Sloane RB: Platelet [^{3}H]imipramine binding in depressed elderly patients. Biol Psychiatry 20:1232–1234, 1985
78. Hrdina PD, Lapierre YD, Horn ER, et al: Platelet [^{3}H]imipramine binding: A possible predictor of response to antidepressant treatment. Prog Neuropsychopharmacol Biol Psychiatry 9:619–623, 1985
79. Tanimoto K, Maeda K, Terada T: Alteration of platelet [^{3}H]imipramine binding in mildly depressed patients correlates with disease severity. Biol Psychiatry 20:340–343, 1985
80. Rehavi M, Weizman R, Carel C, et al: High affinity [^{3}H]imipramine binding in platelets of children and adolescents with major affective disorders. Psychiatry Res 13:31–39, 1984
81. Puig-Antich J, Gittelman R: Depression in childhood and adolescence, in Paykel ES (ed): Handbook of Affective Disorders. London, Churchill, 1981, p. 379–390.
82. Puig-Antich J: Neuroendocrine and sleep correlates of prepubertal major depressive disorder: Current status of the evidence, in Cantwell DP, Carlson GA (eds): Affective Disorders in Childhood and Adolescence: An Update. New York, Spectrum Publications, 1983, p 211–220.
83. Geller B, Rogol AD, Knitter EF: Preliminary data on the dexamethasone suppression test in children with major depressive disorder. Am J Psychiatry 140:620–623, 1983
84. Targum SD, Kapodanno AE: The dexamethasone suppression test in adolescent psychiatric inpatients. Am J Psychiatry 140:589–591, 1983
85. Yaryura-Tobias JA, Bhagavan HN: L-Tryptophan in obsessive-compulsive disorders. Am J Psychiatry 134:1298–1299, 1977
86. Insel TR, Murphy DL, Cohen RM, et al: Obsessive-compulsive disorder: A double blind trial of cloimipramine and clorgyline. Arch Gen Psychiatry 40:605–612, 1983
87. Flament MF, Rapport JL, Berg CJ, et al: Cloimipramine treatment of childhood obsessive-compulsive disorder. Arch Gen Psychiatry 42:977–983, 1985
88. Thoren P, Asberg A, Cronholm B, et al: Chlorimipramine treatment of obsessive-compulsive disorder. Arch Gen Psychiatry 37:1281–1285, 1980
89. Flament MF, Rapoport JL, Murphy DL, et al: Biochemical changes during clomipramine treatment of childhood obsessive-compulsive disorder. Arch Gen Psychiatry 44:219–225, 1987
90. Insel TR, Kalin N, Guttmacher LB, et al: The dexamethasone suppression test in patients with primary obsessive-compulsive disorder. Psychiatry Res 6:153–160, 1980
91. Rapoport J, Elkins R, Langer DH, et al: Childhood obsessive-compulsive disorder. Am J Psychiatry 138:1545–1554, 1981

92. Insel TR, Gillin C, Moore A, et al: The sleep of patients with obsessive compulsive disorder. Arch Gen Psychiatry 39:1372–1377, 1982
93. Siever LJ, Insel TR, Jimerson DC, et al: Growth hormone response to cloidine in obsessive-compulsive patients. Br J Psychiatry 39:1372–1377, 1982
94. Insel TR, Mueller EA, Gillin C: Biological markers in obsessive-compulsive and affective disorders. J Psychiatr Res 18:407–423, 1984
95. Stern RS, Marks IM, Mawson D: Cloimipramine and exposure for compulsive rituals. Plasma levels, side effects, and outcome. Br J Psychiatry 136:161–166, 1980
96. Weizman A, Carmi M, Hermesh H, et al: High affinity imipramine binding and serotonin uptake in platelets of eight adolescent and ten adult obsessive-compulsive patients. Am J Psychiatry 143:335–339, 1986
97. Insel TR, Mueller EA, Alterman I, et al: Obsessive-compulsive disorder and serotonin: Is there a connection? Biol Psychiatry 20:1174–1188, 1985
98. Weizman A, Weizman R, Szekely GA, et al: Abnormal immune response to brain tissue antigen in the syndrome of autism. Am J Psychiatry 139:1462–1465, 1982
99. Van Praag HM: CSF 5-HIAA and suicide in non-depressed schizophrenics. Lancet 2:977–978, 1983
100. Stoff DM, Pollock L, Vitiello B, et al: Reduction of [^{3}H]imipramine binding sites on platelets of conduct-disordered children. Neuropsychopharmacology 1:55–62, 1987
101. Modai I, Rotman A, Munitz H, et al: Serotonin uptake by blood platelets of acute schizophrenic patients. Psychopharmacology 64:193–195
102. Kaplan RD, Mann JJ: Altered platelet serotonin uptake kinetics in schizophrenia and depression. Life Sci 31:583–588, 1982
103. Wood PL, Suranyi-Cadotte BE, Nair NPV, et al: Lack of association between [^{3}H]imipramine binding sites and uptake of serotonin in control, depressed and schizophrenic patients. Neuropharmacology 22:1211–1214, 1983
104. Meltzer HY, Arora RC, Baber R, et al: Serotonin uptake in blood platelets of psychiatric patients. Arch Gen Psychiatry 38:1322–1326, 1981
105. Arora RC, Meltzer HY: Serotonin uptake by blood platelets of schizophrenic patients. Psychiatry Res 6:327–333, 1982
106. Gentesch C, Lichtsteiner M, Gastpar M, et al: [^{3}H]imipramine binding sites in platelets of hospitalized psychiatric patients. Psychiatry Res 14:177–187, 1985
107. Kanof PD, Coccaro EF, Johns CA, et al: Platelet [^{3}H]imipramine binding in psychiatric disorders. Biol Psychiatry 22:278–286, 1986
108. Hanley HG, Stahl SM, Freedman DX: Hyperserotonemia and amine metabolites in autistic and retarded children. Arch Gen Psychiatry 34:521–531, 1977
109. Young JG, Cohen DJ, Shaywitz BA, et al: Molecular pathology in early childhood psychoses, in Wing L. Wing J.K. (eds): Handbook of Psychiatry, Vol. 3: Psychoses of Uncertain Aetiology. Cambridge, Cambridge University Press, 1982, pp 229–235
110. Cohen DJ, Caparulo BK, Shaywitz BA, et al: Dopamine and serotonin metabolism in neuropsychiatrically disturbed children. Arch Gen Psychiatry 34:545–550, 1977
111. Jensen JB, Realmuto GM, Garfinkel BD: The dexamethasone suppression test in infantile autism. J Am Acad Child Psychiatry 3:262–265, 1985
112. Young JG, Cohen DJ, Caparulo BK, et al: Decreased 24-hour urinary MHPG in childhood autism. Am J Psychiatry 136:1055–1057, 1979
113. Geller E, Ritvo ER, Freeman BJ, et al: Preliminary observations on the effect of fenfluramine on blood serotonin and symptoms in three autistic boys. N Engl J Med 307:165–168, 1982
114. August GJ, Raz N, Baird TD: Effects of fenfluramine on behavioral cognitive and affective disturbances in autistic children. J Aut Dev Dis 15:97–107, 1985
115. Yuwiler A, Ritvo ER, Geller E: Uptake and effects of serotonin from platelets of autistic and nonautistic children. J Autism Child Schiz 5:89–98, 1975
116. Siva-Sanker DV: Uptake of 5-hydroxytryptamine by isolated platelets in childhood schizophrenia and autism. Neuropsychobiology 3:234–239, 1977
117. Rotman A, Kaplan R, Szekely GA: Platelets uptake of serotonin in autistic and other psychotic children. Psychopharmacology 67:245–248, 1980
118. Boullin Freeman BJ, Geller E, Ritvo E, et al: Toward the resolution on conflicting finding. J Autism Develop Dis 12:97–101, 1982

119. Anderson GM, Minderaa RB, Van Benthen P-PG, et al: Platelet impramine binding in autistic subjects. Psychiatry Res 11:133–141, 1984
120. Weizman A, Gonen N, Tyano S, et al: Platelet [^{3}H]imipramine binding in autism and schizophrenia. Psychopharmacology 91:101–103, 1987
121. Yehuda R, Meyer JS: A Role for serotonin in the hypothalamic–pituitary–adrenal response to insulin stress. Neuroendocrinology 38:25–32, 1984
122. Field EJ, Caspary EA, Carnegic PR: Lymphocyte sensitization to basic protein of brain in malignant neoplasia: Experiments with serotonin and related compounds. Nature (Lond) 223:284–286, 1971
123. Todd RD, Ciaranello RD: Demonstration of inter and intraspecies differences in serotonin binding sites by antibodies from an autistic child. Proc Natl Acad Sci USA 82:612–616, 1985
124. Traskman-Benz L: CSF 5-HIAA and family history of psychiatric disorder. Am J Psychiatry 140:1257
125. Weizman A, Carel C, Tyano S, et al: Decreased high affinity [^{3}H]imipramine binding in platelets in enuretic children and adolescent. Psychiatry Res 14:39–46, 1985
126. Weizman R, Carmi M, Tyano S, et al: Reduced [^{3}H]imipramine binding but unaltered [^{3}H]serotonin uptake in platelets of adolescent enuretics. Psychiatry Res 19:37–42, 1986

8

Neuroendocrine Abnormalities in Autism and Schizophrenic Disorder of Childhood

Ronit Weizman, Lynn H. Deutsch, and Stephen I. Deutsch

1. INTRODUCTION

Hormonal levels and their alterations in response to provocative stimuli have been measured in children and adolescents with schizophrenic disorder and autism. Ideally, the neuroendocrine strategy would result in the elucidation of subgroups of patients with specific neurotransmitter abnormalities and provide a sensitive and diagnostic laboratory marker specific to a disorder and predictive of treatment response. The neuroendocrine strategy has been applied to the study of autistic and schizophrenic disorders in child psychiatry only to a limited extent. To date, these studies have corroborated the existence of hypothalamic–pituitary abnormalities in subgroups of patients; however, they have not implicated specific neurotransmitter abnormalities. This chapter reviews several studies performed by the authors and other investigators. They were selected to present the problems and potential applications of the neuroendocrine approach.

2. NEUROENDOCRINOLOGY OF SCHIZOPHRENIA

Neuroendocrinologic research in schizophrenia assumes that investigation of alterations in the secretion of specific hormones are indirect indicators of the functional status of neurotransmitters involved in their release. Other goals of the neuroendocrine strategy include the assessment of drug action on neurotransmitter receptor sensitivity and prediction of response to drug treatment. Several central nervous system (CNS) neurotransmitters have been implicated in the regulation of pituitary secretion, mainly through the hypothalamic–median eminence–pituitary axis. Pituitary hormone secretion was studied in schizophrenia because of its rela-

---

Ronit Weizman • Pediatric Department, Hasharon Hospital, Petah Tiqva 49372; and Sackler Faculty of Medicine, Tel Aviv University, Tel Aviv 69978, Israel. *Lynn H. Deutsch and Stephen I. Deutsch* • Psychiatry Service, Veterans Administration Medical Center, and Department of Psychiatry, Georgetown University School of Medicine, Washington, D.C. 20007.

tionship to the dopamine (DA) hypothesis that suggests involvement of overactivity of dopaminergic transmission in the pathophysiology of schizophrenia. This hypothesis is based on the fact that dopaminergic agonists can induce psychotic symptomatology, and neuroleptic drugs act as postsynaptic DA receptor blockers, an effect that correlates with therapeutic potency.[1]

Growth hormone (GH) secretion is regulated by several neurotransmitters, including DA. Prolactin (PRL) release is known to be inhibited by the tuberoinfundibular DA tract. Since a direct assessment of dopaminergic activity in human brain is not possible, neuroendocrine studies have used GH and PRL responses to dopamine agonists as indicators of central dopaminergic activity. Although basal GH and PRL levels in schizophrenic patients were reported to be similar to normal control levels,[2] abnormal GH responses to apomorphine were reported in some adult schizophrenic patients. However, these abnormal GH responses to apomorphine included exaggerated and blunted ones.[2,3]

Unfortunately, only a few neuroendocrine studies have been performed in adolescent schizophrenic patients, in contrast to the large number of clinical studies in adults. This chapter reviews studies of basal PRL and GH levels, as well as their stimulated release in adolescent schizophrenic patients.

In normal subjects, thyrotropin releasing hormone (TRH) and luteinizing release hormone (LRH) do not stimulate GH secretion. However, TRH-induced increases in GH levels have been described in disorders whose pathophysiology are thought to involve specific neurotransmitters. These pathological conditions include endogenous depression,[4] anorexia nervosa,[5] acromegaly,[6] renal failure,[7] and severe liver disease.[8] Aberrant GH responses to LRH have been less commonly observed; they were reported only in acromegaly[9] and in a few cases of anorexia nervosa.[5] In most of these pathological conditions, changes in the metabolism of brain catecholamines and/or serotonin were suggested. Experimentally, abnormal responses of GH to TRH and LRH were demonstrated in hypophysectomized rats bearing an ectopic pituitary under the kidney capsule[10] and in an in vitro preparation of perfused rat pituitary.[11] In these experimental conditions, there is an anatomical or functional disconnection of the anterior pituitary from the hypothalamus.

In a study conducted by our group, the GH response to TRH and LRH in adolescent schizophrenic boys was studied and compared with normal age-matched controls.[12] The study was conducted in 10 patients (aged 15.8 ± 0.4 years) who met the diagnosis of schizophrenia according to the research diagnostic criteria (RDC). The patients were hospitalized and studied during their first psychotic episode. None of the patients received any psychotropic or other medication known to influence GH secretion prior to the study. The control group consisted of nine normal boys. All the schizophrenic patients received an LRH infusion, and six patients underwent a TRH test on the next day. The LRH and TRH tests were repeated 3 months after initiation of neuroleptic treatment in several patients. LH, follicle-stimulating hormone (FSH), PRL, TSH, and GH were measured in each blood sample. The LRH (50-μg/m^2 body surface, i.v.) and TRH (200-μg i.v.) tests were performed between 8 and 9 AM after an overnight fast. Blood samples were collected at baseline and at intervals of 15–30 min during a 90-min period. LRH and TRH induced a marked rise in GH in 8 of 10 and in 4 of 6 untreated patients, respectively. There was no effect on GH secretion in the normal controls. After 3 months of neuroleptic treatment, LRH did not induce a significant rise in GH in three of four patients tested. The mean baseline hormonal levels, and the responses of TSH, FSH, and LH to TRH and LRH administration did not differ from those of the control group. As expected, only PRL was affected (elevated) by the neuroleptic treatment.

The abnormal GH response to TRH demonstrated in newly admitted adolescent schizophrenic patients was subsequently assessed in a different group of chronically treated patients.[13] This study included 10 adolescent schizophrenic subjects (aged 15–18 years) and 10 age- and sex-matched controls. A significant rise in GH response to TRH was demonstrated in 50% of the patients, but in none of the controls. In four of the five patients with abnormal GH responses, there was a positive family history of schizophrenia. By contrast, only one of the GH nonresponding patients had a positive family history of schizophrenia. Thus, the abnormal GH response to TRH in adolescent schizophrenic patients seems to be associated with family history of the disease.

An increase in GH secretion due to TRH provocation in adolescent, but not in adult, schizophrenic patients was reported by DeMilio.[14] An impaired hypothalamic regulatory mechanism or alteration in the cellular membrane of the pituitary GH-secreting cells could have triggered the nonspecific GH secretion after TRH challenge. These data suggest that age of onset might play a role in the development of abnormalities in hypothalamic–pituitary function in schizophrenia.

3. NEUROENDOCRINOLOGY OF AUTISM

The identification of endocrine or endocrine-related abnormalities in autistic children provided evidence in support of the importance of biological factors in the pathogenesis of this disorder. Abnormalities of linear growth were observed in an analysis of the physical characteristics of a carefully diagnosed sample of 101 autistic children.[15] These children fulfilled multiple sets of diagnostic criteria and were diagnosed independently by three research child psychiatrists. All the children were drug free and were evaluated between the ages of 2 and 7 years; thus, the potential confounding effects of maintenance psychoactive medication and institutionalization on the dependent measures were eliminated. The nonautistic, nonretarded siblings served as the comparison group. The autistic children were shorter, and the distribution of percentile heights was abnormal; the autistic children were overrepresented in the lower third of percentiles for height. Bone age was determined radiographically in a subgroup of 84 of the 101 autistic children and did not show a normal distribution about the mean expected for age. The study showed that 6% of the patients were 2 SD below the expected mean, whereas only 1% was 2 SD above the mean. Hormonal assays revealed that triiodothyronine (T_3) and thyroxine (T_4) levels were elevated in about one half to one third of a subsample of these autistic children. The meaningfulness of these findings is increased because repeated measures were obtained for these hormonal values. Interestingly, elevated T_4 levels were positively associated with higher weighted scores of minor physical anomalies in a subsample of 42 autistic children. These data were provocative and suggested a possible relationship among thyroid abnormalities, minor congenital anomalies, and the syndrome of infantile autism. The large sample size of this descriptive study contributed to the importance and validity of these observations.

An earlier study examining the effects of an intravenous infusion of 400 g of synthetic TRH in 10 autistic children on the secretion of TSH by the pituitary, and T_3 over a 2-hr period suggested the existence of hypothalamic–pituitary dysfunction.[16] The group was characterized by its marked individual variability on these measures. There were blunted or delayed T_3 responses and delayed, exaggerated, or subnormal TSH responses. The abnormal kinetics of these responses were consistent with endocrine dysfunction, but the marked variability could not suggest a mechanism or specific neurotransmitter abnormality. Independent of the group's abnormal TSH and T_3 responses to a provocative challenge with TRH, inspection of the clinical and laboratory data obtained on screening supported the existence of possible neurochemical and/or neuroendocrine abnormalities in 9 of the 10 autistic children. These nine children were in a percentile for height that was ≤ 10, showed retarded bone age, and had elevated levels of serotonin in blood. The data were interpreted to show biochemical heterogeneity among patients selected on the basis of uniform and homogeneous descriptive criteria.[16,17]

In order to assess the hypothalamic dysfunction and growth abnormalities of autistic children further, plasma GH levels were measured before and after challenge with two provocative stimuli (oral L-dihydroxyphenylalanine and insulin-induced hypoglycemia).[18,19] These studies had two additional purposes: to implicate an abnormality of a specific neurotransmitter in the pathogenesis of infantile autism, and to see whether short-term treatment with haloperidol altered the plasma GH response. A low-dose administration of haloperidol on an individually titrated basis is an adjunctive and salutary pharmacological approach to the treatment of this disorder.[20] The authors were concerned about possible adverse effects of intervention with haloperidol on the growth and development of these children.

The plasma GH response to an oral challenge of *L*-dihydroxyphenylalanine (L-DOPA) was examined in 22 autistic children.[18] The autistic children were a subsample of a larger group that participated in a study evaluating the effects of haloperidol, behavior therapy, and the interaction of the two on behavioral symptoms and language acquisition. Eight of the autistic children were rechallenged with L-DOPA after 5 weeks of treatment with optimal daily doses of haloperidol (0.5–4.0 mg/day). The opportunity to rechallenge these children provided some preliminary information about the effect of short-term treatment with a relatively selective dopamine receptor antagonist on the regulation of GH release in response to oral L-DOPA administration, and hypothalamic dopamine receptor sensitivity. The L-DOPA-stimulated plasma GH responses of the autistic children were compared with the responses of a "control" series of 42 children who received L-DOPA provocative tests. The majority of the children in the control series were evaluated for short stature and found to have a nonendocrine cause for this complaint. L-DOPA functions as a dopamine agonist subsequent to its uptake into the brain, concentration within dopaminergic nerve terminals, decarboxylation, and release. Dopaminergic agonists stimulate pituitary GH secretion and this effect is thought to be mediated by dopamine receptors in the hypothalamus. Thus, an alteration in the amplitude of kinetics of the pituitary GH response after L-DOPA ingestion in the autistic children would be consistent with a change in the sensitivity of their hypothalamic dopamine receptors. Noradrenergic influences must also be considered when L-DOPA is used as the sole provocative challenge. The dose of L-DOPA was determined according to the weight of the child; children weighing <70 lb received 250 mg, whereas heavier children received a 500-mg dose. The procedure was conducted as described by Weldon et al.[21] The peak GH response of about 30% of this descriptively homogeneous sample of autistic children was below 6 ng/ml over the course of a 135-min period after oral L-DOPA administration. A value below 6 ng/ml is indicative of a blunted or abnormal response. As a group, the autistic children showed blunted responses over the entire 135-min period after L-DOPA administration. The differences between the patients and control series were significant at 90 min ($p = 0.004$) and 120 min ($p = 0.011$) postadministration (Fig. 1). The data imply an impairment in the regulation of GH secretion and a subsensitivity of hypothalamic dopamine receptors. The treatment of eight of the autistic children with optimal daily doses of haloperidol for 5 weeks did not appear to have an effect on the GH response to oral L-DOPA. These data suggest that 5 weeks of treatment with haloperidol may not alter the sensitivity of hypothalamic dopamine receptors, but these conclusions must be viewed tentatively due to the small sample size and the large pre- and post-treatment within-group variance of this measure.

Growth hormone secretion by the pituitary is a highly regulated process and under the influence of several neurotransmitters in addition to dopamine. The blunted plasma GH response to oral L-DOPA administration is suggestive of a dopaminergic abnormality, but these data do not support a neurotransmitter abnormality confined exclusively to the dopaminergic system. Therefore, the plasma GH response to insulin-induced hypoglycemia was examined in

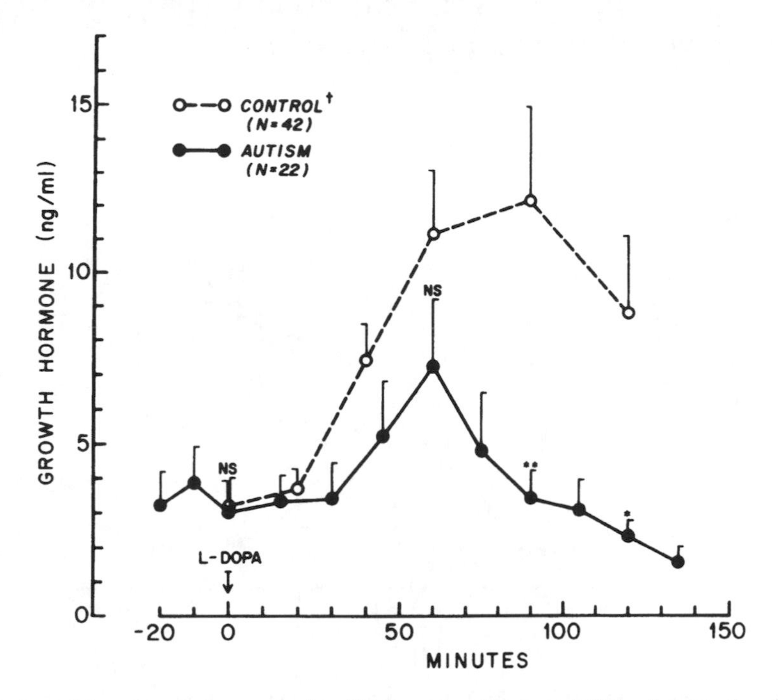

Figure 1. Growth hormone response to oral L-dihydroxyphenylalanine L-DOPA administration. The plasma GH response to oral L-DOPA was examined in 22 autistic children (Deutsch et al.[18]). †The results were compared with a "control" series of published raw data on 42 children who received L-DOPA provocative tests (Weldon et al.).[21] Differences between the autistic and control samples were significant at 90 min ($^{**}p = 0.004$) and 120 min ($^{*}p = 0.011$) after L-DOPA administration. Bars denote standard deviation.

a different sample of eight autistic children.[19] These children represented a subsample of 40 autistic children participating in a study examining the effects of haloperidol on behavioral symptoms and discrimination learning.[20] Six of these eight children were tested on two occasions: on baseline and after 4 weeks of receiving daily pharmacotherapy with haloperidol. In the six children rechallenged with insulin-induced hypoglycemia, the daily haloperidol dosage was individually regulated for the first 2 weeks and then the optimal dose (0.5–3.0 mg/day) maintained for 2 weeks prior to rechallenge. An intravenous bolus of crystalline insulin (0.5 U/kg) was able to produce about a 50% reduction of baseline glucose values in most patients. In this study, no child showed evidence of a blunted plasma GH response to hypoglycemia on repeated testing (i.e., a plasma GH value <7 ng/ml). Short-term treatment with haloperidol did not seem to diminish the magnitude of the GH response to this provocative stimulus. Therefore, if medication status is disregarded, visual inspection of the raw data showed that one half of this sample had persistently elevated plasma GH values with a failure to return to baseline levels over the course of 135 min after insulin infusion on at least one occasion (Fig. 2). The authors attributed the prolonged and exaggerated GH responses to hypothalamic–pituitary dysfunction, rather than a nonspecific effect due to stress. The data emphasized the complexity of GH regulation in autistic children; i.e., they may show a blunted response to one type of stimulus and an exaggerated response to another. The data also argue against an exclusively dopaminergic abnormality in affected children. Unfortunately, a prolonged GH response to insulin-induced hypoglycemia cannot be used to delineate a subgroup of patients with a specific neurotransmitter abnormality.

A failure to suppress plasma cortisol levels in response to the exogenous administration of

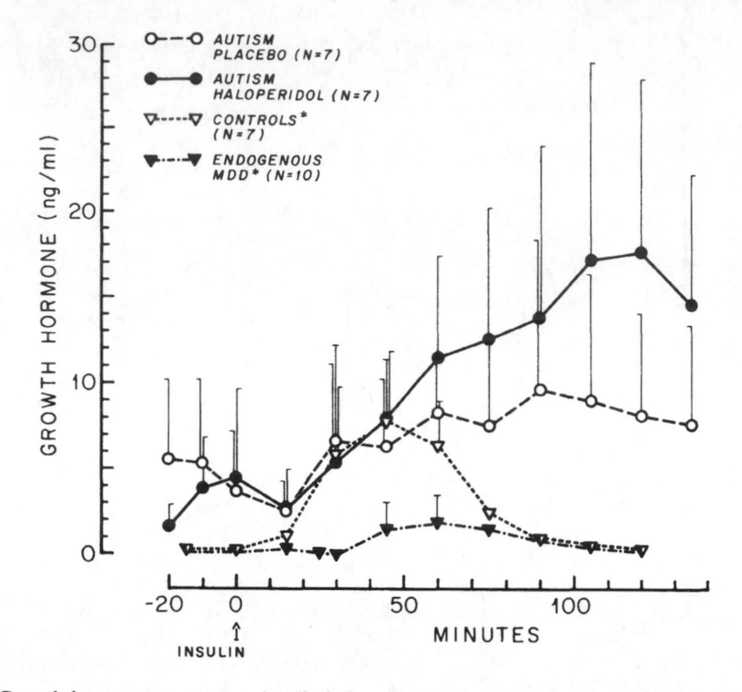

Figure 2. Growth hormone response to insulin-induced hypoglycemia. Plasma growth hormone responses to insulin-induced hypoglycemia were measured in eight autistic children (Deutsch et al.).[19] *The results were compared with those of a group of prepubertal children with nondepressed neurotic disorders and prepubertal children with major depressive disorder endogenous subtype (Puig-Antich et al.).[21a] Bars denote standard deviation. (From Deutsch et al.[19])

dexamethasone, a potent synthetic corticosteroid, is the basis of the dexamethasone suppression test (DST). An abnormal DST result is regarded as evidence of a dysregulation of the hypothalamic–pituitary–adrenal axis.[22] The plasma cortisol values of 11 of 13 autistic children and adolescents (ages 2 years, 6 months to 17 years, 10 months) were elevated above 5 g/dl on at least one occasion on the day following the 11 PM administration of dexamethasone.[23] The dexamethasone dose was determined according to the weight of the child; children weighing below 41 kg received 0.5 mg, whereas children weighing above 41 kg received 1.0 mg. A high percentage of these children (62.9%) continued to show abnormal DST results even after the more rigorous criterion of 8 g/dl was adopted as the cutoff defining nonsuppression. These data argue against the specificity of abnormal DST results to major depressive disorder and for the existence of hypothalamic dysfunction in a significant number of autistic children. However, the abnormal results do not help delineate a specific neurotransmitter abnormality.

4. CONCLUSIONS

The variability in neuroendocrine measures among patients within the same diagnostic group suggests the existence of subgroups among patients with similar descriptive diagnostic features. The significance of these subgroups are uncertain at this time but could reflect different etiologies, circadian variations, sampling problems, or different times in the course of illness. The meaningfulness of these studies would be enhanced by increasing the sample sizes, and following the longitudinal courses of patients differing widely on specific measures.

The stability of a neuroendocrine abnormality on repeat testing at later points in the illness, irrespective of intervention or changes in the severity of symptoms, would suggest that the measure is a *trait* marker of the illness. The identification of a trait marker would be relevant to family studies attempting to define the contribution of *heritability* to the overt manifestation of illness. A neuroendocrine abnormality that changes or reverts to normal in response to an intervention could prove useful as a predictive marker of treatment response. Also, this type of *state* marker might be valuable in exploring the role of a specific neurotransmitter in the pathophysiology of the disorder and the pharmacological mechanism of the intervention. The neuroendocrine strategy is a potentially fruitful one that should be explored further in children and adolescents with autistic and schizophrenic disorders.

REFERENCES

1. Synder SH, Banerjee SP, Yamamura HI et al: Drugs, neurotransmitters and schizophrenia. Science 1984:1243–1253, 1974
2. Meltzer HY, Kolakowska T, Fang VS, et al: Growth hormone and prolactin response to apomorphine in schizophrenia and the major affective disorders: Relation to duration of illness and depressive symptoms. Arch Gen Psychiatry 41:512–519, 1984
3. Meltzer HY, Busch D, Fang VS: Hormone, dopamine receptors and schizophrenia. Psychoneuroendocrinology 6:17–36, 1981
4. Maeda K, Kato, Ohgo S, et al: Growth hormone and prolactin release after injection of thyrotropin releasing hormone in patients with depression. J Clin Endocrinol Metab 40:501–509, 1975
5. Maeda K, Kato Y, Yamaguchi N, et al: Growth hormone release following thyrotropin releasing hormone injection in patients with anorexia nervosa. Acta Endocrinol (Copenh) 81:1–8, 1976
6. Irie M, Trushima T: Increase of serum growth hormone concentration following thyrotropin releasing hormone injection in patients with acromegaly or gigantism. J Clin Endocrinol Metab 35:97–102, 1972
7. Gonzales-Barcena P, Pastin AY, Schalch DS, et al: Response to thyrotropin releasing hormone in patients with renal failure and after infusion in normal man. J Clin Endocrinol Metab 36:117–120, 1973
8. Panerai AE, Salerno F, Maneschi M, et al: Growth hormone and prolactin response to thyrotropin releasing hormone in patients with severe liver disease. J Clin Endocrinol Metab 45:134–140, 1977
9. Faglia G, Beck-Pecoz P, Travaglini P, et al: Elevation in plasma growth hormone concentration after luteinizing hormone releasing hormone (LRH) in patients with acute acromegaly. J Clin Endocrinol Metab 37:338–340, 1973
10. Udeschini G, Cocchi D, Panerai AE, et al: Stimulation of growth hormone release by thyrotropin releasing hormone in the hypophysectomized rat bearing an ectopic pituitary. Endocrinology 98:807–814, 1976
11. Takahara Y, Arimura A, Schally AV: Stimulation of prolactin and growth hormone release by TRH infused with the hypophyseal portal vessel. Proc Soc Exp Biol Med 146:831–835, 1974
12. Gil-Ad I, Dickerman Z, Weizman R, et al: Abnormal growth hormone response to LRH and TRH in adolescent schizophrenic boys. Am J Psychiatry 138:357–360, 1981
13. Weizman R, Weizman A, Gil-Ad I, et al: Abnormal growth hormone response to TRH in chronic adolescent schizophrenic patients. Br J Psychiatry 141:582–585, 1982
14. DeMilio L. TRH response pattern in adolescent schizophrenic males. Br J Psychiatry 145:649–651, 1984
15. Campbell M, Petti TA, Green WH, et al: Some physical parameters of young autistic children. J Am Acad Child Psychiatry 19:193–212, 1980
16. Campbell M, Hollander CS, Ferris S, et al: Response to TRH stimulation in young psychotic children: A pilot study. Psychoneuroendocrinology 3:195–201, 1978
17. Green WH, Deutsch SI, Campbell M: Psychosocial dwarfism, infantile autism, and attention deficit disorder, in Nemeroff CB, Loosen PT (eds): Handbook of Psychoneuroendocrinology. New York, Guilford, 1987, pp 109–142
18. Deutsch SI, Campbell M, Sachar EJ, et al: Plasma growth hormone response to oral L-dopa in infantile autism. J Autism Dev Disord 15:205–212, 1985
19. Deutsch SI, Campbell M, Perry R, et al: Plasma growth hormone response to insulin-induced hypoglycemia in infantile autism: A pilot study. J Autism Dev Disord 16:59–68, 1986
20. Anderson LT, Campbell M, Grega DM, et al: Haloperidol in the treatment of infantile autism: Effects on learning and behavioral symptoms. Am J Psychiatry 141:1195–1202, 1984

21. Weldon VV, Gupta SK, Haymond MW, et al: The use of L-dopa in the diagnosis of hyposomatotropism in children. J Clin Endocrin Metab 36:42–46, 1973
21a. Puig-Antich J, Tabrizi MA, Davies M, et al: Prepubertal endogenous major depressives hyposecrete growth hormone in response to insulin-induced hypoglycemia. *Biol Psychiatry* 16:801–818, 1981
22. Carroll BJ, Feinberg M, Greden JF, et al: A specific laboratory test for the diagnosis of melancholia. Arch Gen Psychiatry 38:15–22, 1981
23. Jensen JB, Realmuto GM, Garfinkel BD: The dexamethasone suppression test in infantile autism. J Am Acad Child Psychiatry 24:263–265, 1985

9

Autoimmune Dysfunction in Neurodevelopmental Disorders of Childhood

Abraham Weizman and Frank J. Vocci, Jr.

1. INTRODUCTION

The syndrome of autism is characterized by a failure to develop normal age-appropriate social relationships, with disturbances of motor, social, adaptive, and cognitive abilities. Etiologic heterogeneity is assumed by most investigators because of the occurrence of autistic syndromes in persons with a variety of disorders (e.g., viral or genetic). As yet, no specific biochemical markers or deviations have been demonstrated. Abnormal responses to sensory stimuli (e.g., auditory, vestibular, visual, tactile, olfactory, and proprioceptive) have been described. Whether these features are due to emotional disturbances or to organic lesion is still controversial. Neurological syndromes such as phenylketonuria (PKU), hypsarrhythmia, and encephalitis caused by fetal rubella infection that may result in autistic features, as well as other brain dysfunctions, raise the possibility that autism is associated with an organic brain lesion(s). Extensive neurological examinations, including EEGs, demonstrated that signs of neurological dysfunction appear more frequently in autistic children than in normal children.[1,2] Damasio and Maurer[3] suggested a neural bilateral dysfunction of the ring of mesolimbic cortex (located in the medial, frontal, and temporal lobes), the neostriatum, and the anterior and medial thalamic nuclei as a result of perinatal viral infection or genetically determined neurochemical abnormalities.

One of the neurobiological approaches in an attempt to elucidate the abnormalities that could be associated with causes or deficits in autism involved the study of possible immunological abnormalities in this disorder. There are only a few investigations in this field; more-

Abraham Weizman • Geha Psychiatric Hospital, Beilinson Medical Center, Petah-Tiqva 49100; and Sackler Faculty of Medicine, Tel Aviv University, Tel Aviv 69978, Israel. *Frank J. Vocci, Jr.* • Medications Development Program, Division of Preclinical Research, National Institute on Drug Abuse, Rockville, Maryland 20857.

over, the result in this area to date require verification from several laboratories to determine the universality of the immune changes noted. Nonetheless, several alterations in the immune system or self/non-self recognition have been reported.

2. ALTERATIONS IN IMMUNE RESPONSIVENESS

The initial report of a defective antibody response after vaccination with rubella autistic children (5 of 13)[4] suggested that a subgroup of autistic children may have an altered immune response. A depressed lymphocyte responsiveness to the T-cell mitogen phytohemoagglutinin (PHA) was observed in autistic children as compared with normal controls.[5] These results were confirmed by a recent study[6] that demonstrated that lymphocytes of autistic patients have reduced responses to the T-cell mitogen concanavalin A (Con A). In addition, Warren et al. reported also a reduced response to B-cell mitogen pokeweed (PWM), as well as a decreased number of T lymphocytes and altered helper/T-suppressor cell ratio. Total immunoglobulin (IgG and IgM) levels in cerebrospinal fluid (CSF) in autistic patients did not differ from controls.[7] A recent study[8] demonstrated reduced natural killer (NK) cell activity in 12 of 31 autistic patients. NK cells are part of the defense mechanism against viruses and malignancy and are involved in immune regulation. Reduced cytotoxic activity of these cells has been found in autoimmune disease, such as lupus erythematosus,[9] multiple sclerosis,[10] rheumatoid arthritis, and Sjögren's syndrome.[11] The relationship of altered immune activity and development of autism is not clear. One is tempted to make several speculations as to the relationship(s) between depressed immune function and the pathophysiology of autism. It is possible that a predisposition to a deficiency in immune function leaves the fetus or the neonate more sensitive to viral infection and secondary neurological damage. Another possibility is that the elevated blood serotonin concentration, demonstrated in 30–40% of autistic patients,[12] suppresses the ability of the lymphocytes to respond to PHA and inhibits the NK activity. An attractive hypothesis suggested by Warren et al.[8] is that immunological damage mediated by autoantibodies to the nervous system also results in a destruction of NK cells. This hypothesis is based on the finding that anti-Leu-7, a mouse monoclonal antibody that is reactive with human NK cells, also binds to myelin-associated glycoprotein. The existence of an epitope shared between cells of the immune system and nervous system may be involved in the autoimmune response toward the two systems.

3. ALTERATIONS IN SELF/NON-SELF RECOGNITION

In a study conducted in our laboratory,[13] evidence for a cell-mediated autoimmune response to human myelin basic protein (inhibition of macrophage migration) in autistic patients was demonstrated. The study included 17 autistic subjects, 6 of whom were drug free for at least 1 year, and 11 patients with similar severe neuropsychiatric disorders who were selected according to the differential diagnosis of autism. The control subjects were treated with drugs identical to the autistic patients and in similar doses. Thirteen of the 17 autistic patients showed a positive migration inhibition factor response to human myelin basic protein, while none of the comparison group responded to this antigen. The results appeared to be independent to drug-treatment effects. It is tempting to speculate that an autoimmune component directed toward myelin basic protein can produce patchy "microlesions" that affect saltatory conduction in the brainstem resulting in prolonged conduction time for sensory information. Other researchers[14] suggest a link between this finding and the hyperserotonemia by pointing out that myelin basic protein contains a serotonin binding site in its tryptophan peptide region; this region is immunologically active and responds by activating lymphocytes. They propose a theory linking myelin basic protein and serotonin in the etiology of autism.

The presence of antiserotonin receptor antibodies in a subgroup of autistic children[15] suggests a novel mechanism for the hyperserotonemia observed in these patients. The study of sera and CSF from a 9-year-old autistic girl revealed autoantibodies against 5-HT binding proteins. These antibodies are of the IgG class, which discriminate between 5-HT_{1A} and other 5-HT binding proteins, and are specific for human 5-HT_{1A} sites. The antibody titer in the CSF of the autistic child was four times greater than in a serum sample drawn at the same time. Thus, elevated 5-HT levels may represent a compensatory response to an apparent decrease in 5-HT_{1A} receptor number.

In order to evaluate further whether antigenic similarities also existed between species, membranes were incubated both with and without added IgG and then assayed for [^{3}H]-5-HT binding.[15] Since no IgG-mediated inhibition of [^{3}H]-5-HT binding was observed for bovine or rat cortex membranes, it seems that differences exist in antigenic domains of the 5-HT_{1A} binding protein across species, despite the pharmacological and physical similarities among 5-HT_{1A} sites. The antigenic domains recognized by this child's antibodies may represent the [^{3}H]-5-HT binding site of the human 5-HT_{1A} receptor itself or other areas that, when occupied by antibody, cause conformational changes in the [^{3}H]-5-HT binding site. A further study of the autistic children revealed circulating antibodies directed against human brain 5-HT receptors in 7 of 13 patients, while none of the 13 normal control had these antibodies. The autoimmune response seems to be restricted to a few specific antigens, since generalized antibrain antibody titers were similar in autistic, depressed, and normal control subjects.[16] In this study, the frontal cortex was used as a source of antigens. Thus, one cannot exclude the possibility that the use of other antigen might give different results. Another possibility is that significant antibody titers were present in early development and are suppressed later. In this case, in vitro stimulation of memory B cells by brain antigens should discriminate autistic children from normal controls.

REFERENCES

1. Knobloch H, Pasamanick B: Some etiologic and prognostic factors in early infantile autism and psychosis. Pediatrics 55:182–191, 1975
2. James AL, Barry RJ: A review of psychophysiology in early onset psychosis. Schizophr Bull 6:506–527, 1980
3. Damasio AR, Maurer RG: A neurological model for childhood autism. Arch Neurol 35:777–786, 1978
4. Stubbs EG: Autistic children exhibit undetectable hemagglutination-inhibition antibody titers despite previous rubella vaccination. J Autism Dev Disord 6:269–274, 1976
5. Stubbs EG, Crawford ML: Depressed lymphocyte responsiveness in autistic children. J Autism Dev Disord 7:49–55, 1977
6. Warren RP, Margaretten NC, Pace NC: Immune abnormalities in patients with autism. J Autism Dev Disord 16:189–197, 1986
7. Young JG, Caparulo BK, Shaywitz BA, et al: Childhood autism: Cerebrospinal fluid examination and immunoglobulin levels. J Child Psychiatry 16:174–179, 1977
8. Warren RP, Foster A, Margaretten NC: Reduced natural killer cell activity in autism. J Am Acad Child Adol Psychol 3:333–335, 1987
9. Goto M, Tanimoto K, Horiuchi Y: Natural cell-mediated cytotoxicity in systemic lupus erythematosus. Arthritis Rheum 23:1274–1281, 1980
10. Waksman BH: Current trends in multiple sclerosis research. Immunol Today 2:87–93, 1981
11. Goto M, Tanimoto K, Chihara T, et al: Natural cell-mediated cytotoxicity in Sjorgen's syndrome and rheumatoid arthritis. Arthritis Rheum 24:1377–1382, 1981
12. Ritvo ER, Yuwiler A, Geller E, et al: Increased blood serotonin and platelets in early infantile autism. Arch Gen Psychiatry 42:129–133, 1970

13. Weizman A, Weizman R, Szekely GA, et al: Abnormal immune response to brain tissue antigen in the syndrome of autism. Am J Psychiatry 139:1462–1465, 1982
14. Westall FC, Root-Bernstein RS: Suggested connection between autism, serotonin and myelin basic protein. Am J Psychiatry 140:1260–1261, 1983
15. Todd RD, Ciaranello RD: Demonstration of inter- and intraspecies differences in serotonin binding sites by antibodies from an autistic child. Proc Natl Acad Sci USA 82:612–616, 1985
16. Todd RD, Hickok JM, Anderson GM, et al: Antibrain antibodies in infantile autism. Biol Psychiatry 23:644–647, 1988

10

Nonketotic Hyperglycinemia

A Paradigm for the Application of Neuroscience to the Understanding and Treatment of a Developmental Disorder

Stephen I. Deutsch, Lynn H. Deutsch, and Ronit Weizman

1. INTRODUCTION

Nonketotic hyperglycinemia is an autosomal-recessive disorder that usually presents during the neonatal period and, in its classic or typical form, is characterized by poor muscle tone, feeding difficulties, intractable seizures, and death due to respiratory insufficiency.[1,2] Phenotypically, the disorder can be quite heterogeneous; atypical presentations include mild mental retardation and expressive language difficulties as the sole features.[3–8] In many ways, nonketotic hyperglycinemia serves as a paradigm for the application of basic neuroscience to the understanding and treatment of a developmental disorder.

The disease is characterized by increased urinary excretion of glycine and elevations of free glycine levels in plasma and cerebrospinal fluid (CSF) in the absence of organic acidemia.[2,9] The diagnosis is based on demonstration of elevated CSF levels with a CSF-to-plasma ratio above 0.03. The hyperglycinemia results from deficient or absent glycine cleavage enzyme system activity in liver and brain. Patients with nonketotic hyperglycinemia show deficits in the formation of $^{14}CO_2$ isolated from expired air after the injection of glycine-1-^{14}C, i.e., glycine labeled in the carboxyl carbon. The glycine cleavage enzyme system is a mitochondrial membrane-associated complex, composed of several polypeptides, requiring tetrahydrofolate (THF), nicotinamide-adenine dinucleotide (NAD), and pyridoxal phosphate as essential cofactors. The complex oxidatively decarboxylates glycine with its α-carbon entering the one-carbon pool as 5,10-methylene THF. Furthermore, glycine contributes to the regula-

Stephen I. Deutsch and Lynn H. Deutsch • Psychiatry Service, Veterans Administration Medical Center, and Department of Psychiatry, Georgetown University School of Medicine, Washington, D.C. 20007. *Ronit Weizman* • Pediatric Department, Hasharon Hospital, Petah Tiqva 49372; and Sackler Faculty of Medicine, Tel Aviv University, Tel Aviv 69978, Israel.

tion of the pool size of available one-carbon units via its metabolic interconversion with serine, a reaction that uses activated THF as the source of hydroxymethyl donation (Fig. 1).

The recognition of nonketotic hyperglycinemia led to an appreciation of the important role played by the glycine cleavage system in maintaining the one-carbon pool size and the prevention of the toxic accumulation of free glycine levels in brain.[2] Glycine is an inhibitory neurotransmitter in the brainstem and spinal cord and an endogenous modulator of glutamatergic transmission. Perry et al.[2] suggested that the neurochemical basis for nonketotic hyperglycinemia is a glycine encephalopathy due to the absence of glycine cleavage activity in brain. A failure to cleave glycine could produce both a reduction in the metabolic one-carbon pool and a toxic accumulation of this neurotransmitter/neuromodulator within the central nervous system (CNS).

Unlike many disorders in pediatric neurology and child psychiatry for which a neurochemical lesion is not known, interventions for the treatment of nonketotic hyperglycinemia are based on the current understanding of the neurochemical consequences of defective glycine cleavage system activity. The fact that many of the intervention trials have been ineffective suggests that this understanding is incomplete and/or the interventions are made after irreversible damage due to the neurotoxic effects of glycine has occurred. To date, the major treatment approaches have been designed to reduce the toxic load of glycine, to supplement the one-carbon pool size with alternative sources of one-carbon units, and to antagonize glycine at postsynaptic receptor sites.[10–23] In the brainstem and spinal cord, glycine acts at specific postsynaptic receptor sites that are structurally coupled to a gated chloride ionophore to promote chloride ion conductance and hyperpolarization.[24–26] This glycine-gated chloride

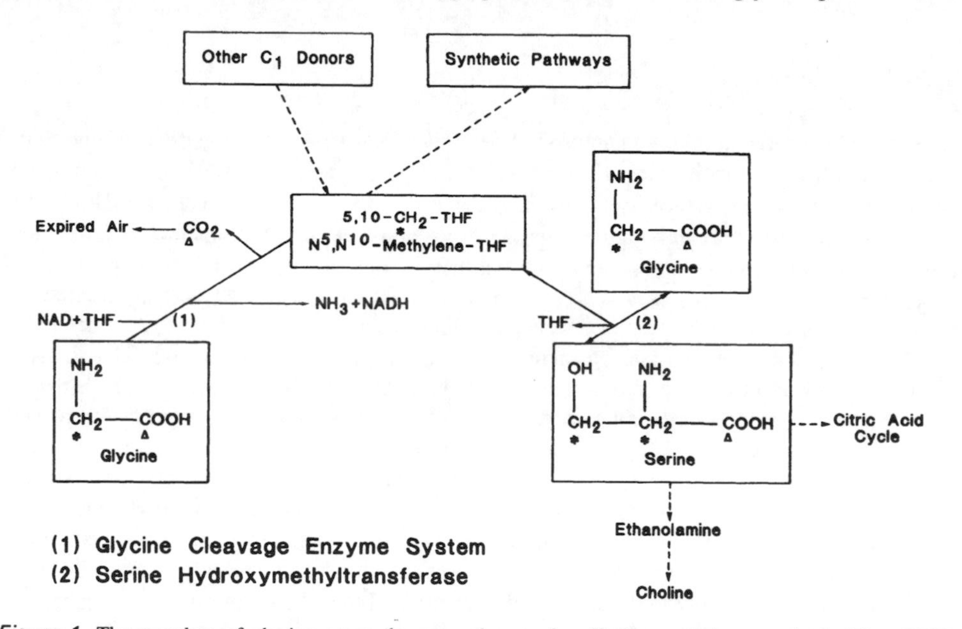

Figure 1. The α-carbon of glycine enters the one-carbon pool as N_5,N_{10}-methylene tetrahydrofolate (THF), and the carboxyl carbon appears in expired air as CO_2 as a result of the glycine cleavage enzyme system (reaction 1). A failure to cleave glycine would result in a decreased availability of N_5,N_{10}-methylene THF and a reduction or absence in the incorporation of the α-carbon of glycine into C-3 of serine by the action of serine hydroxymethyltransferase (reaction 2). The activity of serine hydroxymethyltransferase and the serine–glycine interconversion contributes to the regulation of the pool size of activated THF for one-carbon donation. In nonketotic hyperglycinemia, glycine cleavage enzyme system activity is reduced or absent, resulting in a reduction in the pool size of one-carbon units and a toxic accumulation of glycine, a neurotransmitter/ neuromodulator in the central nervous system (see text for details). (Adapted from Perry et al.[2])

conductance is antagonized in a noncompetitive fashion by strychnine. Recently, glycine has been shown to facilitate glutamatergic-mediated cation flux allosterically at the *N*-methyl-D-aspartate (NMDA) subclass of postsynaptic glutamate–receptor complex.[27–29] Antagonism of this glycine-mediated facilitation of glutamatergic transmission in nonketotic hyperglycinemia may represent a novel intervention strategy for this disorder.

2. BIOCHEMICAL CONSIDERATIONS OF NONKETOTIC HYPERGLYCINEMIA

The glycine cleavage system is the major pathway for the oxidative catabolism of glycine. It is a mitochondrial enzyme system composed of four protein components: P protein (pyridoxal phosphate-containing protein), H protein (a lipoic acid-containing protein), T protein (tetrahydrofolate-dependent protein), and L protein (a lipoamide dehydrogenase).[30,31] Defects in any one of the four protein components can result in typical and atypical presentations of nonketotic hyperglycinemia with absent or severely reduced glycine cleavage system activity in liver and brain. In the atypical presentations, appreciation of a metabolic disorder may be delayed until late infancy, and there is a spectrum of severity with regard to psychomotor development. The activities of the individual components of the complex can be measured and, in an analysis of 10 typical or classic cases, defects in the P and T proteins were shown, whereas defects in the T and H proteins were demonstrated in three atypical cases.[31] In many instances, the clinical phenotype relates to the degree of residual activity for the entire glycine cleavage system. The atypical or more mild presentations are often associated with greater activity and lower, although still significantly elevated, CSF glycine levels than the classic presentations. In at least some atypical cases with greater amounts of residual glycine cleavage system activity there may be a rationale for high-dose cofactor administration.

Diminished activity of the glycine cleavage enzyme system results in elevations of the concentration of this amino acid neurotransmitter in brain tissue and CSF. In two infants with the classic presentation, levels of glycine were elevated two- to fourfold in the frontal and occipital cortex; glycine was also elevated severalfold in the cerebellar cortex, striatum and spinal cord.[2] Interestingly, decarboxylation of radiolabeled glycine-1-^{14}C as measured by the formation of $^{14}CO_2$, was not detectable in autopsied specimens of frontal and cerebellar cortex from these two infants, whereas decarboxylase activity in liver was approximately one third that found in control specimens. Thus, measurement of total glycine cleavage activity in liver biopsies from patients with nonketotic hyperglycinemia may not accurately reflect the severity of the condition in brain. CSF values were about 16–27 times above normal in these two patients with the classic presentation.

In the brainstem, glycine interacts in a high-affinity reversible manner with an agonist recognition site coupled to a chloride ionophore. Toxic accumulation of glycine in this region would result in diminished arousal and respiratory depression. The glycine receptor–chloride ionophore complex may be a member of a family of neurotransmitter-gated ion channels that includes the nicotinic cholinergic receptor and the γ-aminobutyric acid-A ($GABA_A$) receptor complexes.[26] These oligomeric protein complexes seem to have evolved from a common ancestor protein to mediate rapid chemical signal transduction at synapses. The polypeptide that constitutes the gated channel of these receptor complexes shares common structural features: (1) four conserved homologous hydrophobic domains that traverse the membrane, and (2) an extracellular β-structural loop formed by the disulfide bonding of two conserved cysteine residues. The extracellular loop may have a role in agonist recognition. Recently, a 48-kDa polypeptide that contains the [^3H]strychnine binding site has been purified from spinal cord and cloned. The postsynaptic glycine receptor–chloride ionophore complex is the major mediator of synaptic inhibition in the spinal cord and brainstem. Glycine and strychnine, its major pharmacological antagonist, bind to two distinct sites on this receptor complex that interact cooperatively. The [^{3}H]strychnine binding site is structurally associated with the transmembrane hydrophobic domains of the 48-kDa polypeptide and therefore with the chloride channel itself.

Benzodiazepines interact competitively with strychnine at the [^{3}H]strychnine receptor site in crude synaptic membranes prepared from spinal cord; of a series of benzodiazepines that were examined, flunitrazepam was the most effective competitive inhibitor of specific [^{3}H]strychnine binding (ED_{50} = 19 μM).[32] The data suggest that the benzodiazepines interact with the [^{3}H]strychnine site specifically, whereas glycine interacts at a different but cooperative site on the glycine–chloride ionophore complex. Benzodiazepines bind with much higher affinity to a specific receptor site that is structurally associated with the $GABA_A$ receptor and its own distinct chloride ionophore in a tetrameric glycoprotein complex; e.g., the K_D for [^{3}H]flunitrazepam binding to this site is about 1 nM.[33–35] At low concentrations of benzodiazepines, binding to the benzodiazepine–GABA receptor complex would occur preferentially to increase the likelihood that GABA would be effective in promoting chloride ion conductance. However, at the concentrations achieved clinically in the blood and CNS, it is highly probable that significant binding to the [^{3}H]strychnine site occurs interfering with the inhibitory effects of glycine in the spinal cord and brainstem. The abilities of strychnine and benzodiazepines to interfere with glycinergic inhibitory transmission allosterically at the postsynaptic glycine–chloride ionophore complex forms the basis of intervention strategies with these agents.

In addition to its neurotransmitter actions at a specific glycine receptor coupled to a chloride ionophore and a [^{3}H]strychnine binding site, glycine may facilitate the postsynaptic actions of L-glutamate, an excitatory neurotransmitter, in an allosteric fashion at [^{3}H]strychnine-insensitive binding sites.[27–29] L-Glutamate is the principal neurotransmitter of cortical pyramidal cell neurons and is used by these cells to form neocortical efferent and associational pathways.[36] L-Glutamate is also a major neurotransmitter of the entorhinal cortical input to the hippocampus via the perforant pathway, intrahippocampal pathways, and hippocampal efferent pathways. Subclasses of postsynaptic L-glutamate receptors are characterized electrophysiologically and in receptor-binding studies according to their relative sensitivities to three prototypic agonists: kainate, quisqualate, and NMDA. None of these excitants is an endogenously occurring ligand. The postsynaptic L-glutamate binding sites are coupled to cation-selective channels with multiple conductance states responsible for their excitatory actions.

The cation channel coupled to the NMDA subclass shows features of both voltage-sensitive and neurotransmitter receptor-sensitive "gating"; i.e., under conditions of hyperpolarization or quiescence when there is a more negative membrane potential, the channel is relatively insensitive to the effects of NMDA (Fig. 2). This channel insensitivity appears to be due to blockade of its central pore near the intracellular side of the membrane by Mg^{2+} ions at negative membrane potentials. However, as the cell begins to depolarize this Mg^{2+} ion blockade is relieved and NMDA can promote the inward conductance of Ca^{2+} ions. The actions of NMDA to promote Ca^{2+} ion conductance can be antagonized in a noncompetitive allosteric fashion at separate sites located on or within the channel by two potent psychotomimetic agents in man: phencyclidine (PCP) and SKF-10,047. Recently, glycine was shown to be a potent agonist or facilitator of NMDA-mediated neurotransmission; it is effective in nanomolar to low micromolar concentrations. Glycine does not bind to the binding site for NMDA but appears to act at a separate and distinct site that is insensitive to displacement by strychnine. The role of glycine at this site is viewed as analogous to that of benzodiazepines acting at a site distinct from the $GABA_A$ binding site to increase the likelihood that chloride

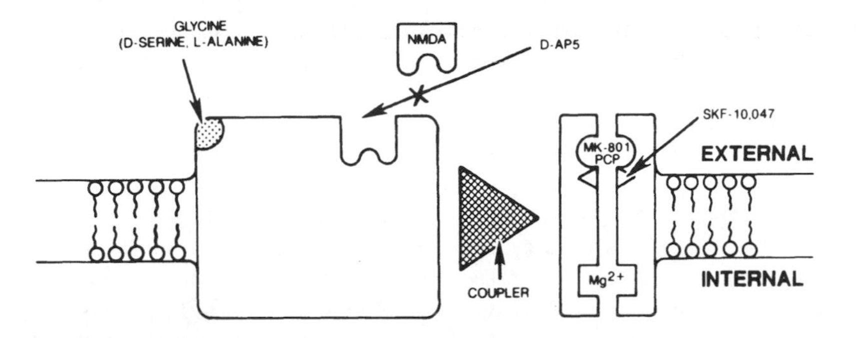

Figure 2. The NMDA receptor complex. The *N*-methyl-D-aspartate (NMDA) receptor is a subclass of postsynaptic glutamate receptor characterized by its affinity for the agonist NMDA. The receptor is coupled to a gated channel that permits cation flux when glutamate is bound to the NMDA agonist recognition site. In the presence of a negative membrane potential, the ability of glutamate to promote cation flux is antagonized by the hydrated magnesium ion occupying a site located near the intracellular side of the channel. As the membrane potential becomes less negative, there is a decreased likelihood that the hydrated magnesium ion will antagonize glutamate-stimulated cation flux. Glycine acts at an allosteric regulatory site that is distinct from the NMDA agonist recognition site to facilitate the ability of glutamate to promote cation conductance. Several compounds have been identified that act at sites located near or within the channel to antagonize the ability of glutamate to promote cation conductance. A therapeutic role for these allosteric antagonists of NMDA-mediated neural transmission may exist in nonketotic hyperglycinemia. Conceivably, a compound from this class may be employed to antagonize the presumed facilitation of glutamatergic transmission by glycine in this condition (see text for additional details). (Adapted from Kemp et al.[29])

ion conductance would be promoted by GABA.[27–29] These recent data on a neuromodulatory role for glycine in promoting NMDA-mediated Ca^{2+} ion conductance raises the intriguing possibility that some of the toxic actions of glycine in nonketotic hyperglycinemia (especially seizures) could be due to a facilitation of glutamatergic transmission. The hypothesis is a testable one in view of the active development of allosteric antagonists of NMDA-mediated transmission for the treatment of seizures and ischemic events in the brain. The prototypic NMDA antagonist under development is MK-801.

There is an impression that neonates with immature development of their blood–brain barriers may be more sensitive to the toxic effects of experimentally induced hyperglycinemia. Newborn and 8-day-old rats were extremely sensitive to the neurotoxic effects of high-dose glycine administration (3 mg/g s.c.); newborn rats treated with equimolar doses of serine did not show neurotoxic effects.[37] Toxicity consisted of a characteristic rotatory behavior (termed *barreling*) that appeared within minutes of injection, inhibition of movement, and respiratory depression. The dose produced glycine concentrations in the forebrain of the newborn rat comparable to concentrations seen in autopsied brain from patients with nonketotic hyperglycinemia. Moreover, more than two thirds of these immature animals receiving high-dose glycine died. Neurotoxic and lethal effects of glycine were not observed in older rats (age 15 and 21 days) with even higher forebrain concentrations. The lethality of exogenous glycine (3 mg/g) in immature animals could be antagonized by a dose of strychnine that was not associated with the induction of rigidity or death when administered alone. These data suggest the existence of a critical period of heightened sensitivity to the neurotoxic effects of glycine in early development. Also, the data are consistent with a receptor-mediated mechanism of neurotoxicity and a salutary role for strychnine in immature animals.

The experimental induction of hyperglycinemia in the CNS of 10-day-old rats was associated with an increase in the maximal density of [^{3}H]strychnine binding sites in spinal cord membranes; the K_D for the binding of [^{3}H]strychnine was not changed.[38] Experimental hyperglycinemia did not affect the maximal density of [^{3}H]strychnine binding sites in spinal cord membranes of rats older than 10 days of age. During the first 10 days of life, the rat shows a rapid increase in the presynaptic uptake sites for, and calcium-stimulated efflux of, glycine and the number of [^{3}H]strychnine binding sites. These data suggest that the ability of hyperglycinemia to alter [^{3}H]strychnine binding is greatest during the postnatal period of rapid development of glycinergic synapses.

However, experimentally induced hyperglycinemia in adult male rabbits was associated with increased levels in CSF and a progression of neurological symptoms that resembled those seen in nonketotic hyperglycinemia.[39] The symptoms included lethargy, hypotonia, ataxia, problems with balance, myoclonic seizures, and respiratory difficulties. In this study, adult animals with fully developed nervous systems and intact blood–brain barriers were sensitive to the neurotoxic effects of glycine. Administration of maximally tolerated doses of strychnine could not alter the progression of symptoms suggesting that at least some of the effects of glycine could be due to its actions at strychnine-insensitive sites.

3. *HETEROGENEITY OF CLINICAL EXPRESSION: NONKETOTIC HYPERGLYCINEMIA AS A CAUSE OF UNEXPLAINED DEVELOPMENTAL DELAY*

Glycine levels in plasma and CSF were measured and the electroencephalogram (EEG) recorded within the first few hours of life in two infants with severe and classic presentations of nonketotic hyperglycinemia.[40] At birth, the plasma glycine levels of both infants were in the normal range; presumably, this reflected maternal clearance of excess plasma levels in the fetus. However, the EEG record of one infant at 30 min and the other at 2 hr of life showed the characteristically abnormal burst-suppression pattern. The abnormal EEG records could be accounted for by elevations of CSF glycine levels in the early neonatal period, despite normal or only midly elevated plasma levels at these times. Glycine values in the CSF were elevated about eightfold in one infant and ninefold in the other infant above the upper limit of normal at 30 and 90 min of life, respectively. The levels in CSF continued to rise during the first day of life. These data emphasize the virtual independence of brain glycine levels from concentrations in the blood; the brain is capable of synthesizing its metabolic requirements for glycine in situ. Also, the brain of the developing fetus with nonketotic hyperglycinemia must express the defect in the glycine cleavage enzyme system with elevations of in situ concentrations to toxic levels. If this is so, the expectations of postnatal therapy for the classic condition must be modest.

The complexity of glycine regulation was demonstrated in the report of three brothers with an atypical presentation of probable nonketotic hyperglycinemia, and upper and lower motor neuron disease.[3] The three brothers showed evidence of spasticity and muscle wasting of the lower extremities and positive Babinski signs; the upper extremities were either unaffected (two cases) or mildly affected relative to the lower extremities. There was no evidence of impaired intellectual function in any of the boys. Glycine levels were elevated in urine, plasma, and CSF; CSF levels were measured in two of the boys and were about twofold higher than the upper limit of normal. Plasma serine levels were measured in two of the boys 1.5 hr after an oral glycine load and did not increase; this finding is consistent with a defect in glycine cleavage system activity and a diminished glycine to serine conversion. Moreover, there was a lower than expected rise in the content of radioactive CO_2 ($^{14}CO_2$) in expired air at 10 and 30 min following intravenous administration of glycine radiolabeled in the C-1 position (glycine-1-^{14}C) to two of the boys. This result is most consistent with a defect in glycine cleavage system activity. If the motor neuron disease is related to the hyperglycinemia and defective glycine cleavage system activity exclusively, these results are difficult to explain. Glycine is an inhibitory neurotransmitter in spinal cord, but spasticity and muscle wasting of the lower extremities are not typical accompaniments of the disorder. Also, these patients showed no evidence of seizure activity or abnormal EEGs. Most strikingly, these patients had normal mentation. In any event, as shown by these cases, atypical presentations of nonketotic hyperglycinemia can occur with mild or no impairment of intellectual function.

Two sibs were reported with glycine elevations in urine, plasma and CSF that were in the range seen in children with classic presentations of nonketotic hyperglycinemia; one of these sibs was only minimally to moderately delayed developmentally, and the other was normal.[4] Neither child showed abnormal elevations of urinary organic acids. At age 23 months, the clinically affected sib showed delays of 5–9 months on the Gesell scales, was slightly hypotonic, and had no spoken language. The children had no history of seizures, and EEG records were normal.

Nonketotic hyperglycinemia may not be as rare a disorder as initially thought and milder presentations may be referred to the child psychiatrist for evaluation. For example, two preschool-age siblings were reported for whom the initial referring complaint was that of a slightly below normal rate of mental and motor development.[5] An aminoacidopathy was only suspected after a developmental evaluation of the younger sibling at age 35 months that included chromatographic analysis of the urine; the analysis revealed an abnormally elevated glycine level. The parents of these children were described as highly intelligent and nonconsanguineous. The oldest child was a boy; his delivery and perinatal period were uneventful. He "sat alone at 7 months, walked alone at 15 months" and linked three words at 22 months. At 22 months, the boy was admitted to a hospital with otitis media and elevated temperature (39.4°C) associated with choreoathetoid movements, visual problems and paroxysmal screaming. A spinal tap was normal (protein and glucose levels and cells were evaluated) and he was discharged after 2 days; a follow-up EEG performed about 6 weeks after discharge was normal. A developmental evaluation was performed at age 30 months; clinically, he showed occasional stumbling and obvious problems with expressive speech. Formal testing with the Bayley Infant Scales and the Stanford–Binet put his mental age at 23 months. Receptive language was at age level, whereas expressive language was at 21–24 months. Perceptual functioning and fine motor development were delayed 6–12 months. A Wechsler scale performed during latency showed a full-scale IQ of 76. His younger sister was reported to reach many of her early motor milestones on time; she sat alone at 7 months, pulled herself to stand at 12 months, and walked alone at 15 months. However, there was some early concern during infancy about head support and ability to roll over. She was hospitalized at age 8 months for 3 days because of a fever followed by a morbilliform rash for 2 days after its resolution. This roseola presentation was accompanied by lethargy, irritability, and a mild dysmetria. Analysis of the CSF and an EEG was normal. At age 35 months, she underwent a developmental evaluation because of both poor coordination and articulation. Gross and fine motor development were delayed (18- and 24-month level, respectively) and dyskinetic rotary movements of forearms, wrists, and fingers were noted. A full-scale IQ during latency was 67. In both children, glycine levels in urine, deproteinized serum, and CSF were elevated significantly. Moreover, glycine levels in the venous blood of these children were persistently elevated for up to 4 hr after an oral glycine load of 150 mg/kg of body weight. The fact that urinary serine levels did not increase in these children after glycine loading suggested the diagnosis of nonketotic hyperglycinemia; in this form of hyperglycinemia, there is a defective conversion of glycine to serine. The baseline EEGs of these children were normal; however, the highest peak levels associated with glycine loading caused slowing of the θ waves. The children were treated with dietary restriction of glycine. Subjectively, this therapy was thought to contribute to a reduction of tantrums and negative behavior. These cases emphasize the importance of including a careful metabolic screening in the evaluation of children with developmental disorders.

The heterogeneity of the clinical presentation was emphasized by the report of a girl studied at age 10 years, one of three affected sisters, with a relatively mild phenotype, despite markedly abnormal biochemical indices.[6] She was able to attend to her feeding and personal hygiene, was sociable with familiar individuals, and able to speak a few words. However, the rate of $^{14}CO_2$ evolution in expired air from an intravenous injection of glycine-1-^{14}C was in the range of severely affected phenotypes. Also, there was a profound reduction in the incorporation of C-2 of glycine-2-^{14}C into the C-3 of serine. Thus, the relatively mild course of her illness could not be explained by the severely abnormal biochemical indices.

The age of onset and clinical course of the disorder are not always uniform or predictable from the biochemical indices. For example, a female infant was described with normal growth and development until age 6 months.[41] Thereafter, she showed a rapidly progressive deterioration with the onset of seizures at 9 months and death due to respiratory arrest at age 15 months. Her rapid deterioration could not be slowed, despite therapeutic trials of strychnine, pyridoxine and citrovorum factor. The ratio of plasma to CSF glycine levels was not in the range associated with the more typical and severe presentations. Thus, there was no basis for predicting her rapidly progressive course.

In another report, three latency-age male siblings showed biochemical abnormalities consistent with an atypical presentation of nonketotic hyperglycinemia.[8] The disorders were typical of those commonly referred to a child psychiatrist for evaluation. In general, the children presented with mild to moderate retardation, speech delay with expressive language more severely affected than receptive, hyperactivity, and poor fine motor coordination. Glycine levels in the CSF were elevated about four- to sevenfold above normal. The three siblings could not clear an oral glycine load (150 mg/kg) normally; plasma glycine levels remained elevated above their abnormally high baseline for up to 8 hr. All three siblings showed a defective conversion of glycine to serine following the oral glycine load. None showed evidence of organic acids in urine.

Neuroimaging studies (computed tomography (CT) and magnetic resonance imaging (MRI) scans) in 23 patients with nonketotic hyperglycinemia showed a range of abnormalities, including parenchymal volume loss, agenesis of the corpus callosum, gyral malformations, cerebellar hypoplasia, delayed myelination, and enlarged lateral ventricles consistent with the fetal configuration, among others.[42,43] Many of these neurostructural abnormalities must have arisen during the first trimester of development. Therefore, nonketotic hyperglycinemia and a potential recurrence risk of 25 percent to future offspring should be considered when a structural abnormality of the brain is recognized in a young infant. Moreover, these data would dampen the expectation of therapeutic efficacy from currently available therapies.

Nonketotic hyperglycinemia can be the cause of unexplained mental retardation in adults. These cases represent survival into adult life of atypical presentations. In one report, two siblings (ages 34 and 17 years) were identified with CSF to plasma glycine ratios elevated above normal (i.e., > 0.03) in the absence of organic acidemia.[7] These patients developed speech that was delayed; they also manifested seizure activity and required institutionalization because of self-mutilatory and aggressive behaviors. In a 22-year-old profoundly retarded, hyperactive man with an atypical presentation (i.e., a relatively mild course of illness characterized by rare seizures, a few simple phrases, and self-care skills), the CSF/plasma glycine ratio was in the range of a mild variant, whereas hepatic activity of the overall glycine cleavage system and P-protein component was in the severely affected range.[44] These data suggest that a dissociation may exist between hepatic and brain glycine cleavage system activity. These data also support the consideration of nonketotic hyperglycinemia as a cause of developmental disorders in childhood.

4. THERAPEUTIC INTERVENTIONS IN NONKETOTIC HYPERGLYCINEMIA

The major approaches to the treatment of nonketotic hyperglycinemia are outlined in Table I. In general, they are designed to reduce the toxic load of glycine, increase the activity of the glycine cleavage enzyme system, supplement the one-carbon pool size, and antagonize glycine at postsynaptic receptor sites. In some instances, therapy has led to a decrease in the frequency and intensity of seizures and improved awareness. The effectiveness of a given intervention seems to depend on the severity of the reduction in glycine cleavage enzyme system activity, the age at which therapy is initiated, and the components of the therapeutic regimen. The classic presentation remains the most refractory to any positive change. This section presents a selective and critical review of the experiences of several investigators.

A 10-day trial of methionine (2.75–3.75 g/day) was added to a regimen of dietary restriction of glycine and serine in a 12-month-old child with nonketotic hyperglycinemia; dietary restriction of glycine and serine alone failed to reduce plasma glycine levels.[10] Methionine was added as a source of one-carbon units, and cystine was also restricted from the diet to maintain sulfur intake within normal limits. The methionine loading reduced plasma glycine levels to the upper limit of normal with a return to elevated values upon termination of the trial. The effect of methionine did not appear to be due to competition with glycine for tubular reabsorption by the kidneys, as there was no increased excretion of glycine in the urine. De Groot et al. speculated that the alternative source of one-carbon units provided by methionine reduced the dependence on the serine to glycine conversion for the provision of one-carbon units. According to this view, the reduced conversion led to a decreased accumulation of glycine. There was no discussion of a change in the clinical state of the infant.

High-dose therapy with N^5-formyl THF (leucovorin) was attempted in a severely affected infant presenting with seizures and apnea that required mechanical ventilation.[11] It was hoped that N^5-formyl THF would replete a deficient one-carbon pool and detoxify glycine by stimulating its conversion to serine by the enzyme serine hydroxymethyltransferase. Also, in the

Table I. Review of Therapeutic Strategies in Nonketotic Hyperglycinemia

Dietary restriction of glycine
Cofactor administration
Folic acid
Pyridoxine
Lipoic acid
Supplementation of one-carbon pool
Leucovorin (N^5-formyl THF)
Formate
Choline
Methionine
Conjugation and accelerated urinary excretion of glycine
Sodium benzoate
Exchange transfusion
Ventricular–peritoneal shunt
Noncompetitive allosteric antagonism of the glycine–chloride ionophore complex
Strychnine
Benzodiazepines
Noncompetitive allosteric antagonism of the NMDA-receptor complex
MK-801

event that the patient had an inborn error that was responsive to high levels of exogenously administered cofactor (THF), this strategy might stimulate a defective apoenzyme to cleave glycine. One day of intramuscular loading (15 mg/kg q6h) was followed by gavage administration of 45 mg/kg q4h. The infant died 3 days after N^5-formyl THF initiation.

The dissociation between the regulation of glycine levels in plasma and CSF was shown dramatically in a reported case of an insertion of a ventricular–peritoneal shunt in a female infant with nonketotic hyperglycinemia.[12] In this patient, CSF levels of glycine remained elevated despite vigorous therapy with sodium benzoate; CSF glycine levels remained 14-fold higher than the upper limit of normal 2 days after the sodium benzoate dose was increased to 250 mg/kg q4h (a more recent study suggests that a 2-week lag may occur before CSF reduction with high-dose benzoate is noted). The patient showed severe developmental delays, including seizures during the waking state, hypotonia, and roving dysconjugate eye movements. The investigators elected to shunt this patient because of their clinical impression in several patients that the severity of brainstem depression was a function of CSF glycine concentration and their inability to lower CSF glycine levels with sodium benzoate administration. At 1 and 5 weeks after the shunt procedure, there was no reduction in glycine levels in ventricular or lumbar fluid. Moreover, the dynamics of glycine resorption from and transport into the CSF must be complex because its levels in lumbar spinal fluid were actually increased by 43% at 5 weeks after the shunt. In this same study, the effect of loading a healthy adult (one of the investigators) with massive doses of glycine (10 g q4h for 9 doses) on plasma and CSF glycine levels was reported; a spinal tap was performed immediately after venipuncture. Plasma levels and CSF values were increased five- and fourfold, respectively. However, in contrast to patients with nonketotic hyperglycinemia, the plasma to CSF ratio was within the normal range. Subjectively, the investigator reported no ill effects from his pharmacologically induced hyperglycinemia.

Beneficial effects of strychnine were first reported in a male infant whose treatment with this noncompetitive allosteric antagonist of glycine was begun at age 6½ months.[13] The presentation of this infant did not include respiratory depression and is thus viewed as less severe than a classic one. Despite hyperglycinuria detected at age 8 days, motor behavior and eye movements were reported as normal when admitted to the hospital at age 16 days. During this admission, the diagnosis of nonketotic hyperglycinemia was made because of elevated CSF glycine levels and the persistent elevation of plasma glycine values with a slow return to the abnormally elevated baseline 24 hr after an oral load (200 mg/kg). At 8 hr after the oral load, EEG changes (generalized sharp wave complexes), generalized cloni and upward deviation of the eyes were recorded. Muscle tone would alternate between hypertonia with rigidity and hypotonicity. On occasion and for brief periods, he would lift his head from the prone position, smile, and make visual contact of a few seconds. Despite treatment with folic acid and vitamin B_6, generalized seizures developed. The vitamins were discontinued and these seizures were poorly responsive to clonazepam (1 mg/day). Therapy with strychnine nitrate alone was initiated at a dose of about 0.1 mg/kg per day in three divided doses. His response to therapy with strychnine alone was complicated and showed fluctuations; on a dose of about 0.4 mg/kg per day, muscle tone, visual contact and head control improved, and he was more alert and active, making "talking noises." However, he had generalized convulsions; therefore, the strychnine dose was reduced and clonazapam (1.2 mg/day) was reintroduced. On this regimen, the patient showed muscular hypotonia, poor head control, less frequent smiling, and an almost absence of following with his eyes. At age 7½ months, the clonazepam was replaced with another benzodiazepine (Ro07-9957) and, on this regimen, dramatic improvement was observed. At 12½ months, a withdrawal strategy was employed to demonstrate the contribution of strychnine to the improved clinical picture; within 2 days of strychnine withdrawal, the patient became extremely floppy, almost motionless with absent tendon reflexes, and severely withdrawn from the environment. In less than 1 day after strychnine reinitiation, he was smiling, socially responsive and babbling, and showed normal muscle tone and tendon refex-es. Although he remained severely developmentally delayed, he reached many developmental milestones and at age 19 months, he was crawling and grasping for desired objects on a dose of strychnine nitrate of about 0.9 mg/kg. At this age, the index patient attained developmental milestones never achieved by a brother age 8½ years with a confirmed diagnosis of nonketotic hyperglycinemia. Clearly, the response of this patient was gratifying, especially in view of the unrelenting deterioration of the older brother. The basis of this dramatic response is somewhat unclear, however, because of the known competitive antagonism between strychnine and benzodiazepines at the [^{3}H]strychnine site on the glycine receptor–chloride ionophore complex. In any event, the beneficial effect of strychnine in this patient was reported to be sustained 4 years after its initiation in a later publication.

Strychnine therapy was reported to have a dramatically beneficial effect in combination with sodium benzoate (150 mg six times/day) and clonazepam (1.5 mg/day) in a severely growth retarded and developmentally delayed 5-month-old black female infant; the sulfated form of strychnine was given in a dosage of 100 μg/kg per day.[14] The infant's presentation was classical with respiratory distress and hypotonicity at 16 hr after birth. Early infancy was characterized by frequent mycolonic and generalized tonic and clonic seizures refractory to medication; irregular respirations; wandering eye movements; vomiting; poor head control, lethargy and hypotonicity. At 5 months of age, the diagnosis of nonketotic hyperglycinemia was made based on a more than 50-fold elevation of CSF glycine levels and the absence of ketones and keto acids in the urine. On the basis of this diagnosis, therapy with benzoate was initiated and followed in 2 weeks by the addition of strychnine; she had been receiving clonazepam before the diagnosis. Although a reduction of plasma glycine levels to near-normal values occurred with benzoate alone, the addition of strychnine was associated with an almost immediate and dramatic improvement of several parameters: alertness, vomiting, involuntary movements, and frequency of clinical seizures. At age 7 months, a normal sleep pattern emerged with daytime alertness. The EEG improved on strychnine therapy, and abnormal epileptiform discharges were not present 3 weeks after therapy was initiated. Empirically, this regimen was remarkably successful for this patient. Theoretically, the possibility of competitive antagonism between strychnine and clonazepam at the specific [^{3}H]strychnine binding site associated with the glycine–chloride ionophore complex could be expected to occur if local clonazepam concentrations were in the micromolar range. Alternatively, the beneficial effects of this combination regimen on seizure production could reflect a synergism between the antagonism of glycine by strychnine and the facilitation of GABA-mediated chloride ion conductance by clonazepam. These data also emphasize the virtual independence of glycine levels in brain from levels in plasma; despite the dramatic reduction in plasma glycine values with benzoate, the levels in CSF remained more than 24-fold elevated.

In two classic presentations with nonketotic hyperglycinemia, vigorous therapy with strychnine sulfate was not associated with significant beneficial effects.[18] The first case described was that of a male infant with CSF glycine levels more than 34 times normal who was referred at age 10 weeks because of almost continuous seizures on phenobarbital, reduced muscle tone, and poor head control. His EEG showed diffuse bilateral asynchronous bursts; spike activity was followed by 1–3 seconds of electrical suppression. Following the diagnosis, leucovorin was initiated for a 1-month trial without improvement. Seizure activity persisted despite combination regimens that included phenobarbital with phenytoin and phenobarbital with carbamazepine. At age 5 months, strychnine sulfate was initiated at a dose of 0.4 mg/kg per day and increased gradually to 0.7 mg/kg per day in divided doses q6h. Hypotonia and seizures persisted, and there was no change in the EEG recording. The patient did not respond to "vigorous physiotherapy and social stimulation." However, feeding improved dramatically in association with strychnine therapy. The strychnine sulfate dose was eventually increased to 1.2 mg/kg per day. On CT scan at 10 months of age, the child showed some evidence of ventricular dilatation and cerebral atrophy. He died at age 10½ months; examination of the brain revealed ischemic changes and a diminution of myelinated fibers within the white matter. The second case was also a male infant transferred at 1 month of age because of constant seizures and marked hypotonia. The EEG was unchanged from that performed at ages 3 and 10 days, which showed the burst-suppression pattern with asynchrony between hemispheres. CSF glycine levels were elevated more than 27-fold above normal values. Strychnine sulfate was initiated at age 6 weeks at a dose of 0.1 mg/kg per day and increased gradually to 0.8 mg/kg per day in divided doses q6h. The EEG pattern at 2 months was unchanged; feeding improved significantly. At age 6 months, the strychnine sulfate dose was 1 mg/kg per day, and there was better seizure control on a regimen that included nitrazepam and phenytoin. A drug-withdrawal strategy for 1 week was used to assess strychnine efficacy, and there was a worsening of hypotonicity and feeding that improved with strychnine reinitiation; there was no change in seizure frequency. The patient remained socially unresponsive, and a study of auditory brainstem-evoked potentials was consistent with a delay in maturation. These authors were not impressed with the value of strychnine therapy.

The failure to influence the course of nonketotic hyperglycinemia positively has often been attributed to the institution of therapy in late infancy, as opposed to the neonatal period. According to this view, the infant's developing nervous system is exposed to the irreversible toxic consequences of glycine accumulation and one-carbon depletion for months before the initiation of therapy. Therefore, it was of considerable interest to examine potential therapeutic effects of vigorous therapy with strychnine in neonates with nonketotic hyperglycinemia.[17] Nonketotic hyperglycinemia was suspected in three neonates presenting with lethargy, hypotonia, and nursing difficulties 7–8 hours after birth and confirmed 15–40 hours after birth. These children were examples of a severe presentation of the disorder. Therapy with strychnine nitrate was initiated within 15–62 hours of birth, and the dosage ranged from 0.2 to 0.9 mg/kg per day in four divided doses; two of the children also received exchange transfusions in an effort to reduce the toxic load of glycine. Despite these efforts, all three children died between the fifth and ninth days of life. The data suggest that, in severe cases, noncompetitive antagonism of glycine (with the dosages of strychnine described in this report) at the glycine receptor–chloride ionophore complex is without obvious beneficial effect.

Strychnine initiation during the first few days of life may be able to reverse the severe brainstem depression in the classic presentation associated with elevated CSF glycine levels, but it may have little effect on the progression of the disorder. A single oral dose of strychnine nitrate (0.2 mg) was effective in permitting the extubation of an 8-day-old infant with respiratory depression within 30 min of its administration.[16] A maintenance regimen of strychnine also had salutary effects on muscle tone and awareness. However, despite the inclusion of strychnine in her regimen, at 7 months of age she required tube feedings and showed no spontaneous motor activity.

Therapy with strychnine nitrate was begun in the 73rd hour of life in dizygotic (DZ) twins with a classic presentation of nonketotic hyperglycinemia.[15] Both twins were severely hypotonic, lacked gag reflexes, and required intubation and mechanical ventilation. The EEGs revealed a burst-suppression pattern with phases of low activity alternating with synchronous high-amplitude paroxysmal discharges. Cofactors of the glycine cleavage enzyme system (pyridoxine, N^5-formyl THF and lipoic acid) and an exogenous source of one-carbon units (N^5-formyl THF) were included in the regimen. A benzodiazepine (Rivotril) was also included to control seizures. The highest daily dose of strychnine nitrate explored was 2.6 mg/kg of body weight. The regimen was associated with improvement in the following areas: muscle tone, spontaneous movement (occurred in one twin and became more apparent in the other twin), periods of hiccuping (either shortened or disappeared), and the appearance of "defense reactions." However, neither twin ever became alert, sucked, or swallowed, and the therapeutic trial was aborted after 60 hr. The clinicians were impressed with what they termed "remarkable" improvement in the areas noted above, but they believed that this was insufficient to justify continued treatment with strychnine. The disorders were not vitamin responsive, as glycine levels in CSF and plasma were not reduced. These data are consistent with several other reports of a less than dramatic effect of strychnine therapy in infants with the early-onset and severe form of nonketotic hyperglycinemia. The results could also suggest that failure to respond to strychnine could reflect irreversible damage to the CNS in utero in infants with a classic presentation.

Strychnine was associated with an improvement in respiratory drive and rate, lethargy, muscle tone, and sucking in two identical male twins with nonketotic hyperglycinemia; the toxic–therapeutic ratio of strychnine dosage was narrow.[19] The diagnosis was made on age 18 days, and therapy with strychnine was initiated soon thereafter. However, strychnine therapy was associated with the induction of myoclonic jerks and generalized seizures; baseline EEGs revealed paroxysmal bursts of sharp wave activity. The dosage of strychnine required frequent adjustment because of the associated exacerbation of the hypsarrhythmic seizure disorder. Clonazepam was added to the regimen to control myoclonic jerks and seizure activity that occurred during treatment with strychnine. The maintenance strychnine dosage on the combined regimen was 2.0 mg/kg per day in divided doses every four hours. The twins died of status epilepticus at age $6\frac{1}{2}$ months; their levels of development were below 1 month. In this report, there was an impression that strychnine acted dichotomously to improve brainstem depression but to exacerbate seizure activity. Warburton et al.[19] speculated that the adverse effect of strychnine could be due to its interaction with sites other than those associated with the glycine receptor–chloride ionophore.

A regimen that resulted in long seizure-free intervals and that increased responsiveness to painful and other environmental stimuli in two female infants consisted of diazepam (blood levels of 180–230 μg/dl), choline (1–4 g/day), sodium benzoate (125 mg/kg), and folic acid (2 mg/day).[20,21] In both infants, levels of glycine in CSF were elevated about 40- to 125-fold greater than normal CSF values; both patients were reported to be hypotonic and have continuous seizure activity. On this regimen, a severely hypotonic infant who showed no interaction with her environment at age 8 months began to follow objects, roll over, and babble after 1 year of therapy. Interestingly, these infants showed an increased state of wakefulness on this regimen, which included diazepam. In both cases, seizures were reported to stop, and the infants became more alert, although they remained severely developmentally delayed. Plasma glycine levels remained elevated despite benzoate administration, and this could reflect an accelerated compensatory interconversion of serine to glycine to maintain the one-carbon pool and the biosynthesis of glycine from choline. Matalon et al.[21] speculated that the failure to observe greater developmental gains could have been due to the institution of therapy at relatively advanced ages (i.e., 5 and 8 months).

The activity of the T-protein component of the glycine cleavage system was about 100th of the lower limit of activity of a control series in a girl with an atypical presentation of nonketotic hyperglycinemia.[23] This patient survived the neonatal period, displayed truncal hypotonia, and developed multifocal myoclonic jerks. At age 5 years, her developmental level was equivalent to that of an infant of 6–12 months. A younger sister of this index patient was shown to have CSF glycine levels elevated more than 10-fold above normal at 36 hr after birth. The younger sister was begun on a regimen that included strychnine (1 mg/kg per day) at age 15 days in an attempt to avert the development of a clinical picture similar to that of her older sibling. On strychnine, she reached all her developmental milestones during the first year of life on time. She showed developmental delays in her second year; she walked at age 21 months. The child was developmentally delayed about 1 full year at age 27 months. The strychnine-treated sister showed no evidence of seizure activity. There was an impression that strychnine contributed to an improvement of the clinical course; the treated sister was less severely retarded and seizure free. However, it is difficult to attribute the milder outcome to strychnine exclusively because there have been cases of discordance with respect to severity of illness in multiply affected sibships.

Glycine can be effectively conjugated with sodium benzoate to form hippurate, an efficiently excreted compound; although this intervention strategy lowers plasma levels of glycine, it has been viewed as inconsistently effective in seizure reduction. A possible reason for the failure to observe consistently beneficial effects with sodium benzoate could be the administration of therapeutically ineffective doses.[22] Plasma glycine levels are an inadequate guide to dosing as these levels are more sensitive to the benzoate-induced reduction than levels in the CSF. The titration of oral administration of sodium benzoate to the lowering of CSF glycine levels in fasting morning samples was associated with a dramatic reduction in the frequency and duration of seizures in three patients. A lag phase of about 2 weeks was observed before the maximal reduction of CSF glycine levels. The relationship between reduction of CSF levels and benzoate dose was linear; however, even with doses approaching 1 g/kg per day, CSF levels were never reduced to normal. Anorexia, vomiting, and renal tubular dysfunction were the adverse effects associated with high-dose benzoate therapy.

5. CONCLUSIONS

The recognition of nonketotic hyperglycinemia led to a greater appreciation of the important metabolic and neurotransmitter functions of glycine in the central nervous system.[2] The clinical manifestations of the disorder are varied and their pathogenesis incompletely understood; they are thought to relate to alterations of one-carbon metabolism and toxic accumulation of glycine, an amino acid neurotransmitter, in the brain. The salutary effect of a therapeutic regimen may depend on its components, the age at which intervention is initiated, and the severity of the reduction in glycine cleavage system activity. Milder presentations of nonketotic hyperglycinemia can account for at least some of the developmental disorders referred to a child psychiatrist for evaluation. Nonketotic hyperglycinemia may be viewed as a paradigm for the application of basic neuroscience to a clinical problem in pediatric neurology and child psychiatry.

ACKNOWLEDGMENT. The editorial assistance and typing of Ms. Denise Reeves are gratefully acknowledged.

REFERENCES

1. Ziter FA, Bray PF, Madsen JA, et al: The clinical findings in a patient with non-ketotic hyperglycinemia. Pediatr Res 2:250–253, 1968
2. Perry TL, Urquhart N, MacLean J, et al: Nonketotic hyperglycinemia. Glycine accumulation due to absence of glycine cleavage in brain. N Engl J Med 292:1269–1273, 1975
3. Bank WJ, Morrow G III: A familial spinal cord disorder with hyperglycinemia. Arch Neurol 27:136–144, 1972
4. Holmgren G, son Blomquist HK: Non-ketotic hyperglycinemia in two sibs with mild psycho-neurological symptoms. Neuropadiatrie 8:67–72, 1977
5. Frazier DM, Summer GK, Chamberlin HR: Hyperglycinuria and hyperglycinemia in two siblings with mild developmental delays. Am J Dis Child 132:777–781, 1978

6. Ando T, Nyhan WL, Bicknell J, et al: Non-ketotic hyperglycinaemia in a family with an unusual phenotype. J Inher Metab Dis 1:79–83, 1978
7. Flannery DB, Pellock J, Bousounis D, et al: Nonketotic hyperglycinemia in two retarded adults: A mild form of infantile nonketotic hyperglycinemia. Neurology (NY) 33:1064–1066, 1983
8. Cole DEC, Meek DC: Juvenile non-ketotic hyperglycinaemia in three siblings. J Inher Metab Dis 8(suppl 2):123–124, 1985
9. Ando T, Nyhan WL, Gerritsen T, et al: Metabolism of glycine in the nonketotic form of hyperglycinemia. Pediatr Res 2:254–263, 1968
10. DeGroot CJ, Troelstra JA, Hommes FA: Nonketotic hyperglycinemia: An *in vitro* study of the glycine–serine conversion in liver of three patients and the effect of dietary methionine. Pediatr Res 4:238–243, 1970
11. Spielberg SP, Lucky AW, Schulman JD, et al: Failure of leucovorin therapy in nonketotic hyperglycinemia. J Pediatr 89:681, 1976
12. Krieger I, Winbaum ES, Eisenbrey AB: Cerebrospinal fluid glycine in nonketotic hyperglycinemia. Effect of treatment with sodium benzoate and a ventricular shunt. Metabolism 26:517–524, 1977
13. Gitzelmann R, Steinmann B, Otten A, et al: Nonketotic hyperglycinemia treated with strychnine, a glycine receptor antagonist. Helv Paediatr Acta 32:517–525, 1977
14. Arneson D, Ch'ien LT, Chance P, et al: Strychnine therapy in nonketotic hyperglycinemia. Pediatrics 63:369–373, 1979
15. Steinmann B, Gitzelmann R: Strychnine treatment attempted in newborn twins with severe nonketotic hyperglycinemia. Helv Paediatr Acta 34:589–599, 1979
16. Melancon SB, Dallaire L, Vincelette P, et al: Early treatment of severe infantile glycine encephalopathy (nonketotic hyperglycinemia) with strychnine and sodium benzoate. Prog Clin Biol Res 34:217–229, 1979
17. von Wendt L, Simila S, Saukkonen A-L, et al: Failure of strychnine treatment during the neonatal period in three Finnish children with nonketotic hyperglycinemia. Pediatrics 65:1166–1169, 1980
18. MacDermot KD, Nelson W, Reichert CM, et al: Attempts at use of strychnine sulfate in the treatment of nonketotic hyperglycinemia. Pediatrics 65:61–64, 1980
19. Warburton D, Boyle RJ, Keats JP, et al: Nonketotic hypreglycinemia. Effects of therapy with strychnine. Am J Dis Child 134:273–275, 1980
20. Matalon R, Michals K, Naidu S, et al: Treatment of non-ketotic hyperglycinaemia with diazepam, choline and folic acid. J Inher Metab Dis 5(suppl 1):3–5, 1982
21. Matalon R, Naidu S, Hughes JR, et al: Nonketotic hyperglycinemia: Treatment with diazepam—A competitor for glycine receptors. Pediatrics 71:581–584, 1983
22. Wolff JA, Kulovich S, Yu AL, et al: The effectiveness of benzoate in the management of seizures in nonketotic hyperglycinemia. Am J Dis Child 140:596–602, 1986
23. Haan EA, Kirby DM, Tada K, et al: Difficulties in assessing the effect of strychnine on the outcome of nonketotic hyperglycinaemia. Observations on sisters with a mild T-protein defect. Eur J Pediatr 145:267–270, 1986
24. Young AB, Snyder SH: Strychnine binding in rat spinal cord membranes associated with the synaptic glycine receptor: Cooperativity of glycine interactions. Mol Pharmacol 10:790–809, 1974
25. Synder SH: The glycine synaptic receptor in the mammalian central nervous system. Br J Pharmacol 53:473–484, 1975
26. Grenningloh G, Rienitz A, Schmitt B, et al: The strychnine-binding subunit of the glycine receptor shows homology with nicotinic acetylcholine receptors. Nature (Lond) 328:215–220, 1987
27. Ascher P, Nowak L: Electrophysiological studies of NMDA receptors. Trends Neurosci 10:284–288, 1987
28. Cotman CW, Iversen LL: Excitatory amino acids in the brain-focus on NMDA receptors. Trends Neurosci 10:263–265, 1987
29. Kemp JA, Foster AC, Wong EHF: Non-competitive antagonists of excitatory amino acid receptors. Trends Neurosci 10:294–298, 1987
30. Hayasaka K, Tada K, Kikuchi G, et al: Nonketotic hyperglycinemia: Two patients with primary defects of P-protein and T-protein, respectively, in the glycine cleavage system. Pediatr Res 17:967–970, 1983
31. Hayasaka K, Tada K, Fueki N, et al: Nonketotic hyperglycinemia: Analyses of glycine cleavage system in typical and atypical cases. J Pediatr 110:873–877, 1987
32. Young AB, Zukin SR, Snyder SH: Interaction of benzodiazepines with central nervous glycine receptors: Possible mechanism of action. Proc Natl Acad Sci USA 71:2246–2250, 1974
33. Mohler H, Okada T: Benzodiazepine receptor: Demonstration in the central nervous system. Science 198:849–851, 1977

34. Squires RF, Braestrup C: Benzodiazepine receptors in rat brain. Nature (Lond) 266:732–734, 1977
35. Olsen RW: GABA–benzodiazepine barbiturate receptor interactions. J Neurochem 37:1–13, 1981
36. Jones EG: Neurotransmitters in the cerebral cortex. J Neurosurg 65:135–153, 1986
37. DeGroot CJ, Boeli Everts V, Touwen BCL, et al: Non-ketotic hyperglycinemia (NKH): An inborn error of metabolism affecting brain function exclusively. Prog Brain Res 48:199–207, 1978
38. Benavides J, Lopez-Lahoya J, Valdivieso F, et al: Postnatal development of synaptic glycine receptors in normal and hyperglycinemic rats. J Neurochem 37:315–320, 1981
39. von Wendt L: Experimental hyperglycinaemia—An evaluation of the efficacy of strychnine therapy in nonketotic hyperglycinaemia. J Ment Defic Res 23:195–205, 1979
40. von Wendt L, Simila S, Saukkonen A-L, et al: Prenatal brain damage in nonketotic hyperglycinemia. Am J Dis Child 135:1072, 1981
41. Trauner DA, Page T, Greco C, et al: Progressive neurodegenerative disorder in a patient with nonketotic hyperglycinemia. J Pediatr 98:272–275, 1981
42. Dobyns WB: Agenesis of the corpus callosum and gyral malformations are frequent manifestations of nonketotic hyperglycinemia. Neurology 39:817–820, 1989
43. Press GA, Barshop BA, Haas RH, et al: Abnormalities of the brain in nonketotic hyperglycinemia: MR manifestations. AJNR 10(2):315–321, 1989
44. Singer HS, Valle D, Hayasaka K, et al: Nonketotic hyperglycinemia: Studies in an atypical variant. Neurology 39:286–288, 1989

II

Application to Specific Clinical Disorders

This is concerned with the application of biological approaches to some of the major clinical problems confronting clinicians in the field. Until recently, pathogenetic mechanisms involved in several clinical disorders of childhood were explained exclusively on the basis of psychodynamic constructs. This was especially true for obsessive-compulsive disorder and suicidality. However, in each of these conditions, biological abnormalities implicating specific neurotransmitters and neuroanatomic regions have been described. It is now apparent that understanding of relevant developments in the basic sciences is essential to the day-to-day practice of clinicians treating children with these, and other, disorders.

11

Genetic Causes of Autism and the Pervasive Developmental Disorders

J. Gerald Young, James R. Brasic, and Len Leven

1. INTRODUCTION: GENETIC RESEARCH ON AUTISM AND THE PERVASIVE DEVELOPMENTAL DISORDERS

Autism is a behaviorally defined syndrome for which many etiologies have been identified, including specific viral and metabolic causes. New evidence confirms that genetic abnormalities cause the disorder in a subgroup of autistic individuals, encouraging further research to differentiate specific genetic etiological mechanisms.

Severe developmental disorders were occasionally observed to cluster in families during the early years of research on autism, but small pilot studies (typically not rigorous in their research design) failed to demonstrate clear chromosomal abnormalities or mendelian patterns of inheritance. Systematic research to assess the presence of genetic influences began during the mid- and late 1970s, leading rapidly to new findings. Several factors converged to stimulate comprehensive clinical genetic research on autism. First, techniques for analysis of chromosomal abnormalities were improved and were applied to autistic patient samples. These methods included new banding techniques and culture methods; the latter led to the techniques demonstrating the presence of the fragile Xq27.3 site. Second, the development of standard diagnostic criteria and application of structured diagnostic assessment methods facilitated an improved understanding of the features of the syndrome. A third, related factor was that clinical observation led to an appreciation of the possibility of variable phenotypic expression of the gene, encompassing a range of type and severity of symptoms among family members. This suggested that behavioral, language, and cognitive symptoms might occur in family members that failed to reach criteria for the presence of the autistic syndrome. The possibility of one or more forms frustes or lesser variants of the disorder began to guide further research. Fourth, renewed attention was given to possible models of transmission of the disorder. For example, pervasive developmental disorders (PDD) are characterized by a preponderance of males among patients in all samples. Investigators began to examine possible mechanisms for

J. Gerald Young, James R. Brasic, and Len Leven • Division of Child and Adolescent Psychiatry, Department of Psychiatry, New York University School of Medicine, New York, New York 10016.

a sex effect on the mode of transmission, including the determination of subgroups with an X-linked mode of inheritance. Fifth, research on familial clustering was facilitated by the availability of more sophisticated clinical genetic methods and mathematical models. A sixth factor was greater emphasis on the concepts of penetrance and expressivity when attempting to interpret familial aggregation data. Finally, an essential ingredient in genetic research on autism was a refinement in methods for obtaining data: a movement from the family history method (one or a few family members give the clinical history for all family members) to the family interview method (each family member is directly interviewed). These improved investigative tactics promise better success in meeting the challenge of unraveling the causes of pervasive developmental disorders.

2. ELEMENTS BASIC TO CLINICAL GENETIC RESEARCH

2.1. Types of Genetic Disorders

Three types of genetic disorders are conventionally described: chromosomal abnormalities, single gene disorders, and polygenic/multifactorial disorders. Chromosomal aberrations are due to abnormalities of whole chromosomes or chromosome segments (either an excess or deficiency). They are common, affecting approximately 7 in 1000 live births and associated with one half of all spontaneous first-trimester abortions. Single gene defects are caused by mutant genes (e.g., due to deletions, additions, or point mutations) whose effects are typically evident on inspection of pedigrees. They are relatively rare (i.e., a frequency of 1 in 2000 or less). Multifactorial disorders are characterized by no single gene error, but, instead, several or many lesser gene alterations (polygenic) that, together with environmental influences (multifactorial), produce a serious abnormality. They tend to recur in families but lack the signature of pedigree patterns characteristic of single gene disorders. They are common among developmental disorders.[1,2]

A genetic contribution to an illness for which no chromosomal abnormality has been determined is classically demonstrated by one or more of the following four methods: adoption studies, twin studies, familial aggregation research, and linkage analysis research.[3] Infrequent reproduction by autistic individuals has removed adoption studies as a practical genetic research method for autism; the other three methods are reviewed.

2.2. Pedigrees and Segregation Analysis: Challenges Guiding Research Design

2.2.1. Diagnosis and Clinical Criteria

Clinical investigators conducting segregation analyses are reliant on clinical phenotypic features in the absence of biological markers of a possible genetic defect. This carries significant risk of unreliable observations due to diagnostic criteria that are only approximations to those for homogeneous subgroups that may later be discovered. This is especially true for behavioral and cognitive disorders. When disorders are well defined by their clinical manifestations, which is rare, this may not be a problem. The need for meticulous assessment and diagnosis is obvious if investigators are to determine segregation patterns accurately within families. The failure of clinical investigators to observe and appreciate spectrum symptoms and diagnoses related to a disorder in its fully expressed form is an example of the potential confusion wrought in clinical genetic research.

At least two results can emerge from these diagnostic problems. First, prevalence rates of autism and the PDDs vary across studies and may be higher than previously estimated; both severity cutoffs for the disorders and inclusion or exclusion of specific related disorders in the PDD group affect prevalence rates. Second, familial aggregation research, even studies using direct interview methods for all family members, are beset by the problem of the need to use best estimate diagnostic procedures that are inherently imperfect; modes of transmission can be proposed inaccurately on the basis of idiosyncratic diagnostic methods.

2.2.2. Availability of Clinical Information

The inability to acquire necessary information on all the family members is a major influence on establishing modes of inheritance. Family members may not be interviewed because they live at a distance, refuse cooperation, have incapacitating illnesses, or have died; for these and other reasons, the data are incomplete and complicate the anticipated segregation analysis.

2.2.3. Application of Optimal Laboratory Methods

Novel laboratory techniques are continually becoming available. Excellent research can be hampered by failing to apply the best (new or old) methods to the problems. Examples are chromosomal banding techniques, culture methods, laboratory and clinical methods for the determination of zygosity, and linkage analysis procedures.

2.2.4. Complexities in the Clinical Expression of the Genotype

Disorders with an apparent genetic origin often have a highly variable clinical expression. The sources of this variability affect the design of each study.

2.2.4a. Genetic Heterogeneity. The phenotypic similarity of children with PDD caused by viral infections, inborn errors of metabolism, or perinatal events is well documented. Similarly, autism with a genetic origin can result from multiple genetic causes. Different mutations can produce the same phenotype, or one so similar as to be easily confused clinically. This genetic heterogeneity may produce pervasive developmental disorders that are superficially alike, but characterized by distinct genetic abnormalities.

2.2.4b. Pleiotropy. A single gene mutation can alter a gene product—an enzyme or structural protein—that will cause multiple phenotypic effects in an individual (pleiotropy). Developmental disorders are characterized by abnormalities that make themselves evident early in development, so that the gene defect(s) can have diverse phenotypic manifestations in later fully differentiated structures. These clinical features may at first appear unrelated; once they are established as components of the clinical disorder, improved understanding of the full breadth of its expressivity enhances analysis of the mode of inheritance.[1,4]

2.2.4c. Penetrance and Expressivity. Variability in the expression of a genetic disorder (including clinical features, age of onset, the influence of environmental factors, and medication effects) complicates clinical genetic research. The mutation of a gene does not necessarily lead to an altered phenotype; when it does, the degree of phenotypic change can be variable. These effects are described as penetrance and expressivity.

Penetrance is an all-or-none concept that indicates whether the genotype is expressed in individuals. It is nonpenetrant in individuals who carry the gene but who do not express it, and there is reduced penetrance in the total group considered. The mathematical representation of penetrance is the percentage of individuals carrying the gene in whom the clinical phenotype is evident. *Expressivity* refers to the variability in form and severity that characterizes the expression of some genes. Individuals in the same family can express a genotype differently, through the expression of different types of abnormalities or varying severity of the abnormalities. Variable expressivity of a genotype is greater across different families than within a family in some disorders, possibly reflecting the effects of other modifying genes or unrecognized genetic heterogeneity. Both variable expression and a failure of penetrance are more common in autosomal dominant disorders, confusing diagnostic assignment and the interpretation of pedigrees.[1,2]

2.2.4d. Sex-Limited and Sex-Influenced Traits. It is assumed that the sex ratio of affected subjects is not 1 : 1 in X-linked disorders, but it is sometimes not appreciated that the sex ratio can be abnormal in autosomal disorders. This might reflect the effects of differing hormonal environments on a gene defect in the developing individual, or other prenatal or perinatal effects. A sex-limited trait is expressed only in a single sex in an autosomal disorder. By contrast, sex-influenced traits are expressed in both sexes, but with markedly different frequencies. Multifactorial traits are often sex influenced.

2.2.4e. Age of Onset. Genes are expressed during fetal life, at birth, and during postnatal development. Consequently, genetic disorders have varying ages of onset, extending even into adulthood; this can be a significant diagnostic characteristic for specific disorders when designating affected or nonaffected status among family members. Determining whether relatives have reached or not reached the age of expression of a disorder can create formidable problems for the investigator, as in pedigree analysis of adult psychiatric disorders. Developmental disorders have different characteristics according to the age of the child, leading to difficult diagnostic research questions, such as the possible relationship between early childhood onset (PDD) and late childhood onset (childhood-onset schizophrenia) severe disorders.

2.2.4f. Gene Interaction and Environmental Effects. The hypothetical perspective that genes act in isolation is convenient for clinical research on genetic disorders. This pretense is temporarily accepted because of the complexity introduced when two major influences are considered: gene interaction and environmental effects. Gene interaction occurs when there is modification of the effects of a gene by its own allele or by genes at other loci. One gene may "turn on" or "turn off" the effects of another gene, or the modulating effects may be more subtle. In either case, modification of the expression of a gene or the metabolism of its genetic product should be assumed to occur. At its most general level, this concept refers to the "genetic background." It can contribute to a variable phenotype that may confuse family studies. Similarly, environmental effects on genetic processes and products are pervasive and must be carefully examined if a genetic disorder is to be illuminated.

3. CONGENITAL ETIOLOGIES OF AUTISM: PRENATAL, PERINATAL, AND POSTNATAL HAZARDS AND OBSTACLES TO THE DIFFERENTIATION OF THEIR GENETIC AND NONGENETIC ORIGINS

Research on many prenatal, perinatal, and neonatal hazards has not identified individual factors that cause autism. While some studies failed to determine a higher rate of prenatal or perinatal insults in autistic individuals, most investigators found specific events occurring more frequently in the autistic group. However, these insults have not been replicated across all studies. The hazards that have been shown in several studies to be more common in autism include bleeding during pregnancy, prematurity/low birthweight or postmaturity, high maternal age, flu-like symptoms during pregnancy, and ingestion of medications during pregnancy.[5–21] In addition, minor physical anomalies are reported to be more common in autistic children, suggesting the possibility of a first-trimester insult.[22] This nonspecific sign, common to several childhood neuropsychiatric disorders, may provide an additional index of the possibility of a pathogenic process already active by the first months of life.

The aggregate conclusion of these studies is that there is an increase in various nonspecific pre-, peri-, and postnatal hazards in the history of autistic individuals when compared to various control groups. This does not identify any single hazard as a specific or frequent cause of autism, but suggests that total reproductive complications are greater. Several hypotheses, not mutually exclusive, can be probed in research. First, there may be an interaction between the reproductive factors and a later environmental influence. For example, genetic predisposition or reproductive complications may lend the predisposing vulnerability of reduced immunological competence that is essential to the development of a viral infection in early childhood and the genesis of the autistic syndrome. Second, clinicians have postulated that an additive mechanism is at work, in which exceeding a threshhold of total reproductive complications markedly increases the likelihood of neuropsychiatric symptoms, modulated by the genetic background, nature of the complications, and the influence of any protective factors. Third, it is possible that most reproductive complications reflect a genetic abnormality that is sufficient to put the pregnancy and delivery at risk but is not lethal.

Several findings are of interest in relationship to these hypotheses. Research at New York University indicated that abnormalities of pregnancy and the neonatal period tend to cluster with a family history of psychiatric illness (nearly one half of the autistic patients having both). Almost all autistic patients had either a reproductive complication or a family history of psychiatric illness.[15] This suggests a genetic predisposition or abnormality. They also showed that the scores of siblings (of autistic children) with minimal brain dysfunction were similar to those of the autistic children, while Gillberg and Gillberg[18] find MBD children to have scores intermediate to autistic and control subjects. These findings are consonant with either an additive or genetic predisposition model.

A study at the University of Goteborg in Sweden[18] utilized data for the autistic and matched control groups obtained by an identical, uniform method that included ratings blind to group membership of the record. They found that reduced optimality in the prenatal, perinatal, and postnatal periods characterized the autistic group in relationship to controls. The reduced optimality was greater in the autistic group during the prenatal period than in the perinatal or neonatal periods. They viewed their findings to indicate that these reproductive complications "signal" a possible problem in the pregnancy or a primary genetic defect. They also found that high maternal age increases the risk of autism, a result compatible with the contribution of a chromosomal abnormality. Finally, research at UCLA suggested that mothers of autistic patients from single incidence autism families had significantly increased frequency of bleeding, flulike symptoms, taking medications during pregnancy, and induced labor when compared to mothers from multiple incidence families.[19]

4. CHROMOSOMAL ABNORMALITIES IN AUTISM

4.1. Chromosomal Aberrations Associated with Autism

Familial clustering of autism has spurred a series of studies reporting chromosomal analyses in autistic patients. Earlier studies employed small numbers of subjects and did not have the advantage of the improved cytogenetic techniques that are now available (Tables I and

Table I. Autosomal Abnormalities Reported in the Pervasive Developmental Disorders

Aneuploidies
Trisomy 6p partial[23]
Trisomy 8 mosaic[24]
Trisomy 21[13,25–27]
Trisomy 22[28,29]
Structural abnormalities
Deletions
del(1)(p35)[30] del(3)[31] del(13)[32] del(17)[31]
Duplications
dup(3)[31] dup(16)[31]
Fragile sites
fra(2)(q13)[33] fra(6)(q26)[27] fra(16)(q23)[27]
Inversions
inv(2)(p11q13)[34] inv(3)[31] inv(5)(p-q+)[35]
inv(9)(p11q12)[33] inv(16)[31]
Translocations
t(4p+;9q−)[36] t(7;20)[36] t(22;13)[37]
Other structural abnormalities
5p+[31] 9h+[27] Gp+[38]

II). They did not uncover compelling evidence for abnormalities in the early karyotype examinations, as the results were negative or the aberrations identified were rare.[28,38,39,42,60,62] For example, a large Y chromosome was reported in a subgroup of autistic patients but thought to be a normal variant.[60] Subsequent research identified a long Y chromosome in nine of 22 autistic cases; a specific associated phenotype was not apparent other than that the autistic patients were higher functioning than usual.[61] Research suggests that an increase in the length of the Y chromosome may be due to duplication of part of the distal part (Yq12) of the long arms.[63–65] The meaning of the finding is unclear. Apart from interindividual and interracial

Table II. Sex Chromosomal Abnormalities Reported in the Pervasive Developmental Disorders

Aneuploidies
xxx[39]
xxy[40]
xxyy[41]
xy mosaicism (long y and short y)[27]
xyy[27,42–46]
Structural abnormalities
Fragile sites
fra(x)(p22)[27]
fra(x)(q27)[27,34,47–62]
fra(x)(q28)[58,59]
Other structural abnormalities
long y (y = F)[27,60,61]

differences in Y chromosome length that have been reported, long Y chromosomes have been reported in various neuropsychiatric and retardation syndromes,[66–73] other diseases, and normal males.

Another study showed increased chromosomal breakage in autistic children when compared to nonautistic children.[74] Autism in a female was associated with a translocation between chromosome 22 and a D group chromosome.[37] A balanced non-robertsonian translocation has also been observed in two autistic/retarded children[35] and a female twin pair with autism/retardation.[36] Again, the significance of these findings has not been determined. A report of a patient with autism was unusual in that this 19-year-old woman also had sporadic retinoblastoma and reduced esterase D activity in association with a deletion on chromosome 13; she had an autistic maternal second cousin.[32] Autism has occurred in association with trisomy 21,[26] trisomy 22,[28,29] a mosaic of trisomy 8,[24] and partial 6p trisomy.[23] A case of autism associated with XYY has been described.[75] Two cases of autism with the heritable folate sensitive fra (2)(q13) site and one case of autism with inv(9)(p11q12) were also identified.[33] Another group reported four different structural autosomal defects: 46,XY,del(17); 46,XY,inv, dup, del(3); 46,XY,5p+; and 46,XX,inv, dup(16).[31]

The changes in cytogenetic methods and the lack of specific diagnostic criteria for autism before 1980 suggest the value of continued systematic investigation of karyotypes of patients with pervasive developmental disorders. This is verified by results of a survey completed in Sweden in which the karyotypes of 66 psychotic children were examined. Of this group, 56 were mentally retarded, 46 (40 boys and 6 girls) were autistic according to strict criteria, and 20 were characterized as other childhood psychoses (15 boys and 5 girls, most fulfilling criteria for childhood onset pervasive developmental disorder in DSM-III). There was no evidence of organicity in early childhood, when the original diagnosis was given, except in a small minority, but at the time of this study 42% had either a known organic syndrome, epilepsy, or cerebral palsy; 48% of the autistic individuals, and 47% of the larger group of pervasive developmental disorder patients, had an associated chromosomal abnormality or marker. These results do not necessarily indicate that the chromosomal aberrations caused the disorders, and this rate of abnormalities is much higher than previously reported studies, so the overall meaning of the results is not yet clear.[27] For those chromosomal abnormalities for which a base rate in the population is not known, it is impossible to judge their significance; for others, a population frequency of less than 1% marks the findings in this study as deviant. The fragile X marker was present in 17% of autistic children, comparable to the rates described below for other studies.[27,76] The findings of this study require replication, but are intriguing in the light of research on chromosomal abnormalities in the childhood neuropsychiatric disorders (Tables I and II).[43,45,77–80]

4.2. Fragile X Syndrome

4.2.1. X-Linked Mental Retardation and Fragile X Syndrome: History, Prevalence, and Clinical Features

The preponderence of males in population samples of the mentally retarded has been known for a century, yet did not lead to systematic investigation of a cause for many years. This probably reflected an assumption that the excess of males was due to differential social expectations for males and females, so that males not reaching a certain level of functional capacities were more likely to be designated as retarded. A more accurate appreciation of the prevalence and importance of X-linked mental retardation began with the studies of Penrose in 1938[81] and Martin and Bell in 1943.[82] There are 25% more mentally retarded males than females in the population. Many rare X-linked disorders are associated with mental retardation, and account for 20–25% of mental retardation. The frequency of all forms is estimated as 1 in 600 male births; about 25% of these are due to the fragile X syndrome.[50]

A *marker X* was identified in four retarded males and four carrier females in a family by Lubs in 1969.[83] Several years later, these findings were corroborated by investigators who identified the marker X in other families. The delay was due to changed, improved conditions for growing cells in the laboratory that were different from those used by Lubs. The marker X is a constriction of the X chromosome causing instability; the fragile site is located on the long arm (Xq27). It was eventually determined by Sutherland[84,85] that special culture conditions were required to enhance expression of the fragile site (a medium deficient in folate and thymidine). Many surveys of families of patients with nonspecific X-linked mental retardation identified individuals with this syndrome; a current estimate of the prevalence of the fragile X syndrome in the general population is 1 in 2000 males and a carrier prevalence in females of 1 in 1000.[86] The prevalence among institutionalized males has been estimated at 2–9%.[50] This makes the fragile X syndrome the most common hereditary form of mental retardation and a close second as the most common genetic cause of mental retardation following Down's syndrome. One methodologically sound study examined 3090 consecutive newborns and found no fragile X positive infants.[87] Other research using indirect methods estimates the incidence at 0.67–0.92 per 1000 male newborns.[88,89]

Previous failure to recognize this syndrome was probably attributable to the fact that phenotypic features are inconsistently present. Many have been identified, among which the most common is macroorchidism, a physical finding present in 80% of boys who have reached puberty. Other physical features commonly observed are prognathism, large ears, a narrow long face, hyperextensible joints and other signs characteristic of connective tissue disorders, dermatoglyphic anomalies,[90–92] and head circumference over the 50th percentile. Physical findings found to be predictive of the fragile X syndrome in institutionalized retarded men include long ears, macroorchidism, and hand calluses or lesions secondary to hand biting. In addition, all the fragile X males identified were ambulatory males in the highest functioning units of a residential treatment center.[129]

Behavioral and cognitive features have also been difficult to classify. They include stereotypies (e.g., hand flapping or hand biting), poor eye contact, seizures (in approximately 25%, with a subgroup possibly having a characteristic paroxysmal EEG pattern, whether seizures are present or not), impaired fine motor coordination, and soft neurological signs.[50,86,94,95] Some investigators report the occurrence of aggressive and self-injurious behavior in individual fragile X subjects. There is increasing recognition of the frequency of attentional dysfunction and hyperactivity, particularly in prepubertal boys. A review of three recent studies[96–98] found 41 of 50 subjects with attentional dysfunction and hyperactivity.[99] Most males have moderate to severe mental retardation, but deficits range from profound to mild retardation; individuals with low normal intelligence and learning disabilities have been reported. Attempts to characterize the cognitive abnormalities have produced varied results, such as deficits in short-term numerical memory and in numerical reasoning and abstraction. However, research using matched controls (non-fragile X nonspecific retarded and nonfragile X autistic groups) failed to find distinctive cognitive differences, indicating the need for caution until additional systematic research has been completed.[100] Language and speech abnormalities range from nonverbal status to minimal language, speech delay, and repetitive speech patterns in severely and profoundly retarded patients. In those patients with mild retardation or borderline or normal IQ, "cluttered" speech has frequently been observed (erratic rhythm in a rapid, disorganized, repetitious speech style).[101] However, comparison of the speech and language characteristics of fragile X mentally retarded males with that of matched groups of subjects with nonspecific retardation or fragile X negative autism showed no distinctive speech and language features of the fragile X group.[102]

While it was initially assumed that female fragile X carriers were clinically normal, it has become apparent that approximately one third of heterozygous females are affected; most are mildly retarded, although occasional individuals are moderately or severely retarded. The intellectual deficit appears to be primarily on the performance score (e.g., visuospatial function and calculations), rather than the verbal score, on IQ testing[103–105]; however, further research is required to clarify these findings.[102] The proportion of affected heterozygous females increases substantially when individuals with normal intelligence accompanied by learning disabilities are included; e.g., 8 of 15 (53%) of nonretarded heterozygotes in one study had specific learning disabilities.[106] The penetrance of fragile X expression and/or cognitive impairment in females is very high for an X-linked disorder, reaching 74% according to the data from one study.[107] The physical features of the syndrome have not been as well documented in females, and appear to be less definitive; they have included a long face, prognathism, abnormal size of ears, palatal abnormalities, malpositioned teeth, hypermobility of finger joints, and other findings.[105,108] Psychiatric disorders are common in subgroups of normal (10%) and retarded (20%) heterozygotes.[86] Schizophrenia spectrum disorders and affective disorders appear to be the major psychiatric disorders appearing in these carriers, occurring at a higher than normal frequency.[109]

The observation that there may be a relationship between fragile X expression and the degree of mental impairment has received continuing support from research findings. One laboratory found that about 90% of the women who are cognitively impaired express the fragile X as contrasted to 26% of carriers with normal intelligence.[107] The association between intelligence and fragile X expression in men has been more controversial. However, a recent protocol found a significant correlation between a lower percentage of fragile X positive cells and higher intelligence in male siblings.[110] There have been reports of males with the fragile X phenotype (the Martin–Bell syndrome) who are fragile X negative, but a careful study of such individuals indicated that they did not fit the full diagnostic criteria and failed to confirm the existence of a fragile X negative Martin–Bell syndrome.[111]

It is especially difficult to recognize phenotypic expression of the fragile X syndrome in early childhood in either males or females, as the adult traits are not reliably evident and other early manifestations have not been determined. However, systematic research is under way,[112] and there is evidence that even in infancy specific physical features, such as a large head and higher than normal birth weight, may suggest that a typical phenotype may later emerge. There is some suggestion that there may be sustained cognitive development of fragile X subjects in childhood, followed by a plateau or decline in adolescence.[113,114] Accurate prenatal diagnosis of the fragile X syndrome is a central goal of research. Amniotic fluid, chorionic villi, and fetal blood samples have been used.[115–117] The occurrence of false-positive and false-negative results has encouraged further efforts to improve the reliability of prenatal diagnosis. Cytogenetic and molecular biological methods complement each other and enhance diagnostic accuracy.[118]

Treatment with folic acid has produced some amelioration of hyperactivity and attentional dysfunction in prepubertal fragile X subjects[119–126] but has not been useful for postpubertal patients. Methylphenidate also appears to have beneficial effects on attentional dysfunction, hyperactivity, and social skills.[93] A single autopsy study reported cerebral dendritic spine abnormalities in an adult fragile X male similar to those found in chromosomal trisomies and associated with synaptic immaturity.[127]

4.4.2. The Fragile X Syndrome and Autism

Approximately 75% of autistic individuals are mentally retarded, and there is a 4 : 1 male-to-female ratio among these patients; this made them a logical group to be surveyed for an X-linked mental retardation disorder. Many centers have described the presence of the fragile X site in rigorously defined autistic individuals following the original reports: 5 of 27 (18.5%) of fragile X male patients had a diagnosis of autism in one study,[45] while another center reported a single case.[49] Prevalence estimates for the fragile X syndrome in the autistic population have ranged from 0 to 17% in different studies. A screening of 183 autistic males found 24 (13.1%) to be positive for the fragile X syndrome. These investigators also reviewed the published research and surveyed 11 other series reporting fragile X marker examinations in autistic males; six series were positive and five were negative. Summing the results of the 12 surveys, about 12% of autistic males were fragile X positive.[50] A more recent review of published research from 10 studies determined 50 of 594 (8.4%) autistic male subjects to have the fragile X syndrome.[95] They also reported on three studies of 46 female autistic patients that found none with the fragile X syndrome.[52,53,128] Two fragile X-positive female autistic patients were reported by Hagerman et al.[31]

The converse is to identify the frequency of the autistic syndrome among fragile X patients. The rate has ranged from 14% to 46% in reports from various centers.[99] One study indicated that 92 of 434 (21.2%) male fragile X subjects had a concurrent diagnosis of autism. Among 150 fragile X positive males in 55 families personally examined by these investigators, 24 (17.3%) were diagnosed as autistic.[50] Their review of the literature accumulated 67 autistic subjects among 284 fragile X patients (24%) studied, consistent with their own findings. A more recent review of the literature found 122 cases of autism reported among 532 fragile X subjects, a rate of 23%.[99] It has been suggested that nearly 50% of fragile X patients may meet criteria for some form of a PDD.[109]

The variable results of studies of the prevalence of the fragile X syndrome with autism have several likely causes. In addition to the problem of different diagnostic procedures[34,51,129] and cytogenetic methods, the sample size has been shown to have a significant effect. Those studies in which no cases of fragile X syndrome were identified were based on small subject samples. These results also support the observation that the proportion of fragile X cases among male subjects with autistic features is greater than among mentally retarded males.[130] The fragile X syndrome has also been reported to be associated with Rett's syndrome.[131,132,125a]

4.4.3. Research on Possible Modes of Inheritance of the Fragile X Syndrome

It was initially assumed that the fragile X syndrome was a typical X-linked disorder, but occasionally manifest in female heterozygotes. If X-linked inheritance occurs, the mother randomly passes one of two X chromosomes to each child; a mother who is a fragile X carrier is therefore expected to have one half of her sons affected and one half of her daughters as unaffected carriers. The father passes an X chromosome to his daughters and a Y chromosome to his sons. The association between the fragile X marker and the clinical signs should be high. However, pedigree analysis showed a surprising variance from this pattern in the fragile X syndrome. While 50% of carrier females express the fragile X marker, one third or more have cognitive impairment; 80% of the latter express the fragile X marker. In addition, there is transmission through normal males to their daughters in many families. These males have normal intelligence and are negative for the fragile X marker: they are nonpenetrant. Their daughters are almost always fragile X negative, but have retarded sons. Examination of pedigrees shows a 20% deficit of fragile X-positive males, indicating 80% penetrance. The sons of daughters of nonpenetrant males are much more likely to be affected in these pedigrees.[133,134] However, a caution in the interpretation of all these findings is that a decline in fragile X expression apparently occurs with age, so that men sampled at an older age (as a grandfather of a proband) may be more likely to be fragile X negative because of this age effect.[110] This mode of inheritance is unique in human genetics and of great interest to investigators; it is clinically crucial because it complicates both genetic counseling and linkage analysis research.

Among the 18 known rare fragile sites (i.e., points on a chromosome susceptible to breakage) on 12 autosomes and the X chromosome, 15 are folate sensitive (inducible in low-folic acid media). The fragile site at Xq27 is the only one associated with mental retardation or other congenital anomaly. Several methods for the induction or enhancement of the sites are used in the laboratory. All seem to affect a postulated molecular basis for fragile sites: a sector of underreplicated DNA or DNA with misincorporated pyrimidine bases formed due to decreased pyrimidine nucleotide triphosphate pools. This causes the chromatin conformation to decondense, so that a gap forms during condensation at mitosis. Research using somatic cell hybrid analysis does not support the hypothesis that there are other autosomal or X chromosomal loci that significantly modify fragile X expression. Rather, fragile X expression apparently is a property intrinsic to the site itself.[135] Another, larger group of fragile sites are observed in all individuals tested and frequently present in both chromosomal homologues; they are known as constitutive or common fragile sites.[136] They are induced by aphidicolin or by thymidylate stress. Recent evidence of a common fragile site at Xq27 raises the question whether the common and rare fragile sites are identical or unrelated. It is possible that the common fragile site is a locus for unequal recombination events that produce the folate-sensitive fragile site and the fragile X syndrome.[137]

It has been proposed that nonpenetrant males inherit and transmit a premutation to their daughters; this DNA alteration is not harmful but predisposes to the fragile X syndrome. The daughters are not clinically retarded, but some of their sons and daughters are. This suggests that the definitive mutation occurs only in ova, apparently generated by interaction between both X chromosomes (possibly a recombination event).[138] Hypotheses have become more specific about possible mechanisms underlying these phenomena. Laird[139–141] postulates that fragile X chromosomes can exist in two states: imprinted and nonimprinted. Chromosome inactivation, a component of the mechanism of dosage compensation occurring in females, makes reactivation necessary; it is proposed that a local mutation blocks the complete reactivation of an X chromosome in females prior to oogenesis in the fragile X syndrome. This "block" is a sustained transcriptional inactivity of the gene(s) at Xq27, and is a product of two sequential states of the mutation: (1) "mutated" (the mutation has the potential to block reactivation of the inactivated X chromosome), or (2) "imprinted" (the mutated fragile X chromosome has blocked full reactivation of the X chromosome, completing the cycle). This hypothesis answers many questions about the unique patterns of inheritance in the fragile X syndrome.[139–141] Other investigators have examined inactivation/reactivation patterns.[142,143] Warren[144] suggests that there is chromosomal breakage at Xq27 with loss of the acentric fragment. This leads to nullisomy for distal genes in the male and functional nullisomy when it occurs on the active X chromosome of female carriers.

5. SINGLE GENE DISORDERS ASSOCIATED WITH THE AUTISTIC SYNDROME

Autism has been reported to occur in association with various single gene disorders (Table III). While it is possible that this represents instances in which a single gene produces the autistic syndrome, it is possible that such an association reflects the additional action of another modifying gene, unusual effects of pleiotropy, environmental influences, or the unre-

Table III. Single-Gene Disorders Associated with the Autistic Syndrome

Autosomal-recessive disorders
- Disorders of purine metabolism[145]
  - Adenylosuccinate lyase deficiency[146]
  - Decreased adenosine deaminase activity[147]
  - Hyperuricosuria[148,149]
  - Increased phosphate 5-phosphoribosyl 1-pyrophosphate synthetase[150–152]
- Histidinemia[153]
- Neurolipidosis[154]
- Phenylketonuria[13,155–162]
- Oculocutaneous albinism[41]

Autosomal-dominant disorders
- Neurofibromatosis[46]
- Noonan syndrome[163]
- Tuberous sclerosis[154,164–172]

cognized coexistence of another disorder. In addition, these single gene disorders have contributed little to the understanding of autism because they are rare causes of autism, and the occurrence of autism is generally a rare event in each of the disorders.

Single gene disorders are often uncovered during screening for an inborn error of metabolism. The most common of the metabolic disorders with which autism is associated is phenylketonuria (PKU). Many centers have reported autistic features (more or less reliably described) in children with PKU.[13,155–160,173] Mental retardation is common in this disorder; while autism is not, it is more frequent in PKU than in other single gene disorders. This disorder is of additional interest because phenylalanine hydroxylase activity is altered as a result of the genetic defect. This enzyme is essential to catecholamine metabolism, which may be affected in autism.[176,177]

Among the other metabolic disorders associated with autistic features are several involving purine metabolism, with one of several enzymatic activities affected and sometimes accompanied by increased uric acid excretion.[146–152] Autism has been observed to occur in association with histidinemia.[153] Improved genetic methods will help clarify the meaning of the association of autism with many single gene disorders (Table III), too numerous to review individually here.

6. SIMPLE MENDELIAN VERSUS POLYGENIC, MULTIFACTORIAL ETIOLOGIES: TWIN STUDIES AND FAMILIAL AGGREGATION RESEARCH ON AUTISM AND THE PERVASIVE DEVELOPMENTAL DISORDERS

While the proportion of cases of autism accounted for by single gene disorders is small, clinicians have continued attempts to determine whether modes of inheritance could be distinguished for subgroups of autistic individuals with an unknown etiology. Evidence began to emerge from two sets of data supporting a genetic contribution to the etiology of autism: multiple incidence families were identified, including monozygotic twins concordant for autism, and the prevalence of autism among siblings of autistic probands was 2%, a rate 50 times that expected.[178]

6.1. Twin Research

Galton[179] first suggested, in 1876, that research using twins could help sort out the relative contributions of heredity and environment to disease, and twin methods are widely applied in psychiatric genetic studies. Improvement of measures used in clinical genetic research disposed investigators to greater confidence in heritability estimates derived from twin studies.[180–183] Nevertheless, twin research, primarily emphasizing the classic twin method in which rates of appearance of a trait in monozygotic (MZ) and dizygotic (DZ) twin pairs are compared, has had to cope with several problems attenuating the validity of results: inadequate diagnostic measures and reliability, unreliable methods used to estimate zygosity, poor sampling methods for the accrual of twin pairs with the disorder, the assumption that the intrafamilial and extrafamilial environments are the same for MZ and DZ twins (ignoring the unique environmental responses to identical twins), and the neglect of differential prenatal, perinatal, and postnatal experiences of twins.[184]

A group of early case reports documented the existence of twin pairs, identical and fraternal, in which both concordance and discordance were observed.[185–190] One series of twins was described.[191] Case reports continue to appear.[192–199] However, several factors made investigators cautious. First, both twins and autistic infants encounter a higher rate of prenatal and perinatal hazards, so that autistic twins may have a "double dose" that will confound twin studies of the heritability of autism.[190] Second, no cases of an autistic child with an autistic parent were known and families with more than one autistic child were rare, creating the illusion that this disorder would not be fertile ground for a familiar aggregation study. This should not be surprising, however, because autistic individuals rarely marry and even more rarely give birth to a child (although a group of autistic individuals who are themselves parents may now have been identified[201]). In addition, the low population prevalence of autism will lead to a low rate of autism in relatives even if heritability is high.[202–204]

A new phase of genetic research on autism was initiated during the late 1970s. Reinterpretation of the discouraging observations above by Folstein and Rutter in 1977[202] gave a new perspective to investigators. First, while only 2% of siblings of autistic children themselves have autism, this rate is 50 times that in the general population. Second, while family histories rarely describe autism in relatives, a family history of speech delays was present in about 25% of cases.[205–207] Third, many methodological deficiencies in prior reports of autistic twins (sampling, diagnostic, zygosity) left their contribution to the understanding of the genetics of autism very doubtful.[202] During this same period, Campbell and colleagues reported results of twin and sibling research from their large cohort of autistic children that further supported a genetic etiology for autism in some families.[200,208–211]

The first study of a relatively large sample of twins evoked a renewed interest in the genetics of autism. A concordance rate for autism of 40% in MZ twin pairs was contrasted with a 10% concordance rate in DZ twins. They also suggested that the inherited factor was not just autism, but a broad vulnerability to cognitive and language disabilities, of which autism would be a more severe form. Their data suggested that an environmental insult was required for full expression of the autistic syndrome. In the group of MZ twins discordant for autism, the "unaffected" twin had cognitive or language impairments in five of the seven pairs and no evidence of perinatal injury. However, there was evidence of perinatal injury in their autistic co-twins. Their model defined a vulnerability to cognitive or language impairment coupled with perinatal injury as factors leading to the autistic syndrome.[202]

The most extensive study of twins and multiple incidence families, by Ritvo and colleagues, has confirmed the heritability of autism in many families and extended our understanding of possible modes of transmission. Concordance rates of 95.7% (22 of 23 pairs) for

MZ twins and 23.5% (4 of 17) for DZ twins were observed in a study of 40 pairs of autistic twins. The 50% concordance rate predicted for DZ twins by a dominant mode of inheritance with full penetrance is quite different from these results; however, they are compatible with autosomal-recessive inheritance, which predicts 100% concordance in MZ twins and 25% concordance in DZ twins. Nevertheless, the findings fail to account for the excess of males in their twin sample (18 of 23 MZ pairs {78.3%} were male; among the DZ pairs, 9 were male, 1 was female, and 7 were mixed sex).[212]

In a related report, Ritvo and colleagues described the ascertainment of 46 families with multiple incidences of autism. There were 41 families with two autistic probands and five with three autistic probands. Segregation analysis calculations gave a maximum likelihood estimate of the segregation ratio (0.19) significantly less than 0.50, the expected value for autosomal-dominant inheritance. Once again, however, it was not significantly different from 0.25, the anticipated value for autosomal-recessive inheritance. Tests of a polygenic threshhold model led to its rejection. Their results were most consistent with an autosomal-recessive mode of inheritance.[213] Families with four autistic siblings and five autistic siblings have now been reported.[214]

6.2. Subject Ascertainment in Genetic Research on Autism: Challenges for Research Design

Ritvo and colleagues emphasized that their results apply only to the specific multiple incidence population ascertained for their study, and the results cannot be generalized to other multiple incidence families or to families with only a single autistic proband. This reflects their awareness of a first difficulty plaguing many genetic studies of autism, the tendency to select a unique sample from a clinic, organization, or other source that biases introduction of subjects into a study. Their ascertainment procedures spawned a unique group that may not reflect the total population of autistic individuals and possibly biases the resulting heritability data (e.g., the autistic subjects had previously been diagnosed and treated at UCLA, had cooperated in research protocols at UCLA, were members of the National Society for Autistic Children, had access to the NSAC newsletter). Second, they confront the well-known problem of ascertainment bias when establishing an autosomal-recessive mode of inheritance: proband families are included, while families which have only unaffected heterozygotes are not.[3,76] Third, rigorous diagnostic methods are essential in a genetic study. They used retrospective chart diagnoses from records of varying quality, not all probands had been tested for the fragile-X syndrome, and there was a strong preponderance of males in the sample. The latter fact makes it possible that the presence of fragile-X autistic subjects or other explanations for the male preponderance were not adequately considered in their research design and interpretation of the data. For example, an X-linked recessive mechanism, polygenic model, or a sex-dependent threshhold effect might be given further scrutiny.

A fourth problem to be considered in familial aggregation research is that disorders that are very severe, imposing major limitations on the child and/or a great burden on the parents, may deter the parents from further childbearing. The family will be limited to a single affected child who is the last child born in the family. Another possibility is that a less severe disorder may have a partial deterrent effect. These effects are represented by the two major childbearing strategies following the birth of a child with a severe disorder: simple termination (no more children) or simple replacement (have one more child, then stop regardless of its affected status). These childbearing behaviors ("stoppage")[215] alter the assumptions of segregation models, because birth order and family size are changed in that the affected child tends to be the last or second to last child in the family because of these "stoppage rules." This generally leads to a serious underestimation of the segregation ratio by most methods of analysis, and this was suggested to have occurred in the Ritvo data in a recent reanalysis considering the effect of stoppage rules.[216] A variety of methods have been employed to take these problems into account in segregation analyses.[217]

Fifth, autism is a disease of delayed onset, so another one or two children may already have been born or conceived, further complicating the analysis. Sixth, the tendency toward smaller families is making sibships with more than one affected child more difficult to find. Seventh, there is data suggesting the possibility that there may be an excess of mothers aged 35 or older among mothers of autistic patients as compared with the general population (see Tsai and Stewart[218] for review of the question). The rate of chromosomal abnormalities in autistic children from these families might be higher. Eighth, many autistic children are products of high-risk pregnancies, potentially confounding twin and familial aggregation studies.[202,218]

6.3. Obtaining a Population Sample for Subject Ascertainment: Previous Epidemiological Research on Autism

Convincing evidence supporting a genetic etiology for an autistic subgroup is reliant on adequate ascertainment procedures within a representative epidemiologic population sample. This also requires the identification of subjects with fragile X syndrome and other known causes of the autistic syndrome, so that analyses can be performed with and without the presence of these subgroups. An unbiased population sample is especially important in relationship to analysis of a possible autosomal-recessive mode of inheritance, where heterozygotes and recessive allele homozygotes may not be identified. This problem is countered by carrying out ascertainment in the context of a full epidemiological survey of a specified population sample, ensuring that the probability of ascertainment of each consecutive sibling is independent of the previous siblings. Efforts should be made to apply a multilevel ascertainment procedure, increasing the chances that each individual subject will be identified. The likelihood of ascertaining a man with a language disorder whose autistic uncle died thirty years previously must be the same as the likelihood of ascertaining a man with a language disorder whose younger brother is autistic or a man with a language disorder with no known autistic relative. Such an epidemiological/familial-genetic survey had not been completed until that by Ritvo and his colleagues in Utah.[219] Comparison of epidemiological data from the research survey with those previously conducted alerts the investigators to possible biases when data are idiosyncratic (e.g., distribution of socioeconomic status or IQ, or the sex ratios). The Utah sample appears to be representative and not unusual in any way. New findings may also emerge; in the Ritvo epidemiological study, 9.7% (20 of 207) of the families had more than one autistic sibling, suggesting that improved ascertainment methods identify a higher rate of multiplex families and adding to the studies suggesting a higher rate of autism among siblings than the 2% previously reported.

Epidemiological studies of the prevalence of autism have been reported on a worldwide basis (Table IV). Two results are evident from these studies. First, the prevalence rates for autism are similar in all countries sampled. Second, to the degree that the rates differ, it appears that the rate of 4 per 10,000 most commonly cited could be an underestimate. This probably reflects diagnostic ambiguities through which a more inclusive definition of autism encompasses more of the individuals with a related PDD and inflates the rate. More thorough ascertainment systems could also produce a higher prevalence rate.

6.4. Diagnostic Criteria and Methods: The Foundation of Clinical Genetic Research on Autism

Establishing concordance or discordance for autism in twin pairs when one twin has autism and another a language disorder that resembles autism can be a difficult diagnostic challenge, and is fundamental to accurate analysis of twin data. Similar problems have been

Table IV. Prevalence Studies of Autism and Related Disorders

Location	Age range (years)	Sex ratio (M : F)	Prevalence per 10,000 population
Canada[220]	6–14	2.5 : 1	10
Denmark[221]	2–14	1.4 : 1	4.3
France[222]	5–9	2.3 : 1	5.1
Ireland[223]	8–10	1.3 : 1	4.31
Japan[224]	5–15	7 : 1	1.1
Japan[225]	0–15	—	1.7
Japan[226]	2–12	—	2.56
Japan[227]	0–18	9 : 1	2.33
Japan[228]	4–12	4 : 1	15.5
Japan[229]	8–14	4 : 1	13.9
Sweden[230,231]	0–20	1.6 : 1	5.6, 6.1
Sweden[75]	4–18	1.8 : 1	4.0
Sweden[232]	0–10	2.9 : 1	6.6
United Kingdom[233]	8–10	2.6 : 1	4.5
United Kingdom[234]	5–14	—	4.8
United Kingdom[235]	0–15	15.7 : 1	4.9
United States[236]	0–12	3.4 : 1	3.1
United States[237]	2–18	2.7 : 1	3.26
United States[219]	0–24	3.7 : 1	4
West Gemany[20]	0–15	2.25 : 1	1.9

discussed in relation to determining rates of autism among fragile X subjects when many have a similar clinical phenotype that merits a PDD diagnosis but may not reach criteria for the full autistic syndrome. These questions are increasingly complex in familial aggregation studies of autism using a broad, inclusive phenotype that encompasses several diagnostic categories. Is the scientific status of the diagnostic procedures and methods for autism sufficient to carry out reliable clinical genetic research on autism and related disorders?

The publication of the third edition of the Diagnostic and Statistical Manual of the American Psychiatric Association (DSM-III)[238] in 1980 was a milestone in the transition of the diagnostic process for autistic children from impressionistic and global to explicit and criterion based. Under this new nomenclature, autism could be reliably diagnosed by investigators in different settings. This was also a necessary prerequisite for validity in diagnosis, a more difficult objective that will be fulfilled only when objective markers for autism are found. However, the DSM-III,[238] DSM-III-R,[239] and ICD-9 systems have accelerated definitions of other aspects of the syndrome, such as the age of onset, prevalence, genetics, natural history, and prognosis, all of which serve to strengthen the validity of the diagnosis.

Unfortunately, autism is not a discrete, well-delineated syndrome comparable to an infection caused by a specific bacterium. Despite advances in defining a "core" autistic syndrome, clinicians agree that autism comprises a broad spectrum of dysfunctions that blends imperceptibly at its edges with several other childhood disorders. This means that a clinical investigator making a diagnostic decision between autism and another disorder in an ambiguous case will have a profound influence on the results of a familial aggregation study, particularly if the decision reflects a consistent idiosyncratic bias. Specific, operationally defined diagnostic criteria are a necessity to reduce this source of variance. In spite of their use, variability in diagnostic decisions continues to influence assignment of affected/nonaffected status in ambiguous cases. This reflects the difficulty encountered in the determination of the presence of the three major behavioral criteria: qualitative impairment in reciprocal social interaction, qualitative impairment in verbal and nonverbal communication and in imaginative activity, and a markedly restricted repertoire of activities and interests. These problems are most evident when considering them in the context of the most common differential diagnoses.

The usual first conditions to be differentiated are the mental retardation syndromes without known etiology. If 75% of autistic children are retarded, and up to 50% of retarded children have serious behavior problems, the stage is set for confusion. A major distinction between the two conditions is the retarded child's capacity for warm social relations in contrast to autistic aloofness and indifference. However, recent studies have shed doubt on this axiom. When separated from their parents, and subsequently reunited under controlled conditions (the "strange situation" paradigm) autistic youngsters resembled normal children of a younger age, or younger mentally retarded children.[240,241] Similarly, a considerable percentage of retarded children have unusual mannerisms, stereotypies, aggressive outbursts, and self-injurious behavior. In addition, the more serious the retardation, the less likely there will be communicative language or meaningful social contact. The distinctions begin to blur.[235,242,243]

Another difficult differential diagnosis is between autism and childhood-onset schizophrenia. Once again, it is easy to differentiate between "pure" forms of both conditions; classically, schizophrenic children present with florid hallucinations, delusions, and thought disorder. They are not likely to be retarded, have had a period of normal development for eight years on the average, and have a fairly even profile of abilities. However, there is a group of children who have odd, eccentric thinking, not quite impaired enough to be childhood-onset schizophrenia but too unusual to be classic autism. It is unclear what their eventual outcome is, and whether they should be classified as schizophrenic.[244] To confuse matters further, debate continues over the natural history of autism.[245] Does it remain unchanged, does it become schizophrenia, or both?

Similar problems of diagnostic discrimination exist with severe developmental language disorders, once again identifiable in their classic forms, but overlapping with autism.[243] Children with developmental language disorders share many features with autistic children: they have diverse etiologies, a similar natural history, nearly half may have some autistic features (including social deficits), and approximately 25% have a family history of speech delay.[206,246,247] Less severe language and cognitive disabilities are not confused with autism but must be meticulously described in familial aggregation research on related disorders. Thus, scrupulous diagnostic procedures are essential for reliable and valid samples that will facilitate accurate data generation in family studies.

6.5. Familial Aggregation Research on Autism, the Pervasive Developmental Disorders, and Related Conditions

Many of the factors discouraging twin studies of autism also led to the neglect of familial aggregation research on autism. In particular, using heritability of the rigorously defined full autistic phenotype as the focus of research, unsystematic surveys by clinical investigators gave results that failed to stimulate interest in more systematic investigation. Probable reasons for these deceptive findings have been discussed above.

Conceptualization of the heritable trait began to be more inclusive, however, as observations suggested that families of autistic individuals were characterized by an increased incidence of several types of psychopathology, even if no other autistic relatives were identified. This was the basis for consideration of a lesser variant or form fruste of the autistic syndrome, centered on the aggregation of cognitive and language impairments. Early studies had been neglected over the years.[190,248–250] Rutter's group found that 25% of the autistic probands had a parent or sibling afflicted with a language disorder or history of delayed speech.[206] Data from another laboratory indicated that 15.5% of siblings of a sample of autistic children had cognitive disabilities, while only 3% of a comparison group of Down's syndrome children had similar diagnoses. They determined that the cognitive disabilities clustered in a few families of autistic probands, and consisted of disturbances in expressive and receptive language, specific learning disabilities, and borderline or more severe mental retardation.[251] Campbell and colleagues reported an excess of subnormal intellectual functioning in siblings of autistic children, and documented significantly lower verbal scores than performance scores on formal intelligence testing.[252]

The importance of thorough clinical assessment was emphasized in findings with another sample of families of autistic children. The incidence of autism in the siblings of autistic probands was 5.9%, and cognitive disabilities were present in 19.6% of siblings. Examination of subgroups showed that the 10 siblings with cognitive disabilities were members of six out of the total of 29 families; 11 siblings from these six families yielded ten siblings with a diagnosis of autism, nonspecific intellectual retardation, or a specific cognitive impairment. The families in which this aggregation occurred were characterized by severely retarded autistic probands. There was no such aggregation of retardation, cognitive disabilities, and autism among siblings of higher-functioning autistic individuals. Their study results suggest that low functioning, severely retarded autistic individuals may carry a specific genetic vulnerability distinctive from the vulnerabilities of other autistic subjects.[253] High-functioning autistic individuals, such as those designated as Asperger's syndrome, might comprise a different subgroup of the autistic syndrome. Further meticulous clinical evaluation research to attempt differentiation of these potential subgroups is a high priority of future research.[254,255] For example, a partial explanation of the earlier inability to identify a genetic contribution to the syndrome of autism might be that earlier studies were more likely to exclude severely retarded autistic children from their samples. It is also possible that fragile X autistic individuals will make up a high percentage of this subgroup. For the severely retarded subgroup of autistic individuals, the heritable factor may be strongly related to a general cognitive impairment.

Completion of the epidemiological sample of autistic probands in Utah by Ritvo and colleagues will be a beginning to unraveling many of these difficult questions. A first result is a surprising lack of replication of previous findings of a broad liability factor (cognitive and communication impairments) transmitted in the families of autistic probands. Scores of parents and siblings on various psychometric tests were distributed in congruence with published normative ranges.[256] Their study had particular strengths in addition to its epidemiological base: a sufficient number of subjects to correct for social class distribution and compare to published normative scores of the tests. Questions will be raised concerning sources of their unexpected results. For example, is the population sample in Utah representative? The authors refer to research data indicating that it is genetically representative.[257,258] Another set of questions will concern the clinical measures of cognitive and language impairments: none are perfect, so did their choice unknowingly bias the results in any way, such as not adequately emphasizing communication/language deficits? For example, Campbell's group has examined metalinguistic performance (e.g., awareness of language as a conventional code, and awareness of language as a rule-governed system) in siblings of autistic children and determined differences from controls.[200] Other investigators have used alternative approaches to distilling out essential language deficits.[102,114,259,260]

6.6. Sex Ratios in Autism and Autistic Subgroups

Multiple studies have confirmed a sex ratio of 1.5–4.5 autistic males per autistic female.[221,233,261–263] There is also a preponderance of males among the mentally retarded, although the excess is less than in autism, ranging from 1.1 to 1.8 males per female.[264,265] A finding replicated in several studies is the high ratio of males to females in the high-IQ autistic group and the low ratio of males to females in the low-IQ autistic group. In other words, more girls are severely impaired, and more boys are among the mildly affected.[263–267]

Several hypotheses have been offered to explain the greater frequency, but lesser severity, of these disorders in males. The first is that there is greater genetic variation in most measurable characteristics in males than in females, so that features in a mild form will be more evident in males; when such features appear in females they reflect more severe pathology.[263,268] Second is a polygenic, multifactorial model of inheritance with a sex-dependent threshold: a continuously distributed liability to the disorder in the population (encompassing genetic and environmental factors) would include a threshhold value for phenotypic manifestation. A higher threshhold for females would be consistent with the observed increased prevalence of autism in males. The trait would appear in fewer females, but with greater severity, in accord with the higher threshhold. The trait would also appear in a larger proportion of relatives of female probands than relatives of male probands, and the female proband's relatives would have a more severe form of the disorder.[76,267,269] A third hypothesis rests on reported sex differences in the normal population bringing a susceptibility to males: boys are more vulnerable to the language and speech problems characteristic of autism and related conditions; lacking these skills, they are reliant upon visuospatial skills which then are exaggerated in the form of various repetitive routines. Females not possessing the compensatory visuospatial skills suffer more severe symptoms when expressing the disorder.[262]

Relevant findings have been reported. Autistic girls have a more deviant form of autism, more neurological abnormalities, and more relatives with autism and autistic spectrum disorders.[263–267,270] Yet no easy explanations emerge, because there is not a linear relationship between an increasing proportion of males and increasing IQ,[267] and a variety of problems interfere with clarifying these questions, such as small sample size, lack of an epidemiological sample, and an absence of necessary comparison groups.[267] The epidemiological sample of autistic patients drawn by Ritvo and colleagues may provide new evidence on these matters. The sex ratio in their findings is similar to those previously reported in autistic cohorts, and contains a suggestion of a linear relationship between the sex ratio and IQ; there is not yet a report of full data analyses in relation to sex ratio.[219]

7. POTENTIAL BIOLOGICAL MARKERS FOR AUTISM

Although many instrumental techniques for examining the neurobiology of autism have yielded suggestions of abnormalities (Table V), no specific biological marker has been validated through replication.[76,177] One exception to this is hyperserotonemia, identified in 30–40% of samples of autistic children in several clinical laboratories.[271] Its meaning is not yet clear, and it does not influence diagnosis or treatment. Imaging techniques hold promise for the near future, and imaginative avenues of research continue to produce new possibilities every year (e.g., blink rate[272] or electroretinogram abnormalities[273]). However, clinical phenomenology remains the basis for diagnostic assignment. Inclusion of potential biological markers in protocols for family aggregation studies may lead to the identification of a useful index.

8. LINKAGE ANALYSIS RESEARCH: GENE MAPPING AND THE PERVASIVE DEVELOPMENTAL DISORDERS

8.1. Molecular Biological Background of Linkage Analysis

The occurrence of alternative forms (alleles) of a gene at the same locus on homologous chromosomes is a basis for genetic variation and the investigation of genetic linkage. When two or more genetically determined alternative phenotypes (and alleles) occur together in a population with appreciable frequency, polymorphism is said to exist. A polymorphism is arbitrarily defined as present in a population when the most common allele at a locus accounts

Table V. Potential Biological Markers for Autism and Related Disorders

Potential marker	Research findings
Imaging	
CT scan	Nonspecific abnormalities[289–293]
MRI	Reduced size of neocerebellar vermal lobules VI and VII[294–298]
	No abnormalities specific to autism[299]
Regional cerebral blood flow	Nonspecific findings[300]
PET scan	Diffusely elevated metabolic rate[301–304]
	No abnormality in mean cerebellar glucose metabolism[302]
Electrophysiological indices	
EEG	Nonspecific abnormalities[305–307]
Brainstem auditory evoked responses	Brainstem transmission time prolonged in some studies, but controversial[308–310]
Event-related potentials	Reduced auditory P300: Possible impaired orienting response[312,313]
Neurochemical measures	
Indoleamines	Hyperserotonemia[314–319,271]
	Decreased CSF 5-HIAA associated with lower global functioning[319,320]
Catecholamines	Higher CSF HVA associated with greater functional impairment[321]
	Decreased urinary free CA and MHPG, but not replicated[319,322]
Neuropeptides	Increased CSF endorphin fraction II[323]
	Reduced blood H-endorphin levels[324]
	Decreased plasma B-endorphin levels[325]
Neuroendocrine indices	
Cortisol	Nonsuppression of plasma cortisol in response to dexamethasone in subgroup[326,327]
Growth hormone	Blunted increase in plasma in response to L-DOPA[326]
Immunological measures	
Autoantibodies to 5-HT1 receptor	Present in CSF and blood of an autistic girl[328,329]
Macrophage migration	Inhibited[330]
Natural killer cell	Reduced activity[331]
HLA	Antigens possibly shared by autistic children and parents more frequently[332,333]
Other measures	
Autonomic responsivity	Impaired[308]
Vestibulo-ocular responses	Abnormal[308,310,334]
Electroretinogram	Subnormal b-wave amplitude[311]
Blink rate	Elevated[335]

for fewer than 99% of the alleles.[1] Polymorphisms are very common: as many as 28% of human genes may show polymorphism and more than 0.2% of the base pairs in DNA play a role in polymorphisms. Approximately 40% of these polymorphisms (or roughly 1 of 1000 of the total base pairs in the human genome) will be detectable by restriction enzyme analysis. If we estimate the human genome to be made up of three billion base pairs, this suggests that several million polymorphisms can be detected as differences between individuals. Most are benign and produce no phenotypic effects.[274–278]

Demonstrating these genetic differences between individuals might seem to require the formidable undertaking of specifying their molecular makeup: determining the sequence of DNA bases in the chromosomes of each. However, another method was recognized long before the discovery that genes are carried in the double helix. Genes occur in a specific linear array on chromosomes, conventionally likened to beads on a string. During meiosis there is an exchange of genetic material between members of a chromosome pair known as crossing over. This crossover between loci causes the formation of new combinations of linked genes.

Several years after the rediscovery of Mendel's laws and the determination that genes are carried on chromosomes (in the first years of this century), violations of mendelian patterns of inheritance were observed. Morgan soon discovered that these unexpected differences were due to crossing over during meiosis. Working from the premise that genes are carried on specific chromosomes, he used experiments involving crossing over to demonstrate that genes are arranged in linear order on the chromosomes and achieved the first mapping of genes on chromosomes in *Drosophila* by measuring the frequency of crossovers: more frequent crossover suggests a greater linear distance between the genes. These techniques were extended to other species, but the inability to control matings, the unknown phase of the alleles, the presence of only partially informative meioses, and small family size, made it improbable that this would be accomplished in humans. However, statistical techniques were devised that have provided useful estimates of linkage from recombinant frequencies.[279]

The invariant base pairing within DNA molecules that determines the genetic code is also the basis for molecular methods for linkage analysis. When the two strands of the DNA double helix are separated, single-stranded DNA will line up in close apposition to ("hybridize" to) another single strand composed of a complementary base sequence. Hybridization will not only occur with the original complementary DNA (cDNA) strand, but with DNA or messenger RNA (mRNA) from any source, as long as the base sequence is complementary. This makes it possible to take DNA, separate the strands, and make multiple complementary copies of a strand that incorporate a radioisotope label (usually 32P), in this manner cloning any genes carried within the strand.

This labeled strand of DNA is known as a *gene probe:* multiple copies of the labeled cloned gene containing a specific base sequence. It can be used to search for a complementary strand or strand fragment, in this way demonstrating that the genes identified by the probe are contained within the DNA of another individual. However, this very useful technical shortcut to finding genes requires either of two pieces of information for construction of a gene probe. A gene probe can be built by knowing the identity of a tissue in which the gene is vigorously expressed, so that extracted messenger RNA (mRNA) can be used as a template to construct the DNA probe that will be used to search for the gene in DNA from another individual. An alternative basis for building a gene probe is knowing the identity of the protein that the gene specifies, so that determination of its amino acid sequence can be used to work backwards to the corresponding base triplets in the mRNA and cDNA.

8.2. Techniques of Linkage Analysis Research: Classic Methods and RFLPs

Synthesis of a gene probe is therefore required for the direct examination of a gene. Paradoxically, it is not necessary to know the specific location of either the gene on a chromosome or even the identity of the chromosome. It is sufficient that we are able to identify the fragment(s) containing the gene of interest, so that we can then clone (differentially amplify) the gene and investigate its structure with restriction enzymes. Regrettably, in 90% of genetic diseases, and most all neuropsychiatric diseases with genetic components, no tissue is known that vigorously expresses the gene, nor has an aberrant protein product of the gene been identified. In order to accomplish our diagnostic purposes within these constraints, the older technique of linkage analysis must be applied.

Linkage analysis is the determination of the tendency for two genetic traits, either genes or polymorphisms, to remain together or recombine during meiosis. Linkage analysis has a long history. Investigators have searched for a trait that is genetically linked to a pathologic trait of interest. This requires that they be located close enough together on the same chromosome that they will be exchanged as a unit in the crossing over of genetic material during meiosis. The traits will be linked 99% of the time if they are separated by only about 1000 kb. This indicates a *recombinant fraction* of 1%. The recombinant fraction refers to the percentage of individuals in which the two traits are not inherited together. A recombinant fraction greater than 30% cannot be distinguished from chance (50%). Most clinical genetic studies use small numbers of subjects, increasing the likelihood of chance association. In order to determine the probability of linkage, the acceptable recombinant fraction is not greater than 10%. This corresponds to a distance of about 10 million base pairs.[274–278]

Clinical scientists had been limited to linking gene to gene, but *restriction enzymes* have dramatically increased the number of traits accessible to linkage analysis. Restriction enzymes are enzymes from bacteria that cut DNA strands at specific "restriction sites" ranging from four to eight bases long. A restriction site is simply a specific base sequence, and is not necessarily pathogenic. Some restriction sites are pathogenic (e.g., if they are large deletions or splicing errors), but others are typically benign (e.g., point mutations affecting only a single base pair). Restriction enzymes are numerous and diverse, so that we can use them to examine approximately 40% of base pairs in any gene. Therefore, if one of the many polymorphisms occurs within a gene, it might be detected by a restriction enzyme that would fail to cut DNA in the usual locus, leaving a longer DNA fragment recognizable on further analysis, or cut the DNA at a new locus, generating smaller DNA fragments. By this means, it is now possible to link a gene with one of the DNA polymorphisms that are common features of the human genome, rather than only to the rare clinical traits whose locus on a chromosome is known. This is the basis for the remarkable surge of linkage studies of diseases whose phenotype can be reliably identified. Phenotype and specific *restriction fragment length polymorphisms* (RFLPs) will travel together in a family pedigree if an appropriate RFLP is selected and they are closely linked on the chromosome.

Linkage studies of psychiatric disorders are not to be undertaken by the faint-hearted: they confront many research problems. Kidd[280] outlined requirements for successful linkage analysis research: standardized reliable diagnostic criteria, systematic symptom assessment methods, extended families with many affected individuals, polymorphic genetic loci distributed throughout the genome, and appropriate powerful statistical methods.

8.3. Linkage Studies of Autism

The first application of linkage analysis to autism was achieved by Spence and Ritvo and colleagues,[281] using 27 of the multiple incidence families of the twin study and seven additional families with a single autistic child. Assuming a fully penetrant autosomal recessive mode of inheritance for the linkage analyses with 30 standard phenotypic gene markers, their results failed to support linkage between the hypothetical autism locus and HLA, and no statistically significant evidence for close linkage could be found for 19 other autosomal markers. They interpreted their results to indicate genetic heterogeneity, even within a sample largely composed of multiplex families. Possible undiagnosed fragile X probands are considered as one possible source of this heterogeneity.

A linkage study of an autistic cohort using X chromosome DNA probes gave no evidence of chromosomal deletions in the fragile X region, nor of any abnormalities on the short arm of X.[56]

8.4. Linkage Analysis Research on the Fragile X Syndrome

Imperfect cytogenetic diagnosis of the fragile X syndrome has encouraged linkage analysis research to identify a closely linked marker to assist with carrier detection and improve prenatal testing methods, as well as further specify the molecular defect and clarify patterns of transmission. Close linkage was first reported between the gene for fragile X expression and a probe for factor IX.[282] Many laboratories have reported a variety of recombinants since that time, leading to controversy whether this linkage heterogeneity indicates two separate loci that can produce the fragile X phenotype, statistical fluctuation, a familial predisposition to crossover, two crossover events for the conversion of a premutation to mutation, or other complex form of heterogeneity.[283–288] Whatever the outcome of further research on these questions, new probes will continue to become available for mapping at sites closely flanking the fragile X loci, carrying the hope of answers to both clinical and molecular questions.

9. PROGRESS ACHIEVED AND ANTICIPATED

While a multitude of questions spur us on to further research, investigators working to help autistic children and their families over the past 15 years can look over a remarkable series of achievements in genetic investigations of autism. Surely research at the clinical, chromosomal, and molecular levels carries realistic hope that we shall reduce the burden of these families in the near future.

ACKNOWLEDGMENT. This work was supported by a grant from the Medical Fellows Program of the Office of Mental Retardation and Developmental Disabilities, State of New York.

REFERENCES

1. Thompson MW: Genetics in Medicine, ed 4. Philadelphia, WB Saunders, 1986
2. Shapiro AK, Shapiro ES, Young JG, et al: Gilles de la Tourette Syndrome, ed 2. New York, Raven, 1988
3. Pauls DL: The familiarity of autism and related disorders: A review of the evidence, in Cohen DJ, Donnellan AM (eds): Handbook of Autism and Pervasive Developmental Disorders. New York, Wiley, 1987, pp 192–198
4. Vogel F, Motulsky AG: Human Genetics: Problems and Approaches, ed 2, revised. Berlin, Springer-Verlag, 1986
5. Taft LT, Goldfarb W: Prenatal and perinatal factors in childhood schizophrenia. Dev Med Child Neurol 6:32–43, 1964
6. Kolvin I: Studies in the childhood psychoses. I. Diagnostic criteria and classification. Br J Psychiatry 118:381–384, 1971
7. Kolvin I, Garside RF, Kidd JSH: IV. Parental personality and attitude and childhood psychoses. Br J Psychiatry 118:403–406, 1971
8. Kolvin I, Humphrey M, McNay A: VI. Cognitive factors in childhood psychoses. Br J Psychiatry 118:415–419, 1971
9. Kolvin I, Ounsted C, Humphrey M, et al: II. The phenomenology of childhood psychoses. Br J Psychiatry 118:385–395, 1971

10. Kolvin I, Ounsted C, Richardson LM, et al: III. The family and social background in childhood psychoses. Br J Psychiatry 118:396–402, 1971
11. Kolvin I, Ounsted C, Roth M: V. Cerebral dysfunction and childhood psychoses. Br J Psychiatry 118:407–414, 1971
12. Rutt CN, Offord DR: Prenatal and perinatal complications in childhood schizophrenia and their sibships. J Nerv Mental Dis 152:324–331, 1971
13. Knobloch H, Pasamanick B: Some etiologic and prognostic factors in early infantile autism and psychosis. Pediatrics 55:182–191, 1975
14. Torrey EF, Hersh SP, McCabe KD: Early childhood psychosis and bleeding during pregnancy: A prospective study of gravid women and their offspring. J Autism Child Schizophr 5:287–297, 1975
15. Campbell M, Hardesty AS, Burdock EI, et al: Demographic and perinatal profile of 105 autistic children: A preliminary report. Psychopharmacol Bull 14(2):36–39, 1978
16. Finegan J, Quarrington B: Pre-, peri- and neonatal factors and infantile autism. J Child Psychol Psychiatry 20:119–128, 1979
17. Deykin EY, MacMahon B: Pregnancy delivery and neonatal complications among autistic children. Am J Dis Child 134:860–864, 1980
18. Gillberg C, Gillberg IC: Infantile autism: A total population study of reduced optimality in the pre-, peri-, and neonatal period. J Autism Dev Disord 13:153–166, 1983
19. Mason-Brothers A, Ritvo ER, Guze B, et al: J Am Acad Child Adol Psychiatry 26:39–42, 1987
20. Steinhausen H-C, Gobel D, Breinlinger M, et al: A community survey of infantile autism. J Am Acad Child Psychiatry 25:186–189, 1986
21. Tsai LY: Pre-, peri-, and neonatal factors in autism, in Schopler E, Mesibov GB (eds): Neurobiological Issues in Autism. New York, Plenum, 1987, pp 179–187
22. Campbell M, Geller B, Small AM, et al: Minor physical anomalies in young psychotic children. Am J Psychiatry 135:573–575, 1978
23. Burd L, Martsolf JT, Kerbeshian J, et al: Partial 6p trisomy associated with infantile autism. Clin Genet 33:356–359, 1988
24. Burd L, Kerbeshian J, Fisher W, et al: A case of autism and mosaic of trisomy 8. J Autism Dev Disord 13:351–352, 1985
25. Maltz AD: Down's syndrome and early infantile autism: diagnostic confusion? J Autism Dev Disord 9:453–455, 1979
26. Wakabayashi S: A case of infantile autism associated with Down's syndrome. J Autism Dev Disord 9:31–36, 1979
27. Gillberg C, Wahlstrom J: Chromosome abnormalities in infantile autism and other childhood psychoses: A population study of 66 cases. Dev Med Child Neurol 27:293–304, 1985
28. Turner B, Jennings AN: Trisomy for chromosome 22. Lancet 2:49–50, 1961
29. Biesele JJ, Schmid W, Lawlis MG: Mentally retarded schizoid twin girls with 47 chromosomes. Lancet 1:403–405, 1962
30. Wenger SL, Steele MW, Becker DJ: Clinical consequences of deletion 1p35. J Med Genet 25:263, 1988
31. Mariner R, Jackson AW III, Levitas A, et al: Autism, mental retardation, and chromosomal abnormalities. J Autism Dev Disord 16:425–440, 1986
32. Ritvo ER, Mason-Brothers A, Menkes JH, et al: Association of autism, retinoblastoma, and reduced esterase D activity. Arch Gen Psychiatry 45:600, 1988
33. Jayakar P, Chudley AE, Ray M, et al: Fra(2) (q13) and inv(9) (p11q12) in autism: Causal relationship? Am J Med Genet 23:381–392, 1986
34. Hagerman RJ, Chudley AE, Knoll JH, et al: Autism in fragile X females. Am J Med Genet 23:375–380, 1986
35. Funderburk SJ, Spence MA, Sparkes RS: Mental retardation associated with "balanced" chromosome rearrangements. Am J Hum Genet 29:136–141, 1977
36. de la Barra FM, Skoknic VC, Alliende AR, et al: Gemelas con autismo y retardo mental asociado a translocación cromósomica balanceada (7;20). Rev Chil Pediatr 57:549–554, 1986
37. Hansen A, Brask BH, Nielsen J, et al: A case report of an autistic girl with an extra bisatellited marker chromosome. J Autism Child Schizophr 7:263–267, 1977
38. Crandell BF, Carrel RE, Sparkes RS: Chromosome findings in 700 children referred to a psychiatric clinic. J Pediatr 80:62–68, 1972
39. Wolraich M, Bzostek B, Neu RL, et al: Lack of chromosome aberrations in autism. N Engl J Med 283:1231, 1970

40. Campbell M, Wolman SR, Breuer H, et al: Klinefelter's syndrome in a three-year old severely disturbed child. J Autism Child Schizophr 2:34–48, 1972
41. Ornitz EM, Guthrie D, Farley AH: The early development of autistic children. J Autism Child Schizophr 7:207–229, 1977
42. Abrams N, Pergament E: Childhood psychosis combined with XYY abnormalities. J Genet Psychol 118:13–16, 1971
43. Forsius H, Kaski U, Schroder J, et al: Is there a common psychopathology of XYY boys? A clinical report on three cases of XYY and one of XY/XYY. Acta Paedopsychiatr 39:28–41, 1972
44. Mallin SR, Walker FA: Effects of the XYY karyotype in one of two brothers with congenital adrenal hyperplasia. Clin Genet 3:490–494, 1972
45. Nielsen J, Christensen KR, Friedrich U, et al: Childhood of males with the XYY syndrome. J Autism Child Schizophr 3:5–26, 1973
46. Gillberg C, and Forsell C: Childhood psychosis and neurofibromatosis—More than a coincidence? J Autism Dev Dis 14:1–8, 1984
47. Turner G, Daniel A, Frost M: X-linked mental retardation, macro-orchidism, and the Xq27 fragile site. J Pediatr 96:837–841, 1980
48. Brown WT, Friedman E, Jenkins EC, et al: Association of fragile X syndrome with autism. Lancet 1:100, 1982
49. Meryash DL, Szymanski LS, Gerald PS: Infantile autism associated with the fragile-X syndrome. J Autism Dev Dis 12:295–301, 1982
50. Brown WT, Jenkins EC, Cohen IL, et al: Fragile X and autism: A multicenter survey. Am J Med Genet 23:341–352, 1986
51. Hagerman RJ, Jackson AW III, Levitas A, et al: An analysis of autism in fifty males with the fragile X syndrome. Am J Med Genet 23:359–374, 1986
52. McGillivray BC, Herbst DS, Dill FJ, et al: Infantile autism: An occasional manifestation of fragile (X) mental retardation. Am J Med Genet 23:353–358, 1986
53. Wahlstrom J, Gillberg C, Gustavson K-H, et al: Infantile autism and the fragile X. A Swedish multicenter study. Am J Med Genet 23:403–408, 1986
54. Wright HH, Young SR, Edwards JG, et al: Fragile X syndrome in a population of autistic children. J Am Acad Child Psychiatry 25:641–644, 1986
55. Bolton P: Autism and the fragile X syndrome. Newsl Assoc Child Psychol Psychiatry 10(3):26–27, 1988
56. Crowe RR, Tsai LY, Murray JC, et al: A study of autism using X chromosome DNA probes. Biol Psychiatry 24:473–479, 1988
57. Gillberg C: The neurobiology of infantile autism. J Child Psychol Psychiatry 29:257–266, 1988
58. August GL: A genetic marker associated with infantile autism. Am J Psychiatry 140:813, 1983
59. Goldfine PE, McPherson PM, Heath GA, et al: Association of fragile X syndrome with autism. Am J Psychiatry 142:108–110, 1985
60. Judd LL, Mandell AJ: Chromosome studies in early infantile autism. Arch Gen Psychiatry 18:450–457, 1968
61. Hoshino Y, Yashima Y, Tachibana R, et al: Sex chromosome abnormalities in autistic children—Long Y chromosome. Fukushima J Med Sci 26:31–42, 1979
62. Book JA, Nichtern S, Gruenberg E: Acta Psychiatr Scand 39:309–323, 1963
63. Sperling K, Lackman I: Large human Y chromosome with two fluorescent bands. Clin Genet 2:352–355, 1971
64. Wahlstrom J: Are variations in length of Y chromosome due to structural changes? Hereditas 69:125–128, 1971
65. Zeuthen E, Nielsen J: Length of the Y chromosome in a general population. Acta Genet Med Gemellol 22:45–49, 1973
66. Bishop A, Blank CA, Hunter H: Heritable variation in the length of the Y chromosome. Lancet 2:18–20, 1962
67. Makino S, Muramoto J: Some observations on the variability of the human Y chromosome. Proc Jpn Acad 40:757–761, 1964
68. Kato T, Takagi N, Morita S: A chromosome study in seven neuropsychiatric patients with special regard to the abnormality of the Y chromosome. Jpn J Genet 40:105–112, 1965
69. Nielsen J: Y chromosomes in male psychiatric patients above 180 cm tall. Br J Psychiatry 114:1589–1590, 1968

70. Tsuchida S, Okamura T, Hayashi T: Chromosome abnormalities in the institute for mentally handicapped children. Jpn J Pediatr 73:1551, 1969
71. Christensen KR, Nielsen J: Incidence of chromosome aberrations in a child psychiatric hospital. Clin Genet 5:205–210, 1974
72. Beltran IC, Robertson FW, Page BM: Human Y chromosome variation in normal and abnormal babies and their fathers. Ann Hum Genet 42:315–325, 1979
73. Ballabio A, Carrozzo R, Gil A, et al: Molecular characterization of human X/Y translocations suggests their aetiology through aberrant exchange between homologous sequences on Xp and Yq. Ann Hum Genet 53:9–14, 1989
74. Sankar Siva DV: Chromosomal breakage in infantile autism. Dev Med Child Neurol 12:572–575, 1970
75. Gillberg C: Infantile autism and other childhood psychoses in a Swedish urban region. Epidemiological aspects. J Child Psychol Psychiatry 25:35–43, 1984
76. Young JG, Leven LI, Newcorn JH, et al: Genetic and neurobiological approaches to the pathophysiology of autism and the pervasive developmental disorders, in Meltzer HY (ed): Psychopharmocology: The Third Generation of Progress. New York, Raven, 1987, pp 825–836
77. Garvey M, Mutton DE: Sex chromosome aberrations and speech development. Arch Dis Child 48:937–941, 1973
78. Leonard MF, Landy G, Ruddle FH, et al: Early development of children with sex chromosome abnormalities: A prospective study. Pediatrics 54:208–212, 1974
79. Bender L, Fry E, Pennington B, et al: Speech and language development in 41 children with sex chromosome abnormalities. Pediatrics 71:262–267, 1983
80. Walzer S: X chromosome abnormalities and cognitive development: Implications for understanding normal human development. J Child Psychol Psychiatry 26:177–184, 1985
81. Penrose LS: Contribution to the psychology and pedagogy of feeble-minded children. Special Report Series No. 229. London, Medical Research Council, 1938
82. Martin JP, Bell J: A pedigree of mental defect showing sex-linkage. J Neurol Neurosurg Psychiatry 6:154–157, 1943
83. Lubs HA: A marker-X chromosome. Am J Hum Genet 21:231, 1969
84. Sutherland GR: Fragile sites on human chromosomes: Demonstration of their dependence on the type of tissue culture medium. Science 197:265–266, 1977
85. Sutherland GR: Marker X chromosomes and mental retardation. N Engl J Med 296:14–15, 1977
86. Chudley AE, Hagerman RJ: Fragile X syndrome. J Pediatr 110:821–831, 1987
87. Sutherland GA, Hecht F: Fragile Sites on Human Chromosomes. New York, Oxford University Press, 1985
88. Gustavson K-H, Blomquist H, K:son, Holmgren G: Prevalence of the fragile-X syndrome in mentally retarded boys in a Swedish county. Am J Med Genet 23:581–587, 1986
89. Herbst DS, Miller J: Nonspecific X-linked mental retardation. II. The frequency in British Columbia. Am J Med Genet 7:461–470, 1980
90. Loesch DZ: Dermatoglyphic findings in fragile X syndrome: A causal hypothesis points to X-Y interchange. Ann Hum Genet 50:385–398, 1986
91. Loesch DZ: Discriminant analysis of dermatoglyphic measurements in fragile X males and females. Clin Genet 33:169–175, 1988
92. Langenbeck U, Varga I, Hansmann I: The predictive value of dermatoglyphic anomalies in the diagnosis of fra(X)-positive Martin–Bell syndrome (MBS). Am J Med Genet 30:169–175, 1988
93. Hagerman RJ, Murphy MA, Wittenberger MD: A controlled trial of stimulant medication in children with the fragile X syndrome. Am J Med Genet 30:377–392, 1988
94. Wisniewski KC, French JH, Fernando S, et al: The fragile X syndrome: Associated neurological abnormalities and developmental disabilities. Ann Neurol 18:665–669, 1985
95. Musumeci SA, Ferri R, Colognola RM, et al: Prevalence of a novel epileptogenic EEG pattern in the Martin–Bell syndrome. Am J Med Genet 30:207–212, 1988
96. Fryns JP, Kleczkowska A, Van den Berghe H: The psychological profile of the fragile X syndrome. Clin Genet 25:131–134, 1984
97. Finelli P, Peuschel SM, Padre-Mendoza T, et al: Neurological findings in patients with the fragile-X syndrome. J Neurol Neurosurg Psychiatr 48:150–153, 1985
98. Largo RH, Schinzel A: Developmental and behavioral disturbances in 13 boys with Fragile-X syndrome. Eur J Pediatr 143:269–275, 1985

99. Bregman JD, Dykens E, Watson M, et al: Fragile-X syndrome; variability of phenotypic expression. J Am Acad Child Adol Psychiatry 26:463–471, 1987
100. Dykens E, Leckman J, Paul R, et al: Cognitive, behavioral, and adaptive functioning in fragile X and non-fragile X retarded men. J Autism Dev Disab 18:41–52, 1988
101. Hanson DM, Jackson AW III, Hagerman RJ: Speech disturbances (cluttering) in mildly impaired males with the Martin–Bell syndrome. Am J Med Genet 23:195–206, 1986
102. Paul R, Dykens E, Leckman JF, et al: A comparison of language characteristics of mentally retarded adults with fragile X syndrome and those with nonspecific mental retardation and autism. J Autism Dev Disord 17:457–468, 1987
103. Miezejeski CM, Jenkins EC, Hill AL, et al: A profile of cognitive deficit in females from fragile X families. Neuropsychology 24:405–409, 1984
104. Kemper MB, Hagerman RJ, Ahmad RS, et al: Cognitive profiles and the spectrum of clinical manifestations in heterozygous fra (X) females. Am J Med Genet 23:139–156, 1986
105. Loesch DZ, Hay DA: Clinical features and reproductive patterns in fragile X female heterozygotes. J Med Genet 25:407–414, 1988
106. Wolff PH, Gardner J, Lappen J, et al: Variable expression of the fragile X syndrome in heterozygous females of normal intelligence. Am J Med Genet 30:213–225, 1988
107. Sherman SL, Turner G, Robinson H, et al: Investigation of the segregation of the fragile X mutation in daughters of obligate carrier women. Am J Med Genet 30:633–639, 1988
108. Fryns J-P: The female and the fragile X. Am J Med Genet 23:157–169, 1986
109. Reiss AL, Hagerman RJ, Vinogradov S, et al: Psychiatric disability in female carriers in the fragile X chromosome. Arch Gen Psychiatry 45:25–30, 1988
110. Turner G, Partington MW: Fragile (X) expression, age and the degree of intellectual handicap in the male. Am J Med Genet 30:423–428, 1988
111. Thode A, Laing S, Partington MW, et al: Is there a fragile (X) negative Martin–Bell syndrome? Am J Med Genet 30:459–471, 1988
112. Hockney A, Crowhurst J: Early manifestations of the Martin–Bell syndrome based on a series of both sexes from infancy. Am J Med Genet 30:61–71, 1988
113. Hagerman RJ, Smith ACM, Mariner R: Clinical features of the fragile X syndrome, in Hagerman RJ, McBogg PM (eds): The Fragile X Syndrome: Diagnosis, Biochemistry, and Intervention. Dillon, Colorado, Spectra, 1983, pp 17–53
114. Paul R, Cohen D, Breg R, et al: Fragile-X syndrome: its relation to speech and language disorders. J Speech Hear Disord 49:328–332, 1984
115. Webb T, Butler D, Insley J, et al: Prenatal diagnosis of Martin–Bell syndrome associated with fragile site at Xq27–28. Lancet 2:1423, 1981
116. Jenkins EC, Brown WT, Duncan C, et al: Feasibility of fragile X chromosome prenatal diagnosis demonstrated. Lancet 2:1292, 1981
117. Shapiro LR, Wilnot PL, Brenholz P, et al: Prenatal diagnosis of fragile X chromosome. Lancet 1:99–100, 1982
118. Shapiro LR, Wilmot PL, Murphy PD, et al: Experience with multiple approaches to the prenatal diagnosis of the fragile X syndrome: Amniotic fluid, chorionic villi, fetal blood and molecular methods. Am J Med Genet 30:347–354, 1988
119. Lejeune J: Metabolisme des monocarbones et syndrome de l'X fragile. Bull Acad Natl Med (Paris) 165:1197–1206, 1981
120. Lejeune J: Is the fragile X syndrome amenable to treatment? Lancet 1:273–274, 1982
121. Carpenter NJ, Barber DH, Jones M, et al: Controlled six-month study of oral folic acid therapy in boys with fragile X-linked mental retardation. Am J Hum Genet 35(suppl):p. 82A, 1983 (abst 243)
122. Brown TW, Jenkins EC, Friedman E, et al: Folic acid therapy in boys with the fragile X syndrome. Am J Med Genet 17:289–297, 1984
123. Brown TW, Cohen IL, Fisch GS, et al: High dose folic acid treatment of fragile (X) males. Am J Med Genet 23:263–271, 1986
124. Froster-Iskenius U, Bodeker K, Oepen T, et al: Folic acid treatment in males and females with fragile-(X)-syndrome. Am J Med Genet 23:273–289, 1986
125. Hagerman RJ, Jackson AW, Levitas A, et al: Oral folic acid versus placebo in the treatment of males with the fragile X syndrome. Am J Med Genet 23:241–262, 1986
126. Wells TE, Madison LS: Assessment of behavior change in a fragile-X syndrome male treated with folic acid. Am J Med Genet 23:291–296, 1986

127. Rudelli RD, Brown WT, Wisniewski K, et al: Adult fragile X syndrome: Clinico-pathologic findings. Acta Neuropathol (Berl) 67:289–295, 1985
128. Venter PA, Hof JO, Coetzee DJ: The Martin–Bell syndrome in South Africa. Am J Med Genet 23:597–610, 1986
129. Hagerman R, Berry R, Jackson AW III, et al: Institutional screening for the fragile X syndrome. Am J Dis Child 142:1216–1221, 1988
130. Fisch GS, Cohen IL, Jenkins EC, et al: Screening developmentally disabled male populations for fragile X: The effect of sample size. Am J Med Genet 30:655–663, 1988
131. Gillberg C, Wahlstrom J, Hagberg B: Infantile autism and Rett's syndrome: Common chromosomal denominator. Lancet 2:1094–1095, 1984
125a. Gillberg C: Autism and Rett syndrome: Some notes on differential diagnosis. Am J Med Genet 24:127–131, 1986
132. Witt-Engerstrom I, Gillberg C: Rett syndrome in Sweden. J Autism Dev Disord 17:149–150, 1987
133. Sherman SL, Morton NE, Jacobs P, et al: The marker (X) syndrome; a cytogenetic and genetic analysis. Ann Hum Genet 48:21–37, 1984
134. Sherman SL, Jacobs PA, Morton NE, et al: Further segregation analysis of the fragile X syndrome with special reference to transmitting males. Hum Genet 69:289–299, 1985
135. Nussbaum RL, Airhart SD, Ledbetter DH, et al: Recombination and amplification of pyrimidine-rich sequences may be responsible for initiation and progression of the Xq27 fragile site: An hypothesis. Am J Med Genet 23:715–721, 1986
136. Berger R, Bloomfield CD, Sutherland GR: Report of the committee on chromosome rearrangements in neoplasia and on fragile sites: 8th international workshop on human gene mapping. Cytogenet Cell Genet 40:490–535, 1985
137. Ledbetter SA, Ledbetter DH: A common fragile site at Xq27: Theoretical and practical implications. Am J Hum Genet 42:694–702, 1988
138. Pembrey ME, Winter RM, Davies KE: A premutation that generates a defect at crossing over explains the inheritance of fragile X mental retardation. Am J Med Genet 23:709–718, 1986
139. Laird CD: Proposed mechanism of inheritance and expression of the human fragile-X syndrome of mental retardation. Genetics 117:587–599, 1987
140. Laird CD: Fragile-X mutation proposed to block complete reactivation in females of an inactive X chromosome. Am J Med Genet 30:693–696, 1988
141. Laird C, Jaffe E, Karpen G, et al: Fragile sites in human chromosomes as regions of late-replication DNA. Trends Genet 3:274–281, 1987
142. Fryns J-P, Van den Berghe H: Inactivation pattern of the fragile X in heterozygous carriers. Am J Med Genet 30:401–406, 1988
143. Wilhelm D, Froster-Iskenius U, Paul J, et al: Fra(X) frequency on the active X-chromosome and phenotype in heterozygous carriers of the fra(X) form of mental retardation. Am J Med Genet 30:407–415, 1988
144. Warren ST: Fragile X syndrome: A hypothesis regarding the molecular mechanism of the phenotype. Am J Med Genet 30:681–688, 1988
145. Rumsey JM, Denckla MB: Neurobiological research priorities in autism, in Schopler E, Mesibov GB (eds): Neurobiological Issues in Autism. New York, Plenum, 1987, pp 43–57
146. Jaeken J, Van den Berghe G: An infantile autistic syndrome characterised by the presence of succinylpurines in body fluids. Lancet 2:1058–1061, 1984
147. Stubbs G, Litt M, Lis E, et al: Adenosine deaminase activity decreased in autism. J Am Acad Child Psychiatry 21:71–74, 1982
148. Coleman M, Landgrebe MA, Landgrebe AR: Purine autism: Hyperuricosuria in autistic children; does this identify a subgroup of autsim? in Coleman M (ed): The Autistic Syndromes. Amsterdam, North-Holland, 1976, pp 183–195
149. Coleman M, Blass JP: Autism and lactic acidosis. J Autism Dev Disord 15:1–8, 1985
150. Nyhan WL, James JA, Teberg AJ, et al: A new disorder of purine metabolism with behavioral manifestations. J Pediatr 74:20–27, 1969
151. Becker MA, Raivo KO, Bakay B, et al: Variant human phosphoribosylpyrophosphate synthetase altered in regulatory and catalytic functions. J Clin Invest 65:109–120, 1980
152. Gruber HE, Jansen I, Willis RC, et al: Alterations of inosinate branchpoint enzymes in cultured human lymphoblasts. Biochim Biophys Acta 846:135–144, 1985
153. Kotsopoulos S, Kutty KM: Histidinemia and infantile autism. J Autism Dev Disord 9:55–60, 1979
154. Creak EM: Childhood psychosis: A review of 100 cases. Br J Psychiatry 109:84–89, 1963

155. Kratter FE: The physiognomic, psychometric, behavioral and neurological aspects of phenylketonuria. J Ment Sci 439:421–427, 1959
156. Lewis E: The development of concepts in a girl after dietary treatment for phenylketonuria. Br J Med Psychol 32:282–287, 1959
157. Hackney IM, Hanley WB, Davidson W, et al: Phenylketonuria: Mental development, behavior, and termination of low phenylalanine diet. J Pediatr 72:646–655, 1968
158. Friedman E: The "autistic syndrome" and phenylketonuria. Schizophrenia 1:249–261, 1969
159. Bliumina MG: A schizophrenic-like variant of phenylketonuria. Zh Nevropatrol Psikhiatr 75:1525–1529, 1975
160. Lowe TL, Tanaka K, Seashore MR, et al: Detection of phenylketonuria in autistic and psychotic children. JAMA 243:126–128, 1980
161. Williams RS, Hauser SL, Purpura DP, et al: Autism and mental retardation: Neuropathologic studies performed in four retarded persons with autistic behavior. Arch Neurol 37:749–753, 1980
162. Pueschel SM, Herman R, Groden G: Brief report: Screening children with autism for fragile-X syndrome and phenylketonuria. J Autism Dev Disord 15:335–338, 1985
163. Paul R, Cohen DJ, Volkmar FR: Autistic behaviors in a boy with Noonan syndrome. J Autism Dev Disord 13:433–434, 1983
164. Taft LT, Cohen HJ: Hypsarrhythmia and infantile autism: A clinical report. J Autism Child Schizophr 1:327–336, 1971
165. Lotter V: Factors related to outcome in autistic children. J Autism Child Schizophr 4:263–277, 1974
166. Mansheim P: Tuberous sclerosis and autistic behavior. J Clin Psychiatry 40:97–98, 1979
167. Hunt A: Tuberous sclerosis: A survey of 97 cases. I. Seizures, pertussis immunisation and handicap. Dev Med Child Neurol 25:346–349, 1983
168. Hunt A: Tuberous sclerosis: A survey of 97 cases. II. Physical findings. Dev Med Child Neurol 25:350–352, 1983
169. Hunt A: Tuberous sclerosis: A survey of 97 cases. III. Family aspects. Dev Med Child Neurol 25:353–357, 1983
170. Curatolo P: Autism and infantile spasms in children with tuberous sclerosis. Dev Med Child Neurol 29:550, 1987
171. Lawlor BA, Maurer RG: Tuberous sclerosis and the autistic syndrome. Br J Psychiatry 150:396–397, 1987
172. Oliver BE: Tuberous sclerosis and the autistic syndrome. Br J Psychiatry 151:560, 1987
173. Leland H: Some psychological characteristics of phenylketonuria. Psychol Rep 3:373–376, 1957
174. Sutherland BS, Berry HK, Shirkey HC: A syndrome of phenylketonuria with normal intelligence and behavior disturbance. J Pediatr 57:521, 1960
175. Bjornson J: Behavior in phenylketonuria: Case with schizophrenia. Arch Gen Psychiatry 10:65–70, 1964
176. Young JG, Cohen DJ, Caparulo BK, et al: Am J Psychiatry 136:1055–1057, 1979
177. Young JG, Cohen DJ, Shaywitz BA, et al: in Maas J (ed): MHPG in Psychopathology. Orlando, Florida, Academic, 1982, pp 193–218
178. Hanson DR, Gottesman II: The genetics, if any, of infantile autism and childhood schizophrenia. J Autism Child Schizophr 6:209–234, 1976
179. Galton F: The history of twins as a criterion of the relative powers of nature and nurture. Popular Science Monthly 8:345–357, 1875
180. Falconer DS: The heritability of liability to certain diseases estimated from the incidence among relatives. Ann Hum Genet 29:51–76, 1965
181. Allen G, Harvald B, Shields J: Measures of twin concordance. Acta Genet Stat Med (Basel) 17:475–481, 1967
182. Smith C: Concordance in twins: Methods and interpretation. Am J Hum Genet 26:454–466, 1974
183. Gottesman II, Carey G: Extracting meaning and direction from twin data. Psychiatr Dev 1:35–50, 1983
184. Stabenau JR: Heredity and environment in schizophrenia: The contribution of twin studies. Arch Gen Psychiatry 18:458–463, 1968
185. Chapman H: Early infantile autism in identical twins. Arch Neurol Psychiatry 78:621–623, 1957
186. Ward TF, Hoddinott BA: Early infantile autism in fraternal twins: A case report. Can Psychiatr Assoc J 7:191–195, 1962
187. Vaillant GE: Twins discordant for early infantile autism. Arch Gen Psychiatry 9:163–167, 1963
188. Rimland B: Infantile Autism. New York, Appleton-Century–Crofts, 1964
189. Kamp LNJ: Autistic Syndrome in one of a pair of monozygotic twins. Psychiatr Neurol Neurochir 67:143–147, 1964

190. Rutter M: Psychotic disorders in early childhood, in Coppen A, Walk A (eds): Recent Developments in Schizophrenia. Br J Psychiatry, Special Publication No. 1, pp 133–155
191. Kallman FJ, Roth B: Genetic aspects of preadolescent schizophrenia. Am J Psychiatry 112:559–606, 1956
192. Kean JM: The development of social skills in autistic twins. NZ Med J 81:204–207, 1975
193. McQuaid PE: Infantile autism in twins. Br J Psychiatry 127:530–534, 1975
194. Kotsopoulos S: Infantile autism in dizygotic twins: A case report. J Autism Child Schizophr 6:133–138, 1976
195. Sloan JL: Differential development of autistic symptoms in a pair of fraternal twins. J Autism Child Schizophr 8:191–202, 1978
196. Wessels WH, Pompe van Meerdervoort M: Monozygotic twins with early infantile autism: A case report. S Afr Med J 55:955–957, 1979
197. Salimi Eshkevari H: Early infantile autism in monozygotic twins. J Autism Dev Disord 9:105–109, 1979
198. Salimi Eshkevari H: Infantile autism in monozygotic twins. J Am Acad Child Psychiatry 24:643–646, 1985
199. Gillberg C, Ohlson V-A, Wahlstrom J, et al: Monozygotic female twins with autism and the fragile-X syndrome (AFRAX). J Child Psychol Psychiatry 29:447–451, 1988
200. Silliman ER, Campbell M, Mitchell RS: Genetic influences in autism and assessment of metalinguistic performance in siblings of autistic children, in Dawson G (ed): Autism. New York, Guilford, 1989, pp 225–259
201. Ritvo ER, Brothers AM, Freeman BJ, et al: Eleven possibly autistic parents. (Letter.) J Autism Dev Disord 18:139–147, 1988
202. Folstein S, Rutter M: Infantile autism: A genetic study of 21 twin pairs. J Child Psychol Psychiatry 18:297–331, 1977
203. Folstein SE: Genetic aspects of infantile autism. Annu Rev Med 36:415–419, 1985
204. Folstein SE, Rutter ML: Autism: Familial aggregation and genetic implications. J Autism Dev Disord 18:3–30, 1988
205. Rutter M, Bartak L, Newman S: Autism—A central disorder of cognition and language? in Rutter M (ed): Infantile autism: Concepts, Characteristics, and Treatment. Edinburgh, Churchill-Livingstone, 1971, pp 148–171
206. Bartak L, Rutter M, Cox A: A comparative study of infantile autism and specific developmental language disorder. I. The children. Br J Psychiatry 126:127–145, 1975
207. Bartak L, Rutter M, Cox A: A comparative study of infantile autism and specific developmental receptive language disorders. III. Discriminant function analysis. J Autism Child Schizophr 7:383–396, 1977
208. Campbell M, Dominijanni C, Schneider B: Monozygotic twins concordant for infantile autism: Follow-up. Br J Psychiatry 131:616–622, 1977
209. Campbell M, Minton J, Green WH, et al: Siblings and twins of autistic children, in Perris C, Struwe G, Jansson B (eds): Biological Psychiatry 1981. Elsevier, North-Holland, 1981, pp 993–996
210. Campbell M, Fish B, Shapiro T, et al: A twin study of preschoolage autistic children (manuscript submitted) Presented at the 27th Annual Meeting of the American Academy of Child Psychiatry, Chicago, Oct. 15–19, 1980
211. Shell J, Campion JF, Minton J, et al: A study of three brothers with infantile autism: A case report with follow-up. J Am Acad Child Psychiatry 26:480–484, 1984
212. Ritvo ER, Freeman BJ, Mason-Brothers A, et al: Concordance for the syndrome of autism in 40 pairs of afflicted twins. Am J Psychiatry 142:74–77, 1985
213. Ritvo ER, Spence MA, Freeman BJ, et al: Evidence for autosomal recessive inheritance in 46 families with multiple incidences of autism. Am J Psychiatry 142:187–192, 1985
214. Ritvo ER, Mason-Brothers A, Jenson WP, et al: A report of one family with four autistic siblings and four families with three autistic siblings. J Am Acad Child Adol Psychiatry 26:339–341, 1987
215. Morton NE: Segregation and linkage, in Burdette WJ (ed): Methodology in Human Genetics. San Francisco, Holden Day, 1962, pp 17–52
216. Jones MB, Szatmari P: Stoppage rules and genetic studies of autism. J Autism Dev Disord 18:31–40, 1988
217. Brookfield JFY, Pollitt RJ, Young ID: Family size limitation: A method for demonstrating recessive inheritance. J Med Genet 25:181–185, 1988
218. Tsai LY, Stewart MA: Etiological implication of maternal age and birth order in infantile autism. J Autism Dev Disord 13:57–65, 1983
219. Ritvo ER, Freeman BJ, Pingree C, et al: The UCLA–University of Utah epidemiologic survey of autism: Prevalence. Am J Psychiatry 146:194–199, 1989

220. Bryson SE, Clark BS, Smith IM: First report of a Canadian epidemiological study of autistic syndromes. J Child Psychol Psychiatry 29:433–445, 1988
221. Brask BH: A prevalence investigation of childhood psychoses, in Nordic Symposium on the Comprehensive Care of the Psychotic Children. Oslo, Norway, Barnespsykiatrisk Forening-Norge, 1972, pp 145–153
222. Cialdella P, Mamelle N: An epidemiological study of infantile autism in a French department (Rhone): A research note. J Child Psychol Psychiatry 30:165–175, 1989
223. McCarthy P, Fitzgerald M, Smith MA: Prevalence of childhood autism in Ireland. Irish Med J 77:129–130, 1984
224. Haga H, Miyamoto H: A survey on the actual state of so-called autistic children in Kyoto prefecture. Jpn J Child Psychiatry 12:160–167, 1971
225. Nakai K: A survey of so-called autistic children in Gifu prefecture. Jpn J Child Psychiatry 12:262–266, 1971
226. Yamazaki K, Yamashita I, Suwa N, et al: Survey on the morbidity rate of "autistic children" in the Hokkaido district. Jpn J Child Psychiatry 12:141–149, 1971
227. Hoshino Y, Kumashiro H, Yashima Y, et al: The epidemiological study of autism in Fukushima-ken. Folia Psychiatr Neurol Jpn 36:115–124, 1982
228. Matsuishi T, Shiotsuki Y, Yoshimura K, et al: High prevalence of infantile autism in Kurume City, Japan. J Child Neurol 2:268–271, 1987
229. Tanoue Y, Oda S, Asano F, et al: Epidemiology of infantile autism in southern Ibaraki, Japan: Differences in prevalence in birth cohorts. J Autism Dev Disord 18:155–166, 1988
230. Bohman M, Björck P-O, Bohman I-L, et al: Barndomspsykoserna—forsummade handikapp? Epidemiologi och habilitering i ett glesbygdslan. Lakartidningen 78:2361–2364, 1981
231. Bohman M, Bohman IL, Björck PO, et al: Childhood psychosis in a northern Swedish county: Some preliminary findings from an epidemiological survey, in Schmidt MH, Remschmidt H (eds): Epidmiological Approaches in Child Psychiatry, Vol. II. International Symposium, Mannheim, 1981. Stuttgart, West Germany, Georg Thieme Verlag, 1983, pp 164–173
232. Steffenburg S, Gillberg C: Autism and autistic-like conditions in Swedish rural and urban areas: A population study. Br J Psychiatry 149:81–87, 1986
233. Lotter V: Epidemiology of autistic conditions in young children. I. Prevalence. Social Psychiatry 1:124–137, 1966
234. Wing L, Yeates SR, Brierley LM, et al: The prevalence of early childhood autism: Comparison of administrative and epidemiological studies. Psychol Med 6:89–100, 1976
235. Wing L, Gould J: Severe impairments of social interaction and associated abnormalities in children: Epidemiology and classification. J Autism Dev Disord 9:11–29, 1979
236. Treffert DA: Epidemiology of infantile autism. Arch Gen Psychiatry 22:431–438, 1970
237. Burd L, Fisher W, Kerbeshian J: A prevalence study of pervasive developmental disorders in North Dakota. J Am Acad Child Adol Psychiatry 26:700–703, 1987
238. American Psychiatric Association: Diagnostic and Statistical Manual, ed 3. Washington, D.C., American Psychiatric Association, 1980
239. American Psychiatric Association: Diagnostic and Statistical Manual, ed 3, revised. Washington, D.C., American Psychiatric Association, 1987
240. Shapiro T, Sherman M, Calamari G, et al: Attachment in Autism and other developmental disorders. J Am Acad Child Adol Psychiatry 28:74–81, 1987
241. Sigman M, Mundy P: Social attachments in autistic children. J Am Acad Child Adol Psychiatry 28:74–81, 1989
242. Wing L: Language, social, and cognitive impairments in autism and severe mental retardation. J Autism Dev Disord 11:31–44, 1981
243. Young JG, Newcorn JH, Leven LI: Pervasive developmental disorders, in Kaplan HI, Sadock BJ (eds): Comprehensive Textbook of Psychiatry, ed 5. Baltimore, Williams & Wilkins, 1989, pp 1772–1787
244. Cohen DJ, Paul R, Volkmar FR: Issues in the classification of pervasive and other developmental disorders: Towards DSM-IV. J Am Acad Child Adol Psychiatry 25:213–220, 1986
245. Tanguay PE, Asarnow R: Schizophrenia in Children, in Michels R, Cavenar JO (eds): Psychiatry, Vol. 2. Philadelphia, JB Lippincott, 1988
246. Cohen DJ, Caparulo B, Shaywitz B: Primary childhood aphasia and childhood autism. J Am Acad Child Psychiatry 15:604–645, 1976
247. Paul R, Cohen DJ, Caparulo BK: A longitudingal study of patients with severe developmental disorders of language learning. J Am Acad Child Psychiatry 22:525–534, 1983

248. Eisenberg L, Kanner L: Early infantile autism 1943–1953. Am J Orthopsychiatry 26:556–566, 1956
249. Whittam H, Simon GB, Mittler PJ: The early development of psychotic children and their sibs. Dev Med Child Neurol 8:552–560, 1966
250. Havelkova M: Abnormalities of siblings of schizophrenic children. Can Psychiatr Assoc J 12:363–367, 1967
251. August GL, Stewart MA, Tsai L: The incidence of cognitive disabilities in the siblings of autistic children. Br J Psychiatry 138:416–422, 1981
252. Minton J, Campbell M, Green WH, et al: Cognitive assessment of siblings of autistic children. J Am Acad Child Psychiatry 21:256–261, 1982
253. Baird TD, August GJ: Familial heterogeneity in infantile autism. J Autism Dev Disord 15:315–321, 1985
254. Wing L: Asperger's syndrome: A clinical account. Psychol Med 11:115–130, 1981
255. Schopler E: Convergence of learning disability, higher-level autism, and Asperger's syndrome. J Autism Dev Disord 15:359–360, 1985
256. Freeman BJ, Ritvo ER, Mason-Brothers A, et al: Psychometric assessment of first-degree relatives of 62 autistic probands in Utah. Am J Psychiatry 146:361–364, 1989
257. Woolf CM, Stephens FE, Mulaik DD, et al: An investigation of the frequency of consanguineous marriages among the Mormons and their relatives in the United States. Am J Hum Genet 8:236–252, 1956
258. McLellan T, Jorde LB, Skolnick MH: Genetic distances between the Utah Mormons and related populations. Am J Hum Genet 36:836–857, 1984
259. Prizant BM: Language acquisition and communicative behavior in autism: Toward an understanding of the "whole" of it. J Speech Hearing Disord 48:296–307, 1983
260. Sherman M, Shaprio T, Glassman M: Play and language in developmentally disordered preschoolers: A new approach to classification. J Am Acad Child Psychiatry 22:511–524, 1983
261. Lotter V: Epidemiology of autistic conditions in young children. II. Some characteristics of the parents and children. Social Psychiatry 1:163–173, 1967
262. Wing L: Sex ratios in early childhood autism and related conditions. Psychiatry Res 5:129–137, 1981
263. Tsai LY, Beisler JM: The development of sex differences in infantile autism. Br J Psychiatry 142:373–378, 1983
264. Lewis EO: Report on an investigation into the incidence of mental deficiency in six areas. 1925–1927, in Report of the Mental Deficiency Committee. Part IV. London, His Majesty's Stationery Office, 1929
265. Abramowicz HK, Richardson SA: Epidemiology of severe mental retardation in children: Community studies. Am J Ment Defic 80:18–39, 1975.
266. Tsai L, Stewart MA, August G: Implication of sex differences in the familial transmission of infantile autism. J Autism Dev Disord 11:165–173, 1981
267. Lord C, Schopler E: Brief report: Differences in sex ratios in autism as a function of measured intelligence. J Autism Dev Disord 15:185–193, 1985
268. Taylor D, Ounsted C: The nature of gender differences explored through ontogenetic analyses of sex ratios in disease, in Ounsted D, Taylor DC (eds): Gender Differences: their Ontogeny and Significance. London, Churchill Livingstone, 1972, pp 215–240
269. Lord C. Schopler E: Neurobiological implications of sex differences in autism, in Schopler E, Mesibov GB (eds): Neurobiological Issues in Autism. New York, Plenum, 1987, pp 191–211
270. Lord C, Schopler E, Revicki D: Sex differences in autism. J Autism Dev Disord 12:317–330, 1982
271. Schain RJ, Freedman DX: Studies on 5-hydroxyindole metabolism in autistic and other mentally retarded children. J Pediatr 58:315–320, 1961
272. Goldberg TE, Maltz A, Bow JN, et al: Blink rate abnormalities in autistic and mentally retarded children: Relationship to dopaminergic activity. J Am Acad Child Adol Psychiatry 26:336–338, 1987
273. Ritvo ER, Creel D, Realmuto G, et al: Electroretinograms in autism: A pilot study of b-wave amplitudes. Am J Psychiatry 145:229–232, 1988
274. White RL: Human genetics. Lancet 2:1257–1262, 1984
275. White RL: Diagnosis when the gene locus is unknown. Hosp Pract 20:103–112, 1985
276. White R: DNA sequence polymorphisms revitalize linkage approaches in human genetics. Trends Genet 1:177–181, 1985
277. White R, Caskey CT: The human as experimental system in molecular genetics. Science 240:1483–1488, 1988
278. White R, Lalouel J-M: Chromosome mapping with DNA markers. Sci Am 258:40–48, 1988
279. Lathrop GM, Lalouel JM, Julier C, et al: Multilocus linkage analysis in humans: Detection of linkage and estimation of recombination. Am J Hum Genet 37:482–498, 1985

280. Kidd KK: Research design considerations for linkage studies of affective disorders using recombinant DNA markers. J Psychiatr Res 21:551–557, 1987
281. Spence MA, Ritvo ER, Marazita ML, et al: Gene mapping studies with the syndrome of autism. Behav Genet 15:1–13, 1985
282. Camerino G, Mattei MG, Mattei JF, et al: Close linkage of fragile X linked mental retardation syndrome to haemophilia B and transmission through a normal male. Nature (Lond) 306:701–707, 1983
283. Winter RM, Pembrey ME: Analysis of linkage relationships between genetic markers around the fragile X locus with special reference to the daughters of transmitting males. Hum Genet 74:93–97, 1986
284. Brown WT, Gross AC, Chan CB, et al: DNA linkage studies in the fragile X syndrome suggest genetic heterogeneity. Am J Med Genet 23:643–664, 1986
285. Brown WT, Jenkins EC, Gross AC, et al: Further evidence for genetic heterogeneity in the fragile X syndrome. Hum Genet 75:311–321, 1987
286. Brown WT, Gross A, Chan C, et al: Multilocus analysis of the fragile X syndrome. Hum Genet 78:201–205, 1988
287. Clayton JF, Gosden CM, Hastie ND, et al: Linkage heterogeneity and fragile X. Hum Genet 78:338–342, 1988
288. Thibodeau SN, Dorkins HR, Faulk KR, et al: Linkage analysis using multiple DNA polymorphic markers in normal families and in families with fragile X syndrome. Hum Genet 79:219–227, 1988
289. Damasio H, Maurer RG, Damasio AR, et al: Computerized tomographic scan findings in patients with autistic behavior. Arch Neurol 37:504–510, 1980
290. Campbell M, Rosenbloom S, Perry R, et al: Computerized axial tomography in young autistic children. Am J Psychiatry 139:510–512, 1982
291. Harcherik DF, Cohen DJ, Ort S, et al: Computed tomographic brain scanning in four neuropsychiatric disorders of childhood. Am J Psychiatry 142:731–734, 1985
292. Jacobson R, Le Couteur A, Howlin P, et al: Selective subcortical abnormalities in autism. Psychol Med 18:39–48, 1988
293. Rumsey JM, Creasey H, Stepanek JS, et al: Hemispheric asymmetries, fourth ventricular size, and cerebellar morphology in autism. J Autism Dev Disord 18:127–137, 1988
294. Gaffney GR, Tsai LY: Brief report: Magnetic resonance imaging of high level autism. J Autism Dev Disord 17:433–438, 1987
295. Gaffney GR, Kuperman S, Tsai LY, et al: Midsagittal magnetic resonance imaging of autism. Br J Psychiatry 151:831–833, 1987
296. Gaffney GR, Tasi LY, Kuperman S, et al: Cerebellar structure in autism. Am J Dis Child 141:1330–1332, 1987
297. Courchesne E, Hesselink JR, Jernigan TL, et al: Abnormal neuroanatomy in a nonretarded person with autism. Arch Neurol 44:335–341, 1987
298. Courchesne E, Yeung-Courchesne R, Press GA, et al: Hypoplasia of cerebellar vermal lobules VI and VII in autism. N Engl J Med 318:1349–1354, 1988
299. Garber HJ, Ritvo ER, Chiu LC, et al: A magnetic resonance imaging study of autism: Normal fourth ventricle size and absence of pathology. Am J Psychiatry 146:532–534, 1989
300. Horowitz B, Rumsey JM, Grady CL, et al: The cerebral metabolic landscape in autism: Intercorrelations of regional glucose utilization. Arch Neurol 45:749–755, 1988
301. Rumsey JM, Duara R, Grady C, et al: Arch Gen Psychiatry 42:448–455, 1985
302. Heh CWC, Smith R, Wu J, et al: Positron emission tomography of the cerebellum in autism. Am J Psychiatry 146:242–245, 1989
303. De Volder A, Bol A, Michel C, et al: Brain glucose metabolism in children with the autistic syndrome: Positron tomography analysis. Brain Dev 9:581–587, 1987
304. De Volder AG, Bol A, Michel C, et al: Metabolisme cérébral du glucose chez les enfants autistes: Etude em tomographie par émission de positrons. Acta Neurol Belg 88:75–90, 1988
305. Ogawa T, Sugiyama A, Ishiwa S, et al: Brain Dev 4:439–449, 1982
306. Martineau J, Bruneau M, Garreau B, et al: Clinical and biological markers in autistic syndromes. Electroencephalogr Clin Neurophysiol 67(suppl):87P, 1987
307. Martineau J, Bruneau N, Garreau, et al: Interêt des marqueurs cliniques et biologiques dans les syndromes autistiques de l'enfant. Rev EEG Neurophysiol Clin 17:159–167, 1987
308. Ornitz EM: Neurophysiology of infantile autism. J Am Acad Child Psychiatry 24:251–262, 1985
309. Garreau B, Bruneau N, Roux S, et al: Study of the auditory function in infantile autism with electrophysiological data. Electroencephalogr Clin Neurophysiol 67(suppl):87p–88P, 1987

310. Rosenhall U, Johansson E, Gillberg C: Oculomotor findings in autistic children. J Laryngol Otol 102:435–439, 1988
311. Ritvo ER, Creel D, Realmuto G, et al: Electroretinograms in autism: A pilot study of b-wave amplitudes. Am J Psychiatry 145:229–232, 1988
312. Courchesne E, Lincoln AJ, Kilman BA, et al: J Autism Dev Disord 15:55–76, 1985
313. Oades RD, Walker MK, Geffen LB, et al: Event-related potentials in autistic and healthy children on an auditory choice reaction time task. Int J Psychophysiol 6:25–37, 1988
314. Kuperman S, Beeghly J, Burns T, et al: Association of serotonin concentration to behavior and IQ in autistic children. J Autism Develop Disord 17:133–140, 1987
315. Badcock NR, Spence JG, Stern LM: Blood serotonin levels in adults, autistic and non-autistic children—With a comparison of different methodologies. Ann Clin Biochem 24:625–634, 1987
316. Yuwiler A, Freedman DX: Neurotransmitter research in autism, in Schopler E, Mesibov GB (eds): Neurobiological Issues in Autism. New York, Plenum, 1987, pp 263–284
317. Geller E, Yuwiler A, Freeman BJ, et al: Platelet size, number, and serotonin content in blood of autistic, childhood schizophrenic, and normal children. J Autism Dev Disord 18:119–126, 1988
318. Anderson GM, Freedman DX, Cohen DJ, et al: Whole blood serotonin in autistic and normal subjects. J Child Psychol Psychiatry 28:885–900, 1987
319. Anderson GM: Monoamines in autism: An update of neurochemical research on a pervasive developmental disorder. Med Biol 65:67–74, 1987
320. Cohen DJ, Shaywitz BA, Caparulo B, et al: Chronic, multiple tics of Gilles de la Tourette's disease: CSF acid monoamine metabolites after probenicid administration. Arch Gen Psychiatry 35:245–250, 1978
321. Gillberg C, Svennerholm L, Hamilton-Hellberg C: Childhood psychosis and monoamine metabolites in spinal fluid. J Autism Dev Disord 13:383–396, 1983
322. Young JG, Cohen DJ, Caparulo BK, et al: Decreased 24-hr urinary MHPG in childhood autism. Am J Psychiatry 136:1055–1057, 1979
323. Gillberg C, Terenius L, Lonnerholm G: Endorphin activity in childhood psychosis: Spinal fluid levels in 24 cases. Arch Gen Psychiatry 42:780–783, 1985
324. Weizman R, Weizman A, Tyano S, et al: Humoral endorphin blood levels in autistic, schizophrenic and healthy subjects. Psychopharmacology 82:368–370, 1984
325. Weizman R, Gil-Ad I, Dick J, et al: Low plasma immunoreactive B-endophin levels in autism. J Am Acad Child Adol Psychiatry 27:430–433, 1988
326. Deutsch SI, Campbell M, Sachar EJ, et al: J Autism Dev Disord 15:205–212, 1985
327. Hoshino Y, Yokoyama F, Watanabe M, et al: The diurnal variation and response to dexamethasone suppression test of saliva cortisol level in autistic children. Jpn J Psychiatr Neurol 41:227–236, 1987
328. Todd RD, Ciaranello RD: Demonstration of inter- and intraspecies differences in serotonin binding sites by antibodies from an autistic child. Proc Natl Acad Sci USA 82:612–616, 1985
329. Todd RD, Hickok JM, Anderson GM, et al: Antibrain antibodies in infantile autism. Biol Psychiatry 23:644–647, 1988
330. Weizman A, Weizman R, Szekely GA, et al: Am J Psychiatry 139:1462–1465, 1982
331. Warren RP, Foster A, Margaretten NC: Reduced natural killer cell activity in autism. J Am Acad Child Adol Psychiatry 26:333–335, 1987
332. Stubbs G: Shared parental HLA antigens and autism. Lancet 2:534, 1981
333. Stubbs EG, Ritvo ER, Mason-Brothers A: Autism and shared parental HLA antigens. J Am Acad Child Psychiatry 24:182–185, 1985
334. Ornitz EM, Atwell CW, Kaplan AR: Brain-stem dysfunction in autism: Results of vestibular stimulation. Arch Gen Psychiatry 42:1018–1025, 1985
335. Goldberg TE, Maltz A, Bow JN, et al: Blink rate abnormalities in autistic and mentally retarded children: Relationship to dopaminergic activity. J Am Acad Child Adol Psychiatry 26:336–338, 1987

12

Biological Studies of Schizophrenia with Childhood Onset

Wayne H. Green and Stephen I. Deutsch

1. INTRODUCTION AND DEFINITION

The literature concerning the biology of schizophrenia pertains almost exclusively to schizophrenia with adolescent or adult onset. Several recent publications on the biology, psychophysiology, and psychoneuroimmunology of schizophrenia in older adolescents and adults are available.[1–4] The relative paucity of biological data on children appears to result from both the rarity of schizophrenia with childhood onset and the increased difficulties in undertaking research with children.

This chapter reviews biological studies of schizophrenia with childhood onset. Other aspects of schizophrenia in childhood have recently been reviewed by Green.[5] The subjects in the studies reviewed meet all DSM-III (1980) criteria,[6] including deterioration from a previous level of functioning, for a schizophrenic disorder as seen in adolescents and adults. As DSM-III-R (1987) criteria[7] for schizophrenia with childhood onset are less stringent than those of DSM-III, the children considered herein fulfill these criteria as well. Although a cogent argument could be made for defining childhood onset physiologically as occurring in a prepubescent youngster, the lack of either Tanner staging or endocrinological status of the children in virtually all published reports necessitates that childhood be defined chronologically for the present. The twelfth birthday, one traditional beginning of adolescence, will be considered the upper limit of childhood; however, in a few instances, pooled data that extend into early adolescence will be reviewed.

Wayne H. Green • Department of Psychiatry, New York University School of Medicine, and Child and Adolescent Psychiatric Clinic, Bellevue Hospital Center, New York, New York 10016. *Stephen I. Deutsch* • Psychiatry Service, Veterans Administration Medical Center, and Department of Psychiatry, Georgetown University Medical Center, Washington, D.C. 20007.

2. EPIDEMIOLOGY

2.1. Prevalence

Schizophrenia with onset during childhood is a rare disorder. A reasonable estimate of prevalence would be on the order of 2–4 per 10,000. The incidence of schizophrenia increases remarkably following puberty and initial diagnosis occurs during late adolescence and early adulthood in most cases.

2.2. Etiology

As this chapter presumes that schizophrenia with onset in childhood is on a continuum with adolescence and adulthood onset schizophrenia, etiological issues overlapping with adolescent and adult-onset schizophrenia will not be addressed here. Most authorities believe schizophrenia to be a diathesis/stress disorder, i.e., having a certain genetic predisposition that may be aggravated by unfavorable psychosocial and/or physical environments. Neither factors that are responsible for nor those that are prophylactic for the development of clinically recognizable schizophrenia in preadolescents have yet been identified.

The data of Kallmann and Roth[8] noting that the onset of childhood schizophrenia was earlier in both male and female singletons than in twins, although as yet unexplained, may be relevant. By contrast, it is possible that schizophrenia is simply noticed and diagnosed earlier in singletons than in twins but actually occurs no earlier.

Kallmann and Roth[8] found no significant etiological relationship between birth order and development of schizophrenia. They did notice, however, that the siblings of their preadolescent index cases seemed more likely to also develop preadolescent schizophrenia than did siblings of adult-onset schizophrenia. Thus, 8.0% of siblings of their index cases had developed schizophrenia before age 15, while only 1.7% of the siblings of adult-onset schizophrenic patients in a previous study had developed schizophrenia before 15 years of age.

2.3. Sex and Age of Onset Differences

Although DSM-III and DSM-III-R note that schizophrenia is apparently equally common in males and in females, this applies to the universe of patients with schizophrenia and does not accurately reflect the situation in the subgroup of schizophrenic patients with childhood onset. The reason for this, reflected in the data of virtually all studies, is that the onset of schizophrenia occurs, on average, at a younger age in males than in females. One recent study of 100 male and 100 female schizophrenic patients diagnosed using strict DSM-III criteria noted times of onset of first treatment, first hospitalization, and the immediate family's initial awareness of psychotic symptomatology.[9] All three events occurred at a mean age of about 5 years younger in males than in females. Of these 200 subjects, only 1 child, a male, had the first psychotic episode noted before age 10. From 10–14 years of age, 8 males and 5 females had their first noted psychotic episodes.

Several studies have found the incidence of first hospitalization for schizophrenia greater in males than in females until about age 30, when the risk of first hospitalization for females surpasses that for males.[10]

Kraepelin[11] noted that of 1054 cases of dementia praecox, 57.4% were men. Of the cases with onset before age 10, however, 70.3% were male. Kallmann and Roth[8] also noted that 72.5% of 102 schizophrenic patients with onset before age 15 were male. They also found that the average age at onset for males was significantly earlier than for females. The average ages of onset of their male singletons was 7 years and of their male twin index cases 8.8 years;

corresponding average age of onset for female singletons was 9.7 years and for female twin index cases 11.1 years. In the study conducted by Kolvin et al.,[12] 33 subjects aged 5–15 years consisted of 24 males and 9 females, a sex ratio of 2.67 : 1.

Using DSM-III diagnostic criteria, in 38 children aged 5 years, 7 months to 11 years, 11 months hospitalized for schizophrenic disorder, Green et al.[13,14] reported that 26 subjects (68.4%) were male and 12 (31.6%) female, a sex ratio of 2.17 : 1. The proportion of males was higher among the 22 children who were diagnosed as having schizophrenic disorder before age 10 (2.67 males per female) compared with those diagnosed age 10 or over (1.67 males per female). Of the 22 cases under 10 years of age, 16 (72.7%) were male and 6 (27.3%) female. This finding is strikingly similar to Kraepelin's data of 70 years ago.[11]

2.4. Genetic Factors

In their important study of the genetic aspects of preadolescent schizophrenia, Kallmann and Roth[8] reported on 102 index cases (52 twins and 50 singletons) with onset of schizophrenia under age 15 years (mean age at onset was 8.6 years). No differences were found in family histories of rates of schizophrenia in parents (8.8%) or siblings (12.2%) or in twin concordance rates between the index cases of preadolescent schizophrenia and a large comparable sample with adult onset. Kallmann and Roth concluded that the heritability of preadolescent schizophrenia was similar to adolescent and adult onset schizophrenia. They noted an increased incidence of early schizophrenia among co-twins and siblings of their early index cases. As there was no direct relationship between an inadequate parental home and preadolescent onset of schizophrenia, nor the converse, these investigators postulated that early-onset schizophrenia was caused by a genetic vulnerability to ordinary environmental stress and by a constitutional decrease in the children's abilities to compensate for this through normal age-appropriate mechanisms.[8]

Kolvin et al.[12] found the rate of schizophrenia among parents of their subjects to be 9.4%. Only one of 56 siblings, however, had an adult type of schizophrenia. Green et al.[14] reported that 6 parents (15.8%) of their 38 schizophrenic children had a history of having been hospitalized with a diagnosis of schizophrenia.

3. STUDIES PERTAINING TO THE CENTRAL NERVOUS SYSTEM

3.1. Pre-, Peri-, and Postnatal Insult to the Central Nervous System

Kolvin et al.[15] obtained perinatal histories for 32 of their 33 subjects. Only 5 (15.6%) had perinatal complications suggestive of possible perinatal brain damage. These were difficult labor in three cases, respiratory distress in one case, and Rh incompatibility in one case. Only one of these five cases, however, showed any collaborating neurological or electroencephalographic (EEG) evidence of brain dysfunction.

Green et al.[14] had historical data concerning pregnancy, delivery, and early infancy for 34 of their 38 subjects. Data were minimal for a few, and complete hospital birth records were available for four children. Pre- and perinatal complications were assessed using the Rochester Research Obstetrical Scale (RROS) developed by Zax and colleagues.[16] The RROS has 27 items in three separately scored areas. The prenatal scale rates maternal age, multiparity, abortions, chronic medication, chronic physical disorder, and maternal infections. The delivery scale contains items on cesarean section, induction of labor, premature rupture of membranes, nonvertex presentations, problems with cord or placenta, abnormalities of amniotic fluid, multiple gestation, forceps, analgesia, and abnormalities of duration of labor. The infant scale rates low birthweight, premature birth, abnormalities of fetal and neonatal heart rates, resuscitation, low Apgar scores, gross physical anomalies, and physical disorders. Average scores on the RROS were prenatal, 1.4; delivery, 1.3; and infancy, 0.5; total, 3.2. These averages were very similar to those of 24 conduct-disordered children and 25 autistic children reported in a previous study from the same ward.[13]

M. Green et al[17] found that 20 schizophrenic patients with onset at or before age 18 years, but mostly during adolescence, had significantly more minor physical anomalies (congenital stigmata) than did schizophrenic patients with later onset. This suggested that first-trimester central nervous system (CNS) insult may play a significant etiological role in increasing the risk of early onset of schizophrenia.

3.2. Developmental Milestones

Kolvin et al.[18] described 33 children and adolescents, aged 5–15 years, diagnosed late-onset psychosis (LOP). Diagnostic criteria consisted of Schneiderian first-rank symptoms and other schizophrenic symptoms in affect, motility, and volition as seen in adult schizophrenic patients and described by F. Fish. Most of these children (52%) had no history of delayed developmental milestones. Fifteen (93.8%) of the 16 children with major delays had delayed use of meaningful three-word phrases.

3.3. Central Nervous System Abnormalities

Kolvin et al.[15] found no evidence of brain dysfunction in 22 (68.8%) of the 32 subjects for whom they had obtained perinatal histories and thorough neurological examination, including EEG in 28 cases. Five subjects (15.6%) had equivocal findings (e.g., an isolated EEG abnormality and a minor motor anomaly on clinical examination) and 5 (15.6%) subjects had clear evidence of brain dysfunction (e.g., seizures or spasticity).

3.4. Electroencephalographic Studies

Kolvin et al.[15] obtained EEGs for 28 of their 33 cases. They found 16 (57.1%) of the recordings to be normal, 3 (10.7%) to be equivocal (including low voltage), and 9 (32.1%) to be abnormal. Four subjects had clinical epilepsy confirmed by EEG. Three of these cases were temporal lobe epilepsy, present for several years before the onset of the psychosis. A fifth subject had a temporal lobe spike focus on EEG but no clinical evidence of epilepsy.

Waldo et al.[19] studied EEGs of 12 males (9.7 ± 2.4 years) and 4 females (11.5 ± 2.4 years) diagnosed as nonautistic childhood psychosis and similar to children diagnosed as suffering from childhood schizophrenia. As a group, 68.8% had normal, 6.3% borderline, and 25% abnormal EEGs.

Green et al.[14] had EEG data available for 24 of their 38 subjects. Seven (29.2%) were read as normal and 17 (70.8%) as abnormal. One child had a diagnosed grand mal seizure disorder.

3.5. Computed Tomography and Magnetic Resonance Imaging

Reiss et al.[20] reported computed tomography (CT) studies on 20 heterogeneous psychiatric patients, 12 males and 8 females (age range 5–15 years, mean age 9.8 years) referred with suspected neurological disease. The ventricular–brain ratios for the psychiatrically referred patients were greater than those of 19 controls, referred from nonpsychiatric sources, whose CT scans were normal. The two largest ratios were for patients diagnosed schizophrenic disorder and schizotypal personality.

Woody et al.[21] performed CT and magnetic resonance imaging (MRI) on a 10-year-old white prepubescent boy diagnosed as having schizophrenic disorder by DSM-III criteria. CT scan performed 2 months after the onset of psychosis showed enlargement of the lateral, third, and fourth ventricles and enlargement of the cisterna magna with left cerebellar atrophy or hypoplasia. The ventricular enlargement and posterior fossa changes were also apparent on MRI. Extensive neurological examinations at that time and over the next 8 months revealed no abnormalities. Woody et al. suggested that these findings in the parenchyma of the brain strongly support the biological hypothesis that neuronal degeneration contributes to the behavioral changes of some schizophrenic patients in both adults and children.[21]

4. LABORATORY STUDIES

4.1. Studies of Peripheral Blood or Serum

The finding of elevated creatine phosphokinase (CPK) levels in adults with acute psychosis prompted Cohen et al.[22] to look at CPK levels in children with severe developmental disturbances. CPK levels were determined for 25 autistic, 10 atypical personality disorders, 5 children with aphasia, 22 children with severe personality disorders, 5 schizophrenic children, 15 normal children, 14 normal siblings, and 14 normal parents. The children with severe personality disorders, a population that is at risk for developing schizophrenia, had significantly higher ($p < 0.005$) levels of CPK (64.8 ± 31.2 IU/liter) than did the normal children and all other children combined. The 5 schizophrenic children had a mean level of CPK (65.0 ± 26.6 IU/liter) almost identical to that of the severe personality disorders but, because of the small number of cases, this did not reach significance. The authors, however, speculated that an elevated CPK level might be a biological marker of vulnerability to schizophrenic spectrum disorders.[22]

Weizman et al.[23] measured basal morning humoral (H)-endorphin blood levels in 12 chronic schizophrenic patients (age range 7–17 years) with childhood onset and compared them with autistic children and normals. Seven of the schizophrenic subjects had subnormal levels, but this did not reach significance. Of note, H-endorphin blood levels were lower in drug-free schizophrenics than in those on antipsychotic medication. This is consistent with the hypothesis of hypofunction of the endorphinergic system in schizophrenia and the enhancement of H-endorphin secretion by neuroleptic medications.

Rogeness et al.[24] compared whole blood serotonin and platelet monoamine oxidase (MAO) in 16 boys diagnosed as schizophrenic or schizophreniform (mean age 10.6 ± 3.1), 8 schizotypal personality-disordered boys (mean age 11.0 ± 1.9), 18 male major depressives (mean age 10.8 ± 2.5) and 20 male controls (mean age 11.2 ± 3.0). There was a nonsignificant elevation of serotonin in the schizophrenic group compared with the controls, consistent with adult studies. Platelet MAO activity was significantly higher in the schizophrenic and schizotypal groups than in the major depression and control groups. This result, the opposite of that found in adults with chronic schizophrenia, was hypothesized as possibly secondary to intense anxiety and stress with resultant high platelet MAO activity or "that elevated MAO is a compensation for overproduction of monoamines or overly sensitive monoamine receptors in children with schizophrenia and that this ability to compensate decreases with age, especially in those who develop a chronic schizophrenic process."[24]

4.2. Immune System Studies

Weizman et al.[25] reported that 13 of 17 autistic subjects (age range 7–22 years, mean 13.35 ± 4.27 years), but none of 5 schizophrenic children (ages 11–17 years), had abnormal immune responses to brain tissue antigen.

5. PSYCHOLOGICAL TESTING

5.1. Intelligence Testing

Kolvin et al.[26] determined full-scale IQs of 30 of their 33 subjects to average 85.9. Twenty-five (83.3%) of these subjects had IQs of 70 or above, as did the 3 children not formerly tested based on school performances and clinical impressions. Only 17 (56.7%) subjects, however, had IQs in the average range or above (90 or higher) whereas 75% of a normal population would be expected to score in these ranges.

Green et al.[14] reported WISC or WISC-R verbal, performance, and full-scale IQs of 36 of their 38 subjects. Both subjects not tested were of at least average intelligence by school report and history prior to the onset of their illnesses. Thirty-one (87.3%) of the children tested had IQs of 70 or above. Only 14 (38.9%) subjects, however, had IQs in the average range or above (90 or higher). The lowest full scale IQ was 59. Performance IQs were higher than verbal IQs in 24 (66.7%) of the children.

Sherman and Asarnow[27] noted that schizophrenic subjects tended to perform consistently better on the verbal than on the performance subtests of the WISC-R. The schizophrenic children did more poorly on timed subtests, such as digit span, which required recruitment of large amounts of attentional capacity and the maintenance of a set of items in working memory. The children did well on the untimed subtests of vocabulary and information, which assess the recall of overlearned facts and do not assess problem-solving abilities.

Asarnow et al.[28] reported WISC-R IQs for 24 nonretarded (full-scale IQs over 70) children mean age 10.1 years (SD = 1.8), diagnosed schizophrenic disorder by DSM-III criteria. All had an onset of schizophrenic symptoms before 11 years of age. Mean full-scale IQ was 87.0 (SD = 9.6); mean verbal IQ was 87.9 (SD = 14.4), and mean performance IQ was 88.3 (SD = 12.0). Following Kaufman's method,[29] the authors computed factor scores in three areas: verbal comprehension (comprised of the information, similarities, vocabulary and comprehension subtests of the WISC-R); a perceptual factor (comprising the picture arrangement, picture completion, block design, and object assembly subtests) and a "freedom from distractibility factor" (composed of arithmetic, coding, and digit span). Scores on the "freedom from distractibility factor," which make significant demands on momentary processing capacity as they are timed and require working memory and attention, were below the range of normal children and significantly lower than scores on the other two factors.[28] These results are consistent with the hypothesis that schizophrenic children have a core cognitive impairment in the recruitment and allocation of their momentary processing capacities.

5.2. Other Psychological Tests

In a study of 11 schizophrenic, 14 autistic, and 28 normal children (age range 7 years, 10 months to 14 years, 4 months), Schneider and Asarnow[30] compared their performances on the Wisconsin Card Sorting Task (WCST), Rey's Tangled Line Task (RTLT), Benton Judgment of Line Orientation (BJLO), Digit Symbol Substitution Test (DSST), and the Peabody Picture Vocabulary Test (PPVT). Mean full-scale (WISC-R) IQs of the schizophrenic children was 92; normal children were estimated as having average intelligence. Schizophrenic subjects scored in the normal range on the PPVT and the BJLO, which require considerable automatic processing of highly ordered or overlearned information. By contrast, the WCST, RTLT, and DSST require greater use of processing capacity. Compared with normal controls, the schizophrenic subjects had more perseverative responses on both halves of the WCST and more nonperseverative errors on the second half of the WCST, which increases the demands on processing capacity relative to the first part because of the necessity of integrating verbal instructions. (Deficits on the WCST are thought to localize to the frontal cortex.) Schneider and Asarnow hypothesized that this paradoxical worsening was the result of providing additional information to the schizophrenic children's already overtaxed momentary processing capacity.[30] Similarly, compared with the normals, the schizophrenic children performed poorly on the DSST and RTLT. Overall, the schizophrenic subjects were impaired on tests which make significant demands on processing capacity, such as categorization, by requiring immediate integration of information from several sources to direct the ongoing action.[30]

6. NEUROPHYSIOLOGICAL AND NEUROPSYCHOLOGICAL STUDIES

6.1. Tests of Information Processing Components or Strategies

Asarnow et al.[31] compared 11 children (mean age 11.6 years ± 1.8 years) diagnosed as having schizophrenic disorder by DSM-III criteria before 9 years of age with two control groups of normal children. One control group was composed of 14 children matched in mental age and the other of 14 chronologically younger children with mental ages approximately 4 years below that of the subjects. Mental ages were determined on the PPVT. Only three of the subjects were receiving neuroleptic medications and their performances did not differ significantly from those of the 8 unmedicated children. Gender differences among the groups did not appear to be responsible for differences in performance.[31]

The groups were compared on three separate experiments. The first measured "partial-report span-of-apprehension," which has been reported to be abnormal in individuals at risk for schizophrenia. The children were asked to note whether a T or an F was among various groups of letters flashed on a screen. As more letters were added to the group, schizophrenic children and the younger control group performed more poorly than the older controls. This was thought to indicate that the schizophrenic children had an impairment in processing this type of information independent of mental age.[31]

The second experiment studied four information processing components/strategies. It was found that the schizophrenic children did not differ significantly from the control groups in (1) stimulus discrimination skills, (2) general response bias, or (3) fatigue/learning/practice effects. The groups differed on the fourth task, which examined information acquisition strategies by studying the relationship between accurate detection of the target stimulus and the quadrant of the screen on which it was projected. Although the older control group performed somewhat better than the schizophrenic group when the stimuli were in the upper two quadrants, both groups performed significantly better when the stimuli were in the upper two quadrants than did the younger control group, for which quadrant location of the stimuli did not influence accuracy. Asarnow et al.[31] concluded the schizophrenic children and the older control group used the same information acquisition strategy. They attributed the schizophrenic children's somewhat poorer performance to a specific deficit in information processing that was unrelated to global intellectual impairment or any general developmental delay.[31]

The third experiment employed the full-report version of the span-of-apprehension test, which requires the subjects to report verbally all the projected letters rather than just the presence of a T or F. This task, a test of iconic and immediate (short-term) memories, requires greater ability to hold the visual stimuli in memory while reporting them than does the partial-report test, which requires subjects to search for critical features and makes greater demands on the preattentive and early attentive processes necessary for critical feature detection. The schizophrenic children performed comparably to the mental age-matched normal controls; both groups performed significantly better than the younger controls. Asarnow et al.[31] concluded that the schizophrenic subjects had adequate iconic and immediate memory abilities and that the poor results on the partial-report test were secondary to a deficit in early attentive processes and that the schizophrenic children were less efficient than normals in the initiation or application of their searches. Together, the three experiments suggested impairments in the mobilization and control of information processing to be a core dysfunction in the schizophrenic children.[31]

6.2. Studies of Visual Event-Related Potentials

Strandburg and Asarnow and colleagues[31–33] also found that the schizophrenic children and mental age-matched normal controls differed on several components of visual event-related potential (ERP) concomitants recorded during their performances on the partial-report span-of-apprehension task described above.

First, the contingent negative variations (CNVs), which are associated with attention, arousal, motivation, expectation, and preparatory set, were slow to develop and resolve in the schizophrenic children, suggesting they had difficulty mobilizing and regulating brain mechanisms underlying preparedness.

Second, the schizophrenic children also failed to show a graded increase in the N1 component of their visual ERPs as task difficulty increased and did not show the enhanced right hemisphere N1 clearly present in the normal children's visual ERPs. As N1 reflects factors such as the complexity of stimulus discrimination, which impinge on selective attention, these data suggest that schizophrenic children fail to regulate mechanisms successfully that determine selective attention.[31–33]

A third visual ERP difference found by Strandburg and colleagues[31–33] was that the P3 component generated by the schizophrenic children was considerably less than that of the controls. The P3 component is thought to reflect processes associated with stimulus discrimination and recognition of stimulus significance. These data suggest less efficient perceptual processing in the schizophrenic children.[31–33]

Fourth, it was noted that the slow-wave (SW) component of the visual ERP, which is believed to indicate "further processing" following the P3 component, did not increase in amplitude with increasing difficulty in the schizophrenic children. The controls, however, showed the normal increase. As this SW component is believed to represent equivocation caused by increasing perceptual difficulty, this was interpreted to show a deficiency in the schizophrenic children's abilities to regulate cognitive activity in resonse to differences in actual information processing demand.[31]

Fifth, it was noted that, unlike the normal controls, schizophrenic children showed no tendency across a number of visual ERP components for electrophysiological activity to reflect changes in information processing demand. The schizophrenic subjects also produced markedly asymmetric CNS activity across the frontal leads during the mobilization of attention. Asarnow et al. suggested that since frontal mechanisms are important for the modulation of the N1 component of the visual ERP, these frontal cortical abnormalities might be reasonsible for the schizophrenic children's weak and unmodulated N1 responses to the visual task in the partial-report span-of-apprehension test.[31–33]

Asarnow et al. speculated that the impaired guided visual search/controlled attentional processes they found in schizophrenic children were caused by abnormalities in late-developing brain systems, specifically, the frontal lobes of the tertiary association cortex.[31]

6.3. Studies of Auditory Event-Related Potentials

In a preliminary study, Erwin et al.[34] recorded auditory event-related potentials (ERP) of five children diagnosed schizophrenic disorder and four schizotypal personality disorder by DSM-III criteria. These nine children were compared (mean age 11.5 years, range 10–13 years) with nine normal controls matched for age (mean age 11.25 years, range 9.75–12.7 years) and with nine younger normal controls (mean age 7.25 years, range 5.8–8.25 years). An "oddball" P300 protocol was used to examine the auditory information-processing abilities and deficits of the children. The P300 potential is present when the subject detects and recognizes a "rare" event, either the presence or absence of a stimulus, that occurs against a background of other frequent events; it is thought to reflect some type of sequential information processing function. The three groups differed significantly in both the structure and latency of the P300 auditory ERP components generated by frequent, target, and rare auditory stimuli. Both control groups showed components that were more positive for rare and target stimuli than for frequent stimuli, but the experimental group had smaller positive responses to rare and target stimuli that were similar to those seen in schizophrenic adults. The failure of the schizophrenic children to produce auditory ERP P300 components that differed across task conditions was thought to reflect some type of inefficient information processing strategy; e.g., they may have included the frequent stimuli in their response set.[34]

Erwin et al.[34] noted that developmental delays did not seem responsible for the experimental subjects' abnormal auditory ERPs, as their various auditory ERP components were similar in latency to the age-matched controls, while the younger controls' auditory ERPs had longer latencies and larger amplitudes typical of younger children.

7. CLINICAL PRESENTATION AND TREATMENT RESPONSE

7.1. Clinical Symptomatology During the Active Phase

Sherman and Asarnow[27] proposed that impairments in the recruitment of attentional capacity explain several of the major behavioral abnormalities of schizophrenic children. According to this conceptualization, schizophrenic children are unable to successfully monitor their own and others' mutual responses to their communicative and behavioral interactions. This type of monitoring demands analyzing and coordinating many complex information-processing tasks, such as the progress of the interaction; the clues (behavioral or verbal) to the others' reactions; and keeping track of their own trains of thought and what has and has not been verbalized. Schizophrenic children who fail in these tasks may have speech that is vague and digressive, and their associations may be loosely connected. The inability to integrate other children's emotional reactions to them into this complex process may contribute to their social ineptness with peers, which, in turn, could result in their being socially isolated.[27]

7.2. Treatment Response to Neuroleptics

Clinically, it has been noted that prepubescent children as a group appear to have poorer therapeutic responses to neuroleptics than do adolescents and adults.[5,14] One possible explanation is that children may metabolize neuroleptics more rapidly than do adolescents and adults.[35]

Green et al.[14] treated 35 of 38 hospitalized schizophrenic children with neuroleptic medications. Response to drug administration was disappointing and the children proved relatively treatment resistant. Six children (17.1%) showed no significant improvement on medication; a seventh child did not have an adequate trial, as her mother refused further medication after the development of untoward effects. Seven children (20%) showed minimal improvement, and 21 (60%) showed modest to moderate improvement. Untoward effects, most frequently sedation or an acute dystonic reaction, usually prevented further increase in dosage or necessitated a change in medication before satisfactory symptom control was achieved. The clinical impression was that these schizophrenic children as a group responded much less satisfactorily to neuroleptic medications than did schizophrenic patients with adolescent or adult onset, and they had a worse immediate prognosis. Corroborating this impression were the relatively poor outcomes as reflected by dispositions after an average three month acute hospitalization. Three (7.9%) children were discharged to long-term hospitalization and 19 (50%) to residential treatment centers. Eleven (28.9%) improved enough to be discharged home with five attending day care centers and 6 outpatient therapy. The other five children were signed out against medical advice; of these, at least four would have been referred for residential treatment or long-term hospitalization.[14]

8. STUDIES OF CHILDREN AT RISK OF DEVELOPING SCHIZOPHRENIA

These studies identify and follow longitudinally children at high risk of developing schizophrenia. Garver[36] emphasized the biological heterogeneity of high-risk populations and pointed out the need to separate high-risk cohorts into those who are biologically/genetically vulnerable and those who are not.

Most high-risk subjects have schizophrenic mothers, suggesting statistically that 8–12% of them will eventually develop schizophrenia. However, the number of cases with onset of schizophrenia before age 12—even in these high-risk cohorts—is exceedingly small, perhaps 1 or 2%. Thus, high-risk studies have not yet shown much about schizophrenia with childhood onset and, with one exception, are not reviewed in this chapter. What they have show of particular interest, however, is that subgroups of at-risk children have abnormalities in attentional processes and event-related potentials that overlap with similar abnormal findings in children, adolescents, and adults with schizophrenia. For a general review of the research on children at risk of schizophrenia, Asarnow[37] and Watt et al.[38] are recommended. More recently, Ferguson and Bawden reviewed psychobiological measures in such children.[39]

In the first longitudinal study of infants at risk for schizophrenia, Fish[40–42] has followed 12 infants of schizophrenic mothers and 12 controls for more than 20 years. Two at-risk children were diagnosed as having schizophrenic disorder at age 10 and a third child at age 17. Most developed moderate to severe personality disorders, including schizotypal personality disorder. Based on this study, Fish hypothesized that transitory pandysmaturation (PDM) in infancy was an early marker for the inheritance of a neurointegrative deficit, the severity of which predicted the risk of a child's later developing clinical symptoms of schizophrenia. Development of schizotypal personality disorder during childhood or adolescence was also thought to indicate increased risk of later schizophrenia. Pandysmaturation is characterized by abnormalities in the timing and integration of neurological maturation. This may be reflected in uneven functioning on examination at a specific time, by fluctuating longitudinal patterns with delays and severe regressions followed by accelerated development in the same area, and by a transitory dysmaturation of CNS organization evidenced by disorders of arousal, physical growth, muscle tone, motility, homeostasis regulation, and higher cognitive functions.

Few schizophrenic children have available such detailed observations of their early development. The proportion of subjects developing schizophrenia and severe psychopathology during childhood and adolescence is unusually large in this sample compared with other at-risk studies. Thus, it is impossible to know whether Fish's interesting findings are characteristic of most schizophrenic children or only a small subgroup.

9. COMMENTS AND SOME DIRECTIONS FOR FUTURE INVESTIGATIONS

Understanding the causes of the similarities and differences in schizophrenia with childhood (prepubertal) onset will undoubtedly shed considerable light on the etiopathogenesis of schizophrenia. This will in turn have an impact on the possible prophylaxis or delay of the clinical onset of schizophrenia, mitigation of the severity of symptoms, and more specific and rational treatments when schizophrenia occurs.

While the preponderance of evidence to date suggests that schizophrenia with childhood onset is on a continuum with adolescent–adult-onset schizophrenia, more research, in particular, longitudinal, follow-up, genetic, and biological study, is necessary before this can be stated with absolute certainty.

Further investigation will also be necessary to clarify such questions as whether schizophrenia with childhood onset can manifest itself with symptomatology different from that of adolescent and adult-onset schizophrenia. If so, what are these other patterns, and how can they be differentiated from other childhood psychopathologies? This issue is of particular relevance for infants and very young, prelatency children whose developmental levels make it difficult to manifest typical schizophrenic symptoms. Whether schizophrenia with childhood onset is a more severe form of schizophrenia with onset during adolescence or adulthood? Is the prognosis worse for childhood onset schizophrenia? Does childhood-onset schizophrenia respond more poorly to neuroleptic therapy and, if so, why?

The consistent finding that schizophrenia has a statistically significant earlier onset in males and the complementary fact that about 70% of schizophrenic children are male is intriguing. What are the factors responsible for this? Is this possibly related to prenatal psychoneurohormonal events that organize the sexual differentiation, development, and maturation of the CNS? Which factors or interaction of factors seem most responsible for developing and conversely for delaying or preventing the clinical expression of schizophrenia during childhood? How, if at all, are these factors related to the biological events of puberty? Do these factors have a significantly different influence on the prepubertal child than on adolescents and adults?

REFERENCES

1. Nasrallah HA (ed): Handbook of Schizophrenia, Vols. 1–5. Amsterdam, Elsevier, 1986–1989
2. Meltzer HY: Biological studies in schizophrenia. Schizophr Bull 13:77–111, 1987
3. Holzman PS: Recent studies of psychophysiology in schizophrenia. Schizophr Bull 13:49–75, 1987
4. Solomon GF: Immunologic abnormalities in mental illness, in Ader R (ed), Psychoneuroimmunology. Orlando, Florida, Academic, 1981, pp 259–278
5. Green WH: Schizophrenia with childhood onset, in Kaplan HI, Sadock BJ (eds), Comprehensive Textbook of Psychiatry, ed 5. Baltimore, Williams & Wilkins, 1989, pp 1975–1981
6. American Psychiatric Association: Diagnostic and Statistical Manual of mental Disorders, ed 3 (DSM-III). Washington, D.C., American Psychiatric Association, 1980
7. American Psychiatric Association: Diagnostic and Statistical Manual of Mental Disorders, ed 3, rev. (DSM-III-R). Washington, D.C., American Psychiatric Association, 1987
8. Kallmann FJ, Roth B: Genetic aspects of preadolescent schizophrenia. Am J Psychiatry 112:599–606, 1956
9. Loranger AW: Sex difference in age at onset of schizophrenia. Arch Gen Psychiatry 41:157–161, 1984

10. Gottesman II, Shields J: Schizophrenia: The Epigenetic Puzzle. Cambridge, Cambridge University Press, 1982, pp 26–27
11. Kraepelin E: Dementia Praecox and Paraphrenia. Translated by RM Barclay from the 8th German Edition of the Textbook of Psychiatry. Edinburgh, Livingstone, 1919
12. Kolvin I, Ounsted C, Richardson LM, et al: Studies in the childhood psychoses. III. The family and social background in childhood. Br J Psychiatry 118:396–402, 1971
13. Green WH, Campbell M, Hardesty AS, et al: A comparison of schizophrenic and autistic children. J Am Acad Child Psychiatry 23:399–409, 1984
14. Green WH, Padron-Gayol M, Hardesty AS, et al: The phenomenology of schizophrenic disorder with childhood onset: A study of 38 cases. In preparation
15. Kolvin I, Ounsted C, Roth M: Studies in the childhood psychoses. V. Cerebral dysfunction and childhood psychoses. Br J Psychiatry 118:407–414, 1971
16. Zax M, Sameraff AJ, Babigian HM: Birth outcomes in the offspring of mentally disordered women. Am J Orthopsychiatry 47:218–230, 1977
17. Green MF, Satz P, Soper HV, et al: Relationship between physical anomalies and age of onset of schizophrenia. Am J Psychiatry 144:666–667, 1987
18. Kolvin I, Ounsted C, Humphrey M, McNay A: Studies in the childhood psychoses. II. The phenomenology of childhood psychoses. Br J Psychiatry 118:385–395, 1971
19. Waldo MC, Cohen DJ, Caparulo BK, et al: EEG profiles of neuropsychiatrically disturbed children. J Am Acad Child Psychiatry 17:656–670, 1978
20. Reiss D, Feinstein C, Weinberger DR, et al: Ventricular enlargement in child psychiatric patients: A controlled study with planimetric measurements. Am J Psychiatry 140:453–456, 1983
21. Woody RC, Bolyard K, Eisenhauer G, et al: CT scan and MRI findings in a child with schizophrenia. J Child Neurol 2:105–110, 1987
22. Cohen DJ, Johnson W, Caparulo BK, et al: Creatine phosphokinase levels in children with severe developmental disturbances. Arch Gen Psychiatry 33:683–686, 1976
23. Weizman R, Weizman A, Tyano S, et al: Humoral–endorphin blood levels in autistic, schizophrenic and healthy subjects. Psychopharmacology 82:368–370, 1984
24. Rogeness GA, Mitchell EL, Custer GJ, et al: Comparison of whole blood serotonin and platelet MAO in children with schizophrenia and major depressive disorder. Biol Psychiatry 20:270–275, 1985
25. Weizman A, Weizman R, Szekely GA, et al: Abnormal immune response to brain tissue antigen in the syndrome of autism. Am J Psychiatry 139:1462–1465, 1982
26. Kolvin I, Humphrey M, McNay A: Studies in the childhood psychoses. VI. Cognitive factors in childhood psychoses. Br J Psychiatry 118:415–419, 1971
27. Sherman T, Asarnow RF: The cognitive disabilities of the schizophrenic child, in M Sigman (ed): Children with Emotional Disorders and Developmental Disabilities. Orlando, Florida, Grune & Stratton, 1985, pp 153–170
28. Asarnow RF, Tanguay PE, Bott L, et al: Patterns of intellectual functioning in non-retarded autistic and schizophrenic children. J Child Psychol Psychiatry 28:273–280, 1987
29. Kaufman AS: Intelligent Testing with the WISC-R. New York, Wiley, 1979
30. Schneider SG, Asarnow RF: A comparison of cognitive/neuropsychological impairments of nonretarded autistic and schizophrenic children. J Abnorm Child Psychol 15:29–45, 1987
31. Asarnow RF, Sherman T, Strandburg R: The search for the psychobiological substrate of childhood onset schizophrenia. J Am Acad Child Psychiatry 25:601–614, 1986
32. Asarnow RF, Sherman T: Studies of visual information processing in schizophrenic children. Child Dev 55:249–261, 1984
33. Strandburg RJ, March JT, Brown WS, et al: Event-related potential concomitants of information processing dysfunction in schizophrenic children. Electroencephalogr Clin Neurophysiol 57:236–253, 1984
34. Erwin RJ, Edwards R, Tanguay PE, et al: Abnormal P300 responses in schizophrenic children. J Am Acad Child Psychiatry 25:615–622, 1986
35. Meyers B, Tune LE, Coyle JT: Clinical response and serum neuroleptic levels in childhood schizophrenia. Am J Psychiatry 137:483–484, 1980
36. Garver DL: Methodological issues facing the interpretation of high-risk studies: Biological heterogeneity. Schizophr Bull 13:525–529, 1987
37. Asarnow RF: Schizophrenia, in RE Tarter (ed), The Child at Psychiatric Risk. New York, Oxford University Press, 1983, pp 150–194

38. Watt NF, Anthony EJ, Wynne LC, et al (eds), Children at Risk for Schizophrenia: A Longitudinal Perspective. Cambridge, Cambridge University Press, 1984
39. Ferguson HB, Bawden HN: Psychobiological measures, in Rutter M, Tuma AH, Lann IS (eds), Assessment and Diagnosis in Child Psychopathology. New York, Guilford, 1988, pp 232–263
40. Fish B: Biologic antecedents of psychosis in children. In Freedman DX (ed), Biology of the Major Psychoses: A Comparative Analysis. New York, Raven, 1975, pp 49–80
41. Fish B: Neurobiologic antecedents of schizophrenia in children: Evidence for an inherited, congenital neurointegrative defect. Arch Gen Psychiatry 34:1297–1313, 1977
42. Fish B: Infant predictors of the longitudinal course of schizophrenic development. Schizophr Bull 13:395–409, 1987

13

Biological Studies of Attention-Deficit Disorder

Ronit Weizman, Abraham Weizman, and Stephen I. Deutsch

1. DOPAMINE THEORY

The hypothesis of dopamine deficiency in the pathophysiology of attention-deficit disorder (ADD) was derived from epidemiologic and pharmacologic studies. The first evidence rises from the analogy between the symptoms of children suffering from attention-deficit disorder with hyperactivity (ADDH) and from children who suffered from the pandemic of von Economo's encephalitis in the beginning of this century. As a result of the viral infection, children exhibited attentional and motor symptoms, while adults developed Parkinson's disease.[1] Since Parkinson's disease is related to dopamine deficiency it suggests[2] that the behavioral disorder observed in ADDH children might also result from a dysfunction of the dopaminergic system. The therapeutic efficacy of stimulants supported the possibility that monoaminergic deficiency may play a role in this disorder.

Determination of amine and metabolite levels in the cerebrospinal fluid (CSF) offered the possibility to assess the function of the dopaminergic system. Basal CSF levels of homovanillic acid (HVA) and 5-hydroxyindoleacetic acid (5-HIAA) did not differ in ADDH children as compared with controls.[3] However, following 2–14 days of treatment with 0.5 mg/kg *d*-amphetamine a reduction of 34% in HVA levels was found in the ADDH children. This reduction correlated with the beneficial effect of the drug. In the same study, only a slight and inconsistent elevation in 5-HIAA was observed. HVA levels in CSF following probenecid loading (100–150 mg orally) was found to be significantly lower in children with ADDH as compared with a control group of children with other neurological disorders. CSF 5-HIAA levels in the children did not differ from that of the control group.[4] Although this study is consistent with the dopamine theory of ADDH, it is worthwhile to mention that only six

Ronit Weizman • Pediatric Department, Hasharon Hospital, Petah Tiqva 49372; and Sackler Faculty of Medicine, Tel Aviv University, Tel Aviv 69978, Israel. *Abraham Weizman* • Geha Psychiatric Hospital, Beilinson Medical Center, Petah Tiqva 49100; and Sackler Faculty of Medicine, Tel Aviv University, Tel Aviv 69978, Israel. *Stephen I. Deutsch* • Psychiatry Service, Veterans Administration Medical Center, and Department of Psychiatry, Georgetown University School of Medicine, Washington, D.C. 20007.

hyperactive children and six controls were included. Two findings are not in accordance with the dopaminergic hypothesis: Levoamphetamine, which is relatively devoid of dopamine agonistic activity, was reported to be efficient in the treatment of ADDH[5]; neuroleptics do not aggravate the disorder, nor do they interfere with the beneficial effect obtained by methylphenidate (MPH) administration.[6,7]

The dopamine deficiency theory of ADDH is further supported by the finding of Shaywitz et al.[8,9] They demonstrated that rats selectively depleted of dopamine during infancy are significantly more active than controls during the 10- to 25-day age range. This 6-OH dopamine-induced hyperactivity is ameliorated by stimulants. Moreover, the lessioned rats show learning deficits later in life. This experimental model of neonatal injection with neurotoxin provides important information about the functional significance of the dopamine system in behavior, cognition, and learning disorder. However, the clinical relevance of this observation to the etiology of ADDH merits further investigation. Moreover, the assumption that dopamine depletion is a prerequisite for the beneficial effect of stimulants is not in accordance with the finding that stimulants improve attention and reduce motor activity in normal children,[10] as well as in children with pure reading disability without attention deficit.[11] Furthermore, dopamine agonists such as piribedel, amantadine and L-DOPA have not proved sufficient in the treatment of this disorder (for review, see Zametkin and Rapoport[12]).

2. NORADRENERGIC HYPOTHESIS

The noradrenergic hypothesis is supported by biochemical and pharmacological studies, although no direct evidence for over- or underactivity of the noradrenergic system is as yet available. No consistent differences in noradrenergic indicators in CSF, blood platelets, plasma, and urine were found between ADDH children and normal controls. Urinary excretion of 3-methoxy-4-hydroxyphenyglycol (MHPG), the major metabolite of norepinephrine (NE), was reported to be reduced,[13–16] unaltered,[17] or increased[18] in ADDH. Prior use of stimulants may account for the reduction in MHPG excretion in ADDH children. As reported previously,[19,20] urinary MHPG levels are suppressed for at least 2 weeks after discontinuation of this medication.

Another approach is to evaluate plasma NE levels and blood pressure in hyperactive children as compared with controls.[21] The investigators reported lack of difference in plasma NE levels and recumbent blood pressure was monitored in the hyperactive children. This increased pressure response on standing was not accompanied by a parallel increase in plasma NE. The authors suggested that these results indicate an increase peripheral adrenergic receptor sensitivity in ADDH children.

The beneficial effect of psychostimulants in ADDH may be related to their effect on the central noradrenergic system.[19] Furthermore, it was found that amphetamine is more potent as a NE uptake inhibitor into synaptosomes than of dopamine.[22] According to the the dopaminergic hypothesis, the *d* and *l*-isomers of amphetamine differ in their effect on dopaminergic function, while they are identical in their effect on the noradrenergic function.[23] Both isomers are effective in the treatment of ADDH, but the *d*-isomer (which possesses dopaminergic agonist activity) is more active in attention improvement.[5] Thus, it seems that at least the beneficial effect of stimulants on attention is mediated by the dopaminergic system rather than by the NE system.

Dextroamphetamine has a suppressive effect on the urine excretion of the NE metabolite MPHG. The reduction in MHPG excretion induced by dextroamphetamine treatment correlated with the beneficial effect of the drug.[24] Decreased MHPG excretion was also found to be associated with other medications that are efficient in the treatment of ADDH, such as desipramine,[25] clorgyline, and tranylcypromine.[12] However, methylphenidate (MPH) and pemoline,[26] which are efficient in the treatment of hyperactivity, do not affect MHPG excretion. Thus, it seems that the peripheral measurement of catecholamine excretion is not associated with a positive response to drug treatment in ADDH.[19]

No consistent differences were found in platelet MAO-B between ADDH children and controls.[27,28] However, it was reported that the normal reduction of MAO activity that occurs at 6–12 years[29] was not obtained in ADDH children.[28] The possibility that increased MAO activity may play a role in ADDH is supported by the beneficial effect of MAO inhibitors. Both, selective MAO-A and MAO-B inhibitors (tranylcypramide), are relatively effective[30] in ADDH, while the selective MAO-B inhibitor (deprenyl) was found to be less effective[12] in the treatment of this disorder. Since the common mechanism of action of the effective MAO inhibitors is in activation of NE metabolism it seems that NE plays a role in the therapeutic effect of these agents in ADDH, although the precise mode of action is as yet unclear.

The possibility of an altered noradrenergic state in ADDH children led to the trial of clonidine treatment in this disorder.[31,32] Clonidine is an α-adrenergic agent that acts preferentially on presynaptic α_2 noradrenergic-receptors.[33] This drug is known to stimulate GH release,[34] to inhibit NE release,[35] and to reduce plasma MHPG concentration.[36] ADDH children exhibited increased GH response to a challenge dose of clonidine in comparison to both children with short stature and patients with Tourette's syndrome.[36] This observation indicates a possible supersensitivity to noradrenergic stimulation. The enhanced sensitivity to noradrenergic provocation may reflect, according to Hunt et al., a combination of diminished presynaptic NE production and increased postsynaptic responsivity. Chronic MPH treatment (12 weeks) reduced the GH response to acute stimulation with clonidine, which may indicate a normalization of the NE postsynaptic receptor in MPH-treated ADDH children.

The possible involvement of the noradrenergic system was further supported by a study that reported a therapeutic benefit from clonidine in a subgroup of ADDH children.[32] The study was conducted in a double-blind placebo-controlled crossover design and included 10 ADDH children aged 8–13. Clonidine was administered for 8 weeks and placebo for 4 weeks. It seems that the inhibitory effect of clonidine on noradrenergic transmission[33] accounts for the improvement in attention and frustration tolerance. Yet chronic clonidine also has a modulatory effect on the firing of dopaminergic neurons.

The noradrenergic hypothesis in ADDH needs further elaboration and clarification via clinical studies and laboratory data using experimental animal models. One of the questions to clarify is the clinical effect of specific agonistic and antagonistic adrenergic agents in this disorder.

3. SEROTONERGIC HYPOTHESIS

Diminished serotonergic activity was suggested to play a role in the pathophysiology of ADDH. The serotonin system is involved in avoidance learning and the acquisition of behavioral inhibitory responses.[37,38] Blood and platelet serotonin levels were reported to be reduced in ADDH children,[39,40] although more recent studies failed to replicate this finding.[41,42] An increase in serotonin blood levels following pyridoxine administration in ADDH children, accompanied by clinical improvement, was reported in one study.[43]

Administration of L-tryptophan, the precursor to serotonin, has shown some beneficial effect in a double-blind placebo-controlled trial conducted on 15 ADDH children.[44] Fenfluramine, an anoretic agent that stimulates the serotonergic system, has been reported to decrease hyperactivity in autistic children.[45] However, clinical trials did not demonstrate any beneficial effect in ADDH children.[46,47] The possible involvement of serotonin in hyperactivity is supported by the beneficial effect of clomipramine[48] and clorgyline[30] in this disorder. Methylphenidate and amphetamine, the drugs of choice in this disorder, also possess serotonergic agonist activity.[49,50]

Serotonergic stimulation in intact neonatal rats as well as in dopamine-depleted neonatal rats (which may represent an animal model of ADDH) was reported to reduce locomotor activity.[51,52] However, a recent study failed to replicate this observation.[53]

The finding that ADDH children do not differ from normal controls in CSF urinary 5-HIAA concentration does not support the serotonergic hypothesis.[4,54] Moreover, high-affinity [^{3}H]imipramine binding to platelets, which labels the serotonin uptake site, did not discriminate between ADDH children and controls.[55] Furthermore, MPH treatment in hyperactive children does not alter platelet serotonin levels,[56] urinary 5-HIAA,[20] and imipramine binding.[55]

In summary, despite some evidence for involvement of the serotonergic system in the ethiopathology of ADDH, it seems that most findings argue against an obvious or major role of serotonin in hyperactivity as well as in the mode of action of MPH in this disorder.

4. OTHER NEUROTRANSMITTER SYSTEMS IMPLICATED IN THE PATHOPHYSIOLOGY OR PHARMACOTHERAPY OF ATTENTION-DEFICIT DISORDER

4.1. Benzodiazepine

Benzodiazepines impair acquisition of new information and can induce an anterograde amnesia. A class of inverse benzodiazepine agonists has been identified with intrinsic pharmacological properties opposite that of the benzodiazepines. These compounds are anxiogenic, proconvulsant, or convulsant. Several of these inverse agonists have been studied in animals and humans and have been shown to possess facilitative effects on learning and memory performance within narrow dose ranges.[57,58] The ideal goal is to identify an inverse agonist that improves cognition within a dosage range devoid of anxiogenic and/or convulsant properties.

Methyl β-carboline-3-carboxylate (β-CCM) is an inverse agonist of the β-carboline group that was studied in three different paradigms used to assess learning and memory and shown to have positive effects.[58] β-CCM facilitated the habituation of food-deprived mice to a novel environment, as reflected in an increase in their food consumption when replaced in this environment (test session) several days later. β-CCM did not increase the food consumption of these mice during their first exposure to this novel environment (training session). The facilitative effects of β-CCM on habituation were observed at doses of 0.2 and 0.3 mg/kg and disappeared at higher doses (i.e., 0.4 and 0.6 mg/kg). Administration of diazepam prior to the training session had opposite effects. Similarly, β-CCM facilitated retention when mice were trained in a single trial passive avoidance procedure. In this paradigm, mice received a low-intensity shock (0.1 mA, 2.0 sec) when they crossed from an illuminated box to a dark box. Treatment with β-CCM (0.2 and 0.3 mg/kg) prior to the training session increased the percentage of mice avoiding the dark box when tested 1 day later. This effect was mediated by the benzodiazepine receptor as it was antagonized by Ro15-1788. By contrast, prior treatment with diazepam impaired the ability of mice to retain the memory of a high-intensity shock (0.6 mA, 2.0 sec). Finally, β-CCM (2.5 mg/kg) administered prior to an "imprinting" session increased the amount of time newly hatched chicks followed a moving decoy when tested 24 hr later. These data support clinical investigations of the salutary effects of inverse agonists on attention, acquisition, and retention.

In a preliminary study with healthy volunteers, there was the suggestion that ZK 93426 (ethyl-5-isopropoxy-4-methyl-β-carboline-3-carboxylate), a member of the β-carboline group, exerted positive effects on measures of attention and concentration.[57] Inverse agonists may have a role in novel pharmacotherapeutic approaches to the treatment of attention deficit disorder and learning disability.

4.2. Opioids

Endogenous opioids may participate in the regulation of selective attention in humans. Naloxone, an opiate antagonist, was shown to increase an electrophysiological index of selective attention in young men.[59] In this electrophysiological paradigm, the men were tested after an intravenous infusion of naloxone (approximately 0.03 mg/kg) and placebo on two separate days. Auditory event-related potentials (AERPs) were recorded as the men listened selectively for randomly delivered discrete tones (target tones) that emanated from one of three spatial locations. The target tones were embedded in a sequence of standard tones and were distinguished by their longer duration. Naloxone did not affect global arousal levels and improved the ability to allocate attention to relevant sources. The mechanism of this naloxone effect could be related to a suppression of attention to distracting stimuli. In any event, opiate antagonists may have a therapeutic role in attention deficit disorders.

In another investigation involving primates, naloxone was shown to improve the performances of four out of five macaque monkeys on a visual recognition test within a narrow dose range.[60] The dose–response relationship appeared to follow an inverted U-shaped function. In this task, the monkeys were required to recognize a previously presented object and to choose a novel object upon their simultaneous presentation in order to receive a food reward. Monkeys differed in their baseline performance on this task and, as a group, performed best on the 0.32 mg/kg intramuscular dose of naloxone hydrochloride. Performance worsened at the highest dose (10.0 mg/kg) examined. The data are consistent with a role for endogenous opiod systems in visual recognition memory. Also, improved performance could have resulted from an enhancement of attention. Examination of potential salutary effects of opiate antagonists on attention and cognition, and their potential therapeutic indications in Child Psychiatry should be pursued.[61]

4.3. Glutamate

The acquisition of a reinforced behavioral pattern could result from an enhancement of attentional mechanisms, a direct effect on learning, or an interaction of the two. Glutamatergic transmission has been implicated in the acquisition of reinforced bar-pressing behavior in rats.[62] Glutamic acid diethyl ester (GDEE), a partial antagonist of glutamate-induced excitations, severely impaired the learning of a correct bar-press response at the two highest intraperitoneally administered doses (240 and 480 mg/kg). GDEE was without any significant effect on the performance of an already learned bar-press response. Glutamate is a prominent neurotransmitter in excitatory afferent hippocampal and intrahippocampal pathways, as well as neocortical efferent pathways. These results are consistent with a role for glutamatergic mechanisms in instrumental learning.

In a variety of preparations, the amino acid L-proline has been shown to suppress glutamate-induced excitations. The pretreatment of brain slices prepared from neonatal chicks with either L-proline or one of its analogs for five minutes reduced the amount of endogenous glutamate released after potassium-induced depolarization.[63] The brain slices were pretreated with 1.0 mM concentrations of L-proline or its analogues. Thus, in this preparation, the antagonistic properties of L-proline occurred as a result of the inhibition of glutamate release. This ability of L-proline and its analogues to inhibit glutamate release could account for the amnestic properties of these compounds in neonatal chicks. Chicks allowed to peck at a target coated with a liquid aversant for 10 sec (training session) showed an impairment of retention when tested 24 hr after the intracerebral injection of test amino acid. That is, chicks treated with L-proline after the training session showed an increased pecking response 24 hr later, as compared with controls. The data suggest that the neonatal chick requires intact glutamatergic transmission in order to learn to avoid pecking at an aversive stimulus.

The intraperitoneal injection of DBA/2 mice with 2-amino-4-phosphonobutyric acid (APB) (500 mg/kg), a glutamate receptor antagonist, impaired their acquisition of two-way avoidance learning in the shuttle box.[64] In this task, the animal must learn to avoid shock by changing compartments after the presentation of a warning signal. The learning of this shuttle-box behavior is thought to involve the hippocampus, a brain region enriched in all subclasses of glutamate receptor. In this study, spatial learning, as evaluated by a water maze task, was unaffected by APB (500 mg/kg, i.p.). Thus, interference with glutamatergic transmission impaired the learning of a behavior that is thought to involve hippocampal mediation.

In contrast to the above study, aminophosphonovaleric acid (AP5), a relatively selective antagonist of the *N*-methyl-D-aspartate (NMDA) subclass of glutamate receptor, impaired the ability of rats to learn a spatial task.[65] In this task, rats placed in a pool of opaque water were required to learn to swim to a quadrant containing an escape platform. The AP5 was administered chronically into the right lateral ventricle via a subcutaneously implanted minipump; the dextrorotatory isomer of AP5 was the active one. The impairment was reflected in longer latencies to reach the escape platform and more indirect routes. The impairment showed some selectivity, as AP5 did not interfere with learning a visual discrimination task. Moreover, chronic infusion of AP5 blocked the induction of long-term potentiation (LTP), an electrophysiological analogue of learning, in the hippocampus. These data suggest that glutamatergic mechanisms mediate LTP in the hippocampus, and spatial learning.

Long-term depression (LTD) is an electrophysiological analogue of motor learning in the cerebellum.[66] LTD is characterized by a decreased firing rate of the cerebellar Purkinje cell in response to excitatory inputs from parallel fibers. It occurs following the conjunctive activation of Purkinje cells by parallel fibers and climbing fibers. The mechanism of LTD appears to be a reduction of the sensitivity of the quisqualate subclass of glutamate receptors located on the dendrites of Purkinje cells. L-Glutamate has been proposed as the neurotransmitter of parallel fibers. These data implicate glutamatergic transmission in LTD, a potential electrophysiological substrate of motor learning processes occurring in the cerebellum.

The maintenance of persistently elevated plasma L-glutamate levels through chronic oral administration can increase cerebral levels of this nonessential amino acid (for review, see Deutsch and Morihisa[67]). Moreover, exogenous administration appears to influence the neurotransmitter pool size. In several poorly controlled clinical trials and pilot studies, L-glutamate was shown to enhance the cognitive and memory performance of samples of demented, learning-disabled, intellectually normal, and retarded individuals (for review, see Deutsch and Morihisa[67]). L-Glutamate was virtually devoid of serious toxicity in these human trials. Clearly, the potential therapeutic efficacy of this excitatory amino acid neurotransmitter in children with disorders of attention, arousal, and learning should be pursued.

REFERENCES

1. Raskin LA, Shaywitz SE, Shaywitz BA, et al: Neurochemical correlates of attention deficit disorder. Pediatr Clin North Am 31:385–395, 1984
2. Wender PH: Some speculations concerning a possible biochemical basis of minimal brain dysfunction. Life Sci 14:1605–1621, 1974

3. Shetty T, Chase TN: Central monoamines and hyperkinesis of childhood. Neurology (NY) 26:1000–1006, 1976
4. Shaywitz BA, Cohen DJ, Bowers MB Jr: CSF monoamine metabolites in children with minimal brain dysfunction: Evidence for alteration of brain dopamine. A preliminary report. J Pediatr 90:67–71, 1977
5. Arnold L, Wender P, McColskey K, et al: Levoamphetamine and dextroamphetamine: Comparative efficacy in the hyperkinetic syndrome: Assessment by target symptoms. Arch Gen Psychiatry 27:816–822, 1972
6. Gittleman R, Klein D, Katz S, et al: Comparative effects of methylphenidate and thioridazine in hyperkinetic children. Arch Gen Psychiatry 33:1217–1231, 1976
7. Weizman A, Weitz R, Szekely GW, et al: Combination of neuroleptic and stimulant treatment in attention deficit disorder with hyperactivity. J Am Acad Child Psychiatry 23:295–298, 1984
8. Shaywitz BA, Klopper JH, Yager RD, et al: Paradoxical response to amphetamine in developing rats treated with 6-hydroxydopamine. Nature (Lond) 261:153–155, 1976
9. Shaywitz BA, Yager RD, Klopper JH: Selective brain dopamine depletion in developing rats. An experimental model of minimal brain dysfunction. Science 191:305–308, 1976
10. Rapoport JL, Buchsbaum M, Zahn T, et al: Dextroamphetamine: Cognitive and behavioral effects in normal prepubertal boys. Science 199:511–514, 1978
11. Gittelman-Klein R, Klein DF: Methylphenidate effects in learning disabilities. I. Psychometric changes. Arch Gen Psychiatry 33:655–664, 1978
12. Zametkin AJ, Rapoport JL: Neurobiology of attention deficit disorder with hyperactivity: Where have we come in 50 years? J Am Acad Child Adol Psychiatry 26:676–686, 1987
13. Shekim WO, Dekirmenjian H, Chapel JI: Urinary catecholamine metabolites in hyperkinetic boys treated with d-amphetamine. Am J Psychiatry 134:1276–1279, 1977
14. Shekim WO, Dekirmenjian H, Chapel JL: Urinary MHPG excretion on minimal brain dysfunction and its modification by d-amphetamine. Am J Psychiatry 136:667–671, 1979
15. Shekim WO, Davis LG, Bylund DB, et al: Platelet MAO, in children with attention deficit disorder and hyperactivity. A pilot study. Am J Psychiatry 139:936–938, 1982
16. Yu-cun A, Yu-Feng W: Urinary 3-methoxy-4-hydroxyphenylglycol sulfate excretion in seventy-three school children with minimal brain dysfunction syndrome. Biol Psychiatry 19:861–870, 1984
17. Rapoport JL, Mikkelsen EJ, Ebert MH, et al: Urinary catecholamine and amphetamine excretion in hyperactive and normal boys. J Nerv Ment Dis 66:731–737, 1978
18. Khan AU, Dekirmenjian H: Urinary excretion of catecholamine metabolites in hyperkinetic child syndrome. Am J Psychiatry 138:108–112, 1981
19. Zametkin AJ, Rapoport JL: Noradrenergic hypothesis of attention deficit disorder with hyperactivity. A critical review, in Meltzer HY (ed), Psychopharmacology: The Third Generation of Progress. New York, Raven, 1987, pp 837–842
20. Zametkin AJ, Karoum FK, Linnoila M, et al: Stimulants, urinary catecholamines, and indoleamines in hyperactivity: A comparison of methylphenidate and dextroamphetamine. Arch Gen Psychiatry 42:251–255, 1985
21. Mikkelsen E, Lake CR, Brown GL, et al: The hyperactive child syndrome. Peripheral sympathetic nervous system function and the effect of d-amphetamine. Psychiatr Res 4:157–169, 1981
22. Kuczenski R: Stimulants, Neurochemical Behavioral and Clinical Perspectives. New York, Raven, 1983
23. Bunney BS, Walters JR, Kuhar MJ, et al: D & L amphetamine stereo-isomers: Comparative potencies in affecting the firing of central dopaminergic and noradrenergic neurons. Psychopharm Commun 1:177–190, 1975
24. Brown GL, Ebert MH, Hunt RD, et al: Urinary 3-methoxy-4-hydroxy-phenylglycol and homovanillic acid response to d-amphetamine in hyperactive children. Biol Psychiatry 16:779–778, 1981
25. Donnelly M, Zametkin AJ, Rapoport JL, et al: Treatment of hyperactivity with desipramine; plasma drug concentration, cardiovascular effects, plasma and urinary catecholamine levels and clinical response. Clin Pharm Ther 39:72–81, 1986
26. Zametkin AJ, Linnoila M, Karoum F, et al: Pemoline and urinary excretion of catecholamines and indolamines in children with attention deficit disorder. Am J Psychiatry 143:359–362, 1986
27. Shekim WO, Davis LG, Bylund DB, et al: Platelet MAO in children with attention deficit disorder and hyperactivity. A pilot study. Am J Psychiatry 139:936–938, 1982
28. Brown GL, Murphy DL, Langer DH, et al: Monoamine enzymes in hyperactivity: Response to d-amphetamine. Annual meeting. Am Acad Child Psychiatry, Toronto, 1984
29. Young JG, Cohen DJ, Waldo MC, et al: Platelet monoamine oxidase activity in children and adolescent with psychiatric disorder. Schizophr Bull 6:324–333, 1984

30. Zametkin AJ, Karoum F, Rapoport JL, et al: Treatment of hyperactive children with monoamine oxidase inhibitors. I. Clinical efficacy. Arch Gen Psychiatry 42:962–969, 1985
31. Hunt RD, Cohen DJ, Shaywitz SE, et al: Strategies for the neurochemistry of attention deficit disorder in children. Schizophr Bull 8:236–251, 1982
32. Hunt RB, Minderaa RB, Cohen DJ: Clonidine benefits children with attention deficit disorder and hyperactivity: Report of a double-blind placebo-crossover therapeutic trial. J Am Acad Child Psychiatry 24:617–629, 1985
33. Aghajanian GK, Bunney BS: Dopamine "autoreceptors": Pharmacological characterization by microintophoretic single cell recording studies. Arch Pharmacol 297:1–7, 1977
34. Lal S, Tolis G, Marin JB, et al: Effects of clonidine on growth hormone, prolactin luteinizing hormone and thyroid stimulating hormone levels in serum of normal men. J Clin Endocrinol Metab 41:703–708, 1975
35. Charney DS, Heninger GR, Sternberg DE, et al: Adrenergic receptor sensitivity in depression: effects of clonidine in depressed patients and healthy subjects. Arch Gen Psychiatry 39:290–294, 1982
36. Hunt RD, Cohen DJ, Anderson G, et al: Possible change in noradrenergic receptor sensitivity following methylphenidate treatment: Growth hormone and MHPG response to clonidine challenge in children in attention deficit disorder and hyperactivity. Life Sci 35:885–897, 1984
37. McElroy J, DuPont A, Feldman R: The effect of fenfluramine and fluoxetine on the acquisition of a conditional avoidance response on rats. Psychopharmacology 77:356–359, 1982
38. Orgen SO: Central serotonin neurons in avoidance learning: Interactions with noradrenaline and dopamine neurons. Pharmacol Biochem Behav 23:107–123, 1985
39. Coleman M: Serotonin concentrations in whole blood of hyperactive children. J Pediatr 78:985–990, 1971
40. Bhagavan HN, Coleman M, Coursin DB: The effect of pyridoxine hydrochloride on blood serotonin and pyridoxal phosphate contents in hyperactive children. Pediatrics 55:437–441, 1975
41. Ferguson HB, Pappas BA, Trites RL, et al: Plasma free and total tryptophan, blood serotonin, and the hyperactivity syndrome: No evidence for the serotonin deficiency hypothesis. Biol Psychiatry 16:231–238, 1981
42. Haslam RHA, Dalby JT: Blood serotonin levels in the attention deficit disorder. N Engl J Med 309:1328, 1983
43. Coleman M, Steinberg G, Tippett J, et al: A preliminary study of the effect of pyridoxine administration in a subgroup of hyperkinetic children: A double-blind crossover. Biol Psychiatry 14:741–751, 1979
44. Nemzer ED, Arnold LA, Votolato NA, et al: Amino acid supplementation as therapy for attention deficit disorder. J Am Acad Child Psychiatry 25:509–513, 1986
45. Ritvo ER, Freeman BJ, Yuwiler A, et al: Fenfluramine treatment of autism: UCLA collaborative study of 81 patients at 9 medical centers. Psychopharmacol Bull 22:133–140, 1986
46. Rapoport JUL, Zametkin A, Donnelly M, et al: New drug trials in attention deficit disorder. Psychopharmacol Bull 21:232–236, 1985
47. Donnelly M, Rapoport JL, Ismond DR: Fenfluramine treatment of childhood attention deficit disorder with hyperactivity: A preliminary report. Psychopharmacol Bull 22:152–154, 1986
48. Garfinfl B, Wender P, Solman L, et al: Tricyclic antidepressant and methylphenidate treatment of attention deficit disorder in children. J Am Acad Child Psychiatry 22:343–348, 1983
49. Shell-Kruger K, Hasselager E: Studies on various amphetamine and clonidine on body temperature and brain 5-hydroxytryptamine metabolism in rats. Psychopharmacologia 36:189, 1974
50. Sloviter RS, Drust EG, Connor JD: Evidence that serotonin mediates some behavioral effects of amphetamine. J Pharmacol Exp Ther 206:348–352, 1978
51. Heffner TG, Seiden LS: Possible involvement of serotonergic neurons in the reduction of locomotor hyperactivity caused by amphetamine in neonatal rated depleted of brain dopamine. Brain Res 244:81–90, 1982
52. Lucot JB, Seiden LS: Effects of serotonergic agonists and antagonists on the locomotor activity of neonatal rats. Pharmacol Biochem Behav 24:537–541, 1986
53. Concannon JT: Serotonin "agonists" in an animal model of attentional deficit disorder (ADD). Soc Neurosci Abs 12(pt 2):1363 (abst 371.18)
54. Wender PH, Epstein RS, Kopin IJ, et al: Urinary monoamine metabolites in children with minimal brain dysfunction. Am J Psychiatry 127:1411–1415, 1971
55. Weizman A, Bernhout E, Weitz R, et al: Imipramine binding to platelets of children with attention deficit disorder with hyperactivity. Biol Psychiatry 23:491–496, 1988
56. Rapoport J, Quinn P, Scirbanu N, et al: Platelet serotonin of hyperactive school boys. Br J Psychiatry 125:138–140, 1974

57. Duka T, Stephens DN, Krause W, et al: Human studies on the benzodiazepine receptor antagonist β-carboline ZK 93426: Preliminary observations on psychotropic activity. Psychopharmacology 93:421–427, 1987
58. Venault P, Chapouthier G, de Carvalho LP, et al: Benzodiazepine impairs and β-carboline enhances performance in learning and memory tasks. Nature (Lond) 321:864–866, 1986
59. Arnsten AFT, Segal DS, Neville HJ, et al: Naloxone augments electrophysiological signs of selective attention in man. Nature (Lond) 304:725–727, 1983
60. Aigner TG, Mishkin M: Improved recognition memory in monkeys following naloxone administration. Psychopharmacology 94:21–23, 1988
61. Deutsch SI: Rationale for the administration of opiate antagonists in treating infantile autism. Am J Ment Defic 90:631–635, 1986
62. Freed WJ, Wyatt RJ: Impairment of instrumental learning in rats by glutamic acid diethyl ester. Pharm Biochem Behav 14:223–226, 1981
63. Pico RM, Keller E, Cherkin A, et al: Brain glutamate inhibition and amnesia: Evidence provided by proline analog action. Dev Brain Res 9:227–230, 1983
64. Sahai S, Buselmaier W, Brussmann A: 2-Amino-4-phosphonobutyric acid selectively blocks two-way avoidance learning in the mouse. Neurosci Lett 56:137–142, 1985
65. Morris RGM, Anderson E, Lynch GS, et al: Selective impairment of learning and blockade of long-term potentiation by an *N*-methyl-D-aspartate receptor antagonist, AP5. Nature (Lond) 319:774–776, 1986
66. Kano M, Kato M: Quisqualate receptors are specifically involved in cerebellar synaptic plasticity. Nature (Lond) 325:276–279, 1987
67. Deutsch SI, Morihisa JM: Glutamatergic abnormalities in Alzheimer's disease and rational for clinical trials with L-glutamate. Clin Neuropharmacol 11:18–35, 1988

14

Biochemical and Genetic Studies of Tourette's Syndrome

Implications for Treatment and Future Research

Phillip B. Chappell, James F. Leckman, David Pauls, and Donald J. Cohen

1. OVERVIEW

Gilles de la Tourette's syndrome (TS) is a chronic, familial neuropsychiatric disorder of unknown etiology characterized by motor and phonic tics that vary in severity and form and a range of complex behavior problems, including some forms of obsessive compulsive disorder (OCD). Although its etiology remains unknown, the vertical transmission of TS within families follows a pattern consistent with an autosomal dominant form of inheritance.[1,2] For nearly a century following its original description in 1885, TS was regarded as a rare medical curiosity. It received relatively little attention until the late 1960s, when interest was renewed largely as a result of the finding that approximately 80% of patients responded well clinically to treatment with haloperidol, a dopamine receptor blocking agent.[3,4] Subsequent expanded research efforts have led to several notable interrelated advances: (1) a greater understanding of the range of phenotypic expression and the possible significance of comorbid conditions, (2) a realization that TS and related conditions are much more common than had previously been considered, (3) a better understanding of the familial transmission of the syndrome, and (4) an emerging body of neurobiologic and pharmacological data concerning the pathophysiology and neuropathological correlates of TS.

The diagnostic criteria included in DSM-IIIR represent the culmination of phenomenological studies conducted throughout the 1970s: (1) onset before age 21; (2) the presence of both multiple motor and one or more vocal tics at some time during the illness, although not necessarily concurrently; (3) the occurrence of tics many times a day (usually in bouts), nearly every day or intermittently throughout a period of more than one year; and (4) variability over

Phillip B. Chappell, James F. Leckman, David Pauls, and Donald J. Cohen • Yale Child Study Center, Yale University School of Medicine, New Haven, Connecticut 06510.

time of the anatomic location, number, frequency, complexity, and severity of tics.[5] These criteria have done much to dispel the historical difficulty in diagnosis, which was attributable in part to the variable manifestations of the syndrome. Recent epidemiological surveys based on these criteria have now estimated the prevalence of TS to be one case per 1000 boys and one case per 10,000 girls, and milder variants of the syndrome are likely to occur in a sizeable percentage of the population.[6]

2. NATURAL HISTORY

While providing a reasonable basis for diagnosis, the DSM-IIIR criteria for TS do not adequately convey a full sense of the varied phenomenology and complex natural history of the disorder, which are in part age-dependent phenomena.[3,7] Problems with attention, hyperactivity, and impulsiveness often occur in young children who later develop motor and phonic tics.[7] Motor tics usually appear before phonic tics and sometimes exhibit a rostral–caudal progression, with tics of the head, face, and shoulders appearing earlier and in a higher proportion of patients than tics of the extremities or trunk.[3,7] The variety of possible motor tics is enormous, ranging from rapid meaningless muscular events such as eye blinking, grimacing, or jerking of the shoulders or arms (simple tics), to more purposive-appearing stereotyped movements such as grooming behaviors, compulsive touching, clapping, arm extension, kicking, and jumping (complex tics).[8,9] In a minority of patients copropraxia and self-abusive behaviors are seen.

Phonic tics, which usually develop after the onset of motor tics, also exhibit a wide range of phenomenology from fast meaningless sounds or noises such as grunting, barking, or sniffing, to more elaborated verbal productions such as syllables, words, and phrases, which can have apparent meaningfulness. Alterations in pitch, volume, and rhythm of speech can also occur as well as echo phenomena, such as echolalia (repeating the words or statements of others) and palilalia (repeating oneself), which are the vocal equivalents of the complex compulsive actions. About 30% of patients develop coprolalia, the sudden and sometimes explosive utterance of obscene or unacceptable words or phrases.[8,9]

Other notable features of tics include their occurrence in bouts; their variability in intensity and frequency in a 24-hr period; their tendency to be less severe during sleep or absorbing activities and more pronounced during stress or fatigue; and the ability of patients to suppress movements voluntarily for minutes to hours.[8–10] Over longer periods tics wax and wane in severity and form, so that a fluctuating pattern of tics and a changing level of severity can occur in an individual patient.[8–11] Periods of increased stress and anxiety are often observed to exacerbate TS symptoms[7,12] and may play a role in the expression of the disorder in genetically vulnerable individuals.[8]

2.1. Relationship between TS and ADHD

While the diagnosis of TS is quite straightforward, the degree to which other behaviors may be associated with the syndrome is not clear. Jaegger et al.[7] reported that symptoms of inattention and hyperactivity were among the first manifestations of the syndrome in a large percentage of patients. Comings and Comings[13] suggested that attention-deficit hyperactivity disorder (ADHD) occurring in the families of TS patients represents a variant expression of the etiologic factors responsible for the manifestation of TS. However, as reviewed in detail below, recent family genetic studies performed at the Yale Child Study Center suggest that ADHD is not etiologically related to TS and that the apparent association between attentional difficulties, hyperactivity and TS reported in the literature may have resulted from ascertainment bias.

2.2. Relationship between TS and OCD

Symptoms of OCD frequently appear in the natural history of TS. In his original 1885 report, Gilles de la Tourette anecdotally described the co-occurrence of chronic motor and phonic tics and OC symptoms in one of his patients.[14] Initial empirical studies based on clinical samples (characterized by uncertain methods and criteria) reported 12 to 35% of patients with TS also had prominent OC symptoms. More recent studies relying on systematic assessment procedures and specified diagnostic criteria indicate that this percentage may be substantially higher, with 45–90% of TS patients displaying OC symptomatology.[11,15,16] Since the available epidemiological data suggest that the population prevalence of OCD is 0.05–2%,[17,18] it is unlikely that this association between TS and OCD is due to chance alone.

The appearance of OC symptoms in TS patients also appears to be an age-dependent phenomenon, with younger children displaying fewer OC symptoms than adult TS patients.[11,15–19] Apart from the presence of complex motor tics, no specific set of OC symptoms has yet been reported to distinguish the OCD seen in TS patients from those observed in OCD patients without a history of tics. Patients display a variety of outcomes; in some cases, the OC symptoms become predominant.

2.3. Relationship between TS and Other Disorders

In contrast to more conservative approaches, Comings and Comings[20,21] have espoused a view of TS as a type of broad disinhibition disorder that may predispose to a wide range of behaviors, including not only ADHD and OCD, but also panic attacks, phobias, depression, mania, conduct disorder, learning disability, sleep disorders, eating disorders, alcohol and drug abuse, and schizophreniform symptoms. However, numerous methodological questions are posed by the studies on which these speculations are based,[22] and further investigations will be required to clarify what relationship, if any, may exist between TS and these disorders.

3. FAMILY AND TWIN STUDIES

Over the past decade, intensive efforts have been made to better understand the role of genetic and environmental factors responsible for the transmission and expression of TS. Specific areas of focus have included (1) examination of the range of expression in monozygotic (MZ) and dizygotic (DZ) twin pairs; (2) the assessment of a large number of TS families using "state-of-the-art" family study methodologies to document the pattern of vertical transmission; (3) determination of whether the same genetic susceptibility gives rise to other behavioral disorders in the relatives of TS patients (e.g., ADHD or OCD); and (4) development of a genetic linkage study in order to provide compelling biological evidence for specific genetic factors in the pathobiology of TS and related disorders.

3.1. Twin Studies

In contrast to studies of the recurrence risk to first degree family members (which may be confounded by forms of "cultural" or nongenetic transmission), studies of identical (MZ) and fraternal (DZ) twin pairs provide direct evidence for the importance of genetic factors in the pathogenesis of TS. In a study of 43 pairs of same-sex twins, Price and co-workers[23] reported a concordance of 53% for MZ twin pairs and 8% for DZ pairs. When the criteria were broadened to include chronic tic disorders as well as TS, the concordance increased to 77% and to 23% for MZ and DZ pairs, respectively. Differences in the concordance of the MZ and DZ pairs are suggestive that genetic factors may play an important role in the etiology of TS. The high incidence of chronic tic disorders in the co-twins also supports the view that these more prevalent tic disorders are etiologically related to TS.

The MZ twin data also indicate that nongenetic factors may influence the expression of the inherited TS diathesis. In this regard, chronic stress and exposure to stimulant medication have been singled out as nongenetic factors, which may be important in the eventual manifestation and severity of TS symptoms.[8] Price et al.[24] studied six pairs of MZ twins who were discordant for stimulant treatment. In all cases, the untreated co-twin also developed TS. However, there were differences in onset and severity of symptoms between the twins, with twins who received central nervous system (CNS) stimulants having a slightly earlier age of onset than twins who were untreated.

The discordant twin data also revealed that twins with TS in the discordant MZ pairs were significantly ($p < 0.006$) lower in birthweight than their unaffected co-twins. These results, which will have to be followed up with data from nontwin pregnancies, suggest that some prenatal effects may be important in the manifestation of the syndrome.

Over the past 2 years, a follow-up study of the original MZ twin pairs studied by Price et al.[23] has been undertaken in an effort to confirm the zygosity and concordance of these twins as well as search for potential nongenetic risk factors that might interact with an inherited vulnerability to produce TS or a related condition. At present, 19 of the original 30 MZ twin pairs have been intensively studied. Following confirmation of zygosity, subjects and parents were interviewed directly to assess current and "worse ever" severity of TS, ADD, and OCD symptoms using the Schedule for TS and Related Disorders,[25] the TS Global Scale,[26] and the Yale–Brown Obsessive Compulsive Scale.[27] The preliminary results suggest that the concordance estimates derived from the Price study may be low. In this follow-up study, there also appears to be a high concordance for OCD among these twin pairs.

3.2. Family Studies

Because a high proportion of TS patients have a positive family history of either TS or chronic multiple tics (CT), the conclusion has frequently been reached that TS shows a familial concentration, especially when CT are regarded as a minor manifestation.[4,11,28] Early work also found a sex difference in the frequency of positive family history, suggesting that the sex difference seen in patient populations might be related to the transmission of the syndrome.[4] Other early studies based on family history data provided strong statistical evidence for genetic transmission, and suggested that the most parsimonious explanation for the pattern of transmission seen in the families of TS patients was a single-major-locus model.[29–33] While these findings pointed to a relatively simple mode of transmission, the specific underlying mode of inheritance remained controversial.

These conclusions, however, were all derived from statistical analyses of data attained by the family history method, in which diagnoses of relatives are based on unstructured interviews of only the proband or one or two close relatives. Because of the inherent limitations of this approach and the need for data of a higher quality, genetic studies at the Yale Child Study Center were initiated based on a methodology of direct interviews of all relatives of TS patients using a fully structured interview form (the Schedule for Tourette's Syndrome and Other Behavioral Disorders[25]). These ongoing studies have provided new information concerning the familial pattern of this disorder and the variation in symptoms that is found in both patients and their relatives.

A series of analyses using the data collected from the first 27 families (of more than 100) ascertained through a TS proband have now been completed. When these data, derived from direct interviews of family members, were compared with family history data previously collected from 52 consecutive patients seen at the Yale Child Study Center Clinic for TS, significant increases in the recurrence risks were observed.[30,34] This was not surprising, since other studies have demonstrated that family history data underestimate the "true" recurrence risks.[35,36] The morbid risks obtained were considerably higher for both TS (10.7% versus 1.9%) and chronic tics (18.5% versus 10.3%). These results suggested that genetic analyses based on family history data should be reevaluated since the parameter estimates obtained from those studies would probably be inaccurate.

Another major difference between the data collected by direct interviews and family history data was that the sex of proband effect disappeared. That is, in the family history data the recurrence risks had been consistent with the hypothesis that the sex difference observed in the population was related to the transmission of the trait, with relatives of females (the less frequently affected sex) being at greater risk for TS and/or CT. The data from this study, in which diagnoses were based on direct interviews, suggested that the risk to relatives was the same for relatives of both male and female probands. While there was still a dramatic difference in the frequency of TS (with the sex ratio about 4:1 in favor of males), that difference did not affect the recurrence risk in the two groups of relatives. This finding suggests that there may not be a difference in genetic loading between males and females and that the observed sex difference may simply reflect a difference in frequency of expression.

To examine whether the same genetic susceptibility gives rise to other behavioral disorders in the relatives of TS patients, the data from these 27 families was scrutinized for any evidence of a possible etiological relationship between TS and ADHD or OCD.

It has been suggested that TS and some forms of ADHD may be etiologically related since these conditions often co-occur in patients.[13] Consistent with this observation, approximately 60% of our group of 27 TS probands satisfied criteria for a diagnosis of ADHD. A comparison of the risks of TS, CT, and ADHD in the families of probands with TS and ADHD (TS + ADHD) with the risk in the families of probands with TS alone (TS − ADHD) showed that the risks of TS and CT were virtually the same in both groups of families. However, the risk of ADHD was found to be significantly higher among the relatives of TS + ADHD probands. The segregation patterns of TS, CT, and ADHD in the families of TS + ADHD probands also suggested that the two disorders were etiologically independent. If the two disorders were etiologically related, a higher than expected number of relatives should show both disorders (i.e., the disorders should be transmitted together in families of TS + ADHD probands). But the number of relatives with both TS and ADHD or CT and ADHD did not differ significantly from the expected values calculated under the assumption that the two disorders were independent. These findings suggest that ascertainment bias of the sort described by Berkson[37] may account for the high association of TS and ADHD observed in clinic populations.

Consistent with previous reports by a number of investigators of a high frequency of OCD symptoms in patients with TS,[11,15,38] 52% of our initial group of 27 TS probands received a diagnosis of OCD. Because of the high frequency of OCD among TS patients, the attractive hypothesis has been advanced that some forms of OCD might be etiologically related to TS. In order to evaluate this possibility, the recurrence risk of OCD in all the first degree relatives was examined and found to be significantly elevated over the population prevalence estimate. While the estimates of the population prevalence for OCD range between 0.5 and 2.0%, the recurrence risk among first-degree relatives for OCD was 23%. Next, the relatives of probands who met criteria for OCD (TS + OCD probands) were examined, and the frequency of OCD among them was compared with the frequency of OCD among the relatives of probands without OCD (TS − OCD probands). The recurrence risks for TS, CT, and OCD were not significantly different in the two groups of relatives, suggesting that the OCD observed in these families may represent an alternate expression of the factors also responsible for the manifestation of either TS or CT, or both.

It should be noted that a dramatic sex difference was observed in the relatives diagnosed as having OCD. Of the 13 first-degree relatives meeting criteria for OCD alone, 10 were female. These data suggest that the sex difference observed for TS and CT may be negated by the preponderance of females who have OCD alone. Thus, there may not be a difference in the frequency of affected individuals, but rather a difference in the number of individuals expressing specific constellations of symptoms (i.e., some with motor and vocal tics, some with motor tics only, some with OCD, and others with various combinations of all), with males showing a preponderance of motor symptoms and females showing a preponderance of obsessive symptoms.

3.3. Segregation Analyses

Another goal of this study was to test specific hypotheses in order to learn about the possible genetic mechanisms which may be responsible for the transmission of the syndrome. Using the computer program POINTER,[39] segregation analyses were performed on data collected from the first 27 families ascertained through a TS proband.[2] Such analyses require that extended pedigrees be separated into nuclear families, and this resulted in the inclusion of 29 nuclear families.

In order to test specific hypotheses, conditional likelihoods were calculated for each sibship. The model incorporated into POINTER is a threshold model requiring population prevalences in order to define liability thresholds. Individuals who exceed these thresholds are affected. Because of the uncertainty of the population prevalence for TS and associated behaviors, the analyses were done using several estimates of this parameter. In addition, based on the findings of the family study, three ways of defining who was affected were tested. In the first set of analyses, only those who met the full criteria for TS were included. In the second set of analyses, those with either TS or CT were counted as affected. In the last set of analyses, individuals were counted as affected if they met criteria for TS, CT, and/or OCD. Specific hypotheses that were examined included a model of no transmission, a mixed model (one that includes a single major genetic locus with polygenic background), a model in which all genetic variance is attributable to a polygenic mechanism, and a single major locus mode.

The results of all these analyses, for all estimates of prevalence and all diagnostic schemes, were consistent with a simple autosomal-dominant mode of transmission. As expected, the estimates of penetrance were dependent on the diagnostic scheme used to assign affected status. It is noteworthy that the results of these analyses were robust over all estimates of population prevalence and that single gene transmission was demonstrated even when only individuals affected with TS were used.

Genetic parameter estimates for males and females were different for all three diagnostic classification schemes. The penetrance for males, for example, was higher for all combinations of diagnostic classifications. However, the penetrance for females was found to increase considerably when individuals with OCD were included as affected. Given the differences between males and females with respect to the frequency of TS, CT, and OCD, this is not surprising. These results were based on best estimates of population prevalences for the three diagnostic schemes; and while there is still a slight sex difference in penetrance, they closely resemble a classic mendelian autosomal-dominant trait.

All the results described above were obtained with data collected from the first 27 families. Preliminary results with data on an additional 38 families are remarkably similar. The recurrence risks of TS, CT, and OCD are not significantly different in the second sample from those risks in the first sample. In addition, preliminary results for ADHD and OCD are also similar. The recurrence risk of OCD is significantly elevated in all families regardless of whether the proband has OCD or not, whereas the recurrence risk of ADHD is elevated in only those families where the proband has ADHD. The patterns within the families of TS + ADHD probands is also similar to that exhibited by the first sample of families. TS and ADHD do not co-occur more frequently than expected by chance. These preliminary findings are consistent with the results of the more complete analyses with data from the first group of families.

In summary, ongoing family and twin studies strongly suggest that some forms of OCD are etiologically related to TS and that the pattern of vertical transmission observed in these families is consistent with an autosomal dominant form of transmission.

3.4. Future Directions: The Search for Chromosomal Linkage

That TS has a significant genetic basis is clearly indicated by the results of the family genetic and twin studies described above. Further clarification of the etiology and pathogenesis of this complex disorder will very likely require a deeper understanding of its genetic basis and of the interactions between relevant genotypes and relevant environmental factors. Documentation of the types of familial patterns that occur was an essential preparation for more basic molecular biological studies of underlying genetic mechanisms. The most compelling evidence for a genetic etiology will be provided when the findings from these family and twin studies have been replicated with biological data, either through the identification of a biological marker or through the demonstration of genetic linkage to a marker locus.

A new class of genetic markers for the human genome, based on DNA sequence polymorphisms, has stimulated renewed interest in linkage approaches to the study of human disorders. Genetic linkage was long recognized as one of the methods useful in clarifying the role of genetic and environmental factors in the expression of complex disorders such as TS. Historically, the method has had limited applicability because of the small number of sufficiently polymorphic genetic markers which were available for study in humans. This situation has changed dramatically over the last several years, however, as advances in recombinant DNA technology have made it possible to detect many highly polymorphic genetic markers. Despite potential problems that might be posed by such factors as genetic heterogeneity, reduced penetrance, and moderate recombination between the disease and marker locus, linkage studies constitute a potentially powerful methodology for identifying the location and verifying the existence of genetic loci important in the expression of complex disorders such as TS.[40,41] Accordingly, extensive linkage mapping of the human genome is now under way, and the locations of several genetic disease loci have already been identified.[42,43]

These advances, along with the identification of appropriate high-density families in Canada[44] and Oregon, have led to intensive ongoing efforts to locate the putative gene responsible for TS using genetic linkage techniques. Should these efforts succeed, a major step forward would be accomplished, and a new era in TS research would be ushered in. It would then become feasible to locate and clone the responsible gene and eventually identify its abnormal product. Increased understanding of the etiology and pathophysiology of the disorder would be likely to follow as well as improved diagnostic techniques and therapeutic strategies. In addition, with the availability of a linked marker, nongenetic risk and protective factors for the illness may be identified through genetic case-control studies, in which affected probands are compared to their genetically identical siblings who have been exposed to similar environmental influences.

4. NEUROANATOMICAL CORRELATES

The etiology and neuropathological correlates of TS remain to be fully established. Earlier hypotheses, based on observations of motor and phonic tics and obsessive-compulsive behaviors (OCB) in patients with encephalitis lethargica, posited abnormalities in dopamine systems of the midbrain tegmentum and periaqueductal gray in the pathophysiology of TS.[45] Alternatively, the neuroanatomical description of a massive amygdalostriatal pathway led Nauta to suggest that this circuit might provide a basis for interactions between the limbic and motor systems and could be the neuropathological site in TS for the association of motor stereotypies and emotionally charged speech production.[46]

The basal ganglia, midbrain regions rich in dopamine as well as a large number of other neurotansmitters and neuromodulators, are widely considered the brain regions associated with a variety of movement disorders, including Parkinson's disease, Huntington's disease, tardive dyskinesia, and encephalitis lethargica. The presence of abnormal movements in TS, suggestive neuropathological findings, and extensive pharmacological and metabolic data implicating midbrain neurochemical systems have led to the hypothesis that the pathophysiology of TS may also involve some dysfunction in this region. In addition, recent positron-emission tomography (PET) studies have implicated the striatum and regions of the frontal cortex in the pathobiology of OCD.[47]

Although an extensive review of the neurobiology of the basal ganglia is beyond the scope of this chapter, certain aspects of recent findings concerning the circuitry and neurochemistry of this brain region seem especially pertinent to clinical features of the natural history of TS and to ongoing and future investigative efforts.[48,49] For example, autoradiographic tracing studies in primates have now demonstrated that the sensorimotor cortex projects exclusively to the putamen, while associational areas of the prefrontal, parietal, temporal, and cingulate cortices project exclusively to the caudate.[50–53] While the putamen is thought to be involved more directly in forebrain motor mechanisms, the caudate may play a role in mediation of complex, associative types of behavior.[48] This segregation of the sensorimotor and associational cortices has been found to be preserved at both pallidal and thalamic levels in nonoverlapping topographic projections.[48,55] Combined with evidence from electrophysiological studies, these observations have led to the suggestion that there exist in the basal ganglia parallel neuronal loops for separate processing of sensorimotor and associational information, each of which is modulated by distinct dopaminergic innervations from the midbrain.[48,54–56]

Moreover, the projection of the sensorimotor cortex onto the putamen is organized as a somatotopic body map of the leg, arm, shoulder, and face in the form of an oblique rostrocaudal strip; this body map is preserved in the projection of the putamen to the ventrocaudal pallidum.[52,56] It is also of interest that electrophysiological studies have demonstrated that ventral pallidal neurons are activated by limb movements, while neurons of the caudolateral pars reticulata of the substantia nigra, which also receives a projection from the putamen, become active during orofacial movements.[56,57]

Certain features of the natural history of TS, such as the preferential involvement with motor tics of the face, head, and shoulder regions, the occurrence of premonitory somatosensory urges preceding tics, and the peripubertal onset of sustained OCD symptoms, suggest that the underlying neuropathological process, whatever its nature, might encroach in some way on the sensorimotor and associational loops of the basal ganglia. For example, the somatotopic sensorimotor territory of the putamen, which persists in the projection of the putamen to the globus pallidus, offers a strikingly suggestive neuroanatomic mirror of the rostrocaudal progression of motor tics, which is seen in some cases of TS. One attractive hypothesis is that a progressive involvement of the two loops might underlie the natural history of this disorder, or alternatively that a lesion of the sensorimotor loop, which might be related to the initial onset of tics, might result in increased compensatory activity that affects both loop systems and that is clinically manifested in the eventual onset of OCD symptoms.

Additional clinical observations that point in this direction include recent imaging studies that have implicated the caudate in childhood-onset OCD[58] and a recent case report of a previously healthy man who developed tics and dyskinesias in association with bilateral pallidostriatal necrosis caused by an anaphylactic reaction to a wasp sting and then, after a delay of 2 years, had onset of severe OCD.[59]

Of particular interest, in view of the intersection of motor symptoms, anxiety, and behavioral disinhibition that can occur in TS, is the anatomical and functional distinction that has now been established between the dorsal and ventral striopallidal complexes.[46,48,49,60] The dorsal complex, which is composed of the dorsal caudate and putamen and their pallidal projections, is believed to be involved mainly with sensorimotor and associative functions.[48,61] By contrast, the ventral complex, which includes the nucleus accumbens, the olfactory tubercle, and the ventral caudoputamen and their projections to a specialized rostroventral extension of the pallidum, is primarily related to components of the limbic system and may provide a site at which the limbic system can interact with the extrapyramidal motor system.[48,61] Interestingly, in functional terms, it has been proposed that the dorsal striopallidal complex may be involved in the initiation of movements related to cognitive activities, while the ventral complex may be involved in motor behaviors related to emotional stimuli.[62] These and other advances in our understanding of the basal ganglia may eventually make it possible to reformulate our hypotheses concerning the underlying neuropathological correlates of TS in much more specific neuroanatomic terms.

5. PATHOPHYSIOLOGIC MECHANISMS

5.1. Dopaminergic Mechanisms

While the principal neurochemical systems implicated in the pathogenesis of TS have been the dopaminergic, noradrenergic, and serotonergic systems, the most established findings concern central dopaminergic mechanisms. Based on neuroanatomic findings described in post mortem studies of encephalitis lethargica, Devinsky[45] suggested that TS is the result of altered dopaminergic function in the midbrain. The substantia nigra and the ventral tegmental area (VTA) are the two adjacent midbrain sites that contain the largest aggregations of dopaminergic neurons in the brain. While the dopamine cell bodies of the substantia nigra give rise to the well-described ascending nigrostriatal pathway, those of the VTA are known to project to various limbic and cortical regions, including the ventral striopallidal complex, the prefrontal cortex, and the cingulate cortex.[63–64] The dopamine neurons of the substantia nigra and the VTA have also been shown to give rise to descending pathways projecting to the dorsal pontine tegmentum in the region of the noradrenergic locus coeruleus.[65] Of great potential importance is the recent demonstration in the primate that the internal segment of the globus pallidus receives both dense dopaminergic and serotonergic innervations from the midbrain.[66]

Other data implicating central dopaminergic mechanisms include clinical trials in which haloperidol and other neuroleptics that block dopaminergic receptors have been found to be effective in the partial suppression of tics in a majority of TS patients.[4,67] Tic suppression has also been reported following administration of α-methyl-*p*-tyrosine, which inhibits dopamine synthesis,[68] and following trials of tetrabenazine, which blocks the accumulation of dopamine in presynaptic storage vesicles.[69] By contrast, TS-like syndromes have appeared following withdrawal of neuroleptics,[70] and tics are often exacerbated following exposure to agents that increase central dopaminergic activity, such as L-DOPA and CNS stimulants.[71]

These pharmacological data, taken together with reports of lowered baseline and postprobenicid mean CSF levels of homovanillic acid (HVA), a major metabolite of brain dopamine, in TS patients compared with available contrast groups,[72–74] have led several groups of investigators to suggest that TS is a disorder in which postsynaptic dopaminergic receptors are hypersensitive.[72,73,75] However, attempts to downregulate putative hypersensitive dopamine receptors with increasing doses of L-DOPA gave equivocal results[75]; also, recent preliminary PET studies of brain dopamine receptors do not support the view that there are increased numbers of these receptors in the few TS patients that have been studied (G. Sedwall, personal communication, 1987). Additional imaging studies are needed to address fully the potential abnormalities of receptor number, affinity, and distribution. It is also important to keep in mind that the reported pharmacological effects are partial. Dopamine receptor-blocking agents suppress tics—they do not eliminate them—and a significant minority of patients do not respond. Taken together, these data suggest that the inferred dysfunction in brain dopaminergic systems may not be the primary pathophysiologic mechanism in tic-disorder patients.

5.2. Noradrenergic Mechanisms

Although retrograde cell-labeling studies in primates have led to the suggestion that the locus coeruleus may be a source of afferent projections to the striatum, this system has remained poorly documented and awaits further confirmation.[48] Nonetheless, noradrenergic mechanisms have figured prominently in several pathophysiological models proposed for TS, and extensive well-documented projections from the locus coeruleus to the thalamic nuclear complexes involved in the sensorimotor and associational loops of the basal ganglia may provide a neuroanatomic basis for them.[76]

Clinical evidence in support of a role for noradrenergic systems has largely been drawn from the reported beneficial effects of clonidine, a specific α_2-adrenergic receptor agonist.[77,78] However, the effectiveness of clonidine in TS is controversial, with some investigators reporting promising results that have not been replicated by others.[77–79] Reduced levels of urinary 3-methoxy-4-hydroxyphenylethylene glycol (MHPG) excretion in TS patients have been reported[80,81]; on the other hand, CSF MHPG levels were found to be normal with the exception of an occasional patient with an elevated level.[74]

To provide a link between central adrenergic and dopaminergic systems, complex hypotheses involving interactions between multiple monaminergic neurotransmitter systems have been advanced[82] based on such findings as altered plasma levels of HVA after chronic clonidine treatment,[83,84] increased plasma HVA levels after acute withdrawal from clonidine,[84] and the concordance of clonidine and haloperidol in a double crossover study involving these agents.[85] However, the reported changes in plasma HVA after chronic clonidine have not been consistently observed, and elevations following acute withdrawal most probably reflect enhanced peripheral sympathetic activity. Other studies of peripheral noradrenergic mechanisms have also been inconclusive.[86]

5.3. Serotonergic Mechanisms

Although serotonergic mechanisms have been repeatedly invoked as potentially playing an important role in the pathophysiology of TS, there is very little hard evidence to support this connection. Serotonergic inputs from the dorsal raphe to the pars reticulata of the substantia nigra, the striatum, and the internal segment of the globus pallidus have been identified and partially characterized, but there are no direct neuropathological data implicating them in TS.[48,66,82] Medications that act by increasing or decreasing serotonergic activity do not consistently alter TS symptoms. Early studies of CSF 5-hydroxyindoleacetic acid (5-HIAA), the principal central metabolite of serotonin, reported a range of low and normal levels in TS patients compared with contrast groups.[72–74] In a recent preliminary study of a small group of TS patients, many with prominent OC symptoms, we have found significantly lower mean CSF 5-HIAA levels as compared with normal controls.[87] Thus, hypotheses for a possible modulatory role for the basal ganglia serotonergic afferent system continue to be attractive. Additional circumstantial evidence includes their neuroanatomic projections and their likely involvement in the pathophysiology of obsessive-compulsive disorder.[88,89]

5.4. Cholinergic and GABAergic Mechanisms

In neurochemical terms, the striatum has classically been viewed as consisting of cholinergic interneurons and γ-aminobutyric acid (GABA)-containing projection neurons, both of which come under the influence of dopaminergic afferents from the midbrain. Because of this close anatomical relationship with dopaminergic systems, it has been suggested that cholinergic and/or GABAergic mechanisms may be implicated in the pathophysiology of TS.

Dysfunction of cholinergic mechanism was proposed based on the effects of cholinergic and anticholinergic agents on symptoms[90] and on findings of increased concentrations of red blood cell (RBC) choline in TS patients and their first-degree relatives in comparison to controls.[91] However, convincing evidence of a central pathophysiological role for cholinergic mechanisms has not been forthcoming. While intravenous physostigmine was reported by some investigators to decrease acutely motor and phonic tics,[92] others found that intramuscular physostigmine exacerbated tics and blocked the beneficial effects of scopolamine, a muscarinic blocking agent.[93] In addition, precursor loading trials with choline and lecithin have been disappointing.[90,94] By contrast, reports that clonazepam was effective in the treatment of TS[95] appear to support a role for GABAergic mechanisms, although other benzodiazepines, including diazepam, were not found to diminish tic symptoms. In a recent open trial, progabide, a direct GABA agonist, reduced tics in two of four patients.[96] Although cumulatively such reports argue for further investigation of both cholinergic and GABAergic mechanisms, a primary defect in either of these systems is probably unlikely in view of the finding that CSF levels of acetylcholinesterase (AChE) and GABA are normal in TS patients as compared with controls.[97,98] However, if central dopaminergic mechanisms are involved in the pathophysiology of tic disorders, as appears likely, secondary alterations in cholinergic and/or GABAergic activity could be expected, given their close neuroanatomical relationship.

5.5. Neuropetidergic Mechanisms: The Endogenous Opioid Peptides and TS

Endogenous opioid peptides (EOP) are localized in structures of the extrapyramidal system,[99] are known to interact with central dopaminergic neurons,[100–102] and may play an important role in the control of motor functions.[103] Two of the three families of opioid peptides, dynorphin and met-enkephalin, have been found to be highly concentrated and similarly distributed in the basal ganglia and substantia nigra.[99] Indeed the highest concentrations of met-enkephalin-immunoreactivity in the brain, as well as the most intense staining by immunohistochemistry, are known to occur in the striatum.[99] Significant levels of opiate receptor binding have also been detected in both primate and human neostriatum and substantia nigra.[104] Not surprisingly, speculations have arisen that alterations in these systems may be etiologically relevant in disorders of the basal ganglia, including TS.

Although no "neuropeptide deficiency" syndrome has yet been described, recent postmortem radioimmunoassay (RIA) and immunohistochemical studies of brains from patients with Parkinson's disease and Huntington's chorea have demonstrated regional decreases in levels of a variety of neuropeptides, including opioid peptides. In Parkinson's disease, Taquet et al.[105] found that dynorphin was normal but that met-enkephalin immunoreactivity was significantly decreased in comparison with controls in the putamen, both segments of the globus pallidus, and the substantia nigra. By contrast, in Huntington's chorea, which can be characterized by both rigidity and hyperkinetic movements, both the dynorphin and met-enkephalin systems appear to be affected. Immunoreactive dynorphin A(1–8), a selective marker for the prodynorphin system, was found to be significantly decreased in the caudate, putamen, external segment of the globus pallidus, and substantia nigra, while the octapeptide met-enkephalin-Arg 6-Gly 7-Leu 8, a selective marker of the proenkephalin system, was decreased in the caudate, putamen, and both segments of the globus pallidus.[106]

A recent immunohistochemical study has provided direct, if limited, neuropathological evidence of an abnormality in dynorphin in TS, a disease that is also characterized by hyperkinetic movements.[107] Haber et al.[107] performed an immunocytochemical survey of the distribution of dynorphin A(1–17)-, met-enkephalin-, and substance P-like immunoreactivity in a single brain from an adult patient who had a lifelong history of severe TS. Distribution of met-enkephalin- and substance P-positive woolly fibers appeared normal in all regions examined, which included the striatum, the globus pallidus, the central gray, and the substantia nigra. By contrast, dynorphin A-like immunoreactivity was notably faint throughout all structures examined and virtually absent in the globus pallidus. By comparison, staining for all three peptides, although diminished, was still present in the globus pallidus of brains attained from patients diagnosed as having Huntington's chorea. These provocative findings, which indicate a marked attenuation of dynorphin A(1–17) in striatal fibers projecting to the globus pallidus, if confirmed and shown not to be due to a technical artifact or to pharmacological effects, would constitute the first evidence for a distinct pathological change in the brain in TS.

The dynorphins constitute the most recently discovered branch of the opioid peptide family. All five members of this group [α-neo-endorphin, β-neo-endorphin, dynorphin A(1–17), dynorphin B, and dynorphin A(1–8)] derive from a common precursor, each containing the amino acid sequence of leu-enkephalin at the *N*-terminal.[99] Widely distributed throughout the central and peripheral nervous system in a distinct pattern, dynorphin is regarded as an extraordinarily potent opiate agonist, which appears to bind selectively to a particular subclass of opiate receptors, the kappa receptors.[108]

Neuroanatomic data have shown that dynorphin is strategically located in brain regions thought to be involved in extrapyramidal motor functions, neuroendocrine regulation, and sensory processing.[99] The highest concentrations of this peptide as determined by RIA have been found in the posterior pituitary, the hypothalamus, the substantia nigra, the caudate, the putamen, and the globus pallidus.[99,109] Immunohistochemical studies also have demonstrated numerous small dynorphin-positive cells in the caudate and putamen; very dense networks of fibers and terminals occur in the medial segment of the globus pallidus, the ventral tegmentum, and the pars reticulata of the substantia nigra.[99,110] These findings provide strong support for the presence of dynorphinergic projections from the striatum to the globus pallidus and substantia nigra.

In animal studies dynorphin and related peptides have been shown to produce a wide range of motor and behavioral effects, some of which are mediated by opiate receptors while others probably are not.[111,112] For example, bilateral injections of dynorphin-related peptides and pharmacological compounds that act specifically at κ-receptors into the substantia nigra of rats caused stereotyped behaviors that could be prevented by pretreatment with naloxone or haloperidol.[103]

Animal studies have also demonstrated functional interactions between dynorphin and the dopaminergic neurotransmitter system, long believed to be involved in TS. Injections of dynorphin or the specific κ-receptor agonist U50,488H into the substantia nigra, pars reticulata, reduced striatal release of dopamine.[103] These effects and results of other studies have led to the conclusion that the dynorphin striatonigral pathway may be involved in negative feedback control of ascending dopaminergic pathways.[103,113] Other evidence for a functional interaction between dynorphin and dopamine includes the demonstration of dose-related increases in dynorphin immunoreactivity in the striatum and substantia nigra of animals chronically treated with apomorphine[102] and the related finding that treatment with haloperidol decreased midbrain levels of dynorphin immunoreactivity.[114] In addition, recent in vivo electrochemical studies have now provided evidence that dynorphin may exert differential modulatory effects on the release of both dopamine and serotonin in the anterior striatum.[115]

The finding by Haber et al. of decreased dynorphin A(1–17) in the globus pallidus of a TS brain, coupled with the neuroanatomic distribution of dynorphin, its broad range of motor and behavioral effects, and its modulatory interactions with striatal dopaminergic and serotonergic neurotransmission suggested that dynorphin might play a role in the pathobiology of TS. In a recent preliminary study of a small group of patients with TS,[87] many of whom had prominent obsessive-compulsive symptoms, we found that CSF levels of dynorphin A(1–8) were increased in comparison with controls, raising the interesting possibility that some aspects of the underlying pathophysiology of TS, such as the mechanisms involved in the obsessive compulsive features of the syndrome, might be associated with regional alterations in processing of the prohormone.

However, the degree to which CSF concentrations of dynorphin A(1–8) might reflect the functional state of neuronal systems in human basal ganglia is far from clear. In addition, cross-sectional studies with small numbers of patients are unable to establish causal mechanisms. Thus, it is impossible at this point to tell whether an elevation in CSF dynorphin, should this finding be substantiated, would represent a primary etiologic factor or a secondary effect of a derangement of some other neurotransmitter system. Such questions can only be clarified with additional pathodynamic studies to elucidate causal mechanisms; larger-scale CSF and postmortem studies will have to examine each of the functional derivatives of prodynorphin simultaneously for evidence of alteration in processing of the prohormone. Although in general CSF studies may lack the precision necessary to address questions of selective involvement of neuronal systems in specific CNS locations, these preliminary findings, which will require replication and extension, are heuristically valuable and suggest a potential new direction for future research concerning the pathoetiology of TS and related disorders.

Additional evidence that opioid peptides may be pathoetiologically implicated in TS is derived from reports of dramatic, although poorly documented, clinical effects of the opiate antagonists naloxone and naltrexone on motor and phonic tics, obsessive-compulsive symptoms, and self-injurious behaviors.[116,117] There are obvious important limitations to uncontrolled brief case reports involving small numbers of patients but, despite their methodological limitations, given the generally unimpressive effects of naloxone in other psychiatric disorders, such dramatic reports, taken together with the recent observations concerning dynorphin, are of interest for their treatment implications as well as for what they suggest concerning the underlying pathoetiology of TS. Future systematic studies should be developed to evaluate the efficacy of the opiate antagonists for the treatment of TS and related disorders in a double-blind placebo-controlled manner.

Another psychopharmacological approach used to examine the role of opioid systems in psychiatric syndromes has been the study of the neuroendocrine effects of challenges with either opiate agonists or antagonists, and such studies have been applied in a preliminary way to TS.[118] It has long been clinically recognized that patients with TS manifest stress-related fluctuations in symptomatology and in response to medication,[7–9] suggesting the presence of an altered threshold for response to stressful stimuli. The importance of a tonic inhibitory modulation of the stress-sensitive hypothalamic–pituitary–adrenal (HPA) axis by endogenous opioid peptide systems has now been well established by numerous studies in animals and humans.[119–121] Taken with the recent findings concerning dynorphin in TS, these observations have raised the interesting possibility that a defect in opioid peptide inhibition of the HPA axis might be related to the enhanced sensitivity to stress, which is frequently characteristic of the syndrome. Such considerations suggest that future applications of neuroendocrine methodologies to the study of TS and related disorders might provide a new source of information concerning the underlying pathophysiology of this disorder.

In summary, recent neuropathological and neurochemical findings in patients with TS suggest that the endogenous opioid peptide dynorphin may play a role in the pathophysiology of this disorder. Such a hypothesis is supported by a wide range of preclinical studies of the distribution of dynorphin, its motor and neuroendocrine effects, and its modulatory interaction with monoaminergic systems long implicated in the pathogenesis of TS. In addition, there have been dramatic, although inconsistent, reports concerning the clinical effects of the opiate antagonists. Future studies should be directed at testing the hypothesis that dynorphin may play a role in the pathophysiology of this disorder and at systematically evaluating the potential clinical value of the opiate antagonists.

6. CONCLUSION

It is clear that a major thrust of future research efforts should be directed at developing and extending the findings from family–genetic and twin studies that suggest that the vertical transmission of TS within families follows an autosomal-dominant mode of inheritance and that some forms of OCD may be etiologically related to TS. A variety of fascinating questions are raised by this finding such as whether the OC symptomatology seen in relatives of TS probands is similar to or different from that which characterizes the relatives of OCD probands, whether there is a sex difference in the frequency of expression of symptoms among the relatives of OCD probands similar to what has been described for TS probands, and whether familial and nonfamilial subtypes of OCD might exist. In order fully to answer such questions, additional families of OCD probands will need to be ascertained using the method of direct interviews of all family members.

Since it is likely that a single locus plays a major role in the transmission of TS, ongoing efforts to establish genetic linkage to a marker locus should be intensified and expanded in view of the enormous potential increment to our understanding of this complex disorder that could follow once linkage is established, the gene cloned, and its abnormal product identified. In the long run, perhaps the most important implications for future TS research to be drawn from our increased understanding of the neurochemistry and neuroanatomy of the basal ganglia will relate to the selection of candidate genes for genetic linkage studies.

Among the most remarkable developments in the neurobiology of the basal ganglia over the past decade are the extensive immunohistochemical studies which have demonstrated the presence of a wide spectrum of differentially distributed classic neurotransmitters, neuromodulators, and neuropeptides.[48–50,60,61,63,66] Neuropeptides have now been identified in most of the major pathways of the basal ganglia (Table I). Genomic and complementary DNA (cDNA) probes are rapidly becoming available for many of these peptides, and restriction fragment length polymorphisms have already been defined for some of them. In the search for linkage, the genes for prodynorphin, proenkephalin, substance P, and other basal ganglia neuropeptides, as well as the genes for regulatory proteins associated with GABAergic, serotonergic, and dopaminergic neurotransmission, will be prime candidates because of the

Table I. Major Neurotransmitters and Neuropeptides of the Basal Ganglia

Pathway	Classic transmitter	Associated neuropeptide(s)
Striatal afferents		
Thalamostriatal	Glutamate/aspartate	Substance P (SP)
Amygdalostriatal	?	Cholecystokinin (CCK), neurotensin (NT)
Corticostriatal	Glutamate/aspartate	CCK
Nigrostriatal	Dopamine	CCK, NT
Ventral tegmentostriatal	Dopamine	CCK, NT
Raphe striatal	Serotonin	
Intrinsic striatal connections	GABA Acetylcholine	Somatostatin, neuropeptide Y
Striatal efferents		
Striatonigral	GABA, taurine	SP, substance K (SK), Dynorphin (DYN), Enkephalin (ENK)
Striatopallidal (external segment)	GABA	ENK (high), SP (low), DYN (low)
Striatopallidal (internal segment)	GABA	SP (high), DYN (high), ENK (low)
Pallidal afferents		
Striatopallidal	GABA	
Nigropallidal (internal segment)	Dopamine	
Raphe pallidal (interal segment)	Serotonin	
Pallidal efferents		
Pallidocortical	Acetylcholine	
Pallidothalamic and pallidosubthalamic	GABA	
Nigral afferents		
Striatonigral	GABA	
Raphe nigral	Serotonin	
Nigral efferents		
Nigrotectal and nigrothalamic	GABA	
Nigrostriatal, nigropallidal, ventral tegmentostriatal	GABA	

accumulating evidence pointing to the basal ganglia as the probable site of the neuroanatomic correlates of TS.

Finally, in view of the increased sophistication of our understanding of the neurochemistry of the basal ganglia and of the availability of many specific and sensitive assays for a wide variety of neuropeptides and monoamine metabolites that were not available a decade ago, it seems timely that a new series of pathodynamic, postmortem, and CSF studies be undertaken in a major new effort to identify and localize the neurobiological mechanisms responsible for this disease. These biochemical and genetic studies should be complemented by a full array of brain-imaging studies using the advanced technologies now available, including PET, single photon emission computed tomography (SPECT), and magnetic resonance imaging (MRI).

REFERENCES

1. Comings DE, Comings BG, Devor EJ, et al: Detection of a major gene for Gilles de la Tourette syndrome. Am J Hum Genet 36:586–600, 1984

2. Pauls DL, Leckman JF: The inheritance of Gilles de la Tourette syndrome and associated behaviors: Evidence for autosomal dominant transmission. N Engl J Med 315:993–997, 1986
3. Shapiro AK, Shapiro E: Treatment of tic disorders with haloperidol, in Cohen DJ, Bruun RD, Leckman JF (eds), Tourette's Syndrome and Tic Disorders: Clinical Understanding and Treatment. New York, Wiley, 1988, pp 267–281
4. Shapiro AK, Shapiro E, Young JG, et al: Gilles de la Tourette Syndrome. New York, Raven, 1978
5. American Psychiatric Association: Diagnostic and Statistical Manual of Mental Disorders, ed 3, revised. Washington, D.C., American Psychiatric Association, 1987, pp 79–80
6. Burd L, Kerbeshian J, Wilkenheiser M, et al: A prevalence study of Gilles de la Tourette's syndrome in North Dakota school aged children. J Am Acad Child Adol Psychiatry 25:552–553, 1986
7. Jaegger J, Prusoff BA, Cohen DJ, et al: The epidemiology of Tourette syndrome: A pilot study. Schizophr Bull 8:267–278, 1982
8. Leckman JF, Cohen DJ, Price RA, et al: The pathogenesis of Tourette syndrome, in Shah NS, Donald AG (eds), Movement Disorders. New York, Plenum, 1986, pp 257–272
9. Leckman JF, Walkup JT, Riddle MA, et al: Tic disorders, in Meltzer HY (ed): Psychopharmacology: The Third Generation of Progress. New York, Raven, 1987, pp 1239–1246
10. Fahn S: The clinical spectrum of motor tics, in Friedhoff AJ, Chase TN (eds), Advances in Neurology: Gilles de la Tourette syndrome, Vol. 35. New York, Raven, 1982, pp 341–344
11. Nee LE, Polinsky RJ, Ebert MH: Tourette syndrome: Clinical and family studies, in Friedhoff AJ, Chase TN (eds), Advances in Neurology: Gilles de la Tourette Syndrome, Vol. 35. New York, Raven, 1982, pp 291–295
12. Surwillo WW, Shafii M, Barrett CL: Gilles de la Tourette: A 20-month study of the effects of stressful life events and haloperidol on symptom frequency. J Nerv Ment Dis 166:812–816, 1978
13. Comings DE, Comings BG: Tourette syndrome and attention deficit disorder with hyperactivity—Are they genetically related? J Am Acad Child Psychiatry 23:138–144, 1984
14. de la Tourette G: Etude sur une affection nerveuse, caracterisée par de l'incoordination motrice, accompagnée d'echolalie et de coprolalia. Arch Neurol 9:19, 1885
15. Montgomery MA, Clayton PJ, Friedhoff AJ: Psychiatric illness in Tourette syndrome patients and first degree relatives, in Chase TN, Friedhoff AJ (eds), Advances in Neurology: Gilles de la Tourette Syndrome, Vol. 35. New York, Raven, 1982, pp 335–339
16. Comings DE, Comings BG: A controlled study of Tourette syndrome. I. Attention-deficit disorder, learning disorders and school problems. Am J Hum Genet 41:701–741, 1987
17. Rudin E: Ein betrage zur frage der zwangsdrankheit, insbesondere inhrere hereditarne beziehugen. Arch Psychiatr Nervenkr 19:14–54, 1953
18. Robins LN, Helzer JE, Weissman MM, et al: Lifetime prevalence of specific psychiatric disorders in three sites. Arch Gen Psychiatry 41:949–958, 1984
19. Grad LR, Pelcovitz D, Olson M, et al: Obsessive-compulsive symptomatology in children with Tourette's syndrome. J Am Acad Child Adol Psychiatry 26:69–74, 1987
20. Comings BG, Comings DE: A controlled study of Tourette syndrome. VII. Summary: A common genetic disorder causing disinhibition of the limbic system. Am J Hum Genet 41:839–866, 1982
21. Comings BG, Comings DE: A controlled study of Tourette syndrome. V. Depression and mania. Am J Hum Genet 41:804–821, 1987
22. Pauls DL, Cohen DJ, Kidd KK, Leckman JF: Tourette syndrome and neuropsychiatric disorders: Is there a genetic relationship? Am J Hum Genet 43:206–209, 1988
23. Price RA, Kidd KK, Cohen DJ, et al: A twin study of Tourette's syndrome. Arch Gen Psychiatry 42:815–820, 1970
24. Price RA, Leckman JF, Pauls DL, et al: Gilles de la Tourette syndrome: Tics and central nervous system stimulants in twins and non-twins with Tourette syndrome. Neurology (NY) 36:232–237, 1986
25. Pauls DL, Hurst C: Schedule for Tourette Syndrome and Other Behavioral Disorders, revised. New Haven, Yale Child Study Center, 1987
26. Harcherik DJ, Leckman JF, Detlor J, et al: A new instrument for clinical studies of Tourette's syndrome. J Am Acad Child Adol Psychiatry 23:153–160, 1984
27. Goodman WK, Rasmussen SA, Price LH, et al: Yale–Brown Obsessive Compulsive Scale. Department of Psychiatry, Yale University School of Medicine and Brown University School of Medicine, 1986
28. Eldridge R, Sweet R, Lake CR, et al: Gilles de la Tourette's syndrome: Further evidence for a major locus mode of transmission. Am J Hum Genet 36:704–709, 1984

29. Kidd KK, Prusoff BA, Cohen DJ: Familial patterns of Gilles de la Tourette syndrome. Arch Gen Psychiatry 37:1336–1339, 1980
30. Pauls DL, Cohen DJ, Heimbuch R, et al: The familial pattern and transmission of Tourette syndrome and multiple tics. Arch Gen Psychiatry 32:1091–1093, 1981
31. Kidd KK, Pauls DL: Genetic hypothesis for Tourette syndrome, in Chase TN, Friedhoff AJ (eds), Advances in Neurology: Gilles de la Tourette Syndrome, Vol. 35. New York, Raven, 1982, pp 243–249
32. Devor EJ: Complex segregation analysis of Gilles de la Tourette syndrome: Further evidence for a major locus mode of transmission. Am J Hum Genet 36:704–709, 1984
33. Comings DE, Comings BG, Devor EJ, et al: Detection of a major gene for Gilles de la Tourette syndrome. Am J Hum Genet 36:586–600, 1984
34. Pauls DL, Kruger SD, Leckman JF, et al: The risk of Tourette syndrome and chronic multiple tics (CMT) among relatives of TS patients obtained by direct interview. J Am Acad Child Psychiatry 23:134–137, 1984
35. Andreason NC, Endicott J, Spitzer RL, et al: The family history method using diagnostic criteria. Arch Gen Psychiatry 34:1229–1235, 1977
36. Orvaschel H, Thompson WD, Belanger A, et al: Comparison of the family history method to direct interview. J Affect Disord 4:49–59, 1982
37. Berkson J: Limitations of the application of fourfold tables analysis to hsopital data. Biometrics 2:47–51, 1946
38. Yargura-Tobias JA: Clinical aspects of Gilles de la Tourette syndrome. Orthomolec Psychiatry 10:263–268, 1981
39. Lalouel JM, Rao DC, Morton NE, et al: A unified model for complex segregation analysis. Am J Hum Genet 35:816–826, 1983
40. Goldin LR, Cox NJ, Pauls DL, et al: The detection of major loci by segregation and linkage analysis: A simulation study. Genet Epidemiol 1:285–296, 1984
41. Gershon ES, Merril CR, Goldin LR, et al: The role of molecular genetics in psychiatry. Biol Psychiatry 22:1388–1405, 1987
42. Gusella J, Wexler N, Conneally P, et al: A polymorphic DNA marker genetically linked to Huntington's disease. Nature (Lond) 306:234–238, 1983
43. Davies K, Pearson P, Harper P, et al: Linkage analysis of two cloned DNA sequences flanking the Duchenne muscular dystrophy locus on the short arm of the human X chromosome. Nucleic Acids Res 11:2303–2312, 1983
44. Kurlan R, Behr J, Medved L, et al: Familial Tourette's syndrome: Report of a large pedigree and potential for linkage analysis. Neurology (NY) 36:772–776, 1986
45. Devinsky O: Neuroanatomy of Gilles de la Tourette's syndrome: Possible midbrain involvement. Arch Neurol 40:508–514, 1983
46. Nauta WJH: Limbic innervation of the striatum, in Friedhoff AJ, Chase TN (eds), Gilles de la Tourette Syndrome. New York, Raven, 1982, pp 41–48
47. Baxter LR, Phelps ME, Mazziota JC, et al: Local cerebral glucose metabolic rates in obsessive-compulsive disorder. Arch Gen Psychiatry 44:211–218, 1987
48. Parent P: Comparative Neurobiology of the Basal Ganglia. New York, Wiley, 1986
49. Graybiel AM: Neurochemically specified subsystems in the basal ganglia, in Evarts EV (chairman), Functions of the Basal Ganglia. London Pitman, (Ciba Foundation Symposium 107), 1984, pp 114–149
50. Graybiel AM, Ragsdale CW: Fiber connections of the basal ganglia. Prog Brain Res 51:239–283, 1979
51. Kunzle H: Bilateral projections from precentral mortor cortex to the putamen and other parts of the basal ganglia. An autoradiographic study in Macaca fascicularis. Brain Res 88:195–209, 1975
52. Percheron G, Yelnik J, François C: A Golgi analysis of the primate globus pallidus. III. Spatial organization of the striato–pallidal complex. J Comp Neurol 227:214–227
53. Ragsdale CW, Graybiel AM: The fronto-striatal projection in the cat and monkey and its relationship to inhomogeneities established by acetylcholinesterase histochemistry. Brain Res 208:259–266, 1981
54. Alexander GE, DeLong MR, Strick PL: Parallel organization of functionally segregated circuits linking basal ganglia and cortex. Annu Rev Neurosci 9:357–381, 1986
55. Delong MR, Georgopoulos AP, Crutcher MD: Cortico-basal ganglia relations and coding of motor performance. Exp Brain Res 7(suppl):30–40, 1983
56. Delong MR, Georgopoulos AP, Crutcher MD, et al: Functional organization of the basal ganglia: Contri-

butions of single-cell recording studies, in Evarts EV (chairman), Functions of the Basal Ganglia. Pitman London, (Ciba Foundation Symposium 107), 1984, pp 64–78

57. DeLong MR, Crutcher MD, Georgopoulos AP: Relations between movement and single cell discharge in the substantia nigra of the behaving monkey. J Neurosci 3:1599–1606, 1983
58. Swedo S, Berg CZ, Luxenberg J, et al: Reduced caudate volume in patients with childhood onset obsessive compulsive disorder (OCD), in Greenhill L, Crosby M (eds), Scientific Proceedings, Presented at the Thirty-Fourth Annual Meeting of the American Academy of Child and Adolescent Psychiatry, Washington, D.C., October 21–25, 1987, p 55
59. Laplane D, Widlocher D, Pillon B, et al: Comportement conpulsif d'allure obsessionnelle par necrose circonscrite bilaterale pallido-striatale. Encephalopathie par piqure de guepe. Rev Neurol 137:269–276, 1981
60. Graybiel AM, Ragsdale CW: Biochemical anatomy of the striatum, in Emson PC (ed), Chemical Neuroanatomy. New York, Raven, 1983, pp 427–504
61. Graybiel AM: Neuropeptides in the basal ganglia, in Martin JB, Barchas JD (eds), Neuropeptides in Neurologic and Psychiatric Disease. New York, Raven, 1986, pp 135–161
62. Heimer L, Switzer RC, Van Hoesen GW: Ventral striatum and ventral pallidum. Components of the motor system? Trends Neurosci 5:83–87, 1982
63. Haber SN: Neurotransmitters in the human and nonhuman primate basal ganglia. Hum Neurobiol 5:159–168, 1986
64. Iverson SD, Alpert JE: Functional organization of the dopamine system in normal and abnormal behavior, in Friedhoff AJ, Chase TN (eds), Advances in Neurology: Gilles de la Tourette Syndrome. New York, Raven, 1982, pp 69–76
65. Deutsch AY, Goldstein M, Roth RH: Activation of locus coeruleus induced by selective stimulation of the ventral tegmental area. Brain Res 363:307–414, 1986
66. Parent A, Smith Y: Differential dopaminergic innervation of the two pallidal segments in the squirrel monkey (*Salmiri sciureus*). Brain Res 426:397–400, 1987
67. Ross MS, Moldofsky H: A comparison of pimozide and haloperidol in the treatment of Gilles de la Tourette syndrome. Am J Psychiatry 135:585–587, 1978
68. Sweet RD, Bruun R, Shapiro E, et al: Presynaptic catecholamine antagonists as treatment for Tourette syndrome: Effects of alpha-methyl-para-tyrosine and tetrabenazine. Arch Gen Psychiatry 31:857–861, 1976
69. Jankovic J, Glaze DG, Frost JD: Effect of tetrabenazine on tics and sleep of Gilles de la Tourette's syndrome. Neurology (NY) 34:688–692, 1984
70. Klawans HL, Falk DK, Nausieda PA: Gilles de la Tourette syndrome after long-term chlorpromazine therapy. Neurology (NY) 28:1064–1068, 1978
71. Golden GS: Gilles de la Tourette syndrome following methylphenidate administration. Dev Med Child Neurol 16:76–78, 1974
72. Butler IJ, Koslow SH, Seifert WE, et al: Biogenic amine metabolism in Tourette syndrome. Ann Neurol 6:37–39, 1979
73. Cohen DJ, Shaywitz BA, Caparulo BK, et al: Chronic, multiple tics of Gilles de la Tourette's disease: CSF acid monoamine metabolites after probenecid administration. Arch Gen Psychiatry 35:245–250, 1978
74. Cohen DJ, Shaywitz BA, Young JG, et al: Central biogenic amine metabolism in children with the syndrome of chronic multiple tics of Gilles de la Tourette: Norepinephrine, serotonin, and dopamine. J Am Acad Child Psychiatry 18:320–341, 1979
75. Friedhoff AJ: Receptor maturation in pathogenesis and treatment of Tourette syndrome, in Friedhoff AJ, Chase TN (eds), Advances in Neurology: Gilles de la Tourette Syndrome, Vol 35. New York, Raven, 1982, pp 133–140
76. Ishikawa M, Tanaka C: Morphological organization of catecholamine terminals in the diencephalon of the rhesus monkey. Brain Res 119:43–55, 1977
77. Cohen DJ, Young JG, Nathanson JA, et al: Clonidine in Tourette's syndrome. Lancet 2:551–553, 1971
78. Cohen DJ, Detlor JG, Lowe T: Clonidine ameliorates Gilles de la Tourette syndrome. Arch Gen Psychiatry 37:1350–1357, 1980
79. Shapiro AK, Shapiro E, Eisenkraft GJ: Treatment of Gilles de la Tourette syndrome with clonidine and neuroleptics. Arch Gen Psychiatry 40:1235–1240, 1988
80. Ang L, Borison R, Dysken M, et al: Reduced excretion of MHPG in Tourette Syndrome, in Friedhoff AJ, Chase TN (eds), Advances in Neurology: Gilles de la Tourette Syndrome, Vol. 35. New York, Raven, 1982, pp 171–175

81. Sweeney RD, Pikar D, REdmond DE, et al: Noradrenergic and dopaminergic mechanisms in Gilles de la Tourette syndrome. Lancet 1:872, 1978
82. Bunny BS, DeRiemer SA: Effects of clonidine on nigral dopamine cell activity: Possible mediation by noradrenergic regulation of serotonergic raphe system, in Friedhoff AJ, Chase TN (eds), Advances in Neurology: Gilles de la Tourette syndrome, Vol. 35. New York, Raven, 1982, pp 99–104
83. Leckman JF, Detlor J, Harcherik DF, et al: Acute and chronic clonidine treatment in Tourette's syndrome: A preliminary report on clinical response and effect on plasma and urinary catecholamine metabolites, growth hormone, and blood pressure. J Am Acad Child Psychiatry 22:433–440, 1983
84. Leckman JF, Ort S, Caruso KA, et al: Rebound phenomenon in Tourette's syndrome after abrupt withdrawal of clonidine. Arch Gen Psychiatry 43:1168–1176, 1986
85. Borison RL, Ang L, Chang S, et al: New pharmacological approaches in the treatment of Tourette's syndrome, in Friedhoff AJ, Chase TN (eds), Advances in Neurology: Gilles de la Tourette Syndrome, Vol. 35. New York, Raven, pp 121–132
86. Shapiro AK, Baron M, Shapiro E, et al: Enzyme activity in Tourette syndrome. Arch Neurol 41:282–285, 1984
87. Leckman, JF, Riddle RA, Berrettini WH, et al: Elevated CSF dynorphin A[1–8] in Tourette's syndrome. Life Sci 43:2015–2023, 1988
88. Cummings JL, Frankel M: Gilles de la Tourette syndrome and the neurological basis of obsessions and compulsions. Biol Psychiatry 20:1117–1126, 1985
89. Insel TS, Mueller EA, Alterman I, et al: Obsessive compulsive disorder and serotonin: Is there a connection. Biol Psychiatry 20:1174–1188, 1985
90. Stahl SM, Berger PA: Cholinergic and dopaminergic mechanisms in Tourette's syndrome, in Friedhoff AJ, Chase TN (eds), Advances in Neurology: Gilles de la Tourette syndrome, Vol. 35. New York, Raven, 1982, pp 141–150
91. Hanin I, Merikangas JR, Merikangas KR, et al: Red-cell choline and Gilles de la Tourette syndrome. N Engl J Med 301:661–662, 1979
92. Stahl SM, Berger PA: Physostigmine in Tourette syndrome: Evidence for cholinergic underactivity. Am J Psychiatry 138:240–241, 1981
93. Tanner CM, Goetz CG, Klawans HL: Cholinergic mechanisms in Tourette syndrome. J Autism Dev Disord 13:207–213, 1983
94. Rosenberg GS, Davis KL: Precursors of acetylcholine: considerations underlying their use in Tourette syndrome, in Friedhoff AJ, Chase TN (eds), Advances in Neurology: Gilles de la Tourette Syndrome, Vol. 35. New York, Raven, 1982, pp 407–412
95. Gonce M, Barbeau A: Seven cases of Gilles de la Tourette's syndrome: Partial relief with clonazepam: A pilot study. Can J Neurol Sci 75:225–241, 1986
96. Mondrup K, Dupont E, Braindgaard H: Progabide in the treatment of hyperkinetic extrapyramidal movement disorders. Acta Neurol Scand 72:341–343, 1985
97. Singer HS, Oshida L, Coyle JT: CSF cholinesterase activity Gilles de la Tourette's syndrome. Arch Neurol 41:756–757, 1984
98. van Woert MH, Rosenbaum D, Enna JJ: Overview of pharmacological approaches for therapy to Tourette syndrome, in Friedhoff AJ, Chase TN (eds), Advances in Neurology: Gilles de la Tourette Syndrome, Vol. 35. New York, Raven, 1982, pp 369–376
99. Nieuwenhuys R: Chemoarchitecture of the Brain. New York, Springer-Verlag, 1985, pp 97–108
100. Buck SH, Yamamura HI: Neuropeptides in normal and pathological basal ganglia, in Friedhoff AJ, Chase TN (eds), Advances in Neurology: Gilles de la Tourette syndrome, Vol. 35. New York, Raven, 1982, pp 121–132
101. Quirion R, Gaudreau P, Martel JC, et al: Possible interactions between dynorphin and dopaminergic systems in rat basal ganglia and substantia nigra. Brain Res 331:358–362, 1985
102. Li S, Sivam SP, Hong JS: Regulation of the concentration of dynorphin A1–8 in the striatonigral pathway by the dopaminergic system. Brain Res 398:390–392, 1986
103. Herrera-Marschitz M, Christensson-Nylander I, Staines W, et al: Striato-nigral dynorphin and substance P pathways in the rat. Exp Brain Res 64:193–207, 1986
104. Kuhar MJ, Pert CB, Snyder SH: Regional distribution of opiate receptor binding in monkey and human brain. Nature (Lond) 245:447–450, 1973
105. Taquet H, Javoy-Agid F, Giraud P, et al: Dynorphin levels in parkinsonian patients: Leu-enkephalin production from either proenkephalin A or prodynorphin in human brain. Brain Res 341:390–392, 1985

106. Seizinger BR, Liebisch DC, Kish SJ: Opioid peptides in Huntington's disease: Alterations in prodynorphin and proenkephalin system. Brain Res 378:405–408, 1986
107. Haber SN, Kowall NW, Vonsattel JP, et al: Gilles de la Tourette's syndrome: A postmortem neuropathological and immunohistochemical study. J Neurol Sci 75:225–241, 1986
108. Goldstein A, Tachibana S, Lowney LI, et al: Dynorphin-(1–13), an extraordinarily potent opioid peptide. Proc Natl Acad Sci USA 76:6666–6670, 1979
109. Gramsch C, Volker H, Pasi A, et al: Immunoreactive dynorphin in human brain and pituitary. Brain Res 233:54–74, 1982
110. Fallon JH, Leslie FM: Distribution of dynorphin and enkephalin peptides in the rat brain. J Comp Neurol 249:293–336, 1986
111. Katz RJ: Behavioral effects of dynorphin—A novel opioid neuropeptide. Neuropharmacology 19:801–803, 1980
112. Walker JM, Moises HC, Coy DH, et al: Non-opiate effects of dynorphin and des-tyr-dynorphin. Science 218:1136–1138, 1982
113. Christensson-Nylander I, Herra-Marschitz M, Staines W, et al: Striato-nigral dynorphin and substance P pathways in the rat. I. Biochemical and immunohistochemical studies. Exp Brain Res 64:169–192, 1986
114. Quirion R, Gaudreau P, Martel J-C, et al: Possible interactions between dynorphin and dopaminergic systems in rat basal ganglia and substantia nigra. Brain Res 331:358–362, 1985
115. Broderick PA: Striatal neurochemistry of dynorphin-(1–13): In vivo electrochemical semidifferential analyses. Neuropeptides 10:369–386, 1987
116. Sandyk R, Iacono RP, Allender J: Naloxone ameliorates compulsive touching behavior and tics in Tourette's syndrome. Ann Neurol 20:437, 1986
117. Sandyk R, Iacono RP, Crinnian CT, et al: Effects of naltrexone in Tourette's syndrome. Ann Neurol 20:437, 1986
118. Sandyk R, Bamford CR, Crinnian CT: Abnormal growth hormone response to naloxone challenge in Tourette's syndrome. Int J Neurosci 37:191–192, 1987
119. Grossman A, Moult PJA, Cunnah D, et al: Different opioid mechanisms are involved in the modulation of ACTH and gonadotrophin release in man. Neuroendocrinology 42:357–360, 1986
120. Morley JE, Baranetsky NG, Wingert TD, et al: Endocrine effects of naloxone-induced opiate receptor blockade. J Clin Endocrinol Metab 50:251–257, 1980
121. Grossman A, Besser GM: Opiates control ACTH through a noradrenergic mechanism. Clin Endocrinol 17:287–290, 1982

15

Obsessive-Compulsive Disorder

Neurologic and Neuroanatomic Perspectives

Elizabeth V. Koby

1. INTRODUCTION

While obsessions and compulsions have been described by clinicians for more than a century,[1] modern nosology distinguishes a pure syndrome of obsessive-compulsive disorder (OCD) from conditions in which obsessive or compulsive features are only part of a wider symptomatology. Currently a diagnosis of OCD is made when intrusive, ego-dystonic thoughts and/or repetitive but stereotyped behaviors are present in the absence of a major neuropsychiatric illness.[2] The lifetime prevalence of this disorder has been estimated at 2%.[3] While the natural history of OCD has not been studied exhaustively, it appears that a substantial fraction of adults with OCD develop their symptoms before age 15.[4] Conversely symptoms of childhood-onset OCD often persist into adulthood.[5,6] By most reports, childhood and adult forms of OCD are clinically similar[7–9] chronic conditions.[10]

2. ASSOCIATION WITH CEREBRAL INSULT

Most patients with childhood or adult OCD have no easily demonstrable neurological impairment nor history of significant brain insult. However, an above-average incidence of obsessions and compulsions has been noted in samples of patients with a variety of serious CNS disturbances, including encephalitis[11–15], head trauma,[16,17] Sydenham's chorea,[18,19] and others.[20] Sporadic OC systems have also been described in patients suffering from gas poisoning,[20] diabetes insipidus,[21] temporal lobe epilepsy,[22] and cerebral tumors.[21,23,24] In several such cases, brain abnormalities have been documented on computed tomography (CT) or electroencephalographic (EEG) studies.[25,26]

Despite the heterogeneity of these reports, some conclusions can be drawn. The central nervous system (CNS) insults described are associated with a variety of psychiatric distur-

Elizabeth V. Koby • Department of Child Psychiatry, Georgetown University School of Medicine, Washington, D.C. 20007.

bances[11,12,16,17,23,27]; obsessions and compulsions, if they occur, can be mild in comparison with other symptoms.[15,16] Cerebral localization has not been possible, at least with respect to head trauma,[16] while the abnormalities described on CT (cortical atrophy, basal ganglia calcification) in the few sporadic cases are nonspecific.[28,29] Collectively, the evidence implies that gross CNS dysfunction may be causally related to OC features in a very small number of patients. On the other hand the heterogeneity and nonspecific sequelae of such cerebral insults militate against a unique etiological process for OC symptoms.

3. EVIDENCE FROM TOURETTE'S SYNDROME

Although the Diagnostic Systems Manual (DSM-III) does not include obsessions and compulsions among the criteria for Tourette's syndrome (TS) and precludes codiagnosing OCD and TS, converging evidence suggests that the two disorders are at times reated. The incidence of significant obsessions and compulsions in TS, originally thought to be as high as 30%,[30–34] may actually be in the 50–90% range.[33–41] Similarities in clinical presentation and course between OCD and TS have been well reviewed recently.[42] Most striking is the familial link between the two disorders. Relatives of TS patients are at increased risk of developing OCD,[35,41,43,44] even if they themselves do not suffer from tics.[32,35,37,44–46] On the basis of such studies, it has been proposed that an autosomal-dominant gene may underlie the expression of some forms of OCD and TS.[44,47]

The increased incidence of nonspecific markers of brain dysfunction, such as hyperactivity, excess sinistrality, soft neurological signs, neuropsychological deficits, EEG and CT scan abnormalities in TS,[42,48,49] are consistent with subtle cerebral pathophysiology. Devinsky[48] reviewed the localizing evidence. Dopamine blocking drugs can suppress, while dopamimetics exacerbate, the tics and vocalizations of TS.[48] Most of the dopaminergic innervation of mammalian brain originates in the brainstem.[51] The highest concentrations of dopamine are found in the caudate nuclei,[50] which have been implicated in several conditions characterized by movement disorders.[52] By inference, it has been posited that aberrant function in the basal ganglia or their connections mediate TS symptoms. The latter is compatible with research data.

Voluntary movements are preceded by cortical electrical potentials that can be detected on EEG. Such potentials have not been detected before the involuntary movements of TS, suggesting that the electrical impulses associated with these abnormal movements are generated subcortically.[53] Even more striking is recent evidence from positron-emission tomography (PET), a functional brain-imaging modality that is briefly discussed later in this chapter. In a preliminary study, a group of TS patients showed increased metabolism in the caudate nuclei and a trend in the same direction in frontal and temporal cortical regions.[54]

In view of the robust genetic and clinical associations between TS and OCD, it might be anticipated that some of these findings might be common to both conditions. Such shared features may however be confined to a subgroup of patients. While OC symptoms appear more often than expected in families of TS probands, an excess incidence of TS has not been found in first-degree relatives of children with OCD.[55] In addition, Tourette's syndrome and OCD appear to have different pharmacological response profiles.[56–60] Accordingly, inferences about OCD based on TS studies must be made cautiously.

4. NEUROLOGICAL AND NEUROPSYCHOLOGICAL FINDINGS

Diverse neurological abnormalities have been described in the rare patient whose OC symptoms developed after significant CNS insult.[11,12,14,16,26] On the other hand, in large sample studies of OCD patients in which a general medical examination was included as part of the evaluation, neurological abnormalities have not been reported,[57,61–63] suggesting that conspicuous neurological signs are generally absent. In the only study in which an adult OCD patient sample was evaluated with attention to subtle neurological signs, an increased incidence of soft[64] signs was found.[65] A preliminary study of children with OCD reported an excess of neurodevelopmental abnormalities[68] (descriptively, immature for age[66,67]). Investigation of a subsequent larger childhood OCD sample showed that 22 of 26 participants had some abnormality[69] as detected on the Physical and Neurological Examination for Soft Signs (PANESS).[70] In particular, subtle choreiform movements could be elicited in several children.[69]

The soft signs found in children with OCD were both more common than, and different from, those in the one adult study where abnormalities were reported as well. The latter may be partially accounted for by the attenuation of some soft signs with development.[71,72] Whatever the case, the origin and implications of soft signs remain controversial.[64] They may be indices of nonspecific, subtle brain pathophysiology.

On the other hand, choreiform movements could have localizing value, given their association with disorders of the basal ganglia.[52] While the findings in both adult and childhood OCD samples must be considered preliminary, it is noteworthy that they would not have been evident without a careful neurodevelopmentally oriented examination, suggesting that if a CNS abnormality were present, it would likely be a subtle one.

Neuropsychological investigations of adult OCD samples have yielded few replicable findings, not surprising given interstudy differences as well as the paucity of systematic studies. While difficulties with decision making tasks,[73] deficits on the Tactual Performance Test, as well as several other WAIS (Wechsler Adult Intelligence Scale) subtests have been reported,[61] other investigators have found no consistent pattern of deficit.[63]

A pilot study of childhood OCD found no evidence of organicity on neuropsychological testing.[7] In a larger sample,[68] the performance of OCD children was reported to be similar to that of controls on tests of attention, memory or decision-making, but significantly worse on several spatioperceptual tasks such as the Money's Road Map Test of Directional Sense[74] and the Stylus Maze Test. These results, confirmed in a subsequent larger childhood OCD sample,[75] imply a relatively selective impairment affecting tasks of mental rotation, a performance profile seen in patients with Huntington's disease (HD), and believed to reflect frontal lobe dysfunction.[76] The OCD patients also displayed mild deficits in the generation and production of new ideas and had more difficulty in adjusting to shifts in test concepts and rules.[75] Moreover, the mild impairment on the Stylus Maze Test and Money's Road Map Test was still detected in a subgroup of patients retested 3–6 years later, suggesting a persistent stable deficit.[75] The overall performance pattern led the investigators to speculate both about frontal lobe and temporoparietal dysfunction.[75]

5. ELECTROENCEPHALOGRAPHY

The estimated incidence of mostly nonspecific EEG abnormalities in patient groups with OC features has ranged widely[77–83]—not surprising, in view of the marked methodological differences between studies. In more recent controlled investigations of unmedicated adult OCD patients, subtle group differences related to the left temporal and right parietal areas were reported by one group;[61] another noted that while 11% of patients had some definite abnormality no consistent pattern was evident.[63] In the only EEG study of children with OCD, 20% of subjects were found to have a variety of mild abnormalities.[68] Thus, controlled studies using modern diagnostic criteria imply that the incidence of abnormalities in OCD detected by conventional EEG techniques may be comparable to that reported for other psychiatric disorders.[84]

In both adult[85] and childhood OCD[7] samples, sleep EEG profiles with similarities (decreases in REM latency, deep sleep, and total sleep time) to those of depressed patients have been described. More recently measurement of brain electrical activity has been paired with sensory stimuli in studies of evoked potentials (EP). Adult OCD patients may have different characteristics in late visual EP components, a finding that appears to be accentuated with increasingly complicated visual stimuli.[62,86] In a somatosensory (median nerve shock) paradigm, the EP waveforms of OCD patients again differed from those of control groups.[87] While the visual EP finding in particular is exciting in that it may be related to cognitive processing, more extensive studies controlling for medication status, attention, and affective state need to be done. No EEG abnormality has emerged as specific for OCD.

6. COMPUTED TOMOGRAPHY

In CT, information about the radiodensity of brain tissue is obtained from multiple small dose X-ray exposures taken around a single axis. Using these data, a computer generates an image, based on radiodensity differences, of a slice of brain tissue. The image provides information about ventricular margins, sulcal markings and other features that can be used to make inferences about brain structure.[29] It must be remembered, however, that the image is a two-dimensional reconstruction of a three-dimensional slice of brain tissue. The CT image will collapse information about the entire thickness (about 5 mm) of the slice in one plane, at times making it difficult to distinguish fluid–parenchyma boundaries.[88] This and other factors contribute to the variations in sensitivity between the quantitative techniques that have been applied to CT scans of the brain.[88]

In the only controlled CT study of adult OCD patients, no significant differences were found in the mean lateral ventricular brain ratio (VBR) or several other indices when compared with those of controls.[63] The sample was deliberately skewed toward patients with possible neuropsychological impairment, a bias that in other psychiatric disorders has been associated with a greater degree of brain structural abnormalities.[29] However, in this study of adult OCD patients, comparison scans were chosen from files and included few true normal controls. It has been suggested that employment of other than normal volunteers for controls in CT scan studies may predispose to a type II error, that is, failure to detect a true difference.[29]

The first childhood OCD CT scan study employed a planimetric measure in which the area of lateral ventricles was compared to that of the area of total brain parenchyma within the same brain slice.[68] Although most patients had an individual VBR with normal limits, the mean VBR was increased compared with the control group. On the other hand, in an investigation in which a volumetric measurement technique was used, no significant differences were found in total intracranial volume or volumes of the third and lateral ventricles between healthy volunteers and young men whose OCD had begun before the age of 18; the patients were reported however to have smaller caudate nuclei than those of controls.[89] Volumetric measurement techniques are thought to be more sensitive than planimetric ones,[90] and it is difficult to conceive that a structural brain change present in childhood, suggested by the first study,[68] would not be detectable later in life. Since different patients were used in the two studies, sampling differences may provide a partial explanation. It is still unclear whether OCD is associated with cerebral ventricular enlargement. At best, such a finding would be nonspecific, probably reflecting dysplasia or brain atrophy.[29] The report of reduced caudate volume[89] is exciting but remains to be replicated in larger samples.

7. POSITRON-EMISSION TOMOGRAPHY

In PET, radionuclides are introduced into the brain. The positrons they emit interact with brain tissue to generate γ-rays. Since the positrons travel a relatively short distance in tissue compared with γ-rays, detection of the latter at the skull surface can provide information about the location and concentration of the radionuclide. In combination with CT techniques similar to that used in CT scanning, a three-dimensional representation of the radionuclide's distribution can be developed. The nature of the PET information depends on the biological properties of the particular radionuclide chosen.

Basic work suggests that the functional activity of brain tissue correlates well with the rate of glucose metabolism; the latter is the brain's main source of energy.[91,92] Accordingly positron-emitting isotopes that enter the glucose metabolic pathway have been used in PET studies, with the intention of gaining information about regional brain activity. Although the complexities of, and controversies surrounding, these techniques are beyond the scope of this chapter and are reviewed elsewhere,[93] in theory PET provides an in-vivo measure of brain function.

Recently, fluorodeoxyglucose (FDG) PET scans of 14 adult OCD patients, 9 of whom also met criteria for major depression, were compared with those of 14 unipolar depressed patients and 14 controls who had no DSM-III diagnosis on axis I.[94] The OCD patients were found to have increased metabolic rates in both caudate nuclei as well as in the left orbital gyrus, while the right orbital gyrus showed a trend to an increased metabolic rate in all comparisons. The metabolic rate in the left orbital gyrus relative to that of the ipsilateral hemisphere (orbital gyrus–hemisphere ratio) was significantly elevated in the OCD group compared with both sets of controls and remained high even in those subjects (8 of 10) who responded to drug treatment. Furthermore, while the caudate/ipsilateral hemisphere metabolic ratio was in the normal range for OCD patients before treatment, it increased significantly, bilaterally and uniformly after treatment, but only in those patients who demonstrated clinical improvement.

These early findings are intriguing, particularly since both normal and psychiatric control groups were used, and since a physiologic change correlated well with successful therapeutic intervention. The authors speculated that OCD symptoms might result when increased cortical activity, particularly in the orbital gyri, overwhelms the ability of the caudate nuclei to process the augmented corticostriatal transmission. The relative increase in caudate metabolism during successful pharmacologic treatment may reflect the reassertion by the caudate nuclei of their capacity to process the increased cortical inputs.[94]

8. INTEGRATING THE EVIDENCE

8.1. Convergent Clinical Findings

The paucity of controlled studies and methodological shortcomings confound many of the data that have emerged from both adult and childhood OCD samples. While the cumulative evidence suggests that subtle CNS dysfunction may be present in at least some OCD subgroups, the implications and nature of this dysfunction remain to be defined. Recognizing the latter, it may still be of heuristic value to draw freely from studies of both children and adults with OCD and speculate on a possible link between the findings.

The increased incidence of choreiform movements reported in one childhood OCD study[69] is compatible with a process that affects basal ganglia function.[52] The neuropsycholo-

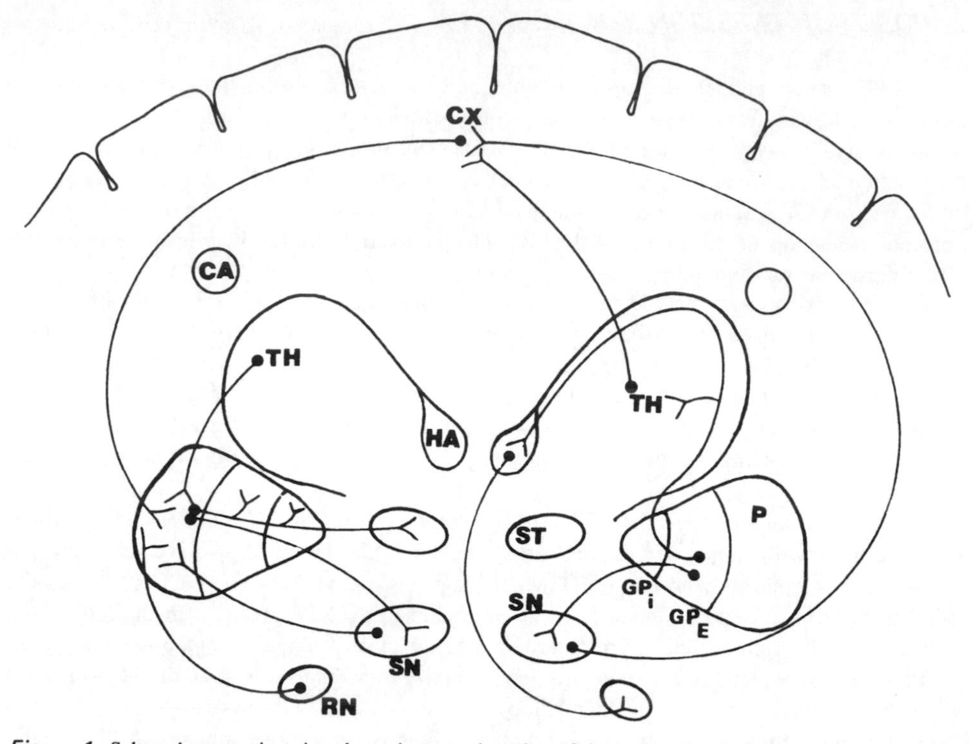

Figure 1. Selected connections in schematic coronal section of the brain. CA, caudate nucleus, CX, cortex; GP_e, external globus pallidus; GP_i, internal globus pallidus; HA, habenular nucleus; P, putamen; RN, raphe nuclei; SN, substantia nigra; SIM, subthalamic nucleus; TH, thalamus.

gical performance profile seen in two samples of children and adolescents with OCD[68,75] has been associated with frontal lobe pathology.[76] In one CT scan study, caudate volume in OCD patients was reduced.[94] It is intriguing, then, that demonstrated differences between patients with OCD and controls in a PET study were limited to the orbital lobes and caudate nuclei.[94] Although the adduced evidence is preliminary, highly selective and based on relatively small samples, it is certainly compatible with the recently proposed hypothesis[94] that regions of the basal ganglia (BG) and frontal cortex (FC) may be involved in mediation of obsessions and compulsions. The latter is also consistent with findings from basic science.

8.2. Anatomy of Implicated Areas

While a thorough discussion of basal ganglia anatomy is beyond the scope of this chapter,[95–97] some generalizations can be made. According to most definitions, the BG includes at least the corpus striatum, the subthalamic nucleus of Luys, and the substantia nigra. The corpus striatum is the largest structure of the BG, consisting of the caudate and lenticular or lentiform nuclei. The latter, in turn, consists of the putamen and globus pallidus (GP). Collectively, the caudate and putamen are often referred to as the striatum, while the globus pallidus, which is subdivided into external (GPe) and internal segments (GPi), is also called the pallidum (Fig. 1). The striatum is the main entrance portal for BG inputs.[95] While the latter come from a variety of brain areas, including much of the limbic system, the most numerous originate in one of three regions: cerebral cortex, intralaminar thalamic nuclei, or substantia nigra.[96] Neocortical projections are the best defined; they are organized topographically but overlap their target areas considerably, so that few if any regions of striatum receive afferents from a single cortical area.[98]

Most efferent fibers from the striatum pass through or synapse in the pallidum. Fibers from the latter project to SN, to the subthalamic region, to the habenula, and to those regions of the thalamus that in turn send afferents to motor and neocortex. The latter is one of several paths by which the basal ganglia can influence the cortex and perhaps modulate corticostriatal input.[96] In particular, a circuit including the orbital frontal cortex and the caudate nucleus has been described.[99] The striatonigral and nigrocortical pathways may form another circuit by which the BG may communicate with cortex. The pallidohabenular pathway is noteworthy because the habenula projects to the median and dorsal raphe nuclei,[100] the major source of serotonergic fibers to striatum.

Neurochemical studies of the BG reveal a degree of organization sufficient to permit fine modulation and processing of inputs and outputs.[101] Identified neurotransmitters include dopamine, serotonin, acetylcholine, glutamate, GABA, substance P, and a host of neuropeptides. In the rat striatum, a mosaic pattern has been described of histochemically distinct islands or striosomes (300–600 μm wide) embedded in a matrix.[102] Similar patterns have been seen in primate and human tissue.[101] This differentiation has been linked to compartmentalization of corticostriatal input and nigrostriatal output systems.[103] For instance, in the dorsolateral striatum, striosomes are innervated by the prefrontal cortex, while sensory and motor cortical areas preferentially project to the matrix. In turn, the striosomes and matrix project to distinct areas of SN. While striosomes and matrix appear to be functionally distinct compartments, intrinsic striatal somatostatin-reactive neurons have been identified that may provide a possible link between them.[103] From both a neurochemical and an anatomic perspective, the basal ganglia appear to be ideally organized to integrate motor output with cortically mediated emotional or cognitive information.

8.3. Physiology of Implicated Areas

Hypotheses about the function of the basal ganglia and connected areas of frontal cortex in humans are largely inferential, based heavily on animal studies. Mettler[104] postulated that the basal ganglia mediate the shifting of attention, that is, allow us to move from one input to another. Accordingly, basal ganglia dysfunction could be manifested as an impaired ability to screen sensory and other inputs, leading to a host of cognitive symptoms including perseveration. It was later suggested that orbital cortex along with the ventral caudate nucleus modulated the probability that an animal's behavior would change in response to environmental alteration.[105] Subsequent studies demonstrated that acaudate cats indeed displayed perseverative behavior addition to other abnormalities.[106]

Cools[107] theorized that the ability to modify previously established behavioral programs, thought to contribute to the wider range of behavioral responses seen in higher animals, is mediated in the neostriatum. Dysfunction of the neostriatum would limit an organism's repertoire of behavioral sets. According to Schneider,[108] the basal ganglia provided a gating mechanism for those sensory inputs that normally triggered behavioral responses. Recently, on the basis of primate studies, Mishkin and co-workers suggested that the basal ganglia participate in one of two coexisting mammalian memory systems.[109,110] In the first, neocortical activation of a limbodiencephalic circuit leads to storage of cognitive, declarative, or associative information. In the second, activation by the frontal cortex of a circuit involving the striatum and other basal ganglia structures stores a noncognitive stimulus response bond, i.e., a habit. In other words, the stimulus–response bond of habit may result when cortically processed sensory inputs become associated with motor inputs generated in the pallidum.[109,110] Dysfunction of the corticostriatal axis could result in decoupling of the original stimulus–response bond and could lead to a behavior outside its original reinforcement contingency.

Of particular interest is the lateral orbitofrontal cortex, which projects to the ventromedial sector of the caudate nucleus; the latter also receives input from auditory and visual association areas of the superior and inferior temporal gyri, respectively.[99] The ventromedial caudate innervates a segment of the GPi and a portion of SN. Projections from the latter reach the mediodorsal and other thalamic nuclei; the circuit is completed by thalamic afferents to lateral orbitofrontal cortex. In primates, bilateral lesions restricted either to the lateral orbitofrontal area or to its projection area in caudate, resulting in perseverative interference with an animal's capacity to make appropriate switches in behavioral set.[111,112] Clinical evidence suggests that the orbital gyri facilitate orderly information processing by inhibiting interfering stimuli.[113] It is most intriguing that preliminary PET studies in adult patients with OCD show metabolic abnormalities in the caudate nucleus and orbitofrontal cortex.[94] It is also worth noting that serotonin, the neurotransmitter most commonly implicated in OCD,[114] is abundant both in the prefrontal cortex[115] and in the caudate nucleus.[101] Experimental depletions of serotonin have been associated both with perseverative behaviors and with impairments in habituation.[116–118]

Considering the variety of animal species and approaches employed, the results have been remarkably consistent. The basal ganglia in conjunction with certain areas of the frontal cortex appear to be critically involved in establishing, facilitating the expression of, and modifying behavioral responses appropriate to, a given context. It is reasonable to speculate whether this system may not also serve as a processor of internally generated neural inputs such as thoughts or emotions, which may be linked to behavioral responses. If dysfunction of connecting areas of the basal ganglia and frontal cortex impairs the ability of subhuman primates to make switches in behavioral sets, the enlargement of the neocortex in humans, its rich connections with the basal ganglia, and the compartmentalization of the latter's inputs, may permit the system to influence processing of the nonmotor or psychic concomitants of these sets as well. Defining the frontal cortical–basal ganglia circuit (FC–BG) to include the orbitofrontal cortex, the caudate nucleus, the brainstem nuclei, and those areas that mediate the connections between them,[99] it is plausible that dysfunction of this circuit could give rise to perseverative behaviors (compulsions) as well as to perseverative thoughts (obsessions).

8.4. Speculation about a Lesion in OCD

The hypothesis that cerebral pathophysiology may underlie OC syndromes was first formulated by von Economo[11] and has since been entertained by many investigators. Proposed regions of dysfunction have included the dominant frontal and cingulate cortices,[61] the temporal and cingulate cortices,[119] as well as the hippocampus.[120] There is relatively little direct clinical evidence implicating these areas. Much of the speculation about the role of the cingulate cortex is based on psychosurgical interventions in OCD,[61,119] and it must be remembered that amelioration after psychosurgery does not necessarily imply antecedent cerebral pathophysiology of the ablated area. Furthermore, since claims of efficacy have been made for surgical lesions of the orbital frontal cortex, anterior cingulum and anterior limb of the internal capsule,[121] it could be argued that interruption of certain frontal cortical connections rather than destruction of the cingulum may be critical for a successful surgical outcome.

The hypothesis that OCD may be associated with dysfunction in regions of the FC and BG in some patient samples is compatible with several preliminary clinical and basic studies. It would be unwarranted and misleading, however, to conclude that a pathognomonic cerebral lesion accompanied by characteristic neuropsychiatric findings invariably underlies the illness. OCD is a syndrome, a constellation of signs and symptoms that "run together." Associations of OCD variously with birth complications,[122] cerebral insults,[11–26] and other neuropsychiatric conditions[30–44] suggest that several etiological processes may lead to a similar phenomenological end state. Whether the similar phenomenology is based in common pathophysiology must be determined empirically. Accordingly, the neurological substrate and accompanying findings in OCD may vary as a result of patient selection. Results from small research studies may not hold for larger cohorts either because the latter represent patients from the less severe end of a continuum, or because the former overrepresent one of several etiologic processes that give rise to OCD.

The classic concept of a neurological lesion, i.e., of a circumscribed cerebral area whose dysfunction entirely mediates clinical presentation, may not be applicable to syndromes such as OCD. Even if it is confirmed that symptomatology in some patient samples is mediated by the FC–BG circuit, it must be remembered that the latter is not a discrete brain region, but a functional unit that includes several brain areas. It is certainly premature to conclude that pathophysiology confined to the orbital cortex, the caudate, the brainstem monoaminergic nuclei, or any other single component of the FC–BG circuit is sufficient to give rise to OC symptoms. It may be that different patterns of dysfunction, involving one or many components, can result in an OCD syndrome. Alternatively, given the complexity of brain organization, pathophysiology may neither be confined to, nor be most conspicuous or primary in, the FC–BG circuit, even in patients with demonstrable dysfunction in those areas. Primary pathophysiology could occur elsewhere yet still affect the function of the FC–BG circuit; the latter might be merely a final common pathway for mediation of OC symptoms. Unitarian models of etiology or pathophysiology should be evaluated with these reservations in mind.

9. FUTURE DIRECTIONS

Obsessive-compulsive disease remains a poorly understood condition with a probable neurobiologic substrate. Its etiology is not known and, while selecting samples of patients with common clinical and/or laboratory indices is advisable for research studies, it must be recognized that clinical uniformity does not guarantee biological homogeneity. Which selection criteria will prove most useful remains to be determined empirically. The use of appropriate normal and psychiatric controls could assist in distinguishing primary features of the illness from secondary features, such as anxiety or depression. Enlisting Tourette's syndrome or Huntington's disease patients in studies could further clarify which findings if any are specific for OCD. It would be of particular value to compare age matched childhood onset and adult onset OCD groups. Similarly, prospective studies of childhood-onset OCD might clarify the relationship between the latter and adult-onset OCD.

Larger-scale specialized neurodevelopmental examinations of OCD samples are certainly indicated. For defining possible structural brain changes, magnetic resonance imaging (MRI) offers a higher degree of resolution and more versatility in selecting the planes of brain slices than does CT. It may be particularly suited to volumetric and densitometric assessments. Neuropsychological evaluation using tests such as the Money Road Map and the Stylus Maze is promising, especially if combined with simultaneous neurophysiological measures. Selective activation of brain regions of interest by neuropsychological tasks has already proved useful in PET[123] and regional cerebral blood flow (rCBF)[124,125] studies of other neuropsychiatric disorders. Some functional imaging studies can be done longitudinally and combined with pharmacological maneuvers. The practicality of such investigative modalities in children may be limited, however, by the radiation exposure involved.

The establishment of a national brain bank would permit neuropathological evaluation of tissue from OCD patients. The complexity of implicated areas and the subtlety of some clinical findings suggest, however, that unless very sensitive techniques are employed, pathology may be overlooked. Finally, etiological studies might consider those processes, whether infectious, traumatic, or genetic, that can affect the development and function of brain regions of interest.

In all these approaches, a few basic facts must be remembered. OCD is a syndrome whose biological homogeneity or heterogeneity is not known. There is considerable evidence that in some patient samples, OCD is associated with subtle nonspecific indices of cerebral dysfunction. Preliminary studies suggest that subgroups of the latter may have demonstrable pathophysiology in a circuit that includes regions of the frontal lobes and components of the basal ganglia. However, the implications of these research findings from small patient samples for the OCD population at large remain to be determined. Most patients with OCD do not have a history of significant cerebral insult, nor do they manifest obvious signs of neurological or neuropsychological impairment. Moreover, demonstration of a pathophysiological substrate does not imply that psychosocial and other factors cannot contribute to the emergence, features, or outcome of a disease process. Provided these facts are not forgotten, the neuroscientific endeavor to define the pathophysiology in OCD promises to be an exciting one.

REFERENCES

1. Tourette de la G: Etude sur une affection nerveuse caracterisee par l'incoordination motrice accompagnee d'echolalie et de coprolalie. Arch Neurol 9:19–42, 1885
2. American Psychiatric Association: Diagnostic and Statistical Manual of Psychiatric Disorders, ed 3, revised. Washington, D.C., APA Press, 1987
3. Robins LN, Helzer JE, Weissman MM, et al: Lifetime prevalence of specific psychiatric disorders in three sites. Arch Gen Psychiatry 41:949–958, 1984
4. Beech HR: Obsessional States. London, Methuen, 1974
5. Hollingsworth C, Tanguey P, Grossman L, et al: Long-term outcome of obsessive compulsive disorder in children. J Am Acad Child Psychiatry 19:134–144, 1980
6. Zeitlin H: The Natural History of Psychiatric Disorder in Children. New York, Oxford University Press, 1986
7. Rapoport J, Elkins R, Langer D: Childhood obsessive-compulsive disorder. Am J Psychiatry 138:1545–1554, 1981
8. Flament M, Rapoport JL: Childhood obsessive-compulsive disorder, in Insel TR (ed), New Findings in Obsessive-Compulsive Disorder. Washington, D.C., APA Press, 1984, pp 24–43
9. Swedo SE, Rapoport, JL, Leonard, H, et al: Obsessive-compulsive disorder in children and adolescents. Arch Gen Psychiatry 46:335–341, 1989
10. Warren W: Some relationships between the psychiatry of children and adults. J Ment Sci 106:815–826, 1960
11. Economo CF von: Encephalitis Lethargica, Its Sequelae and Treatment. London, Oxford University Press, 1931
12. Meyer-Gross W, Steiner G: Encephalitis lethargica in der Selbstbeobachtung. Z Ges Neurol Psychiatry 73:287–289, 1921
13. Jelliffe SE: Psychopathology of forced movements in oculogyric crises. Nerv Ment Dis 55:1–219, 1932
14. Schilder P: The organic background of obsessions and compulsions. Am. J. Psychiatry 94:1397–1413, 1938
15. Wohlfart G, Ingvar DH, Hellberg A-M: Compulsory shouting (Benedek's "klazomania") associated with oculogyric spasms in chronic epidemic encephalitis. Acta Psychiatr Scand 36:369–377, 1961
16. Hillbom E: After-effects of brain injuries. Acta Psych Neurol Scand 35 (suppl 142):82–106, 1960
17. Lishman WA: Brain damage in relation to psychiatric disability after head injury. Br J Psychiatry 114:373–410, 1968
18. Grimshaw L: Obsessional disorder and neurological illness. J Neurol Neurosurg Psychiatry 27:229–231, 1964
19. Swedo S, Rapoport JL, Cheslow D, et al: High prevalence of obsessive-compulsive symptoms in patients with Sydenham's chorea. Am J Psychiatry 146:246–249, 1989
20. Lewis A: Problems of obsessional illness. Proc R Soc Med 29:325–336, 1935

21. Barton R: Diabetes insipidus and obsessional syndrome. Lancet 1:133–135, 1965
22. Hill D, Mitchell W: Epileptic amnesia. Folia Psychiatr 56:718–722, 1953
23. Minski L: The mental symptoms associated with 58 cases of cerebral tumor. J Neurol Psychopathology 13:330–343, 1933
24. Brickner RM, Rosner A, Munro R: Physiological aspects of the obsessive state. Psychosomat Med 11:369–383, 1940
25. McKeon J, McGuffin P, Robinson P: Obsessive-compulsive neurosis following head injury. Br J Psychiatry 144:190–192, 1984
26. Laplane D, Baulac M, Widlocher D, et al: Pure psychic akinesia with bilateral lesions of basal ganglia. J. Neurol Neurosurg Psychiatry 47:377–385, 1984
27. Bear DM, Fedio P: Quantitative analysis of interictal behavior in Temporal lobe epilepsy. Arch Neurol 34:454–467, 1977
28. Cummings JL, Gosenfeld LF, Houlihan JP, et al: Neuropsychiatric disturbances associated with idiopathic calcification of the basal ganglia. Biol Psychiatry 18:591–601, 1983
29. Jaskiw GE, Andreasen NC, Weinberger DR: X-ray computed tomography and magnetic resonance imaging in psychiatry, in Hales RE, Frances AJ (eds), APA Annual Review, Vol 6. Washington, D.C., APA Press, 1987, pp 260–299
30. Kelman DH: Gilles de la Tourette's disease in children: A review of the literature. J Child Psychol Psychiatry 6:219–226, 1965
31. Fernando SJM: Gilles de la Tourette's syndrome. Br J Psychiatry 113:607–617, 1967
32. Morphew JA, Sim M: Gilles de la Tourette's syndrome. A clinical and psychopathological study. Br J Med Psychol 42:293–301, 1969
33. Nee LE, Caine ED, Polinsky RJ et al: Gilles de la Tourette syndrome: Clinical and family study of 50 cases. Ann Neurol 7:41–49, 1980
34. Jaegger J, Prusoff BA, Cohen DJ et al: The epidemiology of Tourette's syndrome: A pilot study. Schizophr Bull 8:267–278, 1982
35. Montgomery MA, Clayton PJ, Freidhoff AJ: Psychiatric illness in Tourette syndrome patients and first-degree relatives, in Friedhoff AJ, Chase TN (eds), Gilles de la Tourette Syndrome. New York, Raven, 1982, pp 335–339
36. Stefl ME: Mental health needs associated with Tourette syndrome. Am J Public Health 74:1310–1313, 1984
37. Nee LE, Polinsky RJ, Ebert MH: Tourette syndrome: clinical and family studies, in Friedhoff AJ, Chase TN (eds), Gilles de la Tourette Syndrome. New York, Raven, 1982, pp 291–295
38. Yaryura-Tobias JA, Neziroglu F, Howard S, et al: Clinical aspects of Gilles de la Tourette syndrome. Orthomol Psychiatry 10:263–268, 1981
39. Comings DE, Comings BG: Tourette syndrome: Clinical, psychological and epidemiological aspects of 250 cases. Am J Hum Genet 3:126–158, 1985
40. Frankel M, Cummings JL, Robertson MM, et al: Obsesions and compulsions in Gilles de la Tourette's syndrome. Neurology (NY) 36:378–382, 1986
41. Pitman RK, Green RC, Jenike MA, et al: Clinical comparison of Tourette's disorder and obsessive-compulsive disorder. Am J Psychiatry 144;1166–1171, 1987
42. Cummings JL, Frankel M: Gilles de la Tourette syndrome and the neurological basis of obsessions and compulsions. Biol Psychiatry 20:1117–1126, 1985
43. Pauls DL, Towbin KE, Leckman JF, et al: Gilles de la Tourette's syndrome and obsessive-compulsive disorder. Arch Gen Psychiatry 43:1180–1182, 1986
44. Comings DE, Comings BG: Hereditary agoraphobia and obsessive-compulsive behavior in relatives of patients with Gilles de la Tourette's Syndrome. Br J Psychiatry 151:1956–199, 1987
45. Walsh PTF: Compulsive shouting in Gilles de la Tourette's disease. Br J Clin Pract 16:651–655, 1962
46. Wassman ER, Eldrgidge R, Abussehab S Sr, et al: Gilles de la Tourette syndrome: Clinical and genetic studies in a midwestern city. Neurology (NY) 28:304–307, 1978
47. Pauls DL, Leckman JF: The inheritance of Gilles de la Tourette's syndrome and associated behaviors. N Engl J Med 315:993–996, 1986
48. Devinsky O: Neuroanatomy of Gilles de la Tourette's Syndrome. Possible midbrain involvement. Arch Neurol 40:508–514, 1983
49. Caine ED: Gilles de la Tourette's Syndrome. Arch Neurol 42:393–397, 1985
50. Moore RY, Bloom FE: Central catecholamine neuron systems: Anatomy and physiology of the dopamine systems. Annu Rev Neurosci 1:129–134, 1978

51. Bacoupolos NJ, Bhatnagar RK: Correlation between tyrosine hydroxylase activity and catecholamine concentration or turnover in brain regions. J Neurochem 29:639–643, 1977
52. Marsden CD: Basal ganglia disease. Lancet 2:1141–1146, 1982
53. Obeso JA, Rothwell JC, Marsden CD: Simple tics in Gille de la Tourette's syndrome are not prefaced by a normal premovement EEG potential. J Neurol Neurosurg Psychiatry 44:735–738, 1981
54. Chase TN, Foster NL, Fedio P, et al: Gilles de la Tourette syndrome: Studies with the fluorine-18-labeled fluorodeoxyglucose positron emission method. Ann Neurol 15(suppl):S175, 1984
55. Lenane M, Swedo S, Leonard H, et al: Obsessive-compulsive disorder in first degree relatives of obsessive-compulsive disordered children. Abstracts, 137th meeting of the American Psychiatric Association, Montreal, May 1988
56. Caine ED, Polinsky RJ, Ebert MH, et al: Trial of chlorimipramine and desipramine for Gilles de la Tourette syndrome. Ann Neurol 5:305–306, 1979
57. Thoren P, Asberg M, Cronholm B, et al: Clomipramine treatment of obsessive-compulsive disorder. I. A controlled clinical trial. Arch Gen Psychiatry 37:1281–1285, 1980
58. Van Woert MH, Rosenbaum D, Enna SJ: Overview of pharmacological approaches to therapy for Tourette syndrome, in Friedhoff AJ, Chase TN (eds), Gilles de la Tourette syndrome. New York, Raven, 1982, pp 369–375
59. Insel TR, Murphy DL, Cohen RM, et al: Obsessive-compulsive disorder: A double-blind trial of clomipramine and clorgyline. Arch Gen Psychiatry 40:605–612, 1983
60. Flament MF, Rapoport JL, Berg CJ, et al: Clomipramine treatment of childhood obsessive-compulsive disorder. A double-blind controlled study. Arch Gen Psychiatry 42:977–983, 1985
61. Flor-Henry P, Yendall LT, Koles ZJ, et al: Neuropsychological and power spectral EEG investigations of the obsessive-compulsive syndrome. Biol Psychiatry 14:119–130, 1979
62. Cieselski KT, Beech HR, Gordon PK: Some electrophysiological observations in obsessional states. Br J Psychiatry 138:479–484, 1981
63. Insel T, Donnelly E, Lalakea L, et al: Neurological and neuropsychological studies of patients with obsessive-compulsive disorder. Biol Psychiatry 18:741–751, 1983
64. Shaffer D, O'Connor PA, Shafer SQ: Neurological soft signs: Their origins and significance for behaviour, in Rutter M (ed), Developmental Neuropsychiatry. New York, Guilford 1983, pp 145–163
65. Hollander E, Schiffman E, Liebowitz MR: Neurologic soft signs in obsessive-compulsive disorders. Abstracts, 140th Annual Meeting of the American Psychiatric Association, Chicago, May 9–14, 1987
66. Denckla MB: Development of speed in repetitive and successive finger movements in normal children. Dev Med Child Neurol 15:635–645, 1973
67. Denckla MB: Development of coordination in normal children. Dev Med Child Neurol 16:729–741, 1974
68. Behar, D, Rapoport JL, Berg CJ, et al: Computerized tomography and neuropsychological test measures in adolescents with obsessive-compulsive disorder. Am J Psychiatry 141:363–369, 1984
69. Koby E, Rapoport JL, Flament M, et al: Basal ganglia pathology in childhood and adolescent obsessive compulsive disorder. Abstracts, 33rd Annual Meeting, Am Acad Child Psychiatry, Los Angeles, 1986
70. Denckla MB: Revised neurological examination for subtle signs. Psychopharm Bull 21:773–789, 1985
71. Denckla MB: Minimal brain dysfunction, in Chall J, Mirsky AF (eds), Education and the Brain. Chicago, University of Chicago Press, 1978, pp 223–268
72. Touwen BL: Examination of the Child with Minor Neurological Dysfunction. London, William Heinemann, 1979
73. Beech HR, Liddell A: Decision-making, mood states and ritualistic behavior among obsessional patients, in Beech HR (ed), Obsessional States. London, Methuen, 1974, pp 143–160
74. Money J, Alexander D, Walker HT: A Standardized Road-Map Test of Direction Sense. Baltimore, Johns Hopkins Press, 1965
75. Cox C, Brouwers P, Berg C, et al: Neuropsychological correlates of obsessive compulsive disorder in adolescents. Abstracts, International Neuropsychological Society, 1986
76. Brouwers P, Cox C, Martin A, Chase T, et al: Differential perceptual-spatial impairment in Huntington's and Alzheimer's dementias. Arch Neurol 41:1073–1076, 1984
77. Pacella BL, Polantin P, Nagler SH: Clinical and EEG studies in obsessive-compulsive states. Am J Psychiatry 100:830–838, 1944
78. Rockwell FV, Simons DJ: The electroencephalogram and personality organization in the obsessive-compulsive reaction. Arch Neurol Psychiatry 57:71–77, 1947
79. Ingram IM, McAdam WA: The electroencephalogram, obsessional illness and obsessional personality. J Ment Sci 106:686–691, 1960

80. Inouye R: Electroencephalographic study in obsessive-compulsive states. Clin Psychiatry 15:1071–1083, 1973
81. Sugiyama T: Clinico-electroencephalographic study on obsessive-compulsive neurosis. Bull Osaka Med School 20:95–114, 1974
82. Bingley T, Persson A: EEG studies on patients with chronic obsessive-compulsive neurosis before and after psychosurgery. Electroencephalogr Clin Neurophysiol 44:691–696, 1978
83. Jenike MA, Brotman AW: The EEG in obsessive-compulsive disorder. J Clin Psychiatry 45:122–124, 1984
84. Solomon, S: Neurological Evaluation, in Kaplan HI, Freedman AM, Sadock BJ (eds), Comprehensive Testbook of Psychiatry, ed 3. Baltimore, Williams & Wilkins, 1980, pp 245–273
85. Insel TR, Gillin JC, Moore A, et al: The sleep of patients with obsessive compulsive disorder. Arch Gen Psychiatry 39:1372–1377, 1982
86. Beech HR, Ciesielski KT, Gordon PK: Further observations of evoked potentials in obsessional patients. Br J Psychiatry 142:605–609, 1983
87. Shagass C, Roemer RA, Straumanis JJ: Distinctive somatosensory evoked potential features in obsessive-compulsive disorder. Biol Psychiatry 19:1507–1524, 1984
88. Shelton RC, Weinberger DR (1986) X-ray computerized tomography studies in schizophrenia: A review and synthesis, in Nasrallah H, Weinberger DR (eds), The Neurology of Schizophrenia. New York, Elsevier, 1986, pp 207–250
89. Luxenberg JS, Swedo SE, Flament MF, et al: Neuroanatomical abnormalities in obsessive-compulsive disorder detected with quantitative X-ray computed tomography. Am J Psychiatry 145:1089–1093, 1988
90. Reveley MA: CT scans in schizophrenia. Br J Psychiatry 146:367–371, 1985
91. Sokoloff L: Relationships among local functional activity, energy metabolism and blood flow in the central nervous system. Fed Proc 40:2311–2316, 1981
92. Sokoloff L: The radioactive deoxyglucose method, in Agranoff BW, Aprison MH (eds), Advances in Neurochemistry, Vol 4. New York, Plenum, 1982, pp 1–82
93. Andreasen NC (ed): Brain Imaging for Psychiatrists. Washington, D.C., APA Press, 1988
94. Baxter LR Jr, Phelps MR, Mazziotta JC, et al: Local cerebral glucose metabolic rates in obsessive-compulsive disorder. Arch Gen Psychiatry 44:211–218, 1987
95. Carpenter MB, Sutin J: Human Neuroanatomy. Baltimore, Waverly Press, 1983
96. Nauta WJH, Domesick VB: Afferent and efferent relationships of the basal ganglia, in Evered, D and O'Connor, M. (eds) Functions of the Basal Ganglia (CIBA Foundation Symposium). London, Pitman, 1984, pp 3–22
97. Pycock CJ, Phillipson OT: A neuroanatomical and neuropharmacological analysis of basal ganglia output, in Iversen LL, Iversen SD, Snyder SH (eds), Handbook of Pharmacopsychiatry, Vol. 18. New York, Plenum, 1985, pp 191–278
98. Kemp JM, Powell TPS: The corticostriate projection in the monkey. Brain 93:525–546, 1970
99. Alexander GE, DeLong MR, Strick PL: Parallel organization of functionally segregated circuits linking basal ganglia and cortex. Annu Rev Neurosci 9:357–381, 1986
100. Herkenham M, Nauta WJH: Efferent connections of the habnular nuclei in the rat. J Comp Neurol 187:19–48, 1979
101. Graybiel, AM: Neurochemically specified subsystems in the basal ganglia, in Evered, D and O'Connor, M (eds) Functions of the Basal Ganglia (Ciba Foundation Symposium). London, Pitman, 1984, pp 114–143
102. Graybiel AM, Ragsdale CW Jr: Histochemically distinct compartments in the striatum of human, monkey and cat demonstrated by acetylcholinesterase staining. Proc Natl Acad Sci USA 75:5723–5726, 1978
103. Gerfen CR: The neostriatal mosaic: Compartmentalization of corticostriatal input and striatonigral output. Nature (Lond) 311:461–464, 1984
104. Mettler FA: Perceptual capacity, functions of the corpus striatum and schizophrenia. Psychiatr Q 29:89, 1955
105. Rosvold HE: The prefrontal cortex and caudate nucleus: a system for effecting correction in response mechanisms, in Rupp C (ed), Mind as a Tissue. New York, Harper & Row, 1968, pp 21–38
106. Villablanca J, Marcus RJ, Olmstead CE: Effects of caudate nuclei and frontal cortical ablation in cats. I. Neurology and gross behavior. Exp Neurol 52:389–420, 1976
107. Cools AR: Role of the neostriatal dopaminergic activity in sequencing and selecting behavioral strategies: Facilitation of processes involved in selecting the best strategy in a stressful situation. Behav Brain Res 1:361–378, 1980

108. Schneider JS: Basal ganglia role in behavior: Importance of sensory gating and its relevance to psychiatry. Biol Psychiatry 19:1693–1710, 1984
109. Mishkin M, Petri HL: Memories and habits: Some implications for the analysis of learning and retention, in Butters N, Squires TL (eds), Neuropsychology of Memory. New York, Guilford, 1985, pp 287–291
110. Mishkin M, Malamut B, Bachevalier J: Memories and Habits: two neural systems, in Lynch G, McGaugh JL, Weinberger NM (eds), Neurobiology of Human Learning and Memory. New York, Guilford, pp 65–77
111. Divac I, Rosvold HE, Szwarcbart MK: Behavioral effects of selective ablation of the caudate nucleus. J Comp Physiol Psychol 63:184–190, 1967
112. Mishkin M, Manning FJ: Nonspatial memory after selective prefrontal lesions in monkeys. Brain Res 143:313–323, 1978
113. Stuss DT, Kaplan EF, Benson DF, et al: Evidence for the involvement of the orbitofrontal cortex in memory functions: An interference effect: J Comp Psychol 96:913–925, 1982
114. Zohar J, Mueller EA, Insel TR, et al: Serotonin responsivity in obsessive-compulsive disorder: Comparison of patients and healthy controls. Arch Gen Psychiatry 44:946–951, 1987
115. Lindvall O, Bjorklund A: General organization of cortical monoamine systems, in Descarries L, Reader TR, Jasper HH (eds), Monoamine Intervention of Cerebral Cortex. New York, A.R. Liss, 1984, pp 9–40
116. Geyer MA, Puerto DB, Menkes DB, et al: Behavioral studies following lesions of mesolimbic and mesostriatal serotonergic pathways. Brain Res 106:257–270, 1976
117. Wirtshafter D, Asin KE, Kent EW: An analysis of open field hyperactivity following electrolytic median raphe lesions in the rat. Soc Neurosci 6:422, 1980
118. Gately PE, Poon SL, Segal DS, et al: Depletion of brain serotonin by 5,7-dyhydroxytryptamine alters the response to amphetamine and the habituation of locomotor activity in rats. Psychopharmacology 87:400–405, 1985
119. Jenike MA: Obsessive-compulsive disorder: A question of a neurologic lesion. Comp Psychiatry 25:298–304, 1984
120. Pitman RK: Neurological etiology of obsessive-compulsive disorders? Am J Psychiatry 139:139–140, 1982
121. Yaryura-Tobias JA, Neziroglu FA: Obsessive-Compulsive Disorders. New York, Marcel Dekker, 1983
122. Capstick N, Seldrup J: Obsessional states: A study in the relationship between abnormalities occurring at the time of birth and the subsequent development of obsessional symptoms. Acta Psychiatr Scand 56:427–431, 1977
123. Cohen RM, Semple WE, Gross M, et al: Dysfunction in a prefrontal substrate of sustained attention in schizophrenia. Life Sci 40:2031–2039, 1987
124. Berman KF: Cortical "stress tests" in schizophrenia: Regional cerebral blood flow studies. Biol Psychiatry 22:1304–1326, 1987
125. Weinberger DR, Berman KF, Zec RF: Physiological dysfunction of the dorsolateral prefrontal cortex in schizophrenia. 1. Regional cerebral blood flow evidence. Arch Gen Psychiatry 43:114–124, 1986

16

Neurochemical and Neuroendocrine Considerations of Obsessive-Compulsive Disorders in Childhood

Susan E. Swedo and Judith L. Rapoport

1. INTRODUCTION

Obsessive-compulsive disorder (OCD) is a common condition in children and adolescents, affecting, at a minimum, 1 in 200 teenagers.[1] The affected individuals are plagued by intrusive obsessive thoughts, and their lives are disrupted by irresistible compulsions. The clinical presentation of childhood-onset OCD is similar to that of adult-onset OCD, and more than one third of adults suffering from OCD report onset of their symptoms during childhood or adolescence.[2] The pre- and post-treatment biochemical profiles and response to pharmacotherapy, particularly treatment with clomipramine, are also similar between adults and children,[3] suggesting that information gained from adult studies can be extrapolated to apply to childhood-onset OCD. This is of particular importance because of the limitations on research investigations in patients under 18 years of age. This chapter presents relevant information from both patient groups, adults and children, in order to examine the neuropathology of childhood OCD more fully.

Several etiologic theories, currently under investigation in OCD, are reviewed in this chapter. These include serotonin dysfunction, as suggested by the selective efficacy of clomipramine, a tricyclic antidepressant (TCA) with particular efficacy in blocking serotonin reuptake, and by challenge tests. Dysfunction of other neurotransmitters, including dopamine, is considered. The compelling evidence for the involvement of hormones, particularly the androgens, in the etiology of OCD is summarized. Finally, a hypothetical model of basal ganglia dysfunction, which incorporates the most widely accepted of these theories, is presented and defended.

Susan E. Swedo and Judith L. Rapoport • Child Psychiatry Branch, National Institute of Mental Health, National Institutes of Health, Bethesda, Maryland 20892.

2. SEROTONIN

Evidence for the serotonergic hypothesis of OCD comes from two sources: the efficacy of drugs with specific serotonergic activity, and challenge tests with metacholorophenylpiperazine (mCPP). The most widely studied of the antiobsessional drugs is chlomipramine hydrochloride (CMI), a relatively selective and potent inhibitor of active serotonin uptake in the brain (It is also a very potent blocker of histamine H_2-receptors. In addition to having significant potency in blocking cholinergic and α-receptors, it has weak antidopaminergic properties.) Its metabolite, desmethylchlomipramine, also effectively blocks serotinin reuptake.[4]

Nine controlled studies of clomipramine have been conducted in adult OCD patients[5–13] and two in children and adolescents.[3,14] In all studies, clomipramine was significantly more effective than placebo. In addition, CMI was superior to clorgyline,[5] imipramine,[6] desipramine, and zimelidine,[7] and in children and adolescents again to desipramine.[10] Although the differences were not statistically significant in two other studies, CMI tended to be more efficacious than nortriptyline[9] and amitriptyline.[12] The antiobsessional effect of the drug is independent of antidepressant effect.

Three other selective serotonin reuptake blockers, fluvoxamine, zimelidine, and fluoxetine, have been reported to be effective in the treatment of OCD. Fluvoxamine, a unicyclic antidepressant with little or no effect on the noradrenergic system, has been shown to be superior to placebo for OCD.[15,16] Zimelidine, withdrawn from the market because of side effects, was superior to placebo in one study,[17] but not in another.[7] Fluoxetine, a bicyclic antidepressant, is currently available in the United States. Although not extensively studied in OCD, fluoxetine seemed effective for OCD in an open trial[18] and in a single-blind trial by Turner et al.[19]

In a study of children and adolescents by Flament et al.,[20] response to clomipramine correlated with pretreatment platelet serotonin concentration. A high pretreatment level of serotonin was a strong predictor of clinical response. However, in this sample of 29 OCD children, there were no differences from age- and sex-matched controls on measures of platelet serotonin, monoamine oxidase (MAO) activity, or plasma epinephrine and norepinephrine concentrations at rest and after a standard orthostatis challenge procedure. In this sample, platelet serotonin concentrations were lower in the more severely ill patients.[20]

Cerebrospinal fluid (CSF) samples from 26 OCD children and adolescents participating in an ongoing National Institute of Mental Health (NIMH) study were analyzed for CSF monoamines. We found that CSF 5-Hydroxy indole acetic acid (5-HIAA) correlated negatively with illness baseline severity, i.e., the most severely ill patients had the lowest CSF 5-HIAA levels. However, the pretreatment 5-HIAA value was not predictive of drug response.

There are several caveats to be considered before accepting a serotonergic hypothesis for OCD. First, CMI is known to effect the norepinephrine, dopamine, and histamine systems, as well as the serotonin system. In addition, if relief of OCD symptoms were the direct result of serotonin reuptake blockade, clinical response should be immediate rather than delayed by several weeks. Finally, if a serotonin imbalance were solely responsible for OCD, one would expect differences in CSF or peripheral measures of serotonin and its metabolites between controls and OCD patients, as well as predictive ability of pretreatment serotonin levels. These discrepancies suggest that the serotonergic hypothesis is too simplistic. Perhaps the efficacy of CMI and other serotonin reuptake blockers is actually the result of an alteration in the balance of serotonin and the other monoamines as suggested by Agren et al.[21] and Murphy et al.[22]

Recent mCPP challenge studies conducted by Zohar et al.[23] suggest that the response of OCD symptoms to CMI treatment is the result of downregulation of serotonergic responsivity. Acute oral administration of mCPP, a serotonin receptor agonist, resulted in increased OC symptoms in untreated OCD patients but no change in OC symptoms or anxiety in those patients on CMI. If down regulation were required for CMI to be effective, initial administration of CMI should result in acute worsening of symptoms. Although Zohar et al.[23] observed this worsening in their patient samples, it has not been reported elsewhere. The lag in response time to CMI administration could be explained by this hypothesis, however.

Charney et al.[24] administered 0.1 mg/kg mCPP intravenously to a group of OCD patients and normal controls. Both groups failed to demonstrate any behavioral change following the challenge. In the group of OCD females, the baseline prolactin levels and prolactin rise (both directly related to serotonergic activity) following mCPP, were significantly reduced compared with the female controls. Tryptophan, a 5-hydroxytryptophan (5-HT) precursor, did not differentiate between groups. Charney et al. conclude that, while serotonin may be involved in the etiology of OCD, it is not solely responsible for the disorder.

3. OTHER NEUROTRANSMITTERS

Dopaminergic dysfunction in OCD is suggested by the presence of obsessive-compulsive symptoms in one third of patients with Tourette's syndrome (TS) and by the increase in OC symptoms following high-dose stimulant administration. Both conditions are thought to be the result of dopaminergic overactivity.[25,26]

The reported incidence of OC symptoms in TS varies from that seen in the general population to 90% of TS patients.[25] In a prospective study of TS and OCD patients, Frankel et al.[25] found that 51% of TS patients met diagnostic criteria for OCD, frequently of greater severity than their primary diagnosis. In the NIMH child study sample, motor tics were reported in one third of OCD patients. In addition, genetic studies in both patient groups link OCD and TS.[27–29] In a family study of TS and chronic multiple tics (CMT), Pauls et al.[29] found the rate of OCD to be significantly increased over estimates for the general population in first-degree relatives of TS probands and OCD patients and frequently supplied the missing generational link between TS patients. Although no increases in TS were found among first-degree relatives of OCD patients, the incidence of OCD was increased. This finding suggests that OCD and TS may represent different manifestations of a similar, autosomal dominant genetic defect.

High-dose stimulant administration has been thought to result in simple stereotypies, rather than in more complex compulsions or obsessions. However, evidence for OC-like behavior following high-dose amphetamines is available.[30,31] In an ongoing study of children with attention-deficit disorder and hyperactivity, B. Borcherding (personal communication, 1987) has noted increased obsessionality and compulsive ritualization with high-dose *d*-amphetamine (1 mg/kg) and methylphenidate (2 mg/kg) administration. In one instance, a 7-year-old boy spent several hours each evening vacuuming the carpet in his home. Another boy was observed to play with Lego blocks for two solid days, stopping only with stiff encouragement to eat and sleep. As is seen in OCD, the children become overly concerned with details and will erase holes in their papers trying to get one letter perfectly shaped or a number just the right darkness.

It is interesting to note, however, that in no stimulant-induced case were the compulsive behaviors ego-dystonic. Similarly, our younger OCD patients, particularly those under 8 years of age, have not reported the compulsions to be ego-dystonic. Rather, they report interference from time wasted or from resulting embarassment. Perhaps compulsions and obsessions result from dopaminergic overactivity and serotonin dysregulation is required for ego-dystonicity. Conversely, TS and OCD might be at opposite ends of a spectrum of dopamine-serotonin dysequilibrium. In TS, dopaminergic overactivity overcomes serotonergic inhibition and results primarily in motor and vocal tics. By contrast, OCD is primarily a serotonergic defect. Here, a primary lack of serotonin results in an inability to inhibit normal dopaminergic activity and fixed action patterns (obsessions and compulsions) are inappropriately released. Ego-dystonicity could then be related to the primary serotonin defect, or secondary to the loss of volitional control.

The role of other neurotransmitters, including the β-adrenergic, α-adrenergic, histaminic and muscarinic systems is uncertain in OCD. Although it is possible that neurotransmitter dysfunction is etiologic in OCD, further investigations are required to elucidate the precise nature of the neuropathology.

4. NEUROENDOCRINE ASPECTS OF OCD

Although most OCD investigations concentrate on hormonal aberrations as secondary rather than primary to the disorder, case reports and anecdotal experience suggest that hormonal dysfunction may be etiologically related to OCD. If one considers OCD from a neuroethological standpoint, obsessions and compulsions represent inappropriately released fixed action patterns.[32] The release of a fixed action pattern depends on presence of a receptive drive state and a key releasing stimulus. The extinction of the fixed action pattern occurs when the drive state is no longer greater than the threshold of response. Hormones determine the internal drive state. In OCD, then, hormonal dysregulation could allow for an inappropriate release of the obsession or compulsion, or a failure to extinguish the behavior. Suggestive evidence for this hypothetical model comes from anecdotal experience, animal studies, and a single drug study.

Several of the NIMH study sample patients have reported an exacerbation of their OCD symptoms during early puberty, while adults with OCD frequently recall the onset occurred during adolescence. Female patients often experience an increase in obsessive thoughts and rituals immediately before their menses, in addition to an inability to resist the OC symptoms. A controlled study of the cyclicity of OCD symptoms in relationship to the menstrual cycle is currently under way. Furthermore, S. Rasmussen has collected a series of 20 OCD cases with postpartum onset (personal communication, 1988).

5. HORMONE TREATMENT

In addition to these postpartum cases of OCD, pilot study data from Casas et al.[33] and from the NIMH study sample also implicate androgens in the etiology of OCD. In the Spanish experience, five out of five patients with OCD experienced a remission in their symptoms following treatment with cyproterone acetate, a potent antiandrogen. Two other patients, one with schizophrenia and the other with a phobic disorder with ritualization, did not respond. At the NIMH, two boys (ages 8 and 15) have been treated with spironolactone, a peripheral antiandrogen, particularly antitestosterone agent, and testolactone, a peripheral antiestrogen medication. Both experienced a temporary reduction of obsessions and compulsions, but relapsed within 3–4 months. Interestingly, although one of the boys had initially failed to respond to clomipramine treatment, he received benefit from a retrial of CMI. The failure to maintain a response to antiandrogen treatment, as well as the greater efficacy of central versus peripheral agents is intriguing and continuing investigations are under way.

6. OTHER HORMONAL SYSTEMS

The hormonal effect is also seen in the sex differences pre- and postpuberty. It has been observed that for children with symptom onset before 10 years of age, the male–female ratio is

7:1, while when the onset of OCD occurs after age 12, the male–female ratio decreases to 1:1.5. This dichotomy could be explained by an androgenic hormone effect, perhaps due to testosterone or estrogen.

In an epidemiologic study of high school students in a rural New Jersey County,[1] it was noted that boys with OCD were smaller and lighter than the control men and boys with other psychiatric illnesses.[34] There were no differences in the size of females with and without OCD. The small size of the OCD males could be due to an effective lack of growth hormone, or to a delay in the pubertal growth spurt, among other possibilities.

Adrenocorticotropic hormone (ACTH) and cortisol have also been linked to obsessive-compulsive behaviors. The increased incidence of psychic complications in Cushing's syndrome, a collection of different disorders having in common elevated ACTH levels, is well documented. Schletter et al.[35] described two Cushingoid children with obesity, short stature, and "submissive personalities and compulsive diligence to their schoolwork," personality traits that disappeared following adrenalectomy. The children are described as "passive, pleasant, nonaggressive children who were extremely compulsive in their schoolwork, achieving extremely high grades and working with an almost exaggerated sense of neatness."

7. ANIMAL GROOMING AND ACTH

Although it is beyond the scope of this chapter to review the extensive literature on hormones and animal grooming behavior, it is important to note the work of Gispen and Isaacson[36] and of Dunn et al.[37] Because novelty and conflict situations are strong releases of ACTH resulting in increased grooming behavior, it was hypothesized that ACTH was involved. Peripheral administration failed to elicit grooming behavior, but intracerebral and intracranial injections of ACTH elicit a dose-dependent grooming response.[36] ACTH (1 μg) injected into the lateral ventricle of a rat causes it to groom for 90% of the subsequent hour, in a pattern indistinguishable from natural grooming.[37] The ACTH response seems to be mediated by a variety of neurotransmitter systems, as dopaminergic antagonists and opiate antagonists prevent the ACTH grooming response. The similarities between this situationally inappropriate ritual and the repetitive behaviors of OCD are striking and worthy of further investigation.

In an attempt to elucidate the etiologic role of ACTH, the androgens, and other hormones in OCD, we are currently analyzing CSF and serum samples from OCD patients, as well as collecting data on variations of OCD in response to the menstrual cycle and pregnancy. Other studies include delineation of male–female differences, comparisons of CNS hormones in OCD patients and controls, and manipulation of the hormonal milieu to induce or alleviate OC symptomatology.

8. BASAL GANGLIA DYSFUNCTION IN OCD

Understanding the etiologic role of the basal ganglia in OCD is probably one of the most exciting challenges in psychiatry because it requires incorporating clues provided by historical reports, clinical presentations, brain-imaging techniques such as computed tomography (CT) and positron-emission tomography (PET) scans, and results of neurotransmitter and hormonal manipulations, into a neuropathologic framework determined by basic scientific research.

The best description of neurologically based OCD comes from Constantin von Economo's treatise on postencephalitic Parkinson's disease, which followed an outbreak of encephalitis lethargica in 1916–1917.[38] Von Economo noted the compulsory nature of the motor tics experienced by his patients. His pathologic examinations revealed only basal ganglia destruction. von Economo's patients, like OCD patients, described "having to" act, while not "wanting to" act—they experienced a neurologically based loss of volitional control.

The association between TS, a known basal ganglia disorder, and OCD has been noted earlier in this chapter. In the NIMH study sample, an increased incidence of tics (30%) has been observed in the children and adolescents with OCD. Choreiform movements, presumably of the same basal ganglia dysfunctional basis as other choreiform disorders, are seen in one third of children and adolescents. The absence of these tics and choreiform movements on follow-up examination 2–5 years later, suggests that they may be the result of the same pathologic process as that causing the child's OCD.

Sydenham's chorea (SC) is a basal ganglia disorder characterized by sudden involuntary, purposeless, jerking movements of the extremities. Husby et al.[39] demonstrated antibodies directed against the cytoplasm of the subthalamic and caudate nuclei in patients with SC, but not in those patients with rheumatic fever (RF) without chorea. The association between OCD and SC was noted by Chapman et al.,[40] who reported OC symptoms in four of eight children with SC, and by Freeman and colleagues[41] in their comparison of adults with and without a history of SC during childhood. Personality disorders, including compulsive personality, were seen in 26 of 40 chorea patients, but in only 8 of 30 controls; "psychoneurosis" (phobic, obsessive-compulsive conversion, and anxiety) was reported in 7 of 40 chorea patients, and no controls.

A recently completed systematic investigation of OC symptoms in 23 children and adolescents with an episode of SC revealed increased OC symptoms and interference as compared with a group of 14 RF controls.[42] SC patients scored significantly higher than did RF patients and the Yes and Interference scales of the Leyton Obsessional Inventory—Child Version. Three of the SC and none of the RF patients had significant obsessional interference and met Diagnostic and Statistical Manual of Mental Disorder-III (DSM-III) criteria for OCD.[42] On the basis of epidemiologic data,[1] a sample of 600 adolescents would be needed to yield three OCD patients. It is particularly noteworthy that the increase in compulsive behaviors and obsessive thoughts occurred selectively in the SC patients. SC differs from RF only in the presence of basal ganglia dysfunction in SC and its absence in RF, once again implicating basal ganglia dysfunction in at least some forms of OCD.

Specific basal ganglia lesions also have resulted in OC behaviors. Laplane et al.[43] reported a case in which OC symptoms developed following a neurotoxic reaction to a wasp sting and two OCD cases following carbon monoxide poisoning. The CT scan of the wasp sting patient demonstrated bilaterial low-density lesions in the internal part of the lentiform nucleus, i.e., the globus pallidus. Clomipramine, but not other drugs, alleviated his OC symptoms.

Surgical lesions that result in the disconnection of the basal ganglia from the frontal cortex are therapeutic in OCD. Currently, the preferred psychosurgeries for OCD are capsulotomy and cingulectomy. In capsulotomy, bilateral basal lesions are made in the anterior limb of the internal capsule in order to interrupt frontal–cingulate projections; however, the surgical target lies within the striatum, near the caudate nuclei. In order to perform a cingulectomy, the surgeon lesions the anterior portion of the cingulate gyrus, interrupting tracks between the cingulate gyrus and the frontal lobes and destroying all of the efferent projections of the anterior cingulate cortex. Both procedures result in significant reduction of obsessions and compulsions. However, the success of these psychosurgeries is not conclusive evidence of a basal ganglia defect in OCD, as the lesion could be anywhere "upstream" from the site of treatment.

Computed tomography data from Luxenberg et al.[44] further implicate the basal ganglia in OCD. In a volumetric analysis of noninfusion head CTs from 10 males with childhood-onset

OCD and 10 healthy controls, the caudate nuclei were found to be significantly smaller in the OCD group. No other differences were seen. A replication study, using magnetic resonance imaging (MRI) techniques is under way. The lack of radioactive exposure in MRI will allow for direct examination of children and adolescents, and comparison with a group of normal controls.

The caudate nuclei were also noted to differ between OCD patients and controls in two 18-FDG PET studies by Baxter et al.[45,46] In their comparison of OCD patients with and without depression, patients with unipolar depression, and normal controls, Baxter et al.[45] found the caudate nuclei and the left orbital frontal gyris to have increased glucose metabolism. In a replication study from the same center,[46] nondepressed OCDs were compared with a group of normal controls. Again, the inferior orbital frontal regions and caudate nuclei were more active in OCD than in controls.

A PET study conducted in the child psychiatry branch of the NIMH compared regional glucose metabolism in 18 adults with childhood-onset OCD and 18 normal controls. Although there was no difference in caudate metabolism between groups, there were significantly elevated metabolic rates in the frontal cortex and cingulate gyri of the OCD patient group. This finding is consistent with the known efficacy of psychosurgery, as well as the hypothesized basal ganglia–frontal lobe dysfunction of OCD.[47]

9. BASAL GANGLIA MODEL

In order to hypothesize that the basal ganglia could be responsible for OCD, it was necessary for Wise and Rapoport[48] to move beyond the traditional view of the basal ganglia as functioning only to control motor activity. Work by Alexander et al.[49] suggests that the basal ganglia are actually a parallel array of loops, including the striatum, pallidum, thalamus, and cerebral cortex. Each loop could function primarily in a motor function, or mediate numerous complex sensorimotor and cognitive functions.

Wise and Rapoport have proposed a model in which the striatum functions as a feature detector/filter that triggers the release of species-typical behaviors.[48] Two sets of input converge on the ventromedial aspect of the caudate nucleus, one from the anterior cingulate and orbital frontal cortex, and the other from cortical "association" areas thought to be involved in the recognition of objects and sounds (superior and inferior temporal areas). The authors postulate that the caudate acts as the stimulus detector, and another portion of the striatum as the internal motivation detector. These striatal assemblies converge to inhibit the pallidal cell assembly, postulated to be tonically discharging. Since the pallidal assembly inhibits thalamic neurons, the inhibition of the pallidal circuits releases, or disinhibits, the thalamic cell assembly, resulting in expression of a fixed action pattern or behavior.

In the second circuit, striatal serotonin receptors potentiate, by excitatory modulation, the inputs to the striatum, including those from the anterior cingulate cortex to the internal motivation detector.[48]

Consider dirty hands as an exemplary model of this hypothetical system. Sensory input to the striatum indicates that the hands are dirty; an innately programmed striatal cell assembly recognizes this input as dirtiness. The striatal cell body then discharges and stops the tonic discharge of the appropriate pallidal cells. This inhibition releases the thalamocortical circuits that lead to the normal behavioral response to dirty hands—washing.

The converging circuit, originating in the anterior cingulate cortex and relaying to the pallidum through different cell bodies within the caudate, provides the signal to perform an act because of internal motivation, i.e., because the animal "wants to do so." If these signals from the cingulate cortex to the striatum disinhibit the same pallidal cell assembly (the hand-washing center), activation of these cingulate neurons causes hand washing to occur, even in the absence of an appropriate sensory stimulus. The hand washing is now triggered by the internal motivation detector and is independent of external justification. In OCD, the behavior is executed compulsively and independently of external justification; i.e., the OCD patient washes and rewashes clean hands, perhaps because of hyperactivity of the cingulate cortex or striatum.

This model is consistent with the results of psychosurgery and with the latest PET scan findings in OCD. It is consistent with the efficacy of clomipramine and other serotonin reuptake blockers, particularly if one considers the pharmacologic effect to be secondary to serotonin receptor downregulation. It is equally consistent with the mCPP challenge studies of Zohar and hypothesized hormonal aberrations of the drive state or receptor receptivity.

The model fails to account for the episodic nature of the disease, i.e., its waxing and waning course. It may or may not be consistent with the behavioral specificity of the disorder; i.e., washers wash and checkers check. If it is considered that only one set of circuits has been disrupted, it is consistent. The washing and checking circuits may have been made vulnerable because of genetic or hormonal influences. Finally, in accepting this model of OCD, it must be assumed that behavior modification alters biochemical circuits in a manner similar to pharmacologic manipulations. Otherwise, the efficacy of this treatment modality cannot be reconciled with this theoretical construct.

Further neuroanatomic investigations of the basal ganglia and frontal cortex and neurophysiologic studies of the role of serotonin and dopamine in OCD are required. In addition, hormonal and genetic inquiries may shed light on the etiology of OCD, while studies of the relationship of OCD to anorexia and to addictive and impulsive behavior disorders could reveal patterns of dysfunction crossing diagnostic categories. For example, ongoing treatment studies of trichotillomania (compulsive hair-pulling) and onchyphagia (pathologic nail-biting) suggest that these two disparate disorders may actually be subtypes of OCD, and more interestingly, suggest that these too are "hard wired" primitive grooming patterns. Further work is underway to delineate the relationship.

Obsessive-compulsive disorder studies of children and adolescents will continue to be particularly valuable because of the opportunity to follow the disorder longitudinally and to analyze CSF and serum samples during the subacute phase of the illness. In addition, children are less likely to have confounding co-morbidity, and parents can provide valuable historical observations. For genetic studies, it is possible to obtain first-person interviews with family members and to perform linkage studies. The question is no longer whether or not OCD represents a neuropsychiatric disorder, but rather, what is the locus and nature of the lesion.

REFERENCES

1. Flament MF, Whitaker A, Rapoport JL, et al: Obsessive compulsive disorder in adolescence: An epidemiologic study. J Am Acad Child Adolesc Psychiatry 27:764–771, 1988
2. Rapoport JL: Annotation: Childhood obsessive compulsive disorder. J Child Psychol Psychiatry 27:289–296, 1986
3. Flament MF, Rapoport JL, Berg CJ, et al: Clomipramine treatment of childhood obsessive compulsive disorder: A double-blind controlled study. Arch Gen Psychiatry 42:977–983, 1986
4. Leonard HL, Rapoport JL: Drug treatment of children and adolescents with obsessive compulsive disorder, in Rapoport JL (ed): Obsessive Compulsive Disorder in Children and Adolescents. Washington, D.C., American Psychiatric Press, 1989, pp 217–236
5. Insel TR, Murphy DL, Cohen RM et al: Obsessive compulsive disorder: A double blind trial of clomipramine and clorgyline. Arch Gen Psychiatry 40:605–612, 1983
6. Volavka J, Neziroglu F, Yaryura-Tobias JA: Clomipramine and imipramine in obsessive compulsive disorder. Psychiatry Res 14:83–91, 1985

7. Insel TR, Mueller EA, Alterman I, et al: Obsessive compulsive disorder and serotonin. Is there a connection? Biol Psychiatry 20:1174–1188, 1985
8. Yaryura-Tobias JA, Neziroglu F, Bergman L: Clomipramine for obsessive compulsive neurosis: An organic approach. Curr Ther Res 20:541–548, 1976
9. Thoren P, Asberg M, Cronholm B, et al: Clomipramine treatment of obsessive compulsive disorder. I. A controlled clinical trial. Arch Gen Psychiatry 37:1281–1285, 1980
10. Marks M, Stern RS, Mawsen D, et al: Clomipramine and exposure for obsessive compulsive rituals. Br J Psychiatry 136:1–25, 1980
11. Montgomery SA: Clomipramine in obsessional neurosis: A placebo controlled trial. Pharm Med 1:189–192, 1980
12. Ananth J, Pecknold JC, VanDenSteen, et al: Double blind comparative study of clomipramine and amitriptyline in obsessive neurosis. Prog Neuropsychopharmacol 5:257–262, 1981
13. Mavissakalian M, Turner SM, Michelson L, et al: Tricyclic antidepressant in obsessive compulsive disorder: Antiobsessional or antidepressant agents? II. Am J Psychiatry 142;572–576, 1985
14. Leonard HL, Swedo S, Rapoport JL, et al: Treatment of childhood obsessive compulsive disorder with clomipramine and desmethylimipramine: A double blind crossover comparison. Psychopharm Bull 24:93–95, 1988
15. Perse TL, Greist JH, Jefferson JW, et al: Fluvoxamine treatment of obsessive compulsive disorder. Am J Psychiatry 144:1543–1548, 1987
16. Price LH, Goodman WK, Charney DS, et al: Treatment of severe obsessive compulsive disorder with fluvoxamine. Am J Psychiatry 144:1059–1061, 1987
17. Prasad A: A double blind study of imipramine versus zimelidine in treatment of obsessive compulsive neurosis. Pharmacopsychiatry 17:61–62, 1984
18. Fontaine R, Chouinard G: An open clinical trial of fluoxetine in the treatment of obsessive compulsive disorder. J Clin Psychopharmacol 6:98–101, 1980
19. Turner SM, Jacob RG, Beidel DC, et al: Fluoxetine treatment of obsessive compulsive disorder. J Clin Psychopharmacol 5:207–212, 1985
20. Flament MF, Rapoport JL, Murphy DL: Biochemical changes during clomipramine treatment of childhood obsessive compulsive disorder. Arch Gen Psychiatry 44:219–225, 1987
21. Agren H, Potter WZ, Nordin C: Antidepressant drug action and CSF monoamine metabolites: New evidence for selective profiles on monoaminergic interactions. Presented at the Sixth Catecholamine Symposium, Jerusalem, 1987
22. Murphy DL, Siever JL, Insel TR: Therapeutic responses to tricyclic antidepressants and related drugs in non-affective disorder patient populations. Prog Neuropsychopharmacol Biol Psychiatry 9:9–13, 1985
23. Zohar J, Insel TR, Zohar-Kadouch RC, et al: Serotonergic responsivity in obsessive-compulsive disorder: Effects of chronic chlomipramine treatment. Arch Gen Psychiatry 45:167–172, 1983
24. Charney DS, Goodman WK, Price LH et al: Serotonin function in obsessive-compulsive disorder: A comparison of the effects of tryptophan and m-chlorophenylpiperazine in patients and healthy subjects. Arch Gen Psychiatry 45:177–185, 1988
25. Frankel M, Cummings JL, Robertson MM, et al: Obsessions and compulsions in Gilles de la Tourette's syndrome. Neurology (NY) 36:378–382, 1986
26. Sharp T, et al: A direct comparison of amphetamine-induced behaviours and regional brain dopamine release in the rat using intracerebral dialysis. Brain Res 401:322–330, 1987
27. Cummings JL, Frankel M: Gilles de la Tourette syndrome and neurological basis of obsessions and compulsions. Biol Psychiatry 20:117–1126, 1985
28. Pauls DL, Leckman JF, Towbin KE, et al: A possible genetic relationship exists between Tourette's syndrome and obsessive-compulsive disorder. Psychopharm Bull 22:730–733, 1986
29. Pauls DL, Leckman JF: The inheritance of Gilles de la Tourette's syndrome and associated behaviors: Evidence for autosomal dominant transmission. N Engl J Med 315;993–997, 1986
30. Koizumi HM: Obsessive-compulsive symptoms following stimulants. (Letter.) Biol Psychiatry 20:1332–1333, 1985
31. Frye P, Arnold L: Persistent amphetamine-induced compulsive rituals: Response to pyridoxine (B6). Biol Psychiatry 16:583–587, 1981
32. Swedo S: Rituals and releasers: An ethological model of OCD, in Rapoport JL (ed): Obsessive Compulsive Disorder in Children and Adolescents. Washington, D.C., American Psychiatric Press, 1989, pp 269–288
33. Casas M, Alvarez E, Duro P, et al: Antiandrogenic treatment of obsessive-compulsive neurosis. Acta Psychiatr Scand 73:221–222, 1986

34. Hamburger SD, Swedo SE, Rapoport JL et al: Growth rate in adolescents with obsessive-compulsive disorder. Am J Psychiatry 146:652–655, 1989
35. Schletter RE, Clift GV, Meyer R, et al: Cushing's syndrome in childhood: Report of two cases with bilateral adrenocortical hyperplasia, showing distinctive clinical features. J Clin Endocrinol 27:22–28, 1967
36. Gispen WH, Isaacson RL: ACTH-induced excessive grooming in the rat. Pharmacol Ther 12:209–240, 1981
37. Dunn AJ, Green EJ, Isaacson RL: Intracerebral adrenocorticotropic hormone mediates novelty-induced grooming in the rat. Science 203:281–283, 1979
38. von Economos C: Encephalitis lethargic, its sequelae and treatment. Oxford University Press, 1931. Translated from Economo: die encephalitis lethargica Deuticke, Vienna 1917–1918 and L'encefalik lethargica Policlinica, Rome 1920
39. Husby G, vandeRyn I, Zabriskie JB, et al: Antibodies reacting with cytoplasm of subthalamic and caudate nuclei neurons in chorea and rheumatic fever. J Exp Med 144:1094–1110, 1976
40. Chapman AH, Pilkey L, Gibbons MJ: A psychosomatic study of eight children with Sydenham's chorea. Pediatrics 21:582–595, 1958
41. Freeman JH, Aron AM, Collard JE, et al: The emotional correlates of Sydenham's chorea. Pediatrics 35:42–49, 1965
42. Swedo SE, Rapoport JL, Cheslow DL, et al: High prevalence of obsessive compulsive symptoms in patients with Sydenham's chorea. Am J Psychiatry 146:246–249, 1989
43. Laplane D, Baulac M, Widlocher D, et al: Pure psychic akinesia with bilateral lesions of basal ganglia. J Neurology Neurosurg Psychiatry 47:377–385, 1984
44. Luxenburg JS, Swedo SE, Flament MF, et al: Neuroanatomical abnormalities in obsessive-compulsive disorder detected with quantitative X-ray computed tomography. Am J Psychiatry 145:1089–1093, 1988
45. Baxter LR, Phelps ME, Mazziotta JC, et al: Local cerebral glucose metabolic rates in obsessive-compulsive disorder. Arch Gen Psychiatry 44:211–218, 1987
46. Baxter LR, Schwartz JM, Mazziotta JC, et al: Cerebral glucose metabolic rates in non-depressed obsessive-compulsives. Am J Psychiatry 145:1560–1563, 1988
47. Swedo SE, Schapiro ME, Grady CL et al: Cerebral glucose metabolism in childhood-onset obsessive compulsive disorder. Arch Gen Psychiat 46:518–523, 1989
48. Wise SP, Rapoport JL: Obsessive-compulsive disorders: Is it basal ganglia dysfunction? in Rapoport JL (ed), Obsessive Compulsive Disorder in Children and Adolescents. Washington, D.C., American Psychiatric Press, 1989, pp 327–346
49. Alexander G, DeLong M, Strick KP: Parallel organization of functionally segregated circuits linking basal ganglia and cortex. Annu Rev Neurosci 9:357–381, 1981

17

The Genetics of Affective Disorder

Nelson Freimer and Myrna M. Weissman

1. INTRODUCTION

Medical genetics had its origin in pediatrics, reflecting its concern with disorders of infancy and development. By contrast, child psychiatry has largely been excluded from psychiatric genetics. Historically, the major psychiatric disorders of adulthood, notably affective disorders and schizophrenia, were not acknowledged to occur in children, particularly before puberty. Recent epidemiologic and clinical studies have demonstrated that the core psychopathology of these disorders typically has its roots before adulthood.[1] During the past decade, consensus agreement on diagnostic criteria for childhood affective disorders was followed by the development of standardized reliable assessment instruments for children, roughly analogous to those used for adults.[2–6] Because of these advances, psychiatric genetic studies are beginning to include children and adolescents as probands and as part of the assessment of relatives. This chapter describes some of the ways in which such advances are important to future clinical practice and research on affective disorders in child psychiatry.

2. EVIDENCE FOR GENETIC ETIOLOGY OF AFFECTIVE DISORDERS

Genetic studies of affective disorders have been limited by uncertainties about the mode of transmission and the unavailability of biochemical disease markers. Many early studies used special populations to assess the probability of genetic transmission, notably registries of twins and adoptees.

Studies of bipolar disorder using a Danish twin registry established increased concordance for monozygotic as compared with dizygotic (DZ) twins.[7] Adoption studies have shown somewhat more equivocal results in support of genetic transmission; in particular, a small New York study found a higher rate of affective disorder in biological parents of bipolar adoptees in

Nelson Freimer • College of Physicians and Surgeons of Columbia University, New York, New York 10032. *Myrna M. Weissman* • Department of Psychiatry, College of Physicians and Surgeons of Columbia University, and Division of Clinical–Genetic Epidemiology, New York State Psychiatric Institute, New York, New York 10032.

comparison to biological or adoptive parents of unaffected adoptees.[8] A somewhat larger Swedish study failed to replicate this finding, although methodological differences, particularly the use of records, may account for this nonreplicability.[9] Current studies suggest that very early environmental influences may play a more important role in infant development than had previously been believed;[10] such evidence suggests that the principal premise of adoption studies may be flawed, i.e., that familial environmental influences can be controlled by studying offspring separated from biological parents in infancy.

Because of their case registry origins, twin and adoption studies effectively excluded children and adolescents; these studies aimed to infer information about disease transmission from prevalence rates among "genetically" selected groups of adults. Most recent studies use techniques that permit identification of early onset pathology for genetic analysis. Two broad types of study are now widely used to explore familial or genetic transmission of psychiatric diseases: family studies, in the wake of reliable diagnostic instruments, and linkage studies, following the discovery that variability in DNA produces large numbers of genetic markers, even in the absence of defined biochemical abnormalities.

3. FAMILY STUDIES

Family studies have had their major impact in demonstrating consistent patterns of disease aggregation within and between families; analysis of these patterns has helped orient subsequent studies of genetic transmission. Until recently, most family studies have studied only adults. Top-down studies begin with an adult proband and focus on disease status of adult relatives, and recently on at-risk children. Bottom-up studies begin with a child proband and assess patterns of illness in relatives of affected children.[11] Top-down studies are generally oriented around probands selected from treatment settings, clinical, and epidemiological studies. In a family study of known types of affective disorder, Weissman et al.[5,12] have included both adults and children in the analysis of relatives. This study has shown that the patterns of risk seen in adult relatives are similar in the relatives who are children. The study drew probands with major depression from two referral clinics and normal controls from a community survey conducted in New Haven, Connecticut.

These studies were primarily designed to assess the degree and special pattern of familial aggregation of illness in relatives of affected individuals. Therefore, the selection of disease-free controls is crucial in order to perform comparative analyses. Generally, this information is obtained using recall of lifetime symptoms, family history information, and medical records to arrive at either RDC or DSM-III/III-R diagnoses.[13] For children and adolescents, the diagnostic process is generally centered on administering a structured interview (e.g., the K-SADS-E[13]) to a parent or other adult about the child and then directly to the child. For adults and children, diagnoses rest on the best estimates by independent, clinically experienced raters using all available information.[4]

The principal findings of these studies are as follows: (1) although major depression aggregates in the adult relatives of probands with MDD or bipolar illnesses, bipolar disorders aggregate in the families of bipolar probands[12,14–16]; (2) although rates of major depression are higher among female relatives, rates of bipolar disorder show a roughly equal sex ratio, and sex of the proband does not affect the degree of familial aggregation[12]; (3) early-onset major depression (age <30 years) is related to increased risk of major depression in adult relatives and children[5]; and (4) onset of major depression in probands before the age of 20 is associated with increased risk of early-onset depression in offspring, beginning at puberty.[16] Although a number of studies have demonstrated increased risk of affective disorders in offspring of depressed parents, these have generally not been designed to permit testing of genetic models.

The results of a number of these studies have been extensively reviewed[17,18] and generally show that children of depressed parents are at increased risk of a host of social, school, and behavioral problems as well as MDD.

Familial studies of bipolar (BP) disorder have also shown consistently higher degree of risk in relatives of probands with early onset, generally defined as prior to age 30. Recently, Strober and colleagues[19] have used bipolar adolescents as probands and have demonstrated that onset of illness in childhood and adolescence is associated with extremely high rates in adult relatives of bipolar and unipolar disorders. They found rates of major affective disorder of more than 50% for first-degree and 15% for second-degree relatives, compared with 10% and 4% among families of schizophrenic controls. In addition, the rate of bipolar I disorder was about four times as high for relatives of probands who had demonstrated prepubertal psychiatric disorders (principally attention deficit and conduct disorders) compared with those with adolescent onset.[19] The inclusion of diverse psychiatric disorders in such analyses is supported by findings of high rates of nonspecific psychopathology in young offspring of BP parents.[20,21] Akiskal et al.[22] examined juvenile and young adult patients in their clinic who had an older sib or parent with a prior diagnosis of BP disorder. They found prepubertal onset of psychiatric symptoms in about one sixth of their sample, with most of the rest showing adolescent onset. Most of these young patients had previously received nonaffective diagnoses in other settings. However, Akiskal et al. identified affective illness in more than 80% of the sample within the first year after onset of psychiatric symptoms. These illnesses included acute depressive or manic episodes, dysthymia, and cyclothymia. The patients were followed for an average of 3 years by which time more than two thirds were classifiable as BP disorder. They confirmed the preliminary suggestion of other investigators that depression is the predominant affective disturbance in child and adolescent offspring of bipolar parents.[23,24]

There is additional evidence that such prodromal syndromes in affected children may be significant for genetic studies. Egeland et al.,[25] in one of several studies of affective disorders among Old Order Amish, found that impairment due to affective symptoms was evident by early adolescence in most individuals who later received a diagnosis of BP disorder.[26] The same Amish population characterized by these findings produced strong linkage between BP disorder and two DNA markers.[26] It has not been reported whether such prediagnosis impairment has been included in further linkage analyses of this population. The studies reviewed above suggest that families highly loaded for BP disorder can be identified through ill children. A number of points must be clarified before embarking on a strategy of selecting early onset probands as a source of families for genetic studies. It is possible that early-onset cases result from assortative mating; i.e., both parents are affected. In such instances, two types of distorting events are conceivable. First, a family with ill maternal and paternal relatives would give the appearance of high genetic loading. By contrast, families with illness occurring in only one parental tree may contain fewer affected members but an equally dense proportion among those who are genetically related. It is also possible that early-onset disorders reflect a genetic "double dose"; e.g., some disease genes with dominant inheritance patterns show varying degrees of expressivity in heterozygotes and homozygotes, notably those associated with hemoglobinopathies.[27]

The results of a study, conducted among a highly selected group of families at the National Institute of Mental Health (NIMH) provide mixed evidence regarding the second explanation.[28] Nurnberger et al.[28] found that vulnerability to developing affective disorder in the children of parents with BP disorder did not differ among those children with one affected parent, compared with those with two. These workers also reported a preliminary finding of enhanced melatonin suppression in response to light in children of affected parents compared with children of controls. However, a greater shift was found in children who had both parents affected compared with those who had one affected parent. Since family studies are organized around pattern of aggregation rather than segregation of traits through genetic transmission, such questions may be difficult to answer from existing data sets.

A major continuing concern of family–genetic studies is resolution of the genetic relationship between MDD and BP disorders as well as panic and other anxiety disorders. Most controversy stems from the presumed heterogeneity of major depression and the hypothesis that some forms of this disorder may be caused entirely by environmental stressors. It has been suggested that several subtypes of major depression belong to the spectrum of BP disorder. These subtypes include early onset, postpartum onset, recurrence, and presence of MDD with delusional features.[29] Aggregation of early-onset depression and BP disorder has been suggested in family and longitudinal studies (M. Weissman, unpublished data) and is an emerging finding of pedigree studies (N. Freimer, unpublished data). Weissman et al.[30] found evidence for a relationship between delusional depression and BP disorder in separate analyses of illness in adult relatives and the children of delusionally depressed probands. They first found that relatives of delusionally depressed probands were more likely to have BP disorder than were relatives of either normal or nondelusionally depressed probands.

In a later study, they found preliminary evidence that children of delusionally depressed probands showed a significantly increased risk of cyclothymia compared with children of nondelusionally depressed probands.[31] Other studies have found associations between familial aggregation of early-onset BP disorder and psychotic depression.[32] The relationship between major depression and panic disorder has been suggested by some,[33,34] but not all,[35,36] family studies. Most studies have ignored the possible occurrence of panic disorder in children. Until recently, the conventional wisdom has held that panic disorder does not exist before adolescence. However, a recent analysis of family data suggests that panic symptoms and disorder can be detected in prepubertal children. (D. M. M. Moreau and M. M. Weissman, unpublished data). Further understanding of prodromal or early onset panic syndromes may come from several current studies of multigenerational families that segregate for panic disorder.

No means of subtyping depressive disorders has been sufficiently consistently demonstrated in family studies to achieve unequivocal acceptance. In addition, recent studies that have successfully shown linkage for BP disorder with chromosomal markers suggest that analysis of large numbers of multigenerational pedigrees may be the most efficient current means of understanding the genetic specificity of subtypes of depressive disorder. These studies demonstrated linkage, despite inclusion of broadly defined depressive disorders. Close analysis of these families suggests that for BP pedigrees, most of those considered genetically affected will carry a diagnosis of BP I, BP II, or cyclothymia. Although inclusion of individuals with unipolar depressive diagnoses has strengthened evidence for linkage in these studies, the number of cases has not been large enough for particular subtypes to emerge.

4. LINKAGE STUDIES

The primary goal of current linkage studies is to provide chromosome localization for a disease gene, thus orienting gene mapping studies aimed at isolating and then characterizing such a gene. Linkage analysis has been applied to the genetics of affective disorders in several studies over a number of years. This section briefly describes the principles of linkage studies and how they are important to child psychiatry.

The basic principle of linkage analysis rests on fundamental characteristics of eukaryotic genetics. Genes are arrayed sequentially along chromosomes; during meiosis, genes in proximity are less likely to recombine, that is, cross over to the homologous chromosome than are genes that are far apart. Therefore, the extent to which two genes display similar recombination patterns is an indicator of their proximity. For a disease gene of unknown chromosomal location, a nearby gene will display a similar pattern across several meioses. Linkage analysis is generally performed using one of several computer programs; these programs assign probabilities of linkage based on a series of algorithms. Linkage is generally expressed in terms of a logarithmic odds ratio or LOD score; generally, a LOD score of 3.0 or more is accepted as evidence in favor of linkage between two genes and a score of −2.0 or less is taken as evidence against linkage.[37]

Until the past few years, disease phenotypes could be analyzed for linkage to other phenotypes (e.g., color blindness) or biochemical/immunologic polymorphisms (e.g., blood types or histocompatibility antigen (HLA) types). As there are very few such available markers and, with the exception of HLA, they are not highly polymorphic, linkage studies had a low success rate, particularly for psychiatric disorders. The advent of recombinant DNA technology during the 1970s opened up the possibility of locating genes more directly. Based on the discovery that human genomic DNA was characterized by immense variation, Botstein et al.[38] proposed that genetic mapping studies could be undertaken to determine the location of any disease gene. Full application of this hypothesis rapidly followed the development of restriction fragment length polymorphism (RFLP) markers on all chromosomes during the early 1980s. The identification and use of RFLPs are based on the knowledge that restriction endonucleases possess specific recognition sequences for cleaving DNA. Sequence changes in genomic DNA led to the appearance or disappearance of cleavage sites, thereby altering the size of such cut pieces, known as restriction fragments. The high degree of genetic variation in the size of these fragments makes them extremely useful as linkage markers.[39] Originally, RFLPs were largely found by chance in mapping or sequencing known genes; more recently, they have been identified through concerted efforts to develop a string of polymorphic markers throughout the genome, in effect forming a genetic linkage map.[40–42]

As part of an ongoing study of Old Order Amish, Egeland et al.[26] used RFLP markers to analyze 81 members of a single multigenerational pedigree heavily loaded for BP disorder. They found convincing evidence of linkage to two markers that had been previously localized to chromosome 11, the insulin gene and the H-*ras* oncogene. This finding represents a landmark in psychiatric genetics in that it demonstrated the possibility of single gene inheritance of a common, biologically complex psychiatric disorder. Moreover, it showed that the rigorous use of operationally derived diagnostic criteria permitted strong genetic inferences, despite continuing concern regarding the validity of such diagnoses. In this study, linkage probabilities remained robust even with a wide range of variations in the assumptions inserted in the linkage model. For example, the LOD score held above 3.0 with maximum penetrance varied between 0.55 and 0.95. Although 14 of the 19 affected individuals had a diagnosis of some sort of bipolar disorder, the linked haplotype was also displayed by individuals with unipolar depression. As the chromosomal region containing the insulin and H-*ras* genes also includes the gene for tyrosine hydroxylase (the rate-limiting enzyme in catecholamine production), there was considerable excitement over the possibility that this gene might be the BP disease gene. Initial analyses showed that tyrosine hydroxylase RFLPs were uninformative in the Amish pedigree. Subsequent analyses using more informative clones have nor erased this possibility but suggest that TH is farther from the BP locus than the H-*ras* marker (i.e., does not represent the BP gene).[43]

Several additional linkage studies of BP disorders have been reported since 1987 and appear to confirm that inheritance of these disorders is heterogeneous; i.e., similar phenotypes result from differing mutations. Baron et al.[44] studied several large Israeli (non-Ashkenazi) pedigrees and found linkage to two conventional (non-DNA) markers on the X chromosome; in these families, major affective disorder was linked to red/green color blindness and to glucose 6-phosphate-dehydrogenase (G6PD) deficiency. There is no report of linkage with DNA markers in these pedigrees. Another group, studying Belgian pedigrees, detected weaker linkage between bipolar disorder and markers at the hemophilia B (factor IX) locus also on the X chromosome.[45] Although the Israeli and Belgian families may share a common disease mutation, current understanding of the genetic distance between the color blindness and factor IX loci suggest that the above findings could also represent linkage to independent mutations. Intensive efforts to map this chromosomal region more completely should resolve this question; the high density of polymorphic markers in this region and the presence of the fragile X site in its midst should permit rapid directional orientation of these bipolar loci.

Hodgkinson et al.[46] and Detera-Wadleigh et al.[47] detected no evidence of linkage between BP disorder and chromosome 11 or X markers in several additional large pedigrees. Thus, it appears as though at least three genes can cause BP disorder; in the absence of corroborating linkage results from additional families, it remains possible that the Israeli and Amish pedigrees represent closed genetic communities in which a founder effect resulted in unique disease mutations. Genealogical documents show that 12,000 Old Order Amish are descended from 30 eighteenth-century progenitors.[26] Although the descent of non-Ashkenazi Israeli populations is less precisely documented, it is clear from studies of a number of genetic diseases that they represent a relatively distinct population. The finding by Baron and colleagues that inclusion of Ashkenazi pedigrees somewhat weakened the strength of linkage may support this possibility.[44] Amish and non-Ashkenazi Israeli families are characterized by large sibships and low rates of alcohol and drug abuse, both of which facilitate the detection of linkage for affective disorders.

None of these recent linkage studies has provided a clear answer as to what subtypes of affective disorder should be grouped for linkage purposes; currently, the best way to maximize detection of linkage is to perform several analyses using a predetermined hierarchical approach, e.g., first using only BP I and II, then including major depressive disorder, and finally including cyclothymia.

Adolescents have not been included in linkage studies of affective disorder. Their exclusion raises important questions about the validity of arbitrary age-of-onset criteria in currently used computer programs for linkage analysis. These programs use preset parameters to define affected and unaffected status for a particular disease, including penetrance, age of onset, and mode of transmission.[48] Age-of-onset parameters are used to establish the likelihood that an unaffected individual has passed through the expected period of risk for developing the disease. This analysis assumes that, as with most nonlethal diseases, the number of cases increases with age. However, a number of studies have demonstrated the existence of a strong cohort effect for unipolar, bipolar, and schizoaffective disorders, with increased prevalence and earlier age of onset for succeeding generations born after 1940.[49] A consistent degree of familial aggregation over the cohorts studied suggests that the effect has acted at least in part through a genetic mechanism.[50,51] It is possible that rising levels of drug and alcohol use have progressively enhanced the degree of expression of predisposition genes. Such a mechanism would strongly affect the penetrance assumption of linkage models. There is still no consensus on how to revise linkage models to include this cohort effect. Until this problem is resolved, it will consequently also be difficult to set expected penetrance levels for analyses that include adolescents. Over the short term, affected adolescents will probably be included in linkage studies through construction of an additional level in hierarchical analyses.

Linkage studies have been successful in suggesting chromosomal location for two BP disease genes. However, this information places each gene within a region of approximately 20 centimorgans (a cM is a unit of genetic distance representing 1% recombination). Based on the estimate that, for the human genome, 1 cM represents approximately 1 million base pairs (mbp) of DNA, single linked markers provide a region of approximately 20 mbp in which to search for a disease gene.[39] In addition, many genes lie in regions of genetic recombination heterogeneity in which the physical distance to be covered is far greater than would be predicted by linkage data; i.e., 1 cM is more than 1 mbp. Such heterogeneity probably characterizes the region containing the Huntington's disease gene,[52] which had not been identified after 5 years of intense search from the discovery of a linkage marker. The elusiveness of the HD gene is particularly striking, in that HD is known to be a genetically homogeneous disease for which large and continually increasing number of families have been made available for study. It is also possible that a gene for X-linked BP disorder lies in a similar region, made more complex by proximity of the breakage point responsible for the fragile X syndrome. Advantages in looking for a gene in this region include the following: (1) exclusion of male-to male transmission for an X-linked disease narrows the number of families that must be screened; (2) hemizygosity of males facilitates analysis of segregation of disease-linked alleles, and (3) a high density of polymorphic markers is mapped to this region. The region surrounding a chromosome 11-linked bipolar gene has not been mapped as thoroughly.

Techniques for moving from a linkage marker to a disease gene are rapidly evolving. A number of these techniques are likely to be directly applicable to BP disease within the next few years. The first step is to try to establish the location of a disease gene relative to markers of known map location; this analysis uses the tightness of linkage and the number of recombinants to examine the likelihood of the disease locus mapping to various positions relative to these markers, using specialized computer programs. With a sufficient number of meiotic events and several informative markers, it should be possible to orient markers that flank the disease gene; such flanking markers define a region that can be approached through physical mapping techniques. Such techniques currently rely on methods for electrophoretically separating large pieces of DNA.[53] Such techniques may be difficult to apply for disease genes that lie near the telomere or tip of the chromosome, and no definite flanking markers have yet been identified for Huntington's disease. No flanking markers have yet been identified for either chromosome 11 or X-linked bipolar disorders.

Identifying and characterizing disease genes or coding sequences are the ultimate goals of linkage and gene-mapping studies. As such studies have proceeded for nervous system disorders, it has become apparent that unique complexities are associated with this task for each disorder.

It has also become apparent that for some nervous system diseases, identifying and cloning the responsible gene will provide enormous amounts of biological information, while for others these steps may provide tantalizingly incomplete understanding. Lesch–Nyhan syndrome and Duchenne muscular dystrophy provide interesting contrasts. Starting with the knowledge that patients with Lesch–Nyhan syndrome are deficient in hypoxanthine ribose transferase (HPRT), the disease gene was cloned and characterized. Subsequent experiments led to development of a mouse strain complete deficient in HPRT; surprisingly, and inexplicably, such mice are phenotypically normal.[54] Localization of the DMD gene has rapidly led to identification of its product, a previously unknown muscle protein termed dystropin; subsequent work has focused on the anatomic localization and physiology of this protein. Thus, the identification of a disease gene through "reverse genetics" has yielded information that may ultimately be important in understanding normal muscle function, as well as the pathogenesis of DMD.[55,56] This set of discoveries may serve as a useful model for those using genetic strategies for psychiatric disorders; they suggest that gene products associated with human diseases are likely to be previously uncharacterized and that substances hypothesized to be associated with pathophysiology may be secondarily involved (e.g., neurotransmitters).

Rapid progress in molecular and mathematical techniques has resulted in near-completion of human genetic linkage maps. One approach has taken several hundred highly polymorphic DNA probes hybridized against several multigenerational reference pedigrees. This approach uses a fast algorithm for assigning maximum likelihood of gene order using many loci at a time.[40] Thus, with a sufficient number of phase-known meioses, any single gene disorder can theoretically be mapped. Another approach has been to use a variety of clone and genomic libraries to saturate regions with markers on a chromosome-by-chromosome basis.[57,42] The availability of comprehensive genetic maps may permit mapping of many disorders with more complex inheritance or in which a predisposition is inherited; nonbipolar affective disorders are likely to be included in these categories. In addition, although it remains hypothetical, it has been suggested that such maps can be used to simultaneously search for linkage markers throughout the genome, thus enabling analysis of polygenic disorders.[58] Progress in constructing genetic maps of nonhuman mammalian genomes may be applicable in developing animal models of affective disorders.[59] For instance, genetic mapping studies of mice bred for behaviorally distinctive traits, such as learned helplessness, could ultimately lead to experimental manipulation of environmental conditions in relation to expression or interaction of the responsible genes.

5. FUTURE DIRECTIONS

Progress toward identification of genes for affective disorder has several important clinical implications for child psychiatry. First, it is likely that debate regarding presymptomatic screening would occur. Since the rediscovery of DNA linkage markers for Huntington's disease (HD), such screening has been cautiously introduced in a few academic centers. The virtual 100% penetrance of the HD gene and devastating clinical course has supported screening of adults, most of whom are deciding whether or not to have children. Psychiatrists have been heavily involved in these screening programs; virtually any situation requires careful assessment, counseling and, often, therapy. For instance, finding out that one is to develop a fatal degenerative disease may lead to severe depression or suicidality, while finding out that one is unaffected while one's sibling has the gene may be accompanied by a different set of psychological responses.[60] In contrast to traditional screening for chromosomal abnormalities, the use of RFLPs, as in HD testing, merely offers a probability range. Thus, the burden of continuing uncertainty after testing also may lead to emotional distress. Different issues are likely to occur for affective disorders. First, affective disorders are of a wide range of severity, and inheritance of a disease gene may have little practical consequence; given the likely incomplete penetrance in most families, "inheriting" individuals may be completely normal. It is therefore hard to conceive a rationale for presymptomatic testing in certain clinical situations, for instance, if a patient is considering termination of pregnancy. As good treatments exist for most forms of affective disorder, there may be greater incentives to make early identification in comparison with HD. Such "preventive" measures pose additional problems; as the majority of cases of familial affective disorders might be expected to show onset by young adulthood, presymptomatic screening would necessarily focus on children and adolescents. Such a focus would require a major deviation from current approaches to presymptomatic screening which are predicated on the principle that only well-informed adults are capable of deciding to be tested.

A potentially serious question is heralded by successful linkage studies of common diseases, such as BP disorder: Who would perform genetic testing and counseling? There is a shortage of medical geneticists and genetic counselors in the United States. Most physicians have a sketchy understanding of human genetics; as of the mid-1980s, approximately one sixth of U.S. medical schools offered no formal course in genetics.[61] Currently, several companies are preparing DNA-based genetic test kits for commercial use; such tests could affect decisions regarding abortion, denial of insurance and employment. Although for HD all cases apparently carry the same mutation, the same is clearly not so for many other disorders, notably BP disorder. For such diseases, linkage markers are inadequate for genetic testing if an individual carries a mutation unlinked to known markers.

In addition, testing programs must take into account problems with laboratory reliability; such problems have plagued screening attempts that use far more straightforward assays. For example, controversies regarding reliability and interpretation have hampered attempts to institute widespread screening for hypercholesterolemia.[62] For genetic diseases, many people will desire testing but may be unable to receive it if their relatives are uninformative (show no heterozygosity at available markers) or are unwilling to be tested. An additional factor that could complicate testing for affective disorders is the possibility that, despite counseling, the procedure could be sufficiently stressful to cause the onset of symptomatic illness. Given the current difficulties in adequately insuring for the long-term intensive treatment often required for these disorders, it is easy to imagine that insurance companies would attempt to force prospective at-risk insurees to undergo predictive testing.

However, a number of opposing factors must be considered. There is evidence suggesting that early intervention and treatment can partially ameliorate the long-term morbidity from BP disorder.[63] Although the efficacy of psychotherapy for mania is uncertain, focused forms of therapy are of demonstrated value in major depression. There is also evidence that even seemingly autonomous mood disorders may be exacerbated by environmental stress.[64] It is conceivable that qualifying the genetic risk of affective disorders could be useful in helping parents identifying which of their children might be most in need of preventive attention, including individual and family therapies. A major clinical issue would be how to prevent such children from becoming iatrogenic psychiatric invalids at an early age. In addition, it is uncertain whether such information would be helpful in its most likely context, a family with several severely ill individuals including parents.

From a scientific standpoint, the identification of genes for affective disorders offers a number of possibilities for examining fundamental questions regarding behavior and personality. In particular, geneticists will collaborate with clinicians, developmental psychologists, and biological psychiatrists to study interactions of genes and environmental factors in normal and abnormal development. Using the presence or absence of disease genes as an independent variable, experimental paradigms could be constructed to study biological factors, such as infection or changes in neurotransmitter systems or cultural, social, and family environments over the course of development.

6. SUMMARY

Molecular genetic techniques have revolutionized psychiatric research. It is now conceivable that at least one gene responsible for BP will be identified in the near future. Evidence from family studies suggests that early age of onset of affective disorders may be associated with higher genetic loading for these disorders. Future progress in elucidating the genetic basis of psychiatric disorders could be facilitated by including selected affected children and adolescents in linkage studies. Child psychiatrists will require additional genetic training to deal with clinical issues that will follow identification of disease genes for affective disorders. Specifically, they should be prepared to assist families in understanding the significance of carrying such a gene and in obtaining more specialized counselling, if appropriate. Finally, research on the expression of disease susceptibility genes will require significant involvement of child psychiatrists, particularly with respect to questions of diagnosis and both physical and psychological development.

ACKNOWLEDGMENTS. This work was supported by grant MH 28274 to MMW from the National Institute of Mental Health, by a NIMH Research Training Fellowship to NF, and by the W. M. Keck Foundation.

REFERENCES

1. Christie KA, Burke JD Jr, Regier DA, et al: Epidemiologic evidence for early onset of mental disorders and higher risk of drug abuse in young adults. Am J Psychiatry 145:971–975, 1988
2. Puig-Antich J: The use of RDC criteria for major depressive disorder in children and adolescent inpatients. J Am Acad Child Adolesc Psychiatry 21:291–293, 1982
3. Puig-Antich, J, Orvaschel H, Tabrizi MA: The Schedule for Affective Disorders and Schizophrenia for School-Age Children—Epidemiologic Version (Kiddie-SADS-E), ed 3. New York, New York State Psychiatric Institute, 1980
4. Weissman MM, Leckman JF, Merikangas KR, et al: Depression and anxiety disorders in parents and children: Results from the Yale family study. Arch Gen Psychiatry 41:845–852, 1984
5. Weissman MM, Merikangas KR, Wickramaratne PF, et al: Understanding the clinical heterogeneity of major depression using family data. Arch Gen Psychiatry 43:430–434, 1986
6. Chambers WJ, Puig-Antich J, Hirsch M, et al: The assessment of affective disorders in children and adolescents by structured interview. Arch Gen Psychiatry 42:696–702, 1985
7. Bertelsen A, Harvald B, Hauge M: A Danish twin study of manic-depressive disorders. Br J Psychiatry 130;330–351, 1977
8. Mendlewicz J, Ranier JD: Adoption study supporting genetic transmission in manic-depressive illness. Nature (London) 268:327–329, 1977
9. Von Knorring AL, Cloninger CR, Bohman M, et al: Adoption study of depressive disorders and substance abuse. Arch Gen Psychiatry 40:943–950, 1983
10. Stern DN: The Interpersonal World of the Infant. New York, Basic Books, 1985
11. Puig-Antich J: Affective disorders in childhood: A review and perspective. Psychiatr Clin North Am 3:403–424, 1980
12. Weissman MM, Gershon ES, Kidd KK, et al: Psychiatric disorder in relatives of probands with affective disorders: The Yale–NIMH collaborative family study. Arch Gen Psychiatry 41:13–21, 1984
13. Weissman MM, Merikangas KR, John K, et al: Family–genetic studies of psychiatric disorders: Developing technologies. Arch Gen Psychiatry 43:1104–1116, 1986
14. Gershon ES, Hamovit J, Guroff J, et al: A family study of schizoaffective, bipolar I, bipolar II, unipolar, and normal controls. Arch Gen Psychiatry 39:1157–1167, 1982
15. Goldin LR, Gershon ES, Targun SD, et al: Segregation and linkage analyses in families of patients with bipolar, unipolar and schizoaffective mood disorders. Am J Hum Genet 35:274–287, 1983
16. Weissman MM, Gammon GD, John K, et al: Children of depressed parents: Increased psychopathology and early onset of major depression. Arch Gen Psychiatry 44:847–853, 1987
17. Orvaschel H: Parental depression and child psychopathology, in Guze SB, Earls FJ, Barrett JE (eds), Childhood Psychopathology and Depression. New York, Raven, 1983, 53–66.
18. Beardslee WR, Bemporad J, Keller MB, et al: Children of parents with major affective disorder: A review. Am J Psychiatry 140:825–832, 1983
19. Strober M, Morell W, Burroughs J, et al: A family study of bipolar I disorder in adolescence: Early onset of symptoms linked to increased familial loading and lithium resistance. J Affective Disord 15:255–269, 1988
20. Gershon ES, McKnew D, Cytryn L et al: Diagnosis in school-age children of bipolar affective disorder patients and normal controls. J Affective Disord 8:283–292, 1985
21. Cytryn L, McKnew DH, Waxler-Zahn C, et al: Developmental issues in risk research: The offspring of affectively ill parents, in Rutter M, Izard CE, Read PE (eds), Depression in Young People: Developmental and Clinical Perspectives. New York, Guilford, 1986, pp 163–189
22. Akiskal HS, Walker P, Puzantian VR, et al: Bipolar outcome in the course of depressive illness: Phenomenologic, familial, and pharmacologic predictors. J Affective Disord 5:115–128, 1983
23. Kuyler PL, Rosenthal L, Igel et al: Psychopathology among children of manic-depressive patients. Biol Psychiatry 15:589–597, 1980
24. Decina P, Kestenbaum CJ, Farber S, et al: Clinical and psychological assessment of children of bipolar probands. Am J Psychiatry 140:548–553, 1983
25. Egeland JA, Blumenthal RL, Nee J, et al: Reliability and relationship of various ages of onset criteria for major affective disorder. J Affective Disord 12:159–165, 1987
26. Egeland JA, Gerhard DS, Pauls DL, et al: Bipolar affective disorders linked to DNA markers on chromosome 11. Nature (Lond) 325:783–787, 1987
27. Weatherall DJ, Clegg JB: The Thalassaemia Syndromes, ed 3. Oxford, Blackwell, 1981
28. Nurnberger J Jr, Hamovit J, Hibbs ED, et al: A high-risk study of primary affective disorder: Selection of

subjects, initial assessment and 1–2-year follow-up in relatives at risk for mental disorders, in Dunner DL, Gershon ES, Barrett JE (eds), Relatives at Risk for Mental Disorder, New York, Raven, 1988
29. Coryell W, Endicott J, Keller MB et al: Phenomenology and family history in DSM-III psychotic depression. J Affective Disord 9:13–18, 1985
30. Weissman MM, Prusoff BA, Merikangas KR: Is delusional depression related to bipolar disorder? Am J Psychiatry 141:892–893, 1984
31. Weissman MM, Warner V, John K, et al: Delusional depression and bipolar spectrum: Evidence for a possible association from a family study of children. Neuropsychopharmacology 1:257–264, 1988
32. Strober M, and Carlson G: Bipolar illness in adolescents: Clinical, genetic and pharmacologic predictors in a three-to-four-year prospective follow-up. Arch Gen Psychiatry 39:549–555, 1982
33. Leckman JF, Weissman MM, Merikangas KR, et al: Panic disorder and major depression: Increased risk of depression, alcoholism, panic and phobic disorders in families of depressed probands with panic disorder. Arch Gen Psychiatry 40:1055–1060, 1983
34. Leckman JF, Weissman MM, Merikangas KR, et al: Major depression and panic disorder: A family study perspective. Psychopharmacol Bull 21:543–545, 1985
35. Crowe R, Noyes R Jr, Pauls DS, et al: A family study of panic disorder. Arch Gen Psychiatry 40:1065–1069, 1983
36. Harris EL, Noyes R Jr, Crowe RR, et al: A family study of agoraphobia. Arch Gen Psychiatry 40:1061–1064, 1983
37. Suarez BK, Cox NJ: Linkage analysis for psychiatric disorders, I. Basic Concepts. Psychiatry Dev 3:219–243, 1985
38. Botstein D, White RL, Skolnick M, et al: Construction of a genetic linkage map using restriction fragment length polymorphisms. Am J Hum Genet 32:314–331, 1980
39. Gusella, JF: DNA polymorphism and human disease. Annu Rev Biochem 55:831–854, 1986
40. Donis-Keller H, Green P, Helms C, et al: A genetic linkage map of the human genome. Cell 51:319–337, 1987
41. Drayna D, Davies K, Hartley D, et al: Genetic mapping of the human X chromosome by using restriction fragment length polymorphisms. Proc Natl Acad Sci USA 81:2836–2839, 1985
42. Nakamura Y, Leppert M, O'Connell P, et al: Variable number of tandem repeat (VNTR) markers for human gene mapping. Science 235:1616–1622, 1987
43. Kidd KK, Kidd JR, Pakstis AJ, et al: Applications of molecular genetic methods to affective disorders, in Molecular Probes: Technology and Medical Applications. New York, Raven, 1988
44. Baron M, Risch N, Hamburger R, et al: Genetic linkage between X-chromosome markers and bipolar affective illness. Nature (Lond) 326;289–292, 1987
45. Mendlewicz J, Simon P, Sery S, et al: Polymorphic DNA marker on X chromosomes and manic depression. Lancet 1:1230–1232, 1987
46. Hodgkinson S, Sherrington R, Gurling H, et al: Molecular genetic evidence for heterogeneity in manic depression. Nature (Lond) 325:805–806, 1987
47. Detera-Wadleigh SC, Berrettini WH, Goldin LR, et al: Close linkage of c-Harvey-ras-1 and the insulin gene to affective disorder is ruled out in three North American pedigrees. Nature (Lond) 325:806–808, 1987
48. Ott, J: Analysis of Human Genetic Linkage. Baltimore, Johns Hopkins University Press, 1985
49. Klerman GL, Lavori PW, Rice J, et al: Birth cohort trends in rates of major depressive disorder among relatives of patients with affective disorder. Arch Gen Psychiatry 42:689–693, 1985
50. Gershon ES, Hamovit JH, Guroff J, et al: Birth cohort changes in manic and depressive disorders in relatives of bipolar and schizoaffective patients. Arch Gen Psychiatry 44:314–319, 1987
51. Rice J, Reich T, Andreasen NC, et al: The familial transmission of bipolar illness. Arch Gen Psychiatry 44:441–450, 1987
52. Gilliam TC, Tanzi RE, Haines JL, et al: Localization of the Huntington's disease gene to a small segment of chromosome 4 flanked by D4S10 and the telomere. Cell 50:565–571, 1987
53. Schwartz D, Cantor CR: Separation of chromosome-sized DNAs by pulsed field gradient gel electrophoresis. Cell 37:67–75, 1984
54. Stout TJ, Caskey, CT: The Lesch–Nyhan syndrome: Clinical, molecular and genetic aspects. Trends Genet 4:175–178, 1988
55. Koenig M, Monaco AP, Kunkel LM: The complete sequence of Dystrophin predicts a rod-shaped cytoskeletal protein. Cell 53:219–228, 1988
56. Bonilla E, et al: Duchenne muscular dystrophy: Deficiency of dystrophin at the muscle cell surface. Cell 54:447–452, 1988

57. White R, Leppert M, Bishop DT, et al: Construction of linkage maps with DNA markers for human chromosomes. Nature (Lond) 313:101–105, 1985
58. Lander ES, Botstein D: Strategies for studying heterogeneous genetic traits in humans by using a linkage map of restriction fragment length polymorphisms. Proc Natl Acad Sci USA 83:7353–7357, 1986
59. Avner P, Amar L, Dandolo L, et al: Genetic analysis of the mouse using interspecific crosses. Trends Genet 4:18–22, 1988
60. Meissen GJ, Myers RH, Mastromauro CA, et al: Predictive testing for Huntington's disease with use of a linked DNA marker. N Engl J Med 318:535–542, 1988
61. Holtzman NA: Recombinant DNA technology, genetic tests and public policy. Am J Hum Genet 42:624–633, 1988
62. Roberts L: Measuring cholesterol is as tricky as lowering it. Science 238:482–483, 1987
63. Goodwin FK, Jamison KR: The natural course of manic-depressive illness, in RM Post, JC Ballenger (eds), The Neurobiology of Mood Disorders. Baltimore, Williams & Wilkins, 1984, pp 20–37
64. Anisman H: Vulnerability to depression: Contribution of stress, in RM Post, JC Ballenger (eds), The Neurobiology of Mood Disorders. Baltimore, Williams & Wilkins, 1984, pp 407–432

18

Biochemical Correlates of Auto-Aggressive Behavior

Inferences from the Lesch–Nyhan Syndrome

Theodore Page and William L. Nyhan

1. INTRODUCTION

The auto-aggressive behavior that characterizes the Lesch–Nyhan syndrome[1] is unique among behavioral abnormalities in that the genetic and molecular basis of the disorder has been firmly established.[2] The syndrome is a consequence of the deficiency of the enzyme hypoxanthine-guanine phosphoribosyl transferase (HPRT; E.C. 2.4.2.8); the deficiency is inherited in a simple X-linked fully recessive manner. However, the exact etiology of the auto-aggressive behavior, the way in which the abnormality in purine metabolism leads to the behavioral phenotype, remains obscure despite an extensive body of research on the metabolic consequences of HPRT deficiency. It is almost certain that neurotransmitter metabolism is involved. Animal models suggest ways in which early neurotransmitter imbalances may cause auto-aggressive behavior. Several mechanisms for an effect of HPRT deficiency on neurotransmitter metabolism seem plausible, including a direct action of the substrates of HPRT on receptors, and the participation of GTP-binding proteins. Experimental approaches to therapy have also contributed to our knowledge of what is—and what is not—involved in auto-aggressive behavior.

2. CLINICAL DESCRIPTION

Patients are always normal at birth,[3] although the enzyme defect is clearly present from the beginning of embryonic development.[4] The first clinical manifestation is often the presence of large amounts of orange crystals of uric acid looking like piles of orange sand in the diaper.[3] All patients overproduce uric acid.[3] The resulting hyperuricemia may lead to nephrolithiasis

Theodore Page and William L. Nyhan • Department of Pediatrics, University of California–San Diego, La Jolla, California 92093.

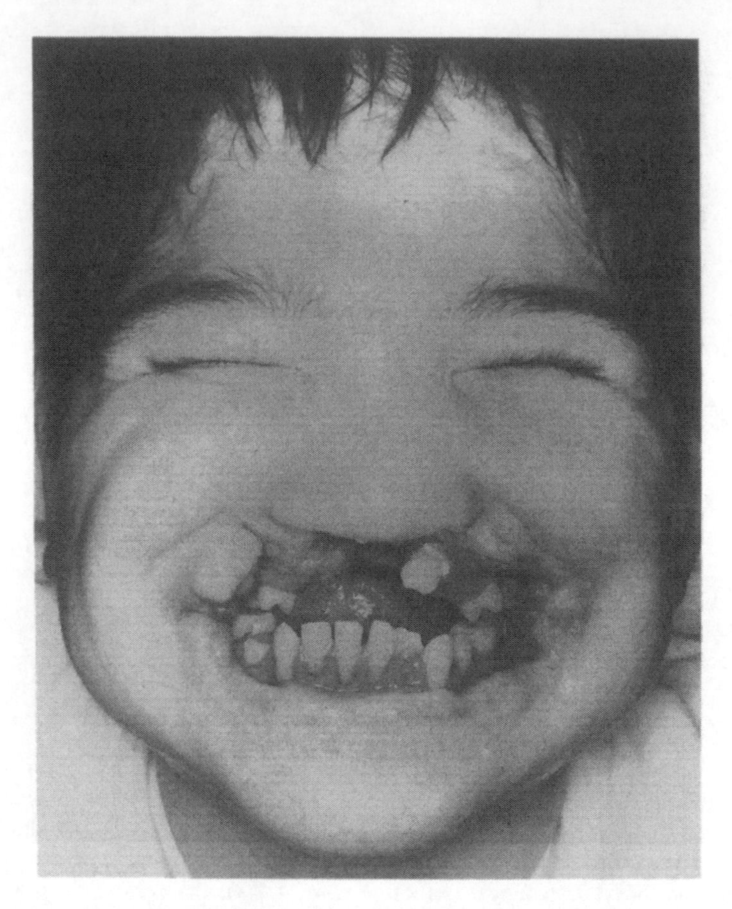

Figure 1. A patient with the Lesch–Nyhan disease in whom there was extensive loss of tissue about the lips and palate.

and renal failure during the first decade. This was virtually always the rule in the untreated patient, but the development of allopurinol therapy has largely prevented the worst of these complications. Delay in motor development is apparent by 6–8 months of age.[3] Abnormal involuntary movements appear that are variably athetoid, choreic, and dystonic. Hyperactive deep tendon reflexes, ankle clonus, scissoring of the legs, and a positive Babinski response are seen regularly after the first year of life. Patients are unable to sit or stand unaided. None learn to walk. Patients with the classic Lesch–Nyhan syndrome are mentally retarded; IQs range from 40 to 70.[5] Variants are known in which intelligence is normal.[6,7]

Among the most distinctive features of the Lesch–Nyhan syndrome is the aggressive, self-mutilative behavior.[8] This behavior is highly stereotyped. It differs from the auto-aggressive or self-injurious behaviors of other conditions in its propensity for actual destruction of tissue. Patients with sensory neuropathies look like pugilists. Their injuries are accidental. The self-injurious behavior commonly seen in nonspecifically retarded patients in institutions is usually low-intensity repetitive and leads to hypertrophy such as cauliflower ears or callus formation around a frequently bitten thumb rather than loss of tissue. By contrast, patients with the Lesch–Nyhan syndrome bite with such ferocity that loss of tissue is the rule (Fig. 1). The hallmark is an absence of tissue around the mouth, and partial amputation of the fingers is common. Patients also bite or pick at the buccal mucosa. One of our patients picked a hole in his ala nasi. Some have amputated the tongue. Others have banged their heads or their chins, caught their legs in wheelchair spokes, or lacerated their legs on braces prescribed for cerebral palsy. Physical restraint is the only effective method of prevention. Extraction of teeth will stop injuries by biting. These patients are not insensitive to pain; they scream in pain when they injure themselves and become agitated and fearful at the anticipation of self-injury. They are quite happy and content when restrained. The behavior of these patients is not solely directed against the patient himself. It is more generally compulsive-aggressive. These patients bite, hit, or kick others. Eyeglasses worn by a doctor, nurse, or attendant may be a favorite target. When they learn speech, they become verbally aggressive. They often spit or vomit in a semivoluntary aggressive manner. Four-letter Anglo-Saxon words are common favorites. Sexually oriented pinchings, touching, or grabbing may turn up with puberty. As with auto-aggression, all these behaviors appear to be compulsive, and patients will often apologize after successfully committing an act against someone else.

Autopsies have been performed on a number of patients with the Lesch–Nyhan syndrome.[5,9–11] Examination of the central nervous system (CNS) including light and electron microscopy has not revealed pathological changes. Those changes that have been described in some patients were likely attributable to uremia. In many of the patients studied most carefully there have been no histological abnormalities.

A number of variants of HPRT deficiency are known in which some of the symptoms characteristic of the Lesch–Nyhan patient are either absent or reduced in severity.[6,7,12,13] These variants have reflected partial deficiency of the enzyme.[14] Thus, if HPRT activity is measured in intact cells, patients with 1.4% or less of the normal activity have the complete Lesch–Nyhan syndrome. We have studied at least five patients whose intelligence was normal but who had otherwise all of the features of the Lesch–Nyhan syndrome. These patients have had 1.4–1.6% of normal activity. Patients in the 1.6–8% range have had normal intelligence and no auto-aggressive behavior, but they have displayed the complete neurological picture of athetoid cerebral palsy. Patients with 8–60% of normal HPRT activity have hyperuricemia and its consequences as their only clinical manifestation; they do not have neurological, cerebral or behavioral abnormalities. Other combinations of symptoms, such as auto-aggressive behavior with no neurological manifestations have never been reported with HPRT deficiency.

3. BIOCHEMICAL STUDIES

3.1. Purine Biochemistry

Purine nucleotides are synthesized in humans in both a de novo pathway and salvage pathways from purine bases and nucleosides (Fig. 2). Purine compounds are degraded ultimately to uric acid, which is excreted in the urine. Thus, the concentrations of purine nucleotides are determined by the rates of de novo synthesis, salvage, and degradation.

The rates of purine salvage, as catalyzed by the enzymes adenine phosphoribosyl transferase (APRT),[15] adenosine kinase,[16] and HPRT[17] are for the most part controlled by the availability of substrates for these enzymes. APRT and HPRT are also regulated by product inhibition.[18] Similarly, the rates of purine degradation catalyzed by 5′ nucleotidase,[19] purine nucleoside phosphorylase,[20] and xanthine oxidase[21] are regulated by the concentrations of the substrates for these reactions.

Purine nucleotide concentrations are controlled largely through de novo synthesis. This 10-step pathway is regulated at several points. The synthesis of the starting compound, 5-phosphoribosyl-1-pyrophosphate (PRPP), catalyzed by PRPP synthetase is inhibited by purine nucleotides, especially ADP, ATP, and GTP.[22] The synthesis of phosphoribosylamine, cata-

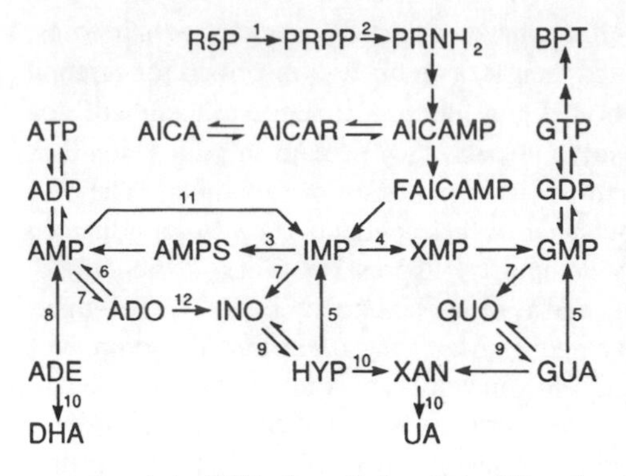

Figure 2. Purine metabolism in the human. ADE, adenine; ADO, adenosine; ADP, adenosine diphosphate; AICA, aminoimidazole carboxamide; AICAMP, aminoimidazole carboxamide ribonucleotide; AICAR, aminoimidazole carboxamide riboside; AMP, adenosine monophosphate; AMPS, adenylosuccinate; ATP, adenosine triphosphate; BPT, biopterin; DHA, dihydroxyadenine; FAICAMP, formamidoimidazole carboxamide ribonucleotide; GDP, guanosine diphosphate; GMP, guanosine monophosphate; GTP, guanosine triphosphate; GUA, guanine; GUO, guanosine; HYP, hypoxanthine; INO, inosine; IMP, inosine monophosphate; $PRNH_2$, phosphoribosylamine; PRPP, phosphoribosyl pyrophosphate; R5P, ribose 5-phosphate; UA, uric acid; XAN, xanthosine; XMP, xanthosine monophosphate. *Enzymes:* 1, PRPP synthetase; 2, PRPP amidotransferase; 3, adenylosuccinate synthetase; 4, IMP dehydrogenase; 5, hypoxanthine-guanine phosphoribosyl transferase; 6, adenosine kinase; 7, 5′ nucleotidase; 8, adenine phosphoribosyltransferase; 9, purine nucleoside phosphorylase; 10, xanthine oxidase; 11, AMP deaminase; 12, adenosine deaminase.

lyzed by PRPP amidotransferase, is the first committed step of the pathway. This reaction is rate-limiting in de novo purine synthesis and is inhibited by AMP and GMP.[23] Finally, the branch point at inosine monophosphate (IMP) is regulated by the availability of GTP; this nucleoside triphosphate is a substrate for adenylosuccinate synthetase,[24] and this is the committed step in adenine nucleotide synthesis; it is an inhibitor of IMP dehydrogenase,[25] the committed step of guanine nucleotide synthesis. The initial steps of the conversion of IMP to either AMP or GMP are also inhibited by the product nucleotides.[24,26]

3.2. Molecular Defect in HPRT and Its Consequences

The enzyme defect in the Lesch–Nyhan disease has been shown to be virtually complete absence of HPRT activity.[2,27] This deficiency was first demonstrated in cultured skin fibroblasts of three patients and subsequently confirmed in many other tissues. The assay is usually carried out in erythrocyte lysates in which the level of activity approximates zero. To date, all patients tested with the complete Lesch–Nyhan syndrome have been HPRT deficient, and the enzyme activity has been found to be deficient in all tissues studied.

The normal enzyme has been extensively studied. It catalyzes the transfer of a phosphoribosyl group to the purine bases hypoxanthine and guanine, as well as to a number of purine analogs such as 8-azaguanine and 6-thioguanine.[28] Only PRPP can serve as the source of the phosphoribosyl group, and magnesium is required for the reaction. For the most part, HPRT occurs as a cytoplasmic enzyme, although it has also been detected in rat brain synaptosomes.[29] In mammals, the activity of HPRT is highest in the central nervous system,[27,29–31] and in humans the activity is highest in the basal ganglia.[27] In the rat brain, there is a large increase in HPRT activity during early embryonic development.[31]

Of the metabolic consequences of HPRT deficiency, the overproduction of purines is the most prominent. Lack of HPRT activity results in the underutilization of hypoxanthine and guanine, which are oxidized to uric acid, and in the underutilization of PRPP, which serves as a substrate for de novo purine synthesis. Studies in which the incorporation of radiolabeled glycine into uric acid was measured in Lesch–Nyhan patients have indicated a 20-fold increase in this incorporation.[32] Since most tissues are capable of de novo synthesis, overproduction may be presumed to occur throughout the body.

Uric acid concentrations in plasma are usually in the range of 7–12 mg/dl. Normal values are usually given as less than 6 mg/dl, but this reflects the fact that hyperuricemia is common in the adult males usually employed to set standards. In children, levels over 4 mg/dl are abnormal. The excretion of uric acid in the urine is enormous and provides an accurate reflection of the increase in purine synthesis. Plasma concentrations of the oxypurines hypoxanthine and xanthine are only slightly elevated because oxidation by the liver is highly efficient. However, CSF concentrations of oxypurines are elevated up to fourfold in Lesch–Nyhan patients and may be threefold higher than their respective plasma concentrations.[27] The fact that the normal CSF/plasma ratio of oxypurines is 0.33 suggests that purine overproduction may be especially severe in the central nervous system (CNS).

Excretion of the compounds 5-aminoimidazole-4-carboxamide (AICA) and its corresponding ribonucleoside (AICAR) is also elevated in Lesch–Nyhan patients.[33,34] The corresponding nucleotide (AICAMP) is an intermediate in de novo purine synthesis. The conversion of AICAMP is formamido-4-imidazole carboxamide ribotide requires formyltetrahydrofolate, which may be rate limiting. Thus, the excretion of AICA and AICAR is a reflection of increased de novo synthesis. In some patients, the excretion of these compounds has been reduced by the administration of folate.[34]

A number of other metabolic abnormalities have been reported in Lesch–Nyhan patients. Many of these seem to be confined to erythrocytes and therefore their significance to the disease process, if any, remains unknown. The activities of two enzymes for which PRPP is a substrate, APRT[2] and orotate phosphoribosyl transferase (OPRT),[35] are increased in the erythrocytes of Lesch–Nyhan patients, but not in cultured fibroblasts or lymphoblasts. An increased activity of IMP dehydrogenase also seems to be confined to erythrocytes.[36] Increased activity of PRPP synthetase has been demonstrated in the cultured fibroblasts of Lesch–Nyhan patients, whereas it is normal in erythrocytes.[37]

One would expect from the fundamental defect that the intracellular concentration of purine nucleotides would be decreased; however this has been difficult to demonstrate. Various studies of nucleotide concentrations in the erythrocytes of Lesch–Nyhan patients have reported normal as well as decreased levels of purine nucleotides.[38,39] However, since erythrocytes are anuclear cells which are incapable of de novo purine synthesis, these findings are probably not representative of other tissues. Several studies of purine nucleotide concentrations in HPRT-cultured cells, including cells of neural origin, have yielded normal values.[40,41] Still, this may not reflect the situation in the living organism.

3.3. Neurotransmitter Abnormalities

There is strong evidence for abnormalities in neurotransmitter metabolism in Lesch–Nyhan patients. Plasma dopamine β-hydroxylase was reported to be significantly decreased in patients with the classic syndrome, as was the magnitude of the norepinephrine increase following postural stress.[42] The cultured fibroblasts of patients have significantly decreased activity of monoamine oxidase.[43,44] Postmortem examination of the brains of four Lesch–Nyhan patients disclosed a number of abnormalities in dopamine neuron function.[45] In areas of the brain with high concentrations of dopamine terminals, tissue concentrations of dopamine and homovanillic acid, and the activities of dopa decarboxylase and tyrosine hydroxylase were significantly decreased. In this same study, tissue concentrations of serotonin, 5-hydroxytryptophan (5-HT), and 5-hydroxyindoleactic acid (5-HIAA) were found to be slightly increased. Lesch–Nyhan patients are known to excrete increased amounts of 5-HIAA.[46] We have reported lowered levels of homovanillic acid (HVA) in the CSF of an infant with Lesch–Nyhan syndrome,[47] and lower levels of homovanillic acid were also reported in the CSF of patients over 16 years of age compared with age-matched controls.[48]

4. BIOCHEMICAL PATHOGENESIS OF THE CENTRAL NERVOUS SYSTEM MANIFESTATIONS

The pathogenesis of the symptoms related to uric acid overproduction is readily understandable in light of the metabolic defect. However, there is no obvious connection between purine metabolism and any aberration of neurotransmitter metabolism, which is thought to underly the neurological and behavioral features of the disease. Any such connection is largely speculative. The fact that no neuroanatomical changes are found in the brains of Lesch–Nyhan patients, that all patients are normal at birth, and that at least one therapy, 5-HT, was capable of bringing about a complete, albeit temporary reversal of the auto-aggressive behavior suggests that the underlying problem is an imbalance of metabolites rather than some type of irreversible structural change, as, for example, is seen in Alzheimer's disease.

The most obvious metabolic effect of HPRT deficiency, the gross overproduction of uric acid, is unlikely to be the cause of CNS manifestations such as auto-aggression, mental retardation, choreoathetosis, and spasticity. The concentration of uric acid in the CSF of Lesch–Nyhan patients is normal.[27,49] Furthermore, antihyperuricemic therapy with allopurinol, which dramatically lowers the plasma uric acid concentration, has no effect on the CNS symptoms, even when initiated at birth.[50] Patients with partial deficiency of HPRT and no CNS abnormalities have concentrations of uric acid in plasma that are just as high as that of patients with the complete Lesch–Nyhan syndrome.[51]

Elevated concentrations of hypoxanthine in plasma and CSF[51] might cause abnormalities of CNS function because hypoxanthine binds to diazapam receptors in neural tissue.[52] However, hypoxanthine levels are also elevated in the CSF of HPRT-deficient patients with no CNS symptoms.[51] Also, the xanthine oxidase inhibitor allopurinol causes a dramatic rise in hypoxanthine concentrations. Yet allopurinol therapy does not aggravate the symptoms of Lesch–Nyhan patients nor does it induce CNS symptoms in HPRT-deficient patients who do not already show them.

The suggestion has been made that the increased de novo synthesis may cause the CNS manifestations of the disease, either by depleting some important cofactor or by the accumulation of a toxic intermediate. Some candidates for the depleted intermediate include glutamine, glycine, and folic acid, but glycine has been found to be increased, rather than decreased, in $HPRT^-$ cells.[53] Patients have been treated with folic acid[34,54] or glutamine[55] with no apparent improvement. As for the toxic intermediate hypothesis, it has been suggested that AICAR may play this role.[39] Lesch–Nyhan patients excrete increased amounts of AICAR and AICA,[33,34] and the erythrocytes of these patients contain significant amounts of AICAMP.[39] However, metabolic studies in which humans subjects ingested up to 14 mg/kg per day of AICAR with no adverse effects suggest that this compound is not toxic.[56] Moreover, other metabolic defects in which there is increased purine synthesis, such as PRPP synthetase superactivity and partial HPRT deficiency, are not normally associated with CNS symptoms.

Although there is little direct evidence that HPRT deficiency causes a decrease in intracellular purine nucleotide concentrations, some circumstantial evidence suggests that just such a nucleotide deficiency may be responsible for the CNS symptoms. In humans,[27] as well as rats[29] and monkeys,[30] HPRT activity is highest in the CNS. The activity of PRPP amidotransferase, the rate-limiting step of de novo purine synthesis, is low,[57] suggesting that purine salvage is an important source of nucleotides for the CNS. In another defect of purine metabolism, purine nucleoside phosphorylase (PNP) deficiency, patients with the most severe enzyme deficiency do have neurological manifestations.[58] These patients do not have self-injurious behavior, but they do have spastic tetraperesis which may resemble some patients with partial HPRT deficiency. The erythrocytes of these patients are deficient in GTP. Since PNP is the only enzyme that is able to produce hypoxanthine and guanine from their respective nucleosides, the enzyme deficiency might result in a deficiency of these purine bases, which would be the functional equivalent of HPRT deficiency.

A metabolic connection between purine metabolism and neurotransmitters is in the synthesis of the cofactor biopterin. This cofactor is produced from GTP and is required for the conversion of tyrosine and tryptophan to dopamine and serotonin, respectively. A deficiency in GTP could theoretically result in a deficiency of biopterin, which might decrease the synthesis of these neurotransmitters, and of norepinephrine and epinephrine as well. Furthermore, biopterin has been shown to serve as a modulator of neurotransmitter synthesis in the rat brain.[59]

Several patients have been reported with defects in the biopterin synthetic pathway.[60–62] These patients have mental retardation, muscular rigidity, and dystonic movements. None has been reported with choreoathetosis or self-injurious behavior. These patients excrete reduced amounts of neurotransmitter metabolites such as HVA and 5-HIAA, whereas in Lesch–Nyhan patients excretions of these metabolites are normal or elevated.[46–48]

Another connection between purine nucleotide metabolism and neurotransmitters is provided by GTP-binding proteins. These proteins are known to act as essential intermediaries between hormone receptors and adenylate cyclase.[63] One model, proposed by Rodbell,[64] operates as follows:

$$\text{H} + \text{R-N} = \text{H-R-N}$$
$$\text{H-R-N} + \text{GTP} = \text{H-R-N-GTP} = \text{H} + \text{R-N-GTP}$$
$$\text{H-R-N-GTP} + \text{C} = \text{H-R-N-GTP-C}$$

In this model, the hormone (H) binds to the receptor (R), which is bound to the nucleotide-binding protein (N) to form an H-R-N complex. GTP does not bind to the R–N complex in the absence of hormone. Without GTP binding the R–N complex has a high affinity for the hormone, but when GTP binds to the H–R–N complex this affinity decreases. It is the binding of the catalytic subunit (C) to the H–R–N–GTP complex that forms the active adenylate cyclase. Two hallmarks of this model are that GTP is required for the hormonal activation of the associated adenylate cyclase, and that GTP decreases the affinity of the receptor for agonists but not antagonists (since antagonists do not cause GTP binding). The dopamine receptor of the corpus striatum of rat brain shows both these properties. GTP is required for dopaminergic stimulation of adenylate cyclase in the corpus striatum. It was found that the concentration of GTP controlled the basal level of adenylate cyclase activity as well as the sensitivity of the enzyme to dopaminergic stimulation.[65] GTP also decreases the affinity of these receptors for agonists such as apomorphine but does not affect the binding of antagonists such as spiroperidol.[66] These effects are specific for GTP and to a lesser extent GDP; neither effect is seen with ATP, ADP, AMP, or GMP. On the basis of this model, one would expect a decrease in guanine nucleotides to render the associated adenylate cyclase less sensitive to dopaminergic stimulation.

Serotonin is also associated with adenylate cyclase stimulation in certain cell types, and this stimulation also requires GTP.[67] It is reasonable to assume that here also, reduced GTP would render these receptors less sensitive to serotonergic stimulation. This would provide a rationale for the success of the serotonin precursor 5-HT in preventing self-injurious behavior in Lesch–Nyhan patients. Administered with a peripheral decarboxylase inhibitor, 5-HT inhibited or completely abolished self-biting behavior in Lesch–Nyhan patients presumably by increasing CNS concentrations of serotonin.[68] Unfortunately, this effect disappeared after a very brief period, despite continued administration of the agents. Lesch–Nyhan patients are not deficient in serotonin, as evidenced by increased excretion of the serotonin metabolite 5-HIAA.[46] Rather, it appears that their serotonergic receptors are less sensitive than normal, and thus require a larger stimulus for normal cellular function, or that there is an imbalance of neurotransmitters.

There is a good deal of evidence that purine compounds act directly as neuromodulators. The best studied of these is adenosine.[69] Adenosine and adenine nucleotides are released by a number of postsynaptic effector cells in response to stimulation by neurotransmitters. The action of adenosine is to inhibit the further release of neurotransmitter presynaptically. The ability of adenosine to inhibit the release of acetylcholine (ACh), dopamine, norepinephrine, serotonin, and γ-aminobutyric acid (GABA) has been demonstrated. Thus adenosine acts as a CNS depressant. This action is potentiated by compounds that inhibit the metabolism of adenosine, such as erythro-9-(2-hydroxy-3-nonyl)adenine (EHNA), and antagonized by theophylline and methylxanthines. The adenosine analogs phenylisopropyladenosine and 2-chloroadenosine mimic the inhibitory action of adenosine. Paradoxically, adenosine also appears to potentiate the action of the neurotransmitters whose release it inhibits by increasing cAMP in the target cell. This effect is also antagonized by theophylline and methylxanthines.

Although adenosine is the best studied of the purine neuromodulators, guanosine, and adenine and guanine nucleotides have similar effects in some cell types.[69] For example, these compounds may be released by smooth muscle in response to nervous stimulation. Their effects are in most cases similar to that of adenosine.

Decreased adenosine in the CNS of Lesch–Nyhan patients has not been reported. A detailed study of urinary purine compounds found that amounts of adenine and adenosine were consistently lower in Lesch–Nyhan disease.[70] It is conceivable that HPRT deficiency could result in adenosine deficiency. Adenosine is normally deaminated to inosine, which undergoes phosphorylytic cleavage to hypoxanthine. Conversion of hypoxanthine to IMP, and then to AMP and adenosine would complete this "adenosine cycle." On the other hand, APRT deficiency would be expected to result in a decrease in AMP and adenosine, and patients with this enzyme deficiency are neurologically normal.[71] With regard to the number and variety of processes adenosine is thought to control, a severe adenosine deficiency might be expected to interfere with a number of autonomic as well as neurological processes.

Research on the possible relationship of neurotransmitter dysfunction to auto-aggression has been done with animal models. Breese has proposed that rats treated neonatally with 6-hydroxydopamine (6-OHDA) serve as a model for the self-mutilation seen in Lesch–Nyhan syndrome.[72] In this model, nerve fibers are destroyed neonatally by 6-OHDA such that later stimulation with L-DOPA or other agonists causes self-biting and other stereotypic behavior in the rats. These behaviors could be blocked by 2-chloroadenosine. This dopamine-induced self-biting did not occur when adult rats were lesioned with 6-OHDA. These observations have led to a developmental hypothesis that even late manifesting self-injurious behavior may be a consequence of early infantile depletion of dopamine. Similar results were obtained with monkeys treated neonatally with 6-OHDA.[73] As with Lesch–Nyhan patients, dopamine in the striatum is reduced and serotonin is increased.

Two populations of dopamine receptors have been characterized in the mammalian brain. D_1-Dopamine receptors are associated with adenylate cyclase and bind thioxanthine compounds, such as piflutixol. The availability of antagonists specific for either the D_1 or D_2 receptors has shown that it is a change in the D_1 receptor, which is responsible for the self-mutilating behavior induced by L-DOPA administration to rats lesioned as neonates.[74] This self-mutilation was prevented by the specific D_1 antagonist SCH-23390, and produced by the specific D_1 agonist SKF-38393. Thus it appears that the dopaminergic D_1 receptors have been rendered functionally supersensitive in these animals.

It is tempting to speculate that in Lesch–Nyhan patients dopaminergic receptors rendered functionally insensitive to dopamine during the neonatal period may be functionally equivalent to 6-OHDA treatment with rats, but there are a number of important differences between Lesch–Nyhan patients and this animal model. First, self-injury in Lesch–Nyhan patients occurs in the absence of any external dopaminergic stimulus; L-DOPA was reported not to produce auto-aggression in these patients and may actually decrease this behavior.[75] Certainly, self-injurious behavior has never been noted in the vast experience of L-DOPA treatment of human patients with Parkinson disease. Finally, D_1 antagonists virtually always protected rats from self-biting,[74] whereas Lesch–Nyhan patients have to date received little or no benefit from these drugs.[76]

Other animal models have been proposed in which self-biting is induced by high doses of caffeine[77] or theophylline.[78] These drugs produce symptoms in rats which are similar to those above, although the neurochemical pathway leading to self-injury is not as well defined as with the dopaminergic model above. Antagonism of the CNS depressant action of adenosine would appear to be one possibility.

5. EXPERIMENTAL APPROACHES TO THERAPY

The hyperuricemia which occurs in the Lesch–Nyhan syndrome can be successfully treated with allopurinol, but no satisfactory treatment exists for the CNS manifestations of the disease; approaches to the treatment of these symptoms are experimental. Treatments include metabolic therapies designed to correct the metabolic defect, and pharmacological therapies designed to reduce or eliminate the various symptoms. Responses to treatment have been almost invariably nonexistent. Despite reports of improvement, this has not usually been our experience.

Although a depletion of purine nucleotides in the cells of the CNS of Lesch–Nyhan patients has not been convincingly demonstrated, a rational metabolic therapy would be designed to correct such a depletion. Administration of IMP,[79] GMP,[79] inosine,[80] or guanosine[51] has not proved beneficial. This might be expected since nucleotides are poorly taken up by intact cells, and no kinase exists in humans to convert the nucleosides to their respective nucleotides.

A more logical choice for nucleotide replacement therapy would be adenine. Adenine is known to cross the blood–brain barrier, and it is readily converted to AMP by ARPT which is present in virtually all tissues. AMP can be converted to IMP by adenylate deaminase and then to guanine nucleotides. Nevertheless, adenine therapy has not proven beneficial, even when initiated at birth.[38,81] The limitation on adenine therapy is due to the nephrotoxicity of 2,8-dihydroxyadenine, which is formed from adenine in a reaction catalyzed by xanthine oxidase. In our experience with the metabolism of radiolabeled purines in intact cells, >99% of the adenine incorporated goes into adenine nucleotides. Although AMP deaminase occurs in most tissues, its activity is probably significant only in skeletal muscle.[82]

The compound 2,6-diaminopurine can be converted to its nucleotide by APRT and then to GMP by AMP deaminase. In one trial, this compound had no effect.[80]

Several attempts have been made to activate de novo purine synthesis by the administration of cofactors which might be ratelimiting. Thus folic acid,[34,54] glutamine,[55] and ribose (together with methylene blue)[5] have been tried without success.

One patient with an unusual variant of HPRT was treated with hypoxanthine.[83] This patient had 7% of normal HPRT activity; he had above-average intelligence and had never evidenced auto-aggression. His enzyme had an increased K_M for hypoxanthine such that at high concentrations of hypoxanthine his enzyme activity was equal to or greater than some HPRT-deficient patients with no neurological dysfunction. During the treatment he reported feeling calmer, with fewer disturbances in equilibrium. His dysarthric speech was reported to improve notably and his excretion of 5-HIAA decreased. His neurological symptoms persisted. The treatment was discontinued because of the consequent hyperuricemia.

Attempts have also been made to replace the HPRT enzyme. Two patient were treated with partial exchange transfusions every two months for 3–4 years, such that their erythrocyte HPRT activity was maintained at 20–70% of normal.[84] Oxypurine levels decreased by 60%. Improvements in self-care and speech were only modest. There was an increase in attempts at self-mutilation after cessation of the program. Another patient was given a bone marrow transplant from a histocompatible sibling.[85] The engraftment was successful and there were no complications. His erythrocyte HPRT activity became normal. Unfortunately there was no improvement in the neurological or behavioral features of the disease. It should be noted that the patient was 22 years old; it remains of interest to assess the effects of early, presymptomatic transplantation. In a disorder such as phenylketonuria (PKU), the CNS effects are caused by a circulating toxic metabolite. Treatment of an adult would not be of any benefit. A presymptomatic transplantation would prove whether or not the CNS manifestations of Lesch-Nyhan syndrome follow the same model as phenylketonuria.

Because of the relationship between purine nucleotides and biopterin, one patient was treated with tetrahydrobiopterin.[83] Neurotransmitter synthesis apparently increased, as evidenced by increased excretion of dopamine, serotonin, and their metabolites. However, this treatment had no effect on the CNS manifestations.

Several pharmacological approaches have been tried with the aim of correcting or compensating for hypothetical imbalances of neurotransmitters. In a followup of the D_1 supersensitivity hypothesis above, two patients were treated with fluphenazine,[76] which is a dopaminergic antagonist but not specific for D_1 receptors. This treatment was reported to result in a decrease in self-biting and improved motor control in a 20-month-old Lesch–Nyhan patient who had begun relatively mild biting behavior at 18 months. Replacement of the drug with a placebo was followed by a considerable worsening in which there was more severe self-biting than had been seen before the treatment. With a 15-year-old patient the treatment produced no improvement, but the frequency and severity of self-biting became worse upon withdrawal of the drug.

Transient success was observed with the serotonin precursor 5-HT.[68] When administered with the peripheral decarboxylase inhibitor carbidopa, several patients showed a decrease in self-biting behavior, and in some the behavior was extinguished altogether. The treatment had no effect on the neurological features of the disease. All of the patients became tolerant to the drug within 1–3 months, and the effect could not be repeated even 1 year later.

One patient was treated with tetrahydrobiopterin, L-DOPA, 5-HT, and carbidopa.[83] His neurological symptoms, particularly his spasticity, were reported to show definite, although temporary, improvement. In another study, patients were treated with L-DOPA alone.[75] This treatment brought about temporary alleviation of self-injury in several patients; in no instance was an improvement in neurological manifestations observed.

Pharmacologic agents that were found to be of no benefit include diazapam, tryptophan, phenobarbitol, thiopropazate, chlorpromazine, and α-methyldopa.[5,81,86]

6. FUTURE DIRECTIONS

The metabolic studies, animals models and experimental therapies have not yet provided a clear rationale for auto-aggressive behavior seen in Lesch–Nyhan syndrome, although they have provided some important findings. It is clear that uric acid is not the cause of the CNS manifestations of the disease. There are definite dopaminergic changes in the brains of Lesch–Nyhan patients. The animal studies suggest that this is the link to auto-aggression; if this is the case, it is potentially treatable. These findings suggest future directions for study.

Animal models have been produced by pharmacological means that simulate some of the symptoms of Lesch–Nyhan syndrome, and work in this area continues. The availability of animal models is of considerable value in the screening of newly discovered agents in order to find those most promising for clinical trials. Similarly, in vitro studies of binding to D_1 and D_2 receptors should yield information on agents best suited for clinical trails. PET imaging studies of D_2 receptors may provide in vivo assessments of agents that effectively alter binding to dopamine receptors. Recently, an HPRT$^-$ mouse was produced by selection or mutation of mouse embryo cells in culture, production of chimeric mice by insertion in developing embryos, and selective mating.[87] These mice have $<1\%$ of the HPRT activity of the parental strain. Surprisingly, these mice are neurologically and behaviorally completely normal. The purine metabolism of both the parental and the HPRT$^-$ strain needs to be studied for quantitative differences. It is possible, for example, that the brains of these mice are less reliant on purine salvage as a source of purine nucleotides. In humans, HPRT activity is highest in the basal ganglia, whereas in mice the distribution is more homogeneous throughout the CNS. Perhaps neurological symptoms could be induced in these mice by treatment with agents which further reduce purine nucleotide synthesis.

Nucleotide replacement strategies will require techniques that actually increase concentrations of purine nucleotides, particularly guanine nucleotides, in the cells of the CNS. Two precursors that have not yet been tried are AICA and AICAR. These compounds are absorbed from the intestine, taken up by cells and converted into AICAMP, which can be converted to both adenine and guanine nucleotides. The metabolism of these precursors requires further study, particularly with regard to which enzymes convert them to AICAMP and how conversion of AICAMP to IMP can be optimized.

More work is needed on the dopaminergic changes in Lesch–Nyhan syndrome. The fact that fluphenazine appeared to be beneficial in a very young patient and not in an adolescent might indicate that therapy must be initiated at an early age to be successful. Positron-emission tomography (PET) scanning of the brains of Lesch–Nyhan patients at different ages with specific D_1 and D_2 ligands might be helpful in identifying changes in dopaminergic receptors. Autopsy data on patients of different ages might also help in this regard.

Work on the transfection of the normal HPRT gene into cells continues. Bone marrow stem cells continue to be the favorite target cells, but work is now in progress on cultured hepatocytes, which dedifferentiate and then differentiate in vitro, and on cells of the CNS such as glial cells. The normal gene has been cloned and sequenced[88] and has been successfully inserted into fibroblasts and lymphoblasts in culture; in the latter cell type, it expresses as much as 20% of the normal activity.[89] The gene has also been successfully inserted into bone marrow stem cells.[90] The strategy would be to follow this transfection by reimplantation of these cells into the patient. First, it must be shown that cells so infected will express the inserted human gene in experimental animals, and so far this has not been the case. Advances in molecular biology are occurring so rapidly that it appears likely that these hurdles will be overcome. When a human trial is possible it should be explored in patients young enough that the experimental assessment will be the prevention of manifestations of the disease rather than their reversal.

REFERENCES

1. Lesch M, Nyhan WL: A familial disorder of uric acid metabolism and central nervous system function. Am J Med 36:561–570, 1964
2. Seegmiller JE, Rosenbloom FM, Kelley WN: Enzyme defect associated with a sex-linked human neurological disorder and excessive purine synthesis. Science 155:1682–1684, 1967
3. Nyhan WL: The Lesch–Nyhan syndrome. Annu Rev Med 24:41–60, 1973
4. Gibbs DA, McFayden IR, Crawford MR, et al: First trimester diagnosis of Lesch–Nyhan syndrome. Lancet 2:1180–1183, 1984

5. Kelley WN, Wyngaarden JB: Clinical syndromes associated with hypoxanthine-guanine phosphoribosyl transferase deficiency, in Stanbury JB, Wyngaarden JB, Fredrickson DS, et al (eds), The Metabolic Basis of Inherited Disease, ed 5. New York, McGraw-Hill, 1983, pp 1115–1143
6. Bakay B, Nissenen E, Sweetman L, et al: Utilization of purines by an HPRT variant in an intelligent nonmutilative patient with features of the Lesch–Nyhan syndrome. Pediatr Res 13:1365–1370, 1979
7. Gottlieb RP, Koppell MM, Nyhan WL, et al: Hyperuricemia and choreoathetois in a child without mental retardation or self-mutilation—A new HPRT variant. J Inher Metab Dis 5:183–186, 1982
8. Nyhan WL: Behavior in the Lesch–Nyhan syndrome. J Autism Child Schizophr 6:235–251, 1976
9. Bassermann R, Gutensohn W, Springmann JS: Pathological and immunological observations in a case of Lesch–Nyhan syndrome. Eur J Pediatr 132:93–104, 1979
10. Crussi FG, Robertson DM, Hiscox JL: The pathological condition of the Lesch–Nyhan syndrome. Am J Dis Child 118:501–506, 1969
11. Seegmiller, JE: Summary: Pathology and pathologic physiology. Fed Proc 27:1042–1046, 1968
12. Emmerson BT, Thompson L: The spectrum of hypoxanthine-guanine phosphoribosyltransferase deficiency. Q J Med 166:1423–1440, 1973
13. Hersh JH, Page T, Hand ME, et al: Clinical correlations in partial hypoxanthine-guanine phosphoribosyltransferase deficiency. Pediatr Neurol 2:302–304, 1986
14. Page T, Bakay B, Nissinen E, et al: Hypoxanthine-guanine phosphoribosyltransferase variants: Correlation of clinical phenotype with enzyme activity. J Inher Metab Dis 4:203–206, 1981
15. Srivastava SK, Beutler E: Purification and kinetic studies of adenine phosphoribosyltransferase from human erythrocytes. Arch Biochem Biophys 142:426–434, 1971
16. Lindberg B, Klenow H, Hansen K: Some properties of partially purified mammalian adenosine kinase. J Biol Chem 242:350–356, 1967
17. Krenitsky TA, Papaioannou R: Human hypoxanthine phosphoribosyltransferase. II. Kinetics and chemical modification. J Biol Chem 244:1271–1277, 1969
18. Henderson JF: Kinetic properties of hypoxanthine-guanine phosphoribosyltransferase. Fed Proc 27:1053–1054, 1968
19. Fox IH, Marchant PJ: Purine metabolism in man: Characterization of microsomal 5′ nucleotidase. Can J Biochem 54:462–469, 1976
20. Wiginton DA, Coleman MS, Hutton JJ: Characterization of purine nucleoside phosphorylase from human granulocytes and its metabolism of deoxyribonucleosides. J Biol Chem 255:6663–6669, 1980
21. Carcassi A, Mercolongo R, Marinello E, et al: Liver xanthine oxidase in gouty patients. Arthritis Rheum 12:17–20, 1969
22. Hershko A, Rain A, Mager J: Regulation of the synthesis of 5-phosphoribosyl-l-pyrophosphte in intact red blood cells and in cell-free preparations. Biochim Biophys Acta 184:64–76, 1969
23. Holmes EW, McDonald JA, McCord JM, et al: Human glutamine phosphoribosylpyrophosphate amidotransferase—Kinetic and regulatory properties. J Biol Chem 248:144–150, 1973
24. Van Der Weyden MB, Kelley WN: Human adenylosuccinic synthetase: Partial purification, kinetic and regulatory properties of the enzyme from placenta. J Biol Chem 249:7242–7281, 1974
25. Holmes EW, Pehlke DM, Kelley WN: The role of human inosinic acid dehydrogenase in the control of purine biosynthesis de novo. Biochim Biophys Acta 364:209–217, 1974
26. Mager J, Magasanik B: Guanosine-5′-phosphate reductase and its role in the interconversion of purine nucleotides. J Biol Chem 235:1474–1478, 1960
27. Rosenbloom FM, Kelley WN, Miller J, et al: Inherited disorder of purine metabolism: Correlation between central nervous system dysfunction and biochemical defects. JAMA 202:175–177, 1967
28. McCollister RJ, Gilbert WR, Ashton DM, et al: Psuedofeedback inhibition of purine synthesis by 6-mercaptopurine and other purine analogs. J Biol Chem 239:1560–1566, 1964
29. Gutensohn W, Guroff G: Hypoxanthine-guanine phosphoribosyltransferase from rat brain (purification, kinetic properties, development and distribution). J Neurochem 19:2139–2150, 1972
30. Krenitzky TA: Tissue distribution of purine ribosyl- and phosphoribosyltransferases in the rhesus monkey. Biochim Biophys Acta 179:506–509, 1969
31. Murray AW: Purine phosphoribosyltransferase activities in rat and mouse tissues and in Ehrlich ascites-tumor cells. Biochem J 100:664–670, 1966
32. Nyhan WL, Oliver WJ, Lesch M: A familial disorder of uric acid metabolism and central nervous system function. J Pediatr 2:257–263, 1965
33. Newcombe DS, Lapes M, Thomson C: Urinary excretion 4-amino-5-imidazole carboxamide in X-linked primary hyperuricemia. Clin Res 15:45, 1967

34. Pignero A, Giliberti P, Tancredi F: Effect of the treatment of folic acid on urinary excretion pattern of amidoimidazole carboxamide in the Lesch–Nyhan syndrome. Perspect Inher Metab Dis 1:42–47, 1973
35. Beardmore TD, Meade JC, Kelley WN: Increased activity of the enzymes of pyrimidine biosynthesis *de novo* in erythrocytes from patients with the Lesch–Nyhan syndrome. J Lab Clin Med 81:43–52, 1973
36. Pehlke DM, McDonald JA, Holmes EW, et al: Inosinic acid dehydrogenase activity in the Lesch–Nyhan syndrome. J Clin Invest 51:1398–1404, 1972
37. Reem GH: Purine metabolism in murine virus-induced erythroleukemic cells during differentiation in vitro. Proc Natl Acad Sci USA 72:1630–1634, 1975
38. Lommen EJP, Bogels GD, Van Der Zee SPM, et al: Concentrations of purine nucleotides in erythrocytes of patients with the Lesch–Nyhan syndrome before and during oral administration of adenine. Acta Paediatr Scand 60:642–646, 1971
39. Sidi Y, Mitchell BS: Z-Nucleotide accumulation in erythrocytes from Lesch–Nyhan patients. J Clin Invest 76:2416–2419, 1985
40. Rosenbloom FM, Henderson JF, Caldwell IC: Biochemical basis of accelerated purine biosynthesis *de novo* in human fibroblasts lacking hypoxanthine-guanine phosphoribosyltransferase. J Biol Chem 243:1166–1173, 1968
41. Snyder FF, Cruikshank MK, Seegmiller JE: A comparison of purine metabolism and nucleotide pools in normal and hypoxanthine-guanine phosphoribosyltransferase-deficient neuroblastoma cells. Biochim Biophys Acta 543:556–569, 1978
42. Lake CR, Ziegler MG: Lesch–Nyhan syndrome: Low dopamanine β-hydroxylase activity and diminished sympathetic response to stress and posture. Science 196:905–906, 1977
43. Breakfield XO, Castiglione CM, Edelstein SB: Monoamine oxidase activity decreased in cells lacking hypoxanthine phosphoribosyltransferase activity. Science 192:1018–1019, 1976
44. Singh S, Willers I, Kluss EM, et al: Monoamine oxidase and catechol-o-methyltransferase activity in cultured fibroblasts from patients with maple syrup urine disease, Lesch–Nyhan syndrome and healthy controls. Clin Genet 15:153–159, 1979
45. Lloyd KG, Hornykiewicz O, Davidson L, et al: Biochemical evidence of dysfunction of brain neurotransmitters in the Lesch–Nyhan syndrome. N Engl J Med 305:1106–1111, 1981
46. Sweetman L, Borden M, Kulovich S, et al: Altered excretion of 5-hydroxyindoleacetic acid and glycine in patients with the Lesch–Nyhan syndrome, in Muller MM, Kaiser E, Seegmiller JE (eds), Purine Metabolism in Man, Vol. II: Regulation of Pathways and Enzyme Defects. New York, Plenum, 1977, pp 398–404
47. Castells S, Chakrabarti C, Winsberg, et al: Effects of L-5-hydroxytryptophan on monoamine and amino acid turnover in the Lesch–Nyhan syndrome. J Autism Devel is 9:95–103, 1979
48. Silverstein FS, Johnston MV, Hutchinson RJ, et al: Lesch–Nyhan syndrome: CSF neurotransmitter abnormalities. Neurology (NY) 35:907–911, 1985
49. Sweetman L: Urinary and cerebrospinal fluid oxypurine levels and allopurinol metabolism in the Lesch–Nyhan syndrome. Fed Proc 27:1055–1059, 1967
50. Marks JF, Baum J, Keele DK, et al: Lesch–Nyhan syndrome treated from the early neonatal period. Pediatrics 42:357–359, 1968
51. Kelley WN, Greene ML, Rosenbloom FM, et al: Hypoxanthine-guanine phosphoribosyltransferase in gout. Ann Intern Med 70:155–206, 1967
52. Skolnick P, Marangos PJ, Goodwin FK, et al: Identification of inosine and hypoxanthine as endogenous inhibitors of [^{3}H]diazepam binding in the central nervous system. Life Sci 23:1473–1480, 1978
53. Skaper SD, Seegmiller JE: Increased concentrations of glycine in hypoxanthine-guanine phosphoribosyltransferase-deficient neuroblastoma cells. J Neurochem 26:689–694, 1976
54. Benke PJ, Herrick N, Smiten L, et al: Adenine and folic acid in the Lesch–Nyhan syndrome. Pediatr Res 7:729–738, 1973
55. Wood MH, Fox RM, Vincent L, et al: The Lesch–Nyhan syndrome: Report of three cases. Aust NZ J Med 2:57–64, 1972
56. Seegmiller JE, Laster L, Stetten D: Incorporation of 4-amino-5-imidazole carboxamide-4-C^{13} into uric acid in the normal human. J Biol Chem 216:653–662, 1955
57. Howard WJ, Kerson LA, Appel SH: Synthesis de novo of purines in slices of rat brain and liver. J Neurochem 17:121–123, 1970
58. Simmonds HA, Fairbanks LD, Morris GS, et al: Erythrocyte GTP depletion in PNP deficiency presenting with hemolytic anemia and hypouricemia, in Nyhan WL, Thompson LF, Watts RWE (eds), Purine and Pyrimidine Metabolism in Man. Vol V: Clinical Aspects Including Molecular Genetics. New York, Plenum, 1986, pp 197–204

59. Miwa S, Watanabe Y, Hayaishi O: 6-*R*-L-Erythro-5,6,7,8-tetrahydrobiopterin as a regulator of dopamine and serotonin biosynthesis in the rat brain. Arch Biochem Biophys 239:234–241, 1985
60. Niederweiser A, Blau N, Wang M, et al: GTP cyclohydrolase I deficiency, a new enzyme defect causing hyperphenylalaninemia with neopterin, biopterin, dopamine, and serotonin deficiencies and muscular hypotonia. Eur J Pediatr 141:208–214, 1984
61. Hoganson G, Berlow S, Kaufman S, et al: Biopterin synthesis defects: Problems in diagnosis. Pediatrics 74:1004–1011, 1984
62. Nyhan WL, Sakati NA: Hyperphenylalaninemia and defective metabolism of tetrahydrobiopterin, in Diagnostic Recognition of Genetic Disease. Philadelphia, Lea & Febiger, 1987, pp 107–112
63. Gilman AG: G Proteins: Transducers of receptor-generated signals. Annu Rev Biochem 56:615–649, 1987
64. Rodbell M: The role of hormone receptors and GTP-regulatory proteins in membrane transduction. Nature (Lond) 284:17–22, 1980
65. Roufogalis BD, Thornton M, Wade DN: Nucleotide requirement of dopamine sensitive adenylate cyclase in the synaptosomal membranes from the striatum of rat brain. J Neurochem 27:1533–1535, 1976
66. Creese I, Usdin TB, Snyder SH: Dopamine receptor binding regulated by guanine nucleotides. Mol Pharmacol 16:69–76, 1979
67. MacDermot J, Higashida H, Wilson SP, et al: Adenylate cyclase and acetylcholine release regulated by separate receptors of somatic cell hybrids. Proc Natl Acad Sci USA 76:1135–11439, 1979
68. Nyhan WL, Johnson AG, Kaufman I, et al: Serotonergic approaches to the modification of behavior in the Lesch–Nyhan syndrome. Appl Res Ment Retard 1:25–40, 1980
69. Fredholm BB, Hedqvist P: Modulation of neurotransmission by purine nucleotides and nucleosides. Biochem Pharmacol 29:1635–1643, 1980
70. Sweetman L, Nyhan WL: Detailed comparison of the urinary excretion of purines in a patient with the Lesch–Nyhan syndrome and a control subject. Biochem Med 4:121–134, 1970
71. Simmonds HA, Van Acker KJ: Adenine phosphoribosyltransferase deficiency, in Stanbury JB, Wyngaarden JB, Fredrickson DS, et al (eds), The Metabolic Basis of Inherited Disease, ed 5. New York, McGraw-Hill, 1983, pp 1144–1156
72. Breese GR, Baumeister AA, McCown TJ, et al: Neonatal-6-hydroxydopamine treatment: Mode of susceptibility for self-mutilation in the Lesch–Nyhan syndrome. Pharmacol Biochem Behav 21:459–1461, 1984
73. Goldstein M, Shegeki K, Kusano N, et al: Dopamine agonist induced self-mutilative biting behavior in monkeys with unilateral ventromedial tegmental lesions of the brainstem: Possible pharmacological model for Lesch–Nyhan syndrome. Brain Res 367:114–120, 1986
74. Breese GR, Baumeister A, Napier TC, et al: Evidence that D-1 dopamine receptors contribute to the supersensitive behavioral response induced by L-dihydroxyphenylalanine in rats treated neonatally with 6-hydroxydopamine. J Pharmacol Exp Ther 235:287–295, 1985
75. Mizuno T, Yugari Y: Prophylactic effect of L-5-hydroxytryptophan on self-mutilation in the Lesch–Nyhan syndrome. Neuropadiatrie 6:13–23, 1975
76. Goldstein M, Anderson LT, Reuben R, et al: Self-mutilation in Lesch–Nyhan disease is caused by dopaminergic denervtion. Lancet 1:338–339, 1985
77. Boyd EM, Dolman M, Knight LM, et al: The chronic oral toxicity of caffeine. Can J Physiol Pharmacol 43:995–1007, 1965
78. Sakata T, Fuchimoto H: Stereotyped and aggressive behavior induced by sustained high doses of theophylline in rats. Jpn J Pharmacol 23:781–785, 1973
79. Rosenberg D, Monnet P, Mamelle JL, et al: Encephalopathie avec troubles du metabolisme des purines. Presse Med 76:2333–2336, 1968
80. Berman PH, Balis ME, Dancis J: Congenital hyperuricemia, an inborn error of purine metabolism associated with psychomotor retardation, athetosis, and self-mutilation. Arch Neurol 20:44–53, 1969
81. Watts RWE, McKeran RO, Brown E, et al: Clinical and biochemical studies on treatment of Lesch–Nyhan syndrome. Arch Dis Child 49:653–702, 1974
82. Lowenstein JM: Ammonia production in muscle and other tissues: The purine nucleotide cycle. Physiol Rev 52:382–414, 1972
83. Manzke H, Gustman H, Koke HG, et al: Hypoxanthine and tetrahydrobiopterin treatment of a patient with features of the Lesch–Nyhan syndrome, in Nyhan WL, Thompson LF, Watts RWE (eds), Purine and Pyrimidine Metabolism in Man. Vol V: Clinical Aspects Including Molecular Genetics. New York, Plenum, 1986, pp 197–204
84. Edwards NL, Jeryc W, Lieberman C, et al: Enzyme replacement in the Lesch–Nyhan syndrome. Clin Res 28:139A, 1980

85. Nyhan WL, Parkman R, Page T, et al: Bone marrow transplantation in Lesch–Nyhan disease, in Nyhan WL, Thompson LF, Watts RWE (eds), Purine and Pyrimidine Metabolism in Man. Vol V: Clinical Aspects Including Molecular Genetics. New York, Plenum, 1986, pp 167–170
86. Mizuno T, Yugari Y: Self-mutilation in Lesch–Nyhan syndrome. Lancet 1:761, 1974
87. Kuehn MR, Bradley A, Robertson EJ, et al: A potential animal model for Lesch–Nyhan syndrome through induction of HPRT mutation into mice. Nature (Lond) 326:295–301, 1987
88. Jolly DJ, Okayama H, Berg P, et al: Isolation and characterization of a full length expressible cDNA for human hypoxanthine phosphoribosyltransferase. Proc Natl Acad Sci USA 80:477–481, 1983
89. Willis RC, Jolly DJ, Miller AD, et al: Partial phenotypic correction of a human Lesch–Nyhan (hypoxanthine-guanine phosphoribosyltransferase-deficient) lymphoblast line with a transmissible retroviral vector. J Biol Chem 259:7842–7849, 1984
90. Gruber HE, Finley KD, Luchtman LA, et al: Insertion of hypoxanthine phosphoribosyltransferase cDNA into human bone marrow by a retrovirus, in Nyhan WL, Thompson LF, Watts RWE (eds), Purine and Pyrimidine Metabolism in Man. Vol V: Clinical Aspects Including Molecular Genetics. New York, Plenum, 1986, pp 171–175

19

Biochemical Studies of Suicide

Boris Birmaher, Laurence L. Greenhill, and Michael Stanley

1. INTRODUCTION

The incidence of suicide in adolescents has been increasing, making it more and more important to arrive at an accurate diagnosis and to mount effective prevention programs.[1] Efforts to construct a behavior profile that predicts risk of attempted or completed suicide have been unsuccessful. Only weak correlations exist at present between suicide and types of premorbid behavior. Current profiles of the child at suicide risk turn up many false-positive results,[2] suggesting that available models have low specificity.

Because behavioral factors by themselves have been of limited clinical value, several investigators have tried to identify a biochemical marker that by itself or in combination with behavioral factors would predict suicide. Many investigations carried out with adults or young adults can serve as a bridge for future research with adolescents and children. This chapter describes the studies that have sought these biological markers.

2. POSTMORTEM FINDINGS IN SUICIDE VICTIMS

2.1. Biogenic Amines and Their Metabolites

Several studies of suicide victims have compared levels of the biogenic amine serotonin, or 5-hydroxytryptophan (5-HT), norepinephrine (NE), and dopamine (DA), and their respective metabolites 5-hydroxyindolacetic acid (5-HIAA), 3-methoxy,4-hydroxyphenylglycol (MHPG), and homovanillic acid (HVA), with levels in nonsuicidal controls. The most consistent finding reported was a decrease in 5-HT or 5-HIAA in the brainstem or hypothalamus.[3–12] Lloyd et al.[3] measured 5-HT and 5-HIAA in raphe nuclei of five suicide victims

Boris Birmaher • Department of Child Psychiatry, College of Physicians and Surgeons of Columbia University, New York, New York 10032. *Laurence L. Greenhill* • Department of Clinical Psychiatry, College of Physicians and Surgeons of Columbia University, New York, New York 10032. *Michael Stanley* • Departments of Psychiatry and Pharmacology, College of Physicians and Surgeons of Columbia University, and Department of Neurochemistry, New York State Psychiatric Institute, New York, New York 10032.

Table I. Postmortem Studies: Neurotransmitters and Metabolites

Investigators	Findings
Shaw et al.[5]	↓ Brainstem 5-HIAA
Beskow et al.[89]	↓ Brainstem 5-HIAA
Bourne et al.[4]	↓ Brainstem 5-HIAA
Pare et al.[3]	↓ Brainstem 5-HT
Lloyd et al.[91]	↓ Brainstem 5-HT
Korpi et al.[13]	↓ Hypothalamus 5-HT ↓ Nucleus accumens 5-HIAA
Cochran et al.[88]	No change in 5-HT in brainstem and frontal cortex
Stanley et al.[10]	No change in 5-HIAA in frontal cortex
Owen et al.[12]	No change in 5-HIAA in frontal cortex
Crow et al.[11]	No change in 5-HT or 5-HIAA in frontal cortex

and five controls. Three of the suicide victims had died by drug overdose. They found no significant difference in 5-HIAA levels between the two groups. There was, however, a significant reduction in brainstem 5-HT levels for the suicide group. Pare et al.[4] determined NE, DA, 5-HT, and 5-HIAA levels in suicide victims who died by carbon monoxide poisoning. These workers found no significant differences between the two groups for NE, DA, and 5-HIAA. They also reported a significant reduction in brainstem levels of 5-HT for the suicide group. Shaw et al.[5] found significantly lower brainstem levels of 5-HT in suicide victims compared with controls. However, it should be noted that approximately one half of the suicide group died of barbiturate overdose and the other half died of carbon monoxide poisoning. More recently, Korpi et al.[6] reported significant decreases in the hypothalamic concentration of 5-HT of suicide victims compared with nonsuicide controls.

Three studies have reported significant reductions in the levels of 5-HIAA in suicide victims. Bourne et al.[8] measured NE, 5-HT, and 5-HIAA in the hindbrain and found significantly lower levels for only 5-HIAA. Beskow et al.[9] measured DA, NE, and 5-HIAA in brainstem areas of suicide victims and controls. They noted significant reductions in 5-HIAA levels for the suicide group. Changes in DA, NE, or their metabolites were either negative or inconsistent (Table I).

2.2. Monoamine Oxidase

Serotonin (5-HT) is a major substrate for the enzyme monoamine oxidase (MAO). Changes in this enzyme may explain previously reported decreases in 5-HT or 5-HIAA that were falsely attributed to abnormal serotonin metabolism in suicide victims. Yet there has been no consistent reporting of MAO enzyme abnormalities. Grote et al.[13] did not find differences between MAO activity in brains of suicide victims and controls. Gottfries et al.[14] found reduced MAO activity in suicidal victims with history of alcoholism but not in suicidal depressive victims without history of alcoholism. Mann and Stanley[15] studied MAO-A and B in the frontal cortex of violent suicide victims and controls matched for age, sex, and postmortem interval; those suicide victims who died from an overdose were excluded to avoid drug effects on MAO activity. They used labeled 5-HT and phenylethylamine (PEA) as substrates for MAO-A and B, respectively. The results of this study showed no significant differences between the groups for either substrates (5-HT or PEA) with respect to MAO V_{max} (maximal

Table II. Receptor Studies

Investigators	Findings
Stanley et al.[20]	↓ [^{3}H]-IMI in cortex
Paul et al.[21]	↓ [^{3}H]-IMI in hypothalamus
Crow et al.[92]	↓ [^{3}H]-IMI in cortex
Perry et al.[93]	↓ [^{3}H]-IMI in cortex
Meyerson et al.[22]	↑ [^{3}H]-IMI in cortex
Stanley and Mann[86]	↑ 5-HT in cortex
Mann et al.[24]	↑ 5-HT in cortex
Meltzer et al.[25]	↑ 5-HT in cortex
Owen et al.[94]	Increase in 5-HT but nonsignificant
Cheetham et al.[26]	↓ 5-HT in hippocampus
Stanley et al.[28]	No change in QNB in cortex
Kauffman et al.[29]	No change in QNB in cortex
Meyerson et al.[22]	↑ In QNB in frontal cortex
Zanko and Biegon[95]	↑ In β in frontal cortex
Biegon[32]	↑ In β in frontal cortex
Mann et al.[96]	↑ In β in frontal cortex
Meyerson[22]	No change in β in frontal cortex

uptake of 5-HT) or K_M (affinity). There was a significant positive correlation between age and MAO (V_{max}) for both groups. These data further suggest that the reported lower brain MAO activity in alcoholic suicides,[14] if confirmed, may be related primarily to alcoholism rather than to suicide per se.

2.3. Receptor Studies

Recently, new assays have been developed to measure the density of receptors associated with pre- and post-synaptic 5-HT neurons[16,17] (Table II). Imipramine binding sites ([^{3}H]-IMI) have been found to be closely associated with 5-HT presynaptic binding sites. The IMI sites have been characterized in platelets and various regions of the brain.[16–18] The idea of studying imipramine binding in suicide victims was derived in part from the clinical findings of Langer et al.,[16] who had reported significantly reduced B_{max} values in platelets of depressed patients compared with controls.[19] The combined association of IMI binding with 5-HT function, as well as the significant reduction in binding density in depressives, suggested the possibility that alterations in imipramine binding might be present in suicide victims. To test this hypothesis, Stanley et al.[20] compared the IMI sites in the brains of suicide victims with controls, matched for age, gender, and postmortem delay. Both groups had experienced sudden violent deaths. They found a significant reduction in the number of [^{3}H]-IMI binding sites in frontal cortex of suicide victims with no differences in binding affinity.

Paul et al.[21] measured [^{3}H]-IMI binding in hypothalamic membranes from suicides and controls. Both groups were matched for age, sex, and postmortem delay. IMI binding was significantly lower in the brains of suicide victims than in the controls. Crow et al.[11] also reported a significant decrease in [^{3}H]-IMI binding in cortex of suicide victims compared with matched controls. Only one study[22] found an increase of [^{3}H]-IMI binding in the brain of suicides compared with a small group of controls.

Another way to assess serotonergic activity in brain is to measure postsynaptic 5-HT-2receptors using ligands such as [^{3}H]spiroperidol or [^{3}H]ketanserin. 5-HT-2 binding in animals has been shown to be downregulated or reduced in response to chronic antidepressant treatment. Lesioning 5-HT nuclei produces upregulation or increased 5-HT-2 receptor binding.[17,23] Stanley et al.[10] measured 5-HT-2 binding in the frontal cortex of suicide victims compared with matched controls. They found a significant increase in the number of 5-HT-2 receptors in the frontal cortex in suicide victims without a change in affinity. The number of binding sites for 5-HT-2 correlated negatively (trend) with the [^{3}-H]-IMI binding sites, suggesting a compensatory increase in postsynaptic binding sites secondary to a reduction in presynaptic input. The increase in 5-HT-2 receptors in suicide victims has been replicated by other investigators.[12,24,25] In contrast, Cheetham and colleagues[26] found decrease in 5-HT-2 receptor binding in cortex of suicide victims.

Muscarinic (QNB) and β-adrenergic binding sites also have been measured in the brains of suicide victims. The rationale for these studies was suggested by the cholinergic–adrenergic imbalance theory postulated by Janowsky et al.[27] and by the known proportion of suicide victims diagnosed as suffering from affective disorder. Stanley et al.[28] measured QNB binding in a study comparing a large number of suicide victims and nonsuicide controls and found no difference between the number of binding sites or in the affinity in the frontal cortex for the two groups. This finding was replicated by Kauffman et al.,[29] who determined QNB binding in three brain regions (including the frontal cortex) in suicide victims. They found no differences in binding parameters for either group in any of the brain regions studied. By contrast, to the findings observed in the two previously described studies, Meyerson et al.[22] reported a significant increase in the QNB binding sites in a small sample group of suicide victims.

In attempt to better understand the functional status of central noradrenergic neurons in suicidal behavior, Mann et al.[24] measured β-adrenergic receptors in suicide victims. It has been suggested that alterations in β-adrenergic receptors might be linked to the therapeutic actions of antidepressant drugs.[30] Mann et al.[24] noted a significant increase in specific β-adrenergic binding in suicide victims compared with controls. Zanko and Beigon[31] and Beigon[32] reported an increase in the number of binding sites with no change in the affinity in a small group of suicide victims compared with matched controls. By contrast, Meyerson et al.[22] did not find changes in dihydroalprenolol (DHA) binding in suicide victims. It should be noted that antemortem use of antidepressants would not explain the receptor alterations observed by Mann et al.[24] and Zanko and Beigon,[31] as data from animal studies indicate that chronic antidepressant treatment causes a downregulation of β-adrenergic receptors.[30]

In addition to those studies that have measured receptors sites for the biogenic amines in suicide victims, a recent study by Nemeroff and colleagues[33] measured binding sites for corticotropin releasing factor (CRF). CRF is the hypothalamic peptide that influences the release of adrenocorticotropic hormone (ACTH) from the anterior pituitary; this peptide may be implicated in the sustained cortisol plasma levels and the "escape" phenomena to the dexamethasone response test seen in some depressive patients. Nemeroff reported low CRF binding sites in the frontal cortex of suicide victims in comparison with nonsuicide controls.

3. CSF STUDIES OF SUICIDE ATTEMPTERS

Many of the investigations into the role of biogenic amines in affective disorders involved measures of CSF metabolites (Table III). One issue that is frequently raised concerning measures derived from CSF relates to the validity of these values as an indicator of central turnover of the parent aminergic system. For example, it has been argued that CSF levels of 5-HIAA may be more indicative of local 5-HT turnover in spinal cord rather than 5-HT turnover in brain.[34] A number of studies have addressed this question using a variety of indirect measures that have yielded equivocal findings.[35,36] In attempt to assess the degree to which 5-HIAA in the CSF reflects brain levels of 5-HIAA directly, Stanley et al.[37] measured 5-HIAA in

Table III. CSF 5-HIAA Studies of Suicide Attempters

Investigators	Diagnosis	CSF 5-HIAA findings
Asberg et al.[38]	Depression	Decreased
Agren[40]	Depression	Decreased
Banki et al.[62]	Depression	Decreased
Traskman et al.[97]	Depression and personality disorders	Decreased
Van Praag[41]	Depression	Decreased
Montgomery et al.[98]	Depression	Decreased
Palanappian et al.[99]	Depression	Decreased
Banki et al.[100]	Depression	Decreased
Perez et al.[101]	Depression	Decreased
Secunda et al.[67]	Depression	No significant difference
Berreteni et al.[102]	Bipolar	No significant difference
Van Praag[102]	Schizophrenia	Decreased
Banki et al.[82]	Schizophrenia	Decreased
Ninan et al.[104]	Schizophrenia	Decreased
Roy et al.[55]	Schizophrenia	No significant difference
Stanley et al.[105]	Schizophrenia	No significant difference
Pickar et al.[56]	Schizophrenia	No significant difference
Banki et al.[106]	Alcoholics	Decreased
Brown et al.[107]	Personality disorders	Decreased

the brain and CSF of the same individual in samples obtained at autopsy. They found that there was a significant positive correlation ($r = 0.78$) between CSF and brain 5-HIAA levels. These findings provide the best proof that CSF metabolites do, in fact, reflect biogenic metabolism in the brain.

The first reports of an association between suicidal behavior and CSF levels of 5-HIAA was published in 1976 by Asberg et al.[38] These workers observed a bimodal distribution of 5-HIAA concentration in the CSF of a group of depressed patients. Within the group that had the lowest levels of CSF 5-HIAA, they found a significant proportion of individuals who had either committed or attempted suicide by violent means. Asberg et al.[39] also reported that persons who attempted suicide and who have low CSF 5-HIAA concentrations are 10 times at greater risk of committing suicide than those with higher 5-HIAA.

Agren et al.[40] compared a group of depressed nonsuicidal patients and a group of depressed suicidal patients and showed that those with history of suicide has lower 5-HIAA CSF levels. Other investigators have reproduced this finding.[41–45]

Low CSF 5-HIAA levels have also been reported for suicidal victims with personality disorders, for schizophrenics, and for persons exhibiting violent behavior.[46–51] Brown et al.[46] studied two groups of personality disorders and were able to identify those with a positive history of suicidal behavior by the presence of lower concentrations of CSF 5-HIAA. In a different study, Brown et al.[47] found an inverse correlation between CSF 5-HIAA concentrations and a history of aggressive behavior. Traskman et al.[48] compared a group of depressed and nondepressed patients who carried diagnoses of personality, anxiety, and manic-depressive disorders with a group of normal volunteers. They found low 5-HIAA levels in CSF of those with suicidal behavior, particularly those who had made violent suicide attempts. There was no difference between depressed and nondepressed suicide attempters.

Linnoila et al.[49] studied violent groups of criminals who had a history of murder or attempted murder and found 5-HIAA concentrations in the CSF of impulsive violent offenders, but no in those offenders who had premeditated their acts. In addition, those violent prisoners who had made suicide attempts had the lowest CSF 5-HIAA concentrations.

Schizophrenics are also a high-risk group for suicide.[52–54] Van Praag et al.[50] Ninan et al.,[51] and Banki et al.[42] reported lower CSF 5-HIAA concentrations in schizophrenic patients who had attempted suicide in comparison with those who never attempted suicide. However, other groups did not replicate this finding.[55,56,105]

Bipolar patients are also at high risk of suicide.[57] However, a comparison of bipolar suicide attempters versus nonattempters has not shown low CSF 5-HIAA concentrations to be present in the attempters.[58,59] It has been suggested that a significant degree of serotonergic dysfunction may be related directly related to the illness itself and therefore obscure any potential suicide–serotonin relationship.[60]

Alcoholic patients are recognized for their high risk of suicide and suicide attempts.[52] Banki et al.[61] found that alcoholics who had made a suicide attempt had lower CSF levels of 5-HIAA than did alcoholics who had not attempted suicide. Other workers have reported an inverse correlation between the length of alcohol abstinence and the levels of CSF 5-HIAA.[62,63] These findings support the hypothesis that alcoholic patients have a lower 5-HT turnover in the central nervous system (CNS) and that acute consumption of alcohol promotes an increase in 5-HT release and turnover. All these findings need to be viewed with caution because alcohol modifies the turnover of biogenic amines.[64–66]

Concentrations of CSF MHPG in suicide attempters have been compared with CSF levels in nonsuicidal controls. While some studies report low CSF MHPG levels[40,67] in the attempters, others found no differences between controls and attempters.[13,47,68] Arato et al.[69] measured CRF levels in the CSF of suicidal and nonsuicidal depressive patients and found no differences between the two groups. Appropriately controlled studies reported elevated CRF levels[33] in the CSF of depressed patients. These findings suggest that CRF changes in CSF may be associated with depression rather than suicidal behavior alone.

4. PERIPHERAL STUDIES IN SUICIDE ATTEMPTERS

Brain and platelets have IMI binding sites that have been shown to be related to 5-HT function. The density of these sites may be reduced in depressive patients.[19] Similar studies are not available for those who have attempted suicide. Stoff et al.[70] measured [^{3}H]-IMI binding sites in platelets of 10 drug-free, impulsive, aggressive conduct disorder children compared with normal controls. Using the Child Behavior Checklist, they showed a significant negative correlation between maximal platelet [^{3}H]-IMI binding and the factor for externalizing behavior (problems directed toward outside world) and aggression.

Shaffer and Bacon[71] reported high rates of aggression and antisocial behavior in adolescent suicide victims and suicide attempters. This finding suggests a biological similarity between aggression toward others and anger turned inward in the form of suicidal behavior.

Platelet MAO has also been found to be low in normal volunteers with a family history of completed or attempted suicide, but not in patients who had attempted suicide.[68,72–74]

5. NEUROENDOCRINE STUDIES

In addition to the study of neurotransmitter metabolites in CSF, depressed and suicidal patients have been the subject of wide ranging neuroendocrine studies. Pituitary hormone secretion, under hypothalamic control, is modulated by neurotransmitter interactions, which have been investigated using sophisticated challenge strategies. Levels of 24-hr, urinary 17-hydroxycorticosteroids and urinary free cortisol has been found to be increased in suicidal patients, or in patients who subsequently committed suicide.[75–77] Other investigations reported normal or low levels of urinary cortisol[78,79] and no significant correlation between hypercortisolemia and suicide attempts.[80] No significant association was found between suicidal behavior and CSF levels of cortisol, ACTH, and CRF.[39,69] Arato et al.[69] found no differences between either ACTH or cortisol concentrations measured in the CSF of suicide victims and controls.

The relationship between dexamethasone suppression of cortisol and suicide behavior is less clear. A positive association has been reported in some instances[80–82] but not in other studies.[80,83] Meltzer et al.[84] administered 5-HT to depressed patients and measured the cortisol response in plasma. Recently, Meltzer et al.[25] reported similar results using the serotonin agonist MK-212, which acts directly on serotonergic receptors. They found an enhanced cortisol response in those patients who made a suicide attempt. This finding is consistent with the reported postsynaptic serotonin supersensitivity in the brain of suicide victims. These data are consistent with report by Stanley and others[10,25,51] reporting an increased number of 5-HT-2 receptors in the brain of suicide victims.

Cocarro et al.[85] reported a negative correlation among severity of suicide behavior, impulsivity and aggressivity, and prolactin response induced by fenfluramine. This drug is an indirect serotonin agonist that releases endogenous stores of 5-HT and inhibits its reuptake. The reduced prolactin release among suicide and impulsive patients suggests hyposerotonergic activity in this group.

6. CONCLUSIONS

There is now a growing body of biochemical studies on adult suicide victims that could serve to direct research and clinical work with children and adolescents at risk for self-injury. While it is true that one must factor in confounding variables such as age, gender, death by overdose or carbon monoxide poisoning, extensive postmortem delay, lack of matched controls, or use of alcohol or other drugs when interpreting postmortem studies of suicide victims,[85] it appears clear that suicidal behavior is correlated with lowered CNS serotonergic activity. This is well supported by a number of methodologically sound papers involving postmortem reports, in vivo CSF protocols, and neuroendocrine studies.

Other important biochemical evidence with implications for children and adolescents has emerged from post mortem studies. Reduced [^{3}H]-IMI receptor binding and reduced prolactin responsivity to fenfluramine has been described, presumably due to low presynaptic serotonergic activity. This may explain the reduced levels of CSF 5-HIAA found in suicide attempters as well as reduced levels of 5-HT and 5-HIAA in the brains of suicide victims. Upregulation of the postsynaptic 5-HT receptors could then occur and has been reported in the postmortem work on suicide victims. Suicidal, depressed patients also show an enhanced response to 5-HT and the serotonin agonist MK-212.

The relationship between suicidal behavior and hypofunctioning of the serotonergic system suggests that a biochemical defect could contribute to a character disorder involving impulsive and/or aggressive behavior, either self-directed or other-directed. Low 5-HT activity has been reported in patients with aggressive or violent behavior and in preclinical investigations on animal aggressive behavior.[87]

In addition to reduced 5-HT activity, some investigators also reported increases in β-adrenergic functioning. Both reduced 5-HT binding and heightened responsiveness in the β-adrenergic system provide additional biochemical clues that could serve as potential "markers" to help the clinician to better identify those children or adolescents at risk for suicidal behavior.[12]

ACKNOWLEDGMENTS. This work was suggested in part by grants MH42242 and MH41847, from the U.S. Public Health Service, the Scottish Rite Schizophrenia Research Program and the Lowenstein Foundation, and MH38838-04 from the National Institute of Mental Health. Portions of the material in this review have been presented at the following meetings: American College of Neuropsychopharmacology, December, 1987; American Psychiatric Association, May, 1987; and Biological Psychiatry, May, 1989.

REFERENCES

1. Curran DK: Adolescent Suicidal Behavior. Hagerstown, Maryland, Harper & Row, 1987, pp 14
2. Cohen J: Statistical approaches to suicidal risk factor analysis, in Stanley M, Mann JJ (eds), Psychobiology of Suicidal behavior. New York, New York Academy of Sciences, 1986, pp 34–41 ·
3. Lloyd KG, Farley IJ, Deck JHN, et al: Serotonin and 5-hydroxyindoleacetic acid in discrete areas of the brainstem of suicide victims and control patients, in Advances in Biochemical Psychopharmacology, Vol. II. New York, Raven Press, 1974, pp 387–397
4. Pare CMB, Yeung DPH, Price K, et al: 5-Hydroxytryptamine, noradrenaline and dopamine in brain steam, hypothalamus and caudate nucleus of controls of patients committing suicide by coal-gas poising. Lancet 1:131–135, 1969
5. Shaw DM, Camps FE, Eccleston EG: 5-hydroxytryptamine in the hind brain of depressive suicides. Br J Psychiatry 113:1407–1411, 1967
6. Korpi ER, Kleinman JE, Goodman SI, et al: Serotonin and 5-hydroxyindolacetic acid concentrations in different brain regions of suicide victims: Comparison in chronic schizophrenic patients with suicide as cause of death. Presented at the International Society of Neurochemistry, Vancouver, Canada 1983
7. Gillin JC, Kelsoe JR, Kaufman CA, et al: Muscarinic receptor density in skin fibroblasts and autopsied brain tissue in affective disorder. Presented at Psychobiology of Suicide Behavior. New York, New York Academy of Sciences, September 1985, pp 143–147
8. Bourne HR, Bunney WE Jr, Colburn RW, et al: Noradrenaline, 5-hydroxytryptamine, and 5-indoleacetic acid in the human brain: Post mortem studies in a group of suicides and in a control group. Lancet 2:805–808, 1968
9. Beskow J, Gottfriess GC, Roos BE, et al: Determination of monoamine and monoamine metabolites in the human brain: post mortem studies in a group of suicides and in a control group. Acta Psychiatr Scand 53:7–20, 1976
10. Stanley M, Mann JJ: Increased serotonin-2 binding sites in frontal cortex of suicide victims. Lancet 2:214–216, 1983
11. Crow TJ, Cross AJ, Cooper SJ, et al: Neurotransmitter receptors and monoamine metabolites in the patient with Alzheimer-type dementia and depression and suicides. Neuropharmacology 23:1561–1569, 1984
12. Owen F, Cross AJ, Crow TJ, et al: Brain 5-HT_2 receptors and suicide. Lancet 2:1256, 1963
13. Grote SS, Moses SG, Robins, E, et al: A study of selected catecholamine metabolizing enzymes: A comparison of depressive suicides and alcoholic suicides with controls. J Neurochem 23:791–236, 1974
14. Gottfries, CG, Knorring LV, Oreland L: Platelet monamine oxidase activity in brains from alcoholic. J Neurochem, 25:667–673, 1975
15. Mann JJ, Stanley M: Postmortem monoamine oxidase enzyme kinetics in the frontal cortex of suicide victims and controls. Acta Psychiatr Scand 69:135–139, 1984
16. Langer SZ, Moret C, Raisman R, et al; High-affinity [^{3}H]imipramine binding in rat hypothalamus: Association with uptake of serotonin but not of norepinephrine. Science 210:1133–1135, 1980
17. Brunello N, Chuang DM, Costa E: Different synaptic location of mianserin and imipramine binding sites. Science 215:1112–1115, 1982
18. Rehavi M, Skolnick P, Paul SM: Solubilization and partial purification of the high affinity [^{3}H]imipramine binding site from human platelets. FEBS Lett 150:514–518, 1982
19. Raisman R, Sechter D, Briley MS, et al: High affinity ^{3}H-imipramine binding in platelets from untreated and treated depressed patients compared to healthy volunteers. Psychopharmacology 75:368–371, 1981
20. Stanley M, Virgilio J, Gershon S: Tritiated imipramine binding sites are decreased in the frontal cortex of suicides. Science 216:1337–1339, 1982
21. Paul SM, Rehavi M, Skolnick P, et al: High affinity binding of antidepressants to a biogenic amine transport site in human brain and platelet: Studies in depression, in Post RM, Ballenger JC (eds), Neurobiology and Mood Disorders. Baltimore, William & Wilkins, 1984, pp 846–853

22. Meyerson LR, Wennogle LP, Abel MS, et al: Human brain receptor alterations in suicide victims. Pharmacol Biochem Behav 17:159–163, 1982
23. Peroutka SJ, Snyder SH: Regulation of serotonin (5-HT_2) labeled with [^{3}H] spiroperidol by chronic treatment with antidepressant amitriptyline. Pharmacol Exp Therap 215:582–587, 1980
24. Mann JJ, Stanley M, McBride AP, et al: Increased serotonin and beta-adrenergic receptor binding in the frontal cortices of suicide victims. Arch Gen Psychiatry 43:954–959, 1986
25. Meltzer HY, Nash JF, Ohmori T, et al: Neuroendocrine and biochemical studies of serotonin and dopamine in depression and suicide. International Conference on New Directions in Affective Disorders, S60, 1987
26. Cheetham SC, Crompton MR, Katona CLE, Horton RW: Brain 5-HT_2 receptor binding sites in depressed suicide victims. Brain Res. 443:272–280, 1988
27. Janowsky DS, El-Yousef MK, David JM, et al: A cholinergic-adrenergic hypothesis of mania and depression. Lancet 1:632, 1972
28. Stanley M: Cholinergic receptor binding in the frontal cortex of suicide victims. Am J Psychiatry 141:1432–1436, 1984
29. Kaufman CA, Gillin JC, Hill B, et al: Muscarinic binding sites in suicides. Psychiatry Res 12:47–55, 1984
30. Sulser F, Robonson SE: Clinical implications of pharmacological differences among antipsychotic drugs, In Lipton MA, DiMascio A, Killam KF (eds), Psychopharmacology: A Generation of Progress. New York, Raven, 1978, pp 943–954
31. Zanko MT, Biegon A: Increased β-adrenergic receptor binding in human frontal cortex of suicide victims. Abstract of the Annual Meeting, Society of Neuroscience, Boston, 1983
32. Biegon A: Findings in beta receptors in suicide victims. Grand Rounds, Rockefeller University, June 1987
33. Nemeroff CB, Widerlow E, Bissett G, et al: Arch Gen Psychiatry, 45:577–579, 1988
34. Ashby P, Verrier M, Wash JJ, et al: Spinal reflexes and the concentrations of 5-HIAA, MHPG and HVA in lumbar cerebrospinal fluid after spinal lesions in man. J Neurol Neurosurg Psychiatry 39:1191–1200, 1976
35. Curzon G, Gumpert EJ, Sharpe DM: Amine metabolites in the lumbar cerebrospinal fluid of humans with restricted flow of cerebrospinal fluid. Nature New Biol 231:189–191, 1971
36. Post RM, Goodwin FK, Gordon E, et al: Amine metabolites in human cerebrospinal fluid: Effects of cord transection and spinal fluid block. Science 179:897–899, 1973
37. Stanley M, Traskman-Bendz L, Dorovini-Zis K: Correlations between aminergic metabolites simultaneously obtained from human CSF and brain. Life Sci 37:1279–1286, 1985
38. Asberg M, Thoren P, Traskman L, et al: Serotonin depression: A biochemical subgroup within the affective disorders? Science 191:478–480, 1976
39. Asberg M, Traskman L, Birtilson L, et al: Studies of CSF 5-HIAA in depression and suicidal behavior. Exp Med Biol 133:739–752, 1981
40. Agren H: Symptom patterns in unipolar and bipolar depression correlating with monoamine metabolites in the cerebrospinal fluid. II. Suicide Psychiatry Res 3:225–236, 1980
41. Van Praag HM: Depression, suicide and the metabolism of serotonin in the brain. J Affective Dis 4:275–290, 1982
42. Banki CM, Arato M, Papp Z, et al: Biochemical markers in suicidal patients: Investigations with cerebrospinal fluid amine metabolites and neuroendocrine tests. J Affective Dis 6:341–350, 1984
43. Palanappian V, Ramachandran V, Somasundaram O: Suicidal ideation and biogenic amines in depression. Indian J Psychiatry 25:268–292, 1983
44. Montgomery SA, Montgomery D: Pharmacological prevention of suicidal behavior. J Affective Dis 4:291–298, 1982
45. Perez de los Cobos JZ, Lopez-Ibor Alino JJ, Saiz Ruiz J: Correlatos biologicaos del suicido y la agresivivad en depressiones mayores (con melancholia): 5-HIAA en LCR, DST, y respuesta terapeutica a 5-HTP. Presented at the First Congress of the Spanish Society for Biological Psychiatry, Barcelona, 1984
46. Brown GL, Goodwin FK, Ballenger JC, et al: Aggression in humans correlates with cerebrospinal fluid amine metabolism. Psychiatry Res 1:131–139, 1979
47. Brown GL, Ebert MH, Goyer PF, et al: Aggression, suicide and serotonin; relationships to CSF amine metabolites. Am J Psychiatry 139:741–746, 1982
48. Traskman L, Asberg M, Bertillsson K, et al: Monoamine metabolites in CSF and suicidal behavior. Arch Gen Psychiatry 38:631–636, 1981
49. Linnoila M, Virkkunen M, Scheinin M, et al: Low cerebrospinal fluid 5-hydroxyindoleacetic acid concentrations differentiates impulsive violent behavior. Life Sci 33:2609–2614, 1983
50. Van Praag HM: CSF 5-HIAA and suicide in non-depressed schizophrenics. Lancet 2:977–978, 1984
51. Ninan PT, Van Kammen DP, Scheinin M, et al: CSF 5-hydroxyindoleacetic acid in suicidal schizophrenic patients. Am J Psychiatry 141:566–569, 1984

52. Miles CL: Conditions predisposing to suicide. J Nerv Ment Dis 164:231–246, 1977
53. Tsuang MT: Suicide in schizophrenics, manics, depressives, and surgical controls. Arch Gen Psychiatry 35:153–155, 1978
54. Roy A: Suicide in chronic schizophrenics. Br J Psychiatry 144:171–177, 1982
55. Roy A, Ninam P, Mazonson A, et al: CSF monoamine metabolite in chronic schizophrenia patients who attempt suicide. Psychol Med 15:335–340, 1985
56. Pickar D, Roy A, Brier A, et al: Suicide and aggression in schizophrenia: neurobiologic correlates, in Mann JJ, Stanley M (eds), Psychobiology of Suicidal Behavior. New York, New York Academy of Sciences, 1986, pp 189–196
57. Jamison KR: Suicide and bipolar disorders, in Mann JJ, Stanley M (eds), Psychobiology of Suicidal Behavior. New York, New York Academy of Sciences, 1986, pp 301–315
58. Roy-Byrne P, Post RM, Rubinow DR, et al: CSF 5-HIAA and personal and family history of suicide in affectively ill patients: a negative study. Psychiatry Res 10:263–274, 1983
59. Berrettini W, Nurenberger J, Narrow W, et al: Cerebrospinal fluid studies of bipolar patients with and without a history of suicide attempts, in Mann JJ, Stanley M (eds): Psychobiology of Suicidal Behavior. New York, New York Academy of Sciences, 1986, pp 197–201
60. Goodwin FK: Suicide aggression and depression: A theoretical framework for future research, in Mann JJ, Stanley M (eds), Psychobiology of Suicidal Behavior. New York, New York Academy of Sciences, 1986, pp 351–356
61. Banki C, Arato M, Kilts C: Aminergic studies and cerebrospinal fluid cautions in suicide, in Mann JJ, Stanley M (eds), Psychobiology of Suicidal Behavior. New York, New York Academy of Sciences, 1986, pp 221–230
62. Banki C: Factors influencing monoamine metabolites and tryptophan in patients with alcohol dependence. J Neural Transm 50:98–101, 1981
63. Ballenger JC, Goodwin FK, Major LF, et al: Alcohol and central serotonin metabolism in man. Arch Gen Psychiatry 36:224–227, 1979
64. Tabakoff B, Ritzmann R: Inhibition of the transport of 5-hydroxyindoleacetic acid from brain by ethanol. J Neurochem 24:1043–1051, 1975
65. Ellingboe J: Effect of alcohol on neurochemical processes, in Lipton M, DiMascio A, Killman K (eds), Psychopharmacology: A Generation of Progress. New York, Raven, 1978, pp 1653–1654
66. Herrero E: Monoamine metabolism in rat brain regions following long-term alcohol treatment. J Neural Transm 47:227–236, 1980
67. Secunda JA, Cross CK, Koslow K, et al: Biochemistry and suicidal behavior in depressed patients. Biol Psychiatry 21:756–767, 1986
68. Oreland L, Wilberg A, Asberg M, et al: Platelet MAO activity and monoamine metabolites in cerebrospinal fluid in depressed and suicidal patients and in healthy controls. Psychiatry Res 4:21–29, 1981
69. Arato M, Banki CM, Nemeroff CB, et al: Hypothalamic–pituitary–adrenal axis and suicide, in Mann JJ, Stanley M, (eds), Psychobiology of Suicidal Behavior. New York, New York Academy of Sciences, 1986, pp 263–270
70. Stoff D, Pollack L, Bridger WL: Platelet imipramine binding sites correlate with aggression in adolescents. Biol Psychiatry 12:180–182, 1986
71. Shaffer D, Bacon K: A critical review of prevention intervention efforts in suicide with particular preference to youth suicide. Prepared for the Prevention and Intervention Work Group of HHS Task Force on Youth Suicide. Presented in Oakland, California, June 11–13, 1986
72. Grottfries CG, Knorring LV, Oreland L: Platelet monamine oxidase activity in mental disorders. Neuropsychopharmacology 4:185–192, 1980
73. Buchsbaum MS, Hairr RJ, Murphy DL: Suicide attempts, platelet monoamine oxidase and the average evoked response. Acta Psychiatry Scand 56:69–77, 1979
74. Meltzer HY, Arora RC: Platelet markers of suicidality, in Mann JJ, Stanley M (eds), Psychobiology of Suicidal Behavior. New York, New York Academy of Sciences, 1986, pp 271–280
75. Bunney WE Jr, Fawcett JA, Davis JM, et al: Possibility of a biochemical test for suicidal potential. Arch Gen Psychiatry 13:232–239, 1965
76. Bunney WE Jr, Fawcett JA, Davis JM, et al: Further evaluation of urinary 17-hydroxycorticosteroids in suicidal patients. Arch Gen Psychiatry 21:138–150, 1969
77. Ostroff R, Giller E, Bonese K, et al: Neuroendocrine risk factors of suicide. Am J Psychiatry 139:1323–1325, 1982

78. Levy B, Hensen E: Failure of the urinary test for suicidal potential. Arch Gen Psychiatry 20:415–418, 1969
79. Carroll BJ, Greden JF, Feinberg M: Suicide, neuroendocrine dysfunction and CSF 5-HIAA concentrations in depression, in Angrist B (ed), Recent Advances in Neuropsychopharmacology. Oxford, Pergamon, 1981, pp 307–313
80. Kocsis JH, Kennedy S, Brown RP, et al: Neuroendocrine studies in depression: Relationship to suicidal behavior, in Mann JJ, Stanley M (eds), Psychobiology of Suicidal Behavior. New York, New York Academy of Sciences, 1986, pp 256–262
81. Targum SD, Rosen L, Capodanno AE: The dexamethasone suppression test in suicidal patients with unipolar depression. Am J Psychiatry 140:877–879, 1983
82. Banki CM, Aroto M: Amine metabolites and neuroendocrine responses related to depression and suicide. J Affective Disord 5:223–232, 1983
83. Van Waltere JP, Charles G, Wilmotte J: Tests de function à la dexamethasome et suicide. Acta Psychiatr Scand 83:569–578, 1983
84. Meltzer HY, Perline R, Tricou BJ, et al: Affect of 5-hydroxytryptophan on serum cortisol levels in major affective disorders. II. Relation to suicide, psychosis and depressive symptoms. Arch Gen Psychiatry 41:379–387, 1984
85. Cocarro EF, Siever LJ, Klar H, et al: Serotonergic studies in patients with affective and personality disorders. Arch Gen Psychiatry 46:587–599, 1989
86. Stanley MM, Mann JJ: Biological factors associated with suicide, in Frances AJ and Hales RE, APA Review of Psychiatry, Vol. 7, 1987, pp 334–352
87. Valzelli L: Psychobiology of Aggression and Violence. New York, Raven, 1981
88. Cochran E, Robins E, Grote S: Regional serotonin levels in brain: A comparison of depressive suicides and alcoholic suicides with controls. Biol Psychiatry 11:283–295, 1976
89. Beskow J, Gottfries CG, Roos BE, et al: Determination of monoamine and monoamine metabolites in the human brain: Postmortem studies in a group of suicides and in a control group. Acta Psychiatr Scand 53:7–20, 1976
90. Bourse HR, Bunney WE Jr, Colburn RW, et al: Noradrenadine, 5-hydroxyindoleacetic acid of suicidal patients. Lancet ii:805–808, 1968
91. Korpi ER, Kleinman JE, Goodman SI, et al: Serotonin and 5-hydroxyindoleacetic acid concentrations in different brain regions of suicide victims: Comparison in chronic Schizophrenic patients with suicide as cause of death. Presented at the International Society for Neurochemistry, Vancouver, Canada, 1983
92. Crow TJ, Cross AJ, Cooper SJ, et al: Neurotransmitter receptors and monoamine metabolites in the patients with alzheimer-type dementia and depression and suicides. Neuropharmacology 23:1561–1569, 1984
93. Perry EK, Marshall EF, Blessed G, et al: Decreased imipramine binding in the brains of patients with depresseve illness. Br J Psychiatry 142:188–192, 1983
94. Owens F, Cross AJ, Crow TJ, et al: Brain 5-HT receptors and suicide. Lancet 2:1283, 1983
95. Zanko MT, Biegon A: Increased B-adrenergic receptor binding in human frontal cortex of suicide victims. Abstr. of the Annual Meeting, Society of Neuroscience, Boston, 1983
96. Mann JJ, Stanley M, McBride AP, et al: Increased serotonin and beta-adrenergic receptor binding in the frontal orifices of suicide victims. Arch Gen Psychiatry 43:954–959, 1986
97. Traskman L, Asberg M, Bertilsson K, et al: Monoamine metabolites in CSF and suicidal behavior. Arch Gen Psychiatry 38:631–636, 1987
98. Montgomery SA, Montgomery D: Pharmacological prevention of suicidal behavior. J Affective Disord 4:291–298, 1982
99. Palanappian V, Rauachandler V, Somosundaran O: Suicidal ideation and biogenic amines in depression. Indian J Psychiatry 25:268–292, 1983
100. Banki CM, Arato M, Papp Z, et al: Biochemical markers in suicidal patients: Investigations with cerebrospinal fluid amine metabolites and neuroendocrine tests. J Affective Disord 6:341–350, 1984
101. Perez de los Cobos JZ, Lopez-Ibor Alino JJ, Saiz Ruiz J: Correlatos biologicos del suicido y la agresividad en depressiones mayores (con melancolia): 5-HIAA en LCR, DST, y respuesto terapeutica a 5-HTp. Presented to the first Congress of the Spanish Society for Biological Psychiatry, Barcelona, 1984
102. Berrettini W, Nurenberger J, Narrow W, et al.: Cerebrospinal fluid studies of bipolar patients with and without a history of suicide attempts, in Mann JJ, Stanley M (eds), Psychobiology of Suicidal Behavior, New York, New York Academy of Sciences, 1986
103. Van Praag HM: CSF 5-HIAA and suicide in nondepressed schizophrenics. Lancet 2:977–978, 1983

104. Ninan PT, Van Kammen DP, Scheinin M, et al: CSF 5-hydroxyindoleacetic acid in suicidal schizophrenic patients. Am J Psychiatry 141:566–569, 1984
105. Stanley M, Stanley B, Traskman-Bendz L, Winchel R, and Jones JS: An assessment of the biochemical findings in schizophrenic patients who attempt suicide. Athens, Greece, VIII World Congress of Psychiatry, Abstract, 1989
106. Banki CM, Alato M: Amine metabolites and neuroendocrine responses related to depression and suicide. J. Affective Disord 5:225–232, 1983
107. Brown GL, Goodwin FK, Ballenger, JC, et al: Ag ression in humans correlates with cerebrospinal fluid amine metabolites. Psychiatry Res. 1:131–139, 1979

20

Relationship of Down's Syndrome to Alzheimer's Disease

Ausma Rabe, Krystyna E. Wisniewski, Nicole Schupf, and Henryk M. Wisniewski

1. INTRODUCTION

Down's syndrome (DS), or trisomy 21, is a major known cause of mental retardation, occurring in 1 of every 1000 live births.[1] Alzheimer's disease (AD) is a progressive dementing disorder, with characteristic brain pathology, that affects about 10% of people over 65 years of age.[2,3] The presence of Alzheimer-type pathology in the brains of almost all people 35 years and older with DS has been the cornerstone for the widely held view that people with DS will develop AD not only at a much younger age, but also in much larger numbers than will people without DS.[4–7] Because of this relationship between DS and AD, it has been widely assumed that knowledge about almost any aspect of one of these conditions will illuminate the other.[8–11] For example, the similarity of the morphological and neurochemical abnormalities in AD and DS has led to the proposition that aging DS brains may serve as a model for the pathogenesis of brain abnormalities in AD.[12,13] In addition, since DS is a genetic disorder with many features of premature aging,[14] it has been expected that the study of DS will contribute important information about the role of genetics and aging in the etiology of AD.[15–19]

2. DELAYED AND ABNORMAL BRAIN DEVELOPMENT IN DOWN'S SYNDROME

It is assumed that the prenatal and early postnatal brain abnormalities are responsible for mental retardation in DS. Similarly, the occurrence of Alzheimer neuropathology later in life must be related to the premorbid abnormalities of brain structure and function. However, in either case the critical variables are not yet known.[20]

The brains of people with DS are smaller than normal.[21] The hippocampus also appears to

Ausma Rabe, Krystyna E. Wisniewski, Nicole Schupf, and Henryk M. Wisniewski • New York State Institute for Basic Research in Developmental Disabilities, Staten Island, New York 10314.

be reduced by as much as 50%.[22] The differences in brain size relative to normal are smallest at birth and increase during infancy and early childhood,[23] suggesting defects in the developmental program. The shape of the DS brain is roundish, largely because of the reduced frontal pole; the superior temporal gyrus is frequently quite narrow.[21,23] Myelination of cerebral white matter is delayed.[24] The anterior commissure is considerably (50% or more) smaller than normal in many (but not all) DS brains.[25]

A number of cytoarchitectonic and synaptic abnormalities are revealed by light and electron microscopic studies, most of which suggest arrested and/or delayed development.[26,27] The cerebral cortex of the DS brain has a reduced number of neurons that is particularly marked in the number of granule cells of cortical layers II and IV of the cortex.[23] Precocious cessation in the development of dendritic spines in the visual cortex has also been reported.[28]

There is no consistent agreement among investigations on whether fetal DS brains are less developed than normal brains of the same gestational age. For example, Sylvester[22] found that the hippocampus in a DS fetus was not only smaller but also less mature than that of an age-matched normal fetus. By contrast, K. Wisniewski et al.,[29] found the brains of a 15- to 22-week-old DS fetuses to be as well developed as those of normal fetuses. Petit et al.[30] reported that synapses in the cerebral cortex of the fetal DS brain had reduced synaptic parameters and formed more primitive contacts. These findings are consistent with the observations that synaptic length is shortened in cortical neurons of the DS brain.[23] By contrast, synaptic density of cerebrocortical neurons in DS is reduced very little.[30,31] A reduction in the number of spines on pyramidal dendrites from the hippocampus and cingulate gyrus of young DS brains had been observed earlier,[32] but a more recent study failed to confirm this finding.[33]

Nothing appears to have been reported on the development and status of glial cells in the DS brain. In view of our recent findings about the possible role of glia in brain amyloidosis in AD, one wonders whether there are congenital abnormalities in the glia and/or macrophage cells of the DS brain that contribute to the formation of the Alzheimer-type lesions that develop in all people with DS later in life.

3. DEMENTIA AND NEUROPATHOLOGY IN ALZHEIMER'S DISEASE

Alzheimer's disease is characterized by progressive dementia with onset in middle life (early onset or presenile dementia) or late in life (late onset or senile dementia of the Alzheimer type).[34] In addition to variations in the time of onset, there are considerable individual differences in the behavioral manifestations that suggest that AD may be a heterogeneous disease.[35–37]

Alzheimer's disease usually begins as a mild impairment of memory, decline in spatial competence, difficulties with language, and/or other cognitive functions. During the course of several years, the dementia progresses to a total cognitive and physical incompetence and eventual death. Detailed descriptions of the various phases of Alzheimer dementia in nonretarded persons are available.[34–38] Since dementia in the elderly is often caused also by other conditions (e.g., multiple strokes, depression), differential diagnosis of Alzheimer dementia is important.[39,40] Proper diagnosis early in the disease process is important for treatable conditions and will be essential for AD when curative or palliative procedures become available.[41] Psychometric tests capable of assessing the nature and degree of impairment are critical for the evaluation of the effectiveness of curative measures.[42,43] Since dementia can have many different causes, a definite diagnosis of AD requires a neuropathologic confirmation.[44–46] Neuropathologically, AD is characterized by brain atrophy[47,48] and the presence of extracellular senile (neuritic and amyloid) plaques[49] and intracellular neurofibrillary tangles[50] in the neocortical association areas, hippocampus, and other brain regions.[51] While plaques and tangles have been considered the definitive lesions of AD, neuron loss in the neocortex and other regions is a conspicuous feature in all cases.[48,52,53] Lipofuscin granules in neurons, granulovacuolar degeneration and Hirano bodies in the hippocampus are other less frequently described morphological abnormalities associated with AD.[47]

As people age, they develop at least some plaques in the cerebral cortex and tangles in the hippocampus without developing dementia.[54–56] However, brains of people with Alzheimer dementia usually have many more plaques and tangles than the brains from age-matched nondemented elderly persons.[36,57,58]

These facts suggest that there may be a threshold number of plaques and tangles that must be exceeded before the appearance of clinical signs of dementia. (The idea of a threshold also implies a preclinical stage of AD.[36]) Age-adjusted quantitative guidelines for threshold numbers of plaques and tangles for a neuropathological diagnosis of AD have been proposed,[45,50,59] but it is likely that they will have to be revised since some neuropathologists consider the proposed threshold values too low, and a recent survey indicates the need for differential quantitative standards for different tissue stains.[61]

A quantitative relationship between the degree of neuropathology, especially plaque and tangle counts, and the degree of dementia has been widely assumed.[57] A few clinicopathological studies have shown that a positive correlation exists between plaque counts (after death) and the degree of dementia near the end of life, as well as between tangle counts and dementia.[58] The correlations, highly significant statistically but always considerably less than unity, suggest that plaques and tangles may be reasonable semiquantitative markers of the pathological processes underlying dementia.[60]

Several recent reports, however, indicate a more complicated picture. Terry et al.[62] reported that as many as 30% of aged people with Alzheimer dementia may have only neocortical plaques and hippocampal tangles, but no neocortical tangles. Others have shown that neocortical plaques do not discriminate well between aged persons with and without Alzheimer dementia, whereas neocortical tangles do.[63–66] Still others demonstrate that neither plaques nor tangles may be adequate markers for dementia, but that loss of pyramidal neurons in the neocortex may provide a good quantitative index of Alzheimer dementia.[67] Mann et al.[53] confirm this by demonstrating that the number of cortical pyramidal neurons decreased during a 3- to 7-year course of Alzheimer dementia in five patients (between biopsy and necropsy), whereas plaque and tangle counts did not change consistently. An integration of these apparently disparate findings regarding the relationship of plaques and tangles to dementia, we believe, will be achieved only through a better understanding of the pathogenesis of plaques and tangles and of their relationship to each other and to the death of neurons.

Current research at our institute is addressing the problem of plaque and tangle pathogenesis in AD and has led to the hypothesis of AD as a form of brain amyloidosis.[68] The use of highly sensitive methods of immunocytochemistry in the study of brains of Alzheimer patients has demonstrated several stages in the development of neurofibrillary tangles[69] and has also shown that amyloid (the main component of plaques) and paired helical filaments (PHF) (the main components of tangles) are more diffusely distributed in the brain than previously thought. PHFs are not confined to the soma of tangle-bearing neurons and the dystrophic neurites associated with plaques but also appear in the neuropil.[66] Moreover, amyloid is not confined to plaque cores and blood vessel walls as heretofore believed but is diffusely distributed in the cortical neuropil, subpial layer, and white matter, as well as several other brain regions, suggesting that brain amyloidosis in AD is much grater than so far thought and may be at the root of AD pathology.[70] It appears that amyloid formation may produce trauma to the brain and that particular neurons in particular brain regions may react to this trauma by forming PHFs (tangles) and dying thus leading to neuron loss.

What leads to amyloidosis in the brain of people susceptible to AD? Recently scientists here have identified cells which are associated with the formation of amyloid fibers in brains of AD patients. These cells appear to belong to the reticuloendothelial system: microglia, pericytes, macrophages, and fibroblast-like cells in the walls of medium and large vessels of the meninges. A subset of these cells may be defective in such a way that the amyloid precursor protein is abnormally cleaved to generate the β- peptide that polymerizes into amyloid fibers.[68]

4. ALZHEIMER NEUROPATHOLOGY IN DOWN'S SYNDROME

The similarity between DS and AD has been defined by neuropathology, not by generally accepted clinical criteria of dementia.[60,71] The brain of almost every person with DS who has come to autopsy and is over 35 years of age has had numerous plaques and tangles.[4,7,13,33] The lesion densities in DS brains, even in the 30- to 40-year age group, are usually in the range of those found in the brains of nonretarded people with AD. Very few persons with DS have been shown to have plaques and/or tangles before the age of 20, and almost none who have come to autopsy after the age of 50 have been free of these lesions.[13] The distribution of these two classic AD lesions in the brain is similar in AD and DS,[13] and the lesions are usually considered identical at the light-microscopic as well as the electron microscopic level.[72] However, a few recent reports have described differences between the AD and DS plaques.[73,74]

Loss of neurons may be quite pronounced.[10] The evaluation of neuron loss due to aging and/or AD in DS brains is complicated by the fact that several regions of their brains may already have fewer neurons from the very start. Unfortunately, investigators have not always recognized this fact. The other less frequently studied neuropathological changes associated with AD—atrophy,[75] granulovacuolar degeneration, and Hirano bodies in the hippocampus,[76] amyloid angiopathy,[77] lipofuscin granules in neurons,[13] and basal ganglia calcification[78,79] are also quite pronounced in aging DS brains.

We have shown previously that in DS brains large numbers of plaques appear 10–20 years before a large increase in the number of tangles.[7,71] The temporal primacy of plaques in DS brains may be a reflection of the triple dose of the gene for β-amyloid on chromosome 21 and is consistent with the idea that amyloid deposits may be instrumental in, or contribute to, the formation of tangles and neuron loss. Moreover, Mann and Esiri[80] recently reported that, using immunocytochemical staining methods with tissue from brains of DS persons aged 30–50 years, they observed diffuse amyloid deposits without any PHF (i.e., tangles) in the youngest brains. Several of the older brains in addition to the amyloid deposits showed neuritic changes (typical of plaque formation) and the presence of PHF. Some of the other older brains showed both abundant plaques and tangles. These findings are consistent with the idea that brain amyloidosis may be responsible for the Alzheimer neuropathology in DS brain.[68]

What stimulates early amyloidosis in DS brains is not known, but it appears plausible to assume that the triple gene does for beta amyloid is involved.[81–84] The DS brain may also possess an excess of cells of the reticuloendothelial system recently linked to amyloidosis in the brains of AD patients perhaps as a congenital abnormality and/or as a result of a compromised immune system or some other premature degenerative process. The idea of a congenital excess of such cells finds indirect support in the recent findings of Kornguth and Bersu,[85] who demonstrated a marked increase in macrophages in the colliculi and cerebellum of fetal trisomy 16 mice, a model system for the study of DS. (The mouse chromosome 16 and human chromosome 21 are syntenic for several genes, among them superoxide dismutase and precursor for the amyloid β-protein.)

5. DEMENTIA IN DOWN'S SYNDROME

The clinical signs of Alzheimer dementia and its course in people with DS have been largely unknown until very recently.[86–88] It is difficult to detect intellectual decline in the mentally retarded. We will briefly review two informative accounts. Dalton and Crapper–

McLachlan[89] reviewed 33 DS cases from the published literature for whom both at least some clinical description and neuropathological data were available. These workers found that no single clinical or functional feature had been reported as a certain sign of dementia in DS; the most frequently reported feature was the occurrence of seizures (in 88% of the cases), followed by focal neurological signs and personality changes (in 46% of cases). By contrast, memory loss (one of the most notable signs of Alzheimer dementia in the general population) was the least frequently reported feature, and there was no mention of changes in any language-based functions, probably reflecting the difficulty of detecting these subtler functional changes against the background of mental retardation. In presenting their current longitudinal study, Lai and Williams[90] list memory impairment, temporal disorientation, and reduced verbal output as the earliest signs of dementia for the higher functioning DS persons; for the lower-functioning DS individuals, apathy, inattention, and decreased social interaction are among the early indications of dementia. As dementia progresses, loss of self-help skills, motor impairment, and seizures can be observed. They characterize the terminal phase of dementia by incontinence and pathological reflexes. Lai and Williams[90] also note that seizures are a distinctive feature of dementing persons with DS: they were observed in 84% of their cases within 2 years of the beginning of mental decline; according to them, seizures are rare in nondemented DS persons (about 5%) and in AD in the general population (about 10%). This pattern of seizure frequencies suggests that in aging DS the degenerative processes responsible for dementia interact with the congenital abnormalities of their brains.

6. RELATIONSHIP BETWEEN NEUROPATHOLOGY AND DEMENTIA IN DOWN'S SYNDROME

The significant positive correlation between plaques and tangles and dementia in AD in the general population has led to the expectation that all elderly people with DS should have dementia.[60,71] Contrary to the expectation, the presumed near ubiquitous presence of many plaques and tangles in DS brains may not be accompanied by progressive dementia in many mature and elderly persons with DS.[7,13,33,91,92] While the demented DS persons may have somewhat more plaques and tangles than the DS individuals without dementia,[7,60] the lesion numbers for the nondemented DS people are clearly in the range of the numbers associated with dementia in the general population.[7,10,33,93] Although this discrepancy between neuropathology and clinical symptoms has been recognized,[13,33,60,71] there has not been the same recognition that the universal presence of plaques and tangles in all brains of aging persons with DS, whether demented or not, precludes the use of these lesions as diagnostic criteria of AD in DS.[94] Moreover, there have been very few attempts to investigate why some people do not develop dementia, and how demented persons with DS differ from the nondemented elderly people with DS.[60,75,95–97]

The prevalence of Alzheimer-type dementia in mature and aging persons with DS is not known with any precision. The reported frequency of dementia in mature and elderly DS ranges from 6% to 75%, reflecting significant inconsistencies that require explanation. We have reviewed these earlier studies in detail elsewhere.[60] Two longitudinal studies with aging DS persons show that dementia may indeed be more frequent in the aging DS than in the general population. Dalton and Crapper-McLachlan[86] demonstrated that 12 (or 24%) persons of the 49 studied over an 8-year period (the annual incidence was 9%) showed a decline in visual short-term memory, which they interpreted as an early sign of dementia. Full clinical and neuropathological description of AD has been published for seven of these 12 persons.[77] The average onset of dementia was 53 years, and its duration to death ranged from 3.5 to 10.5 years. In a recent report of another prospective study of 96 DS persons over 35, Lai and Williams[90] describe that 49 (51%) were demented, with an average age of onset at about 55 years of age (range 45–65). Twenty-three of the demented patients have died; for these, the average duration of dementia was about 5 years. Twelve of these were autopsied, and abundant AD neuropathology was found in all of these brains. They also report that the percentage of demented DS cases increased with age: 13% of persons between 35–49 years, 55% of people in their 50s, and 75% of those over age 60 were demented. This progressive increase in the number of demented people would seem to suggest that all persons with DS might eventually develop Alzheimer dementia if they lived long enough, a view expressed earlier by Dalton and Crapper-McLachlan.[89] These percentages may, however, considerably overestimate AD as the cause of dementia, since they do not take into account the possibility that some of the decline in function may have been due to factors related to the well-documented early mortality in DS rather than AD. Schupf and collaborators (unpublished data from our institute) have shown that there is decline in adaptive behavior skills before death in 10–25% of retarded persons without DS.

Both longitudinal studies reviewed above found that the onset of dementia was much later than one would expect on the basis of the presumed pervasive presence of plaques and tangles in DS brains by the age of 40. Dementia appears at about the same age as the increase in tangles,[7,71] suggesting a possible relationship. This later-than-expected decline in function has been confirmed by other studies. A recent cross-sectional study from our institute by Zigman et al.,[91] with more than 2000 DS people ranging in age from 20 to 69 years, showed a decline in adaptive competence only after the age of 50. These results are in essential agreement with those reported by Silverstein et al.,[92] who detected a functional decline in aging persons with DS only after 60 years of age. In a prospective study by K. Wisniewski and Hill (now in its second year at our Institute), visual and verbal memory of 51 mildly and moderately retarded persons with DS is being monitored over time (and compared with a retarded non-DS group matched for age and functional level). To date, the subjects have been tested twice, 1 year apart. Of particular interest within the present context is the finding that the older DS subjects (18 persons 40 and older) performed as well as the younger DS subjects (20 persons under 40) under all test conditions. This was true even of the seven oldest persons who were 45–57 years of age.

Are people with DS more or less likely to developing AD dementia than the general population? The answer depends on the frame of reference. When compared with persons without DS, both normal and retarded, people with DS appear to be much more susceptible to dementia: in addition to almost everyone developing large numbers of plaques and tangles at 35–40 years of age, as many as 50% may develop Alzheimer dementia in their 50s and 60s. In sharp contrast, only relatively few (perhaps 0.05%) of the people without DS develop AD in their 50s (early-onset dementia), and the prevalence after 65 is 10% (late-onset dementia).[98] By contrast, when we consider only people with DS and then compare time of onset of dementia with the age of appearance of plaques and tangles, aging DS persons appear to be resistant to developing Alzheimer dementia; i.e., far fewer than 100% appear to become demented in their lifetime, and most of those who develop dementia do so in their 50s or later, at least some 10 years after the presumed appearance of neuropathology severe enough to be consistent with a diagnosis of AD. Several recent cliniconeuropathological reports confirm that many elderly people with DS have numerous plaques and tangles but appear not to be demented.[13,33]

We have discussed elsewhere some of the possible reasons for this vexing discrepancy.[71] A resolution of this disparity between the clinical and neuropathological signs of AD in DS is important for several reasons: If the discrepancy is genuine, either (1) the people with DS are in some unknown way protected against developing dementia, or (2) AD in people with DS is a different disease from that in people without DS. More attention should be paid to all measures that might differentiate between those who become demented and those who do not. The degree of brain atrophy[75] and metabolic activity (assessed through brain imaging),[99] and superoxide dismutase[95] and vitamin E levels in plasma[97] are promising recent examples.

It is possible, however, that the discrepancy is only apparent and has arisen from sampling errors, differences in measurement, and variations in diagnostic procedures. Neuropathological criteria based on plaque and tangle counts may simply not differentiate well between demented and nondemented people with DS, although they may be better markers for AD dementia in the general population. From a clinical point of view, valid assessment of declining intellectual function and/or dementia in a mentally retarded population has been thwarted by many problems. Among them are the lack of baseline data for a given individual,[19] lack of adequate assessment instruments, especially for the lower levels of functioning, as well as lack of proper criteria for what constitutes progressive dementia in the mentally retarded.[60,71]

Only longitudinal (prospective) studies of a sufficient number of retarded persons with DS can resolve the disparity between neuropathological evidence and reports of dementia. After death, quantitative neuropathological investigation should also include nondemented individuals. Psychometric procedures are needed for the early detection of decline in intellectual and behavioral functions and monitoring course of dementia at all levels of retardation. Some successful efforts have been made.[86] In addition, a group of scientists in Edinburgh[15] have found a neurophysiological measure (the auditory P300 potential) that may be useful for detecting the onset of dementia in persons with DS,[100,101] and we are developing additional behavioral nonverbal tests of cognitive function for use with DS people at the lower level of functioning.

7. NEUROCHEMICAL ABNORMALITIES AND DEMENTIA

7.1. Alzheimer's Disease

Brain neurotransmitter chemistry is abnormal in AD, and it is very difficult to single out any particular neurotransmitter deficit as a causative factor.[102–105] Deficits in all neurotransmitters appear to be more severe in younger cases, again reminding one of the possible heterogeneity of AD.[106]

A severe acetylcholine (ACh) deficiency has been consistently demonstrated by many different groups of investigators.[107] There is a loss of neurons in several basal forebrain nuclei (nucleus basalis of Meynert, diagonal band of Broca and medial septum) that send cholinergic projection fibers to the cerebral cortex.[108–110] A reduction in the enzyme that is needed for the synthesis of ACh, choline acetyltransferase (ChAT), is invariably reduced in the neocortex and other brain regions.[111–113] Several investigators have reported significant correlations between cortical ChAT levels and plaque and/or tangle counts, and ChAT levels and the degrees of dementia.[114–116] Although it is widely held that the cholinergic deficiency may be responsible for the memory deficit characteristic of AD, its relationship to dementia is not understood.[103]

Other neurotransmitter systems are also impaired in AD.[117] There is loss of neurons in the locus coeruleus (site of noradrenergic projection neurons to the neocortex),[118,119] and reduced level of norepinephrine in the brain.[120–122] By contrast, it does not appear to be clearly established that the level of cortical dopamine is reduced. Serotonin level is also reduced in the neocortex of AD brains,[123,124] and loss of neurons and numerous tangles have been reported in the raphe nucleus, which sends serotonergic projections to the cortex.[8,125] Of the amino acid neurotransmitters, γ-aminobutyric acid (GABA) levels may be reduced in some areas of the cortex.[112,116] Of the neuropeptide neurotransmitters, somatostatin alone is consistently reduced in the cortex of brains with AD.[126] Immunoreactive staining has recently shown cortical plaques[127] and tangles[128] that stain somatostatin-positive, suggesting that somatostatin containing cortical neurons are susceptible to AD pathology.

7.2. Down's Syndrome

The brains of mature and aging persons with DS appear to show abnormalities in the same neurotransmitter systems as observed in AD in the general population.[129,130] These deficits appear to be the largest in brains that also have substantial numbers of plaques and tangles.[8] No reports have shown correlations between any neurotransmitter deficits in DS brains and plaque and tangle counts, nor with presence or absence of dementia. Some investigators have reported small deficits in neurotransmitter levels[130–132] in younger brains without any AD neuropathology, suggesting that there may be congenital neurotransmitter deficits in DS. The number of young DS brains studied systematically is still too small for a proper evaluation of how much of the deficits in neurotransmitter systems is due to trisomy 21 and how much is a result of aging and/or AD. Fortunately, investigators are beginning to recognize the need to separate congenital contributions from those of degenerative processes of AD. A brief review of specific neurotransmitter systems in DS is presented below.

Changes in cholinergic structures of DS brains are quite pronounced, both as cortical deficits in ChAT concentration and as reduced cell counts in the nucleus basalis of Meynert.[133,134] Norepinephrine levels and neuron number in locus coeruleus are also reduced.[121,122,130,133] Similarly, serotonin levels are lower than normal in regions commonly rich in this neurotransmitter[130,132] and, correspondingly, cell counts in the raphe nuclei are also reduced.[133] Dopamine levels, in contrast, are not affected[122] although there may be a reduction in neuron numbers in one of the brain stem nuclei (A10 or ventral tegmental area) that sends dopaminergic fibers to frontal and limbic areas of the cortex.[11] The nigrostriatal dopamine system does not appear to be affected.[11] The somatostatin level has also been reported to be lowered in DS brains.[136]

8. METABOLIC AND OTHER ABNORMALITIES

Brain blood flow and glucose metabolism may be abnormal in persons with DS, even before the appearance of AD neuropathology. Cerebral blood flow was found to be reduced in a relatively young (29 ± 8 years) group of DS subjects.[137] Glucose metabolism was found to be elevated in a group of young adult DS individuals, but it declined with age and declined still further in those who were demented.[99] Reduced blood flow and reduced glucose metabolism have also been reported for persons with AD from the general population.[138,139] These measures of brain activity are taken from living individuals and may provide useful physiological markers of dementia.

Three other recent studies that have attempted to distinguish between demented and nondemented living individuals with DS may be of considerable interest if they can be replicated. In one study,[95] the enzyme Cu,Zn superoxide dismutase was found to be significantly reduced in those who were demented, relative to the nondemented DS persons in whom the enzyme showed the expected elevation resulting from the triple gene dosage. This result suggests a relationship between a decreased antioxidant potential and dementia. The same demented DS individuals also displayed a different pattern of measures of thyroid function than those without dementia.[96] The third study by Jackson et al.[97] found that demented aging persons with DS had lower levels of vitamin E than did an age-matched nondemented DS group. These investigators have proposed that Alzheimer neuropathology is a characteristic of DS, but that some other factor determines the age of onset of dementia. Such a factor might be vitamin E; since vitamin E protects against oxidative damage to cells, it might also protect against development of dementia in persons with DS.

9. AGING

Alzheimer's disease is clearly an age-related disease. Since 90% of all AD cases are 65 years or older,[2] age is the single best predictor of the probability whether a given person will develop AD. Although the large majority of old people do not have AD, their brains nevertheless show a mild degree of the same pathological changes as seen in AD (plaques, tangles, cell loss, and deficits in some neurotransmitters).[36] This overlap in brain pathology between normally aging persons and those with AD makes one wonder whether all people would eventually develop AD if they lived long enough. Thus, for example, Wright and Whalley[15] argue eloquently that AD is premature aging, probably at a genetically determined rate. There are others who believe that AD is a disease and not the outcome of normal aging. Roth[36] has carefully catalogued the brain changes reported to occur with normal aging and compared them with those reported for AD. He concludes that aging and early-onset AD are clearly different and that even late-onset AD, although more similar to aging, still differs from it.

Our view has been that aging, as a gradual loss of reserve, provides an environment within the body in which certain pathological changes can readily develop.[56] Thus, the aging of the brain and the disease process(es) responsible for AD must be intricately related.

The early appearance of Alzheimer neuropathology in people with DS may be related to premature aging but, despite the considerable attention the idea has received, the relationship is not understood.[15,18,19,140] It is not clear now AD neuropathology in DS is related to aging genes. Although DS is prominent among the various genetic syndromes that show accelerated aging with, by far, the highest number of aging features,[14] there are many other syndromes of premature aging without plaques and tangles.[141] These provide dramatic evidence that premature aging does not necessarily produce AD neuropathology.

10. GENETIC FACTORS

Is the risk of developing AD inherited? Genetic factors appear to play a significant role, but the mode of transmission, autosomal-dominant or polygenic, has not been established. The available data on the cumulative incidence of AD among first-degree relatives vary so widely that, for all practical purposes, we do not yet know the magnitude of the risk of inheriting AD.[142–147]

There is also evidence for a familial link between AD and DS: families with members who have died with AD have shown an excess rate of births with DS (3.5/1000 instead of 1/1000).[143] That early-onset AD and DS cluster together was recently confirmed by a study of the frequency of AD in families with DS: the frequency of early-onset, but not late-onset, AD was significantly increased in these families.[148]

There is no evidence for consistent chromosomal abnormalities in persons with AD.[149] On the other hand, the chromosomal basis of DS is well known: 96% have chromosome 21 in triplicate; a small portion of DS with the characteristic features of DS indicate that the critical portion is 21q22. How this imbalance produces DS with its congenitally abnormal brain and AD neuropathology later in life is not known. Several gene loci on chromosome 21 have been known for some time through increased gene products in DS, among them Cu,Zn superoxide dismutase.[150]

The last couple of years have been particularly exciting in the area of molecular genetics of AD and DS. Several studies have mapped a gene that codes β-amyloid to human chromosome 21.[81–84] This gene was seen to be of direct relevance to the production of excessive senile or amyloid plaques in DS and AD, and was seen as providing a genetic link between AD and DS. This perception was further strengthened by reports suggesting that a gene responsible for familial AD also maps on chromosome 21 and that the β-amyloid gene on chromosome 21[151] is duplicated in AD.[152,153]

Subsequent reports, however, have not replicated either finding linking chromosome 21 to AD. A study with AD families failed to establish linkage to chromosome 21 for both early-onset and late-onset AD, including the region where early-onset AD had been localized by St.

George-Hyslop and associates.[154] Other studies failed to find three copies of the β-amyloid gene in their AD patients.[155–157] Correspondingly, no elevation in β-amyloid gene dose has been found in AD, while the expected 1.5 times gene dosage has been observed in DS.[156,157] The apparent inconsistencies in the chromosomal abnormalities in non-DS cases may be a reflection of heterogeneity of AD.

11. CONCLUSIONS

The strong similarities among the morphological and neurochemical abnormalities found in the brains of AD patients and those of aging persons with DS remain unchallenged. Consequently, the status of the DS brain as a natural model for a pathogenesis of Alzheimer neuropathology appears to be firmly established. The cross-fertilization of knowledge derived from studying the two conditions is well illustrated by the hypothesis of AD being a cerebral form of amyloidosis in both DS and non-DS populations. However, neglect of the role played by congenital abnormalities of the DS brain in the development of the precocious and ubiquitous Alzheimer neuropathology in the aging DS brain has been a serious shortcoming in most attempts to relate AD and DS pathophysiologically.

The clinical picture of aging persons with DS and of those with AD but without DS is incongruent. On the basis of the established relationship in the general population between Alzheimer brain pathology and dementia, almost all persons with DS should show signs of dementia by about 40 years of age, since by then their brains have large numbers of Alzheimer-type lesions. However, recent research data show that those DS persons who develop dementia do so in their 50s and that a significant number may not become demented at all. This discrepancy must be resolved, if we are to continue considering aging DS individuals as having the same disease as those with AD in the general population. Moreover, the time may have come to consider the neuropathological criteria of AD in the general population as insufficient for a similar diagnosis in DS.

In our opinion, the study of DS has not clarified the role of aging in AD. Until we understand better the process(es) involved in aging, this state of affairs is unlikely to improve. Similarly, knowledge about the genetic basis of DS has contributed little to knowledge about the risk of inheriting AD by people without DS. Improvement in the methodology of inheritance studies will play a major role here. By contrast, work in molecular genetics has shown a great deal of fruitful interplay among studies of DS and AD and, even though no consistent chromosomal abnormalities for AD have yet been found, perhaps because of the heterogeneity of the disease, we should be constantly watching for new developments.

ACKNOWLEDGMENTS. We wish to thank Dr. Arthur Dalton for editorial comments. Our recent and current research referred to in this chapter is supported, in part, by the New York State Office of Mental Retardation and Developmental Disabilities and by a NIH Program Project P01 HD 22634.

REFERENCES

1. Adams NM, Erickson JD, Layde PM, et al: Down's syndrome: Recent trends in the United States. JAMA 246:758–760, 1981
2. Mortimer JA: Alzheimer's disease and senile dementia: Prevalence and incidence, in Reisberg B (ed), Alzheimer's Disease. New York, Free Press, 1983, p 141
3. Henderson AS: The epidemiology of Alzheimer's disease. Brit Med Bull 42:3–10, 1986

4. Malamud N: Neuropathology of organic brain syndromes associated with aging, in Gaitz CM (ed), Aging and the Brain. New York, Plenum, 1972, p 63
5. Burger PC, Vogel FS: The development of the pathologic changes of Alzheimer's disease and senile dementia in patients with Down's syndrome. Am J Pathol 73:457–468, 1973
6. Whalley LJ: The dementia of Down's syndrome of Alzheimer's disease and its relevance to aetiological studies of Alzheimer's disease. Ann NY Acad Sci 396:39–53, 1982
7. Wisniewski KE, Wisniewski HM, Wen GY: Occurrence of neuropathological changes and dementia of Alzheimer's disease in Down's syndrome. Ann Neurol 17:278–282, 1985
8. Mann DMA, Yates PO, Marcyniuk B: Alzheimer's presenile dementia, senile dementia of Alzheimer type and Down's syndrome in middle age form an age related continuum of pathological changes. Neuropath Appl Neurobiol 10:185–207, 1984
9. Mann DMA, Yates PO, Marcyniuk B: Some morphometric observations on the cerebral cortex and hippocampus in presenile Alzheimer's disease, senile dementia of Alzheimer type and Down's syndrome in middle age. J Neurol Sci 69:139–159, 1985
10. Mann DMA, Yates PO, Marcyniuk B, et al: Loss of neurones from cortical and subcortical areas in Down's syndrome patients at middle age. Quantitative comparisons with younger Down's patients and patients with Alzheimer's disease. J. Neurol Sci 80:79–89, 1987
11. Mann DMA, Yates PO, Marcyniuk B: Dopaminergic neurotransmitter systems in Alzheimer's disease and in Down's syndrome at middle age. J. Neurol Neurosurg Psychiatr 50:341–344, 1987
12. Fishman MA: Will the study of Down syndrome solve the riddle of Alzheimer disease? J Ped 108:627–629, 1986
13. Mann DMA: The pathological association between Down syndrome and Alzheimer disease. Mech Aging Devel 43:99–136, 1988
14. Martin GM: Genetic syndromes in man with potential relevance to pathobiology of ageing. Birth Defects Orig Artic Ser 14(1):5–39, 1978
15. Wright AF, Whalley LF: Genetics, ageing, and dementia. Br J Psychiatry 145:20–38, 1984
16. Sylvester PE: Ageing in the mentally retarded, in Dobbing J (ed), Scientific Studies in Mental Retardation. London, Royal Society of Medicine, 1984, p 259
17. Editorial: Aging in Down's syndrome. Lancet II:885, 1985
18. Hewitt KE, Carter G, Jancar J: Ageing in Down's syndrome. Br J Psychiatry 147:58–62, 1985
19. Oliver C, Holland AJ: Down's syndrome and Alzheimer's disease: a review. Psychol Med 16:307–322, 1986
20. Coyle JT, Oster-Granite ML, Gearhart JD: The neurobiologic consequences of Down syndrome. Brain Res Bull 16:773–787, 1986
21. Crome L, Stern J: Pathology of Mental Retardation. London, Churchill Livingstone, 1972
22. Sylvester PE: The hippocampus in Down's syndrome. J Ment Defic Res 27:227–236, 1983
23. Wisniewski KE, Laure-Kamionowska M, Connell F, et al: Neuronal density and synaptogenesis in the postnatal stage of brain maturation in Down syndrome, in Epstein CJ (ed): The Neurobiology of Down Syndrome. New York, Raven, 1986, p 29
24. Wisniewski KE, Schmidt-Sidor B: Myelination in Down's syndrome brains (pre-and postnatal maturation) and some clinical–pathological correlations. Ann Neurol 20:429–430, 1986
25. Sylvester PE: The anterior commissure in Down's syndrome. J Ment Defic Res 30:19–26, 1986
26. Ross MH, Galaburda AM, Kemper TL: Down's syndrome: Is there a decreased population of neurons? Neurol 34:909–916, 1984
27. Wisniewski KE, Laure-Kamionowska M, Wisniewski HM: Evidence of arrest of neurogenesis and synaptogenesis in brains of patients with Down's syndrome. N Engl J Med 311:1187–1188, 1984
28. Takashima S, Becker LE, Armstrong DL, et al: Abnormal neuronal development in the visual cortex of the human fetus and infant with Down's syndrome: A quantitative and qualitative Golgi study. Brain Res 225:1–21, 1983
29. Wisniewski K, Schmidt-Sidor B, Shepard TH: Normal brain growth and maturation in Down's syndrome fetuses aged 15 to 22 weeks. Ann Neurol 22:430–431, 1987
30. Petit TL, Le Boutilier JC, Alfano DP, et al: Synaptic development in the human fetus: A morphometric analysis of normal and Down's syndrome neocortex. Exp Neurol 83:13–23, 1984
31. Cragg BG: The density of synapses and neurons in normal, mentally defective, and aging human brains. Brain 98:81–90, 1975
32. Suetsugu M, Mehraein P: Spine distribution along the apical dendrites of the pyramidal neurons in Down's syndrome. Acta Neuropathol (Berl) 50:207–210, 1980

33. Williams RS, Matthysse S: Age-related changes in Down syndrome brain and the cellular pathology of Alzheimer disease. Prog Brain Res 70:49–67, 1986
34. Reisberg B: Clinical presentation, diagnosis, and symptomatology of age-associated cognitive decline and Alzheimer's disease, in: Reisberg B (ed), Alzheimer's Disease. New York, Free Press, 1983, p 173
35. Katzman R: Alzheimer's disease. N Eng J Med 314:964–973, 1986
36. Roth M: The association of clinical and neurobiological findings and its bearing on the classification and aetiology of Alzheimer's disease. Brit Med Bull 42:42–50, 1986
37. Friedland RP: Alzheimer disease: Clinical and biological heterogeneity. Ann Int Med 109:298–311, 1988
38. Folstein MF, Whitehouse PF: Cognitive impairment of Alzheimer disease. Neurobehav Toxicol Teratol 5:631–634, 1983
39. Hachinski VC: Differential diagnosis of Alzheimer's dementia: Multi-infarct dementia, in Reisberg B (ed), Alzheimer's Disease. New York, Free Press, 1983, p 188
40. Wells CE: Differential diagnosis of Alzheimer's dementia: Affective disorder, in Reisberg B (ed), Alzheimer's Disease. New York, Free Press, 1983, p 193
41. Branconnier RJ, DeVitt DR: Early detection of incipient Alzheimer's disease: Some methodological considerations on computerized diagnosis, in Reisberg B (ed), Alzheimer's Disease. New York, Free Press, 1983, p 214
42. Fuld PA: Psychometric differentiation of the dementias: An overview, in Reisberg B (ed), Alzheimer's Disease. New York, Free Press, 1983, p 201
43. Crook T: Psychometric assessment in Alzheimer's disease, in Reisberg B (ed), Alzheimer's Disease. New York, Free Press, 1983, p 211
44. McKhann G, Drachman D, Folstein M, et al: Clinical diagnosis of Alzheimer's disease: Report of the NINCDS-ADRDA work group under the auspices of Department of Health and Human Services Task Force on Alzheimer's disease. Neurol 34:939–944, 1984
45. Khachaturian ZS: Diagnosis of Alzheimer's disease. Arch Neurol 42:1097–1105, 1985
46. Joachim CL, Morris JH, Selkoe DJ: Clinically diagnosed Alzheimer's disease: Autopsy results in 150 cases. Ann Neurol 24:50–56, 1988
47. Brun A: An overview of light and electron microscopic changes, in Reisberg B (ed), Alzheimer's Disease. New York, Free Press, 1983, p 37
48. Perry RH: Recent advances in neuropathology. Brit Med Bull 42:34–41, 1986
49. Wisniewski HM: Neuritic (senile) and amyloid plaques, in Reisberg B (ed), Alzheimer's Disease. New York, Free Press, 1983, p 57
50. Iqbal K, Wisniewski HM: Neurofibrillary tangles, in Reisberg B (ed), Alzheimer's Disease. New York, Free Press, 1983, p 48
51. Price DL: New perspectives on Alzheimer's disease. Ann Rev Neurosci 9:489–512, 1986
52. Hubbard BM, Anderson JM: Age-related variations in the neuron content of the cerebral cortex in senile dementia of Alzheimer type. Neuropath Appl Neurobiol 11:369–382, 1985
53. Mann DMA, Marcyniuk B, Yates PO, et al: The progression of the pathological changes of Alzheimer's disease in frontal and temporal neocortex examined both at biopsy and at autopsy. Neuropath Appl Neurobiol 14:177–195, 1988
54. Tomlinson BE, Blessed G, Roth M: Observations on the brains of non-demented old people. J Neurol Sci 7:331–356, 1968
55. Tomlinson BE, Henderson G: Some quantitative cerebral findings in normal and demented old people, in Terry RD, Gershon S (eds), Neurobiology of Aging. New York, Raven, 1976, p 183
56. Wisniewski HM, Merz GS: Aging, Alzheimer's disease, and developmental disabilities, in Janicki MP, Wisniewski HM (eds), Aging and Developmental Disabilities. Baltimore, Brookes, 1984, p 177
57. Blessed G, Tomlinson BE, Roth M: The association between quantitative measures of dementia and of senile changes in the cerebral grey matter of elderly subjects. Br J Psychiatry 114:797–817, 1968
58. Wilcock GK, Esiri MM: Plaques, tangles, and dementia: A quantitative study. J Neurol Sci 56:343–356, 1982
59. Wisniewski HM, Merz GS: Neuropathology of the aging brain and dementia of the Alzheimer type, in Gaitz CM, Samorajski T (eds), Aging 2000: Our Health Care Destiny: Biomedical Issues, Vol. I. New York, Springer, 1985, p 231
60. Wisniewski HM, Rabe A: Discrepancy between Alzheimer-type neuropathology and dementia in persons with Down's syndrome. Ann NY Acad Sci 477:247–259, 1986
61. Wisniewski HM, Rabe A, Silverman W, et al: Neuropathological diagnosis of Alzheimer disease. J Neuropath Exper Neurol 48:606–609, 1989

62. Terry RD, Lawrence MD, Hansen A, et al: Senile dementia of the Alzheimer type without neocortical neurofibrillary tangles. J Neuropath Exper Neurol 46:262–268, 1987
63. Katzman R, Terry R, DeTeresa R, et al: Clinical, pathological, and neurochemical changes in dementia: A subgroup with preserved mental status and numerous neocortical plaques. Ann Neurol 23:138–144, 1988
64. Crystal H, Dickson D, Fuld P, et al: Clinico-pathologic studies in dementia: Nondemented subjects with pathologically confirmed Alzheimer's disease. Neurology 38:1682–1687, 1988
65. Dickson DW, Farlo J, Davies P, et al: Alzheimer's disease. A double-labeling immunohistochemical study of senile plaques. Am J Pathol 132:86–101, 1988
66. Barcikowska M, Wisniewski HM, Bancher C, et al: About the presence of paired helical filaments in dystrophic neurites participating in the plaque formation, Acta Neuropathol 78:225–231, 1989
67. Neary D, Snowden JS, Mann DMA, et al: Alzheimer's disease: A correlative study. J Neurol Neurosurg Psychiatry 49:229–237, 1986
68. Wisniewski HM, Currie JR, Barcikowska M, et al: Alzheimer's disease, a cerebral form of amyloidosis, in Pouplard-Barthelaix A, Emile J, Cristen (eds), Immunology and Alzheimer's Disease. Berlin, Springer, 1988, p 1
69. Bancher C, Brunner C, Lassmann H, et al: Accumulation of abnormally phosphorylated tau precedes the formation of Alzheimer neurofibrillary tangles in Alzheimer's disease. Brain Res 477:90–99, 1989
70. Wisniewski HM, Bancher C, Barcikowska M, et al: Spectrum of morphological appearance of amyloid deposits in Alzheimer's disease, Acta Neuropathol 78:337–347, 1989
71. Wisniewski HM, Rabe A, Wisniewski KE: Neuropathology and dementia in people with Down's syndrome, in Banbury Report 27: Molecular Neuropathology of Aging. Cold Spring Harbor Laboratory, 1987, p 399
72. Schochet SS, Lampert PW, McCormack WF: Neurofibrillary tangles in patients with Down's syndrome: A light and electron microscopic study. Acta Neuropathol (Berl) 23:342–346, 1973
73. Allsop D, Kidd M, Landon M, et al: Isolated senile plaque cores in Alzheimer's disease and Down's syndrome show differences in morphology. J Neurol Neurosurg Psychiatry 49:886–892, 1986
74. Szumanska G, Vorbrodt AW, Mandybur TI, et al: Lectin histochemistry of plaques and tangles in Alzheimer disease. Acta Neuropathol 73:1–11, 1987
75. Shapiro M: Alzheimer disease in premorbidly normal persons with the Down syndrome: Disconnection of neocortical brain regions, in Friedland RP (moderator), Alzheimer disease: Clinical and biological heterogeneity. Ann Intern Med 109:298–311, 1988
76. Ball MJ, Nuttall K: Neurofibrillary tangles, granulovacuolar degeneration, and neuron loss in Down syndrome: Quantitative comparison with Alzheimer dementia. Ann Neurol 7:462–265, 1980
77. Wisniewski KE, Dalton AJ, McLachlan DR, et al: Alzheimer disease in Down syndrome: Prospective clinico-pathological studies. Neurology 35:957–961, 1985
78. Wisniewski KE, French JH, Rosen JF: Basal ganglia calcification (BGC) in Down's syndrome (DS)—another manifestation of premature aging. Ann NY Acad Sci 396:179–189, 1982
79. Mann DMA: Calcification of the basal ganglia in Down's syndrome and Alzheimer's disease. Acta Neuropathol 76:595–598, 1988
80. Mann, DMA, Esiri MM: The pattern of acquisition of plaques and tangles in the brains of patients under 50 years of age with Down's syndrome. J Neurol Sci, 89:169–179, 1989
81. Robakis NK, Wisniewski HM, Jenkins EC, et al: Chromosome 21q21 sublocalisation of gene encoding beta-amyloid peptide in cerebral vessels and neuritic (senile) plaques of people with Alzheimer's disease and Down syndrome. Lancet I:384–385, 1987
82. Goldgaber D, Lerman MI, McBride OW, et al: Characterization and chromosomal localization of a cDNA encoding brain amyloid of Alzheimer's disease. Science 235:877–880, 1987
83. Kang J, Lemaire HG, Unterbeck A, et al: The precursor of Alzheimer's disease amyloid A4 protein resembles a cell-surface receptor. Nature 325:733–736, 1987
84. Tanzi RE, Gusella JF, Watkins PC, et al: Amyloid beta protein gene: cDNA, mRNA distribution, and genetic linkage near the Alzheimer locus. Science 235:880–884, 1987
85. Kornguth S, Bersu E: Cerebellar and collicular development in murine trisomy 16 animals: Morphological changes, monocyte infiltration, and increased expression of MHC H-2Kk. Brain Dysfunction 1:255–271, 1988
86. Dalton AJ, Crapper-McLachlan DR: Incidence of memory deterioration in aging persons with Down's syndrome, in Berg JM (ed), Perspectives and Progress in Mental Retardation, Vol. II: Biomedical Aspects. Baltimore, University Park Press, 1984, p 55
87. Wisniewski KE, Hill AL: Clinical aspects of dementia in mental retardation and developmental dis-

abilities, in Janicki MP, Wisniewski HM (eds): Aging and Developmental Disabilities. Baltimore, Brookes, 1985, p 195
88. Barcikowska M, Silverman W, Zigman W, et al: Alzheimer type neuropathology and clinical symptoms of dementia in mentally retarded people without Down syndrome. 1989, submitted for publication
89. Dalton AJ, Crapper-McLachlan DR: The clinical expression of Alzheimer's disease in Down's syndrome. Psychiatric Clinics of North America 9:659–670, 1986
90. Lai F, Williams R: Alzheimer's dementia in Down's syndrome. Neurology 37:332, 1987
91. Zigman WB, Schupf N, Lubin RA, et al: Premature regression of adults with Down syndrome. Am J Ment Def 92:161–168, 1987
92. Silverstein AB, Herbs D, Miller TJ, et al: Effects of age on the adaptive behavior of institutionalized and noninstitutionalized individuals with Down syndrome. Am J Ment Ret 92:455–460, 1988
93. Mann DMA, Yates PO, Marcyniuk B, et al: The topography of plaques and tangles in Down's syndrome patients of different ages. Neuropath Appl Neurobiol 12:447–457, 1986
94. Ropper AH, Williams RS: Relationship between plaques, tangles, and dementia in Down's syndrome. Neurology 30:639–644, 1980
95. Percy ME, Dalton AJ, Markovic VD, et al: Alzheimer's disease in Down syndrome: Red cell superoxide dismutase, glutathione peroxidase and catalase. Am J Med Genetics, in press
96. Percy ME, Dalton AJ, Markovic VD, et al: Thyroid function and autoantibodies to thyroid in Down syndrome: Relationship with Alzheimer's disease. Am J Med Genetics, in press
97. Jackson CVE, Holland AJ, Williams CA, et al: Vitamin E and Alzheimer's disease in subjects with Down's syndrome. J Ment Def Res 32:479–484, 1988
98. Molsa PK, Marttila RJ, Rinne UK: Epidemiology of dementia in a Finnish population. Acta Neurol Scand 65:541–552, 1982
99. Cutler NR: Cerebral metabolism as measured with positron emission tomography (PET) and [^{18}F]2-deoxy-D-glucose: Healthy aging, Alzheimer's disease and Down syndrome. Prog Neuro-Psychopharmacol & Biol Psychiat 10:909–321, 1986
100. Blackwood DHR, St Clair DM, Muir WJ, et al: The development of Alzheimer's disease in Down's syndrome assessed by auditory event-related potentials. J Ment Def Res 32:439–453, 1988
101. Muir WJ, Squire I, Blackwood DHR, et al: Auditory P300 response in the assessment of Alzheimer's disease in Down's syndrome: A 2-year follow-up study. J Ment Def Res 32:455–463, 1988
102. Mann, DMA: Neuropathological and neurochemical aspects of Alzheimer's disease, in Iversen LL, Iversen SD, Snyder SH (eds): Handbook of Psychopharmacology. New York, Plenum, 1988, p 1
103. Hohmann C, Antuono P, Coyle JT: Basal forebrain cholinergic neurons and Alzheimer's disease, in Iversen LL, Iversen SD, Snyder SH (eds): Handbook of Psychopharmacology. New York, Plenum, 1988, p 69
104. Rossor M: Neurochemical studies in dementia, in Iversen LL, Iversen SD, Snyder SH (eds): Handbook of Psychopharmacology. New York, Plenum, 1988, p 107
105. Saper CB: Chemical neuroanatomy of Alzheimer's disease, in Iversen LL, Iversen SD, Snyder SH (eds): Handbook of Psychopharmacology. New York, Plenum, 1988, p 131
106. Rossor MN, Iversen LL, Reynolds GP, et al: Neurochemical characteristics of early and late onset types of Alzheimer's disease. Brit Med J 288:961–964, 1984
107. Perry EK: The cholinergic hypothesis—ten years on. Brit Med Bull 42:63–69, 1986
108. Wilcock GK, Esiri MM, Bowen DM, et al: The nucleus basalis in Alzheimer's disease: Cell counts and cortical biochemistry. Neuropath Appl Neurobiol 9:175–179, 1983
109. Rogers JD, Brogan D, Mirra SS: The nucleus basalis of Meynert in neurological disease: A quantitative morphological study. Ann Neurol 17:163–170, 1985
110. Arendt T, Bigl V, Tennstedt A, et al: Neuronal loss in different parts of the nucleus basalis is related to neuritic plaque formation in cortical target areas in Alzheimer's disease. Neurosci 14:1–14, 1985
111. Bowen DM, Benton JS, Spillane JA, et al: Choline acetyltransferase activity and histopathology of frontal neocortex from biopsies of demented patients. J Neurol Sci 57:191–202, 1982
112. Rossor MN, Garrett NJ, Johnson AL, et al: A postmortem study of the cholinergic and GABA systems in senile dementia. Brain 105:313–330, 1982
113. Nagai T, McGreer PL, Peng JH, et al: Choline acetyltransferase immunohistochemistry in brains of Alzheimer's disease patients and controls. Neurosci Lett 36:195–199, 1983
114. Perry EK, Tomlinson BE, Blessed G, et al: Correlations of cholinergic abnormalities with senile plaques and mental tests scores in senile dementia. Br Med J 2:1457–1459, 1978

115. Wilcock GK, Esiri MM, Bowen DM, et al: Alzheimer's disease: Correlation of cortical choline acetyltransferase activity with the severity of dementia and histological abnormalities. J Neurol Sci 57:407–417, 1982
116. Mountjoy CQ, Rossor MN, Iversen LL, et al: Correlation of cortical cholinergic and GABA deficits with quantitative neuropathological findings in senile sementia. Brain 107:507–518, 1984
117. Rossor M, Iversen LL: Non-cholinergic neurotransmitter abnormalities in Alzheimer's disease. Br Med Bull 42:70–74, 1986
118. Bondareff W, Mountjoy CQ, Roth M: Selective loss of neurons of origin of adrenergic projection to cerebral cortex (nucleus locus coeruleus) in senile dementia. Lancet I:783–784, 1981
119. Tomlinson BE, Irving D, Blessed G: Cell loss in locus coeruleus in senile dementia of Alzheimer type. J Neurol Sci 49:419–428, 1981
120. Mann DMA, Lincoln J, Yates PO, et al: Changes in the monoamine-containing neurones of the human CNS in senile dementia. Br J Psychiatry 136:533–541, 1980
121. Yates CM, Ritchie IM, Simpson J, et al: Noradrenaline in Alzheimer-type dementia and Down syndrome. Lancet II:39–40, 1981
122. Yates CM, Simpson J, Gordon A, et al: Catecholamines and cholinergic enzymes in pre-senile and senile Alzheimer-type dementia and Down's syndrome. Brain Res 280:119–126, 1983
123. Adolfsson R, Gottfries CG, Roos BE, et al: Changes in the brain catecholamines in patients with dementia of Alzheimer type. Br J Psychiatry 135:216–223, 1979
124. Arai H, Kosaka K, Iizuka T: Changes of biogenic amines and their metabolites in postmortem brains from patients with Alzheimer-type dementia. J Neurochem 43:388–393, 1984
125. Curcio CA, Kemper T: Nucleus raphe dorsalis in dementia of the Alzheimer type: Neurofibrillary changes and neuronal packing density. J Neuropath Exp Neurol 43:359–368, 1984
126. Davies P, Katzman R, Terry RD: Reduced somatostatin-like immunoreactivity in cerebral cortex from cases of Alzheimer disease and Alzheimer senile dementia. Nature 288:279–280, 1980
127. Morrison JH, Rogers J, Scherr S, et al: Somatostatin immunoreactivity in neuritic plaques of Alzheimer patients. Nature 314:90–92, 1985
128. Roberts GW, Crow TJ, Polak JM: Location of neuronal tangles in somatostatin neurones in Alzheimer's disease. Nature 314:92–94, 1985
129. Mann DMA, Yates PO, Marcyniuk B, et al: Pathological evidence for neurotransmitter deficits in Down's syndrome of middle age. J Ment Defic Res 29:125–135, 1985
130. Godridge H, Reynolds GP, Czudek C, et al: Alzheimer-like neurotransmitter deficits in adult Down's syndrome brain tissue. J Neurol Neurosurg Psychiatry 50:775–778, 1987
131. McGeer EG, Norman M, Boyes B, et al: Acetylcholine and aromatic amine systems in postmortem brain of an infant with Down's syndrome. Exper Neurol 87:557–570, 1985
132. Yates CM, Simpson J, Gordon A: Regional brain 5-hydroxytryptamine levels are reduced in senile Down's syndrome as in Alzheimer's disease Neurosci Lett 65:189–192, 1986
133. Shortridge BA, Vogel FS, Burger PC: Topographic relationship between neurofibrillary change and acetylcholinesterase-rich neurons in the upper brain stem of patients with senile dementia of the Alzheimer's type and Down's syndrome. Clin Neuropath 4:227–237, 1985
134. Casanova MF, Walker LC, Whitehouse PJ, et al: Abnormalities of the nucleus basalis in Down's syndrome. Ann Neurol 18:310–313, 1985
135. Yates CM, Simpson J, Maloney AFJ, et al: Alzheimer-like cholinergic deficiency in Down syndrome. Lancet II:979, 1980
136. Pierotti AR, Harmar AJ, Simpson J, et al: High-molecular-weight forms of somatostatin are reduced in Alzheimer's disease and Down's syndrome. Neurosci Lett 63:141–146, 1986
137. Melamed E, Mildworf B, Sharav T, et al: Regional cerebral blood flow in Down's syndrome. Ann Neurol 22:275–278, 1987
138. Yamaguchi F, Meyer JS, Yamamoto M, et al: Noninvasive regional cerebral blood flow measurements in dementia. Arch Neurol 37:410–418, 1981
139. Duara R, Grady C, Haxby J, et al: Position emission tomography in Alzheimer's disease. Neurology 36:879–887, 1986
140. Wisniewski KE, Wisniewski HM: Age-associated changes and dementia in Down's syndrome, in Reisberg B (ed), Alzheimer's Disease. New York, Free Press, 1983, p 319
141. Brown WT, Wisniewski HM: Genetics of human aging. Rev Biol Res Aging 1:81–99, 1983
142. Larsson T, Sjogren T, Jacobson G: Senile dementia: A clinical, sociometical, and genetic study. Acta Psychiatr Scand 39(Suppl 167)1–259, 1963

143. Heston LL, Mastri AR, Anderson VE, et al: Dementia of the Alzheimer type. Clinical genetics, natural history and associated conditions. Arch Gen Psychiatry 38:1085–1091, 1981
144. Heyman A, Wilkinson WE, Hurwitz BJ, et al: Alzheimer's disease: Genetic aspects and associated clinical disorders. Ann Neurol 14:507–515, 1983
145. Breitner JCS, Folstein MF: Familial Alzheimer dementia: A prevalent disorder with specific clinical features. Psychol Med 14:63–80, 1984
146. Martin RL, Gerteis G, Gabrielli WF: A family-genetic study of dementia of the Alzheimer type. Arch Gen Psychiatry 45:894–900, 1988
147. Schupf N, Zigman WB, Silverman WP, et al: Genetic epidemiology of Alzheimer's disease, in Battistin L (ed), Aging Brain and Dementia: New Trends in Diagnosis and Therapy. New York, Alan R. Liss, 1990, p. 57
148. Yatham LN, McHale PH, Kinsella A: Down's syndrome and its association with Alzheimer's disease. Acta Psychiatr Scand 77:38–41, 1988
149. Glenner GG: Alzheimer's disease: Its proteins and genes. Cell 52:307–308, 1988
150. Epstein CJ: Trisomy 21 and the nervous system: From cause to cure, in Epstein CJ (ed): The Neurobiology of Down Syndrome. New York, Raven, 1986, p 1
151. St. George-Hyslop PH, Tanzi RE, Polinsky RJ: The genetic defect causing familial Alzheimer's disease maps on chromosome 21, Science 235:885–890, 1987
152. Delebar JM, Goldgaber D, Lamour Y, et al: Beta amyloid gene triplication in Alzheimer's disease and karyotypically normal Down syndrome. Science 235:1390–1392, 1987
153. Schweber M: A possible unitary genetic hypothesis for Alzheimer's disease and Down syndrome. Ann NY Acad Sci 450:223–238, 1985
154. Schellenberg GD, Bird TD, Wijsman EM, et al: Absence of linkage of chromosome 21q21 markers to familial Alzheimer's disease. Science 241:1507–1501, 1988
155. Warren AC, Robakis NK, Ramakrishna N, et al: Beta-amyloid gene is not present in three copies in autopsy-validated Alzheimer's disease. Genomics 1:307–312, 1987
156. Furuya H, Sasaki H, Goto I, et al: Amyloid beta-protein gene duplication is not common in Alzheimer's disease: Analysis by polymorphic restriction fragments. Biochem Biophys Res Commun 150:75–81, 1988
157. Podlisny MB, Lee G, Selkoe DJ: Gene dosage of the amyloid beta precursor protein in Alzheimer's disease. Science 238:669–671, 1987

III

Evaluation, Drug Development, and Ethical Considerations

The final section of the book considers very practical issues facing clinicians and investigators concerned with behavioral disorders of children. A comprehensive psychological evaluation, including specialized neuropsychological testing procedures, is often an indispensable component of the diagnostic assessment. Moreover, research in the discipline of neuropsychology has provided clinicians with a large data base of "deficits" referable to specific neuroanatomic regions in specific disorders. The interpretation of these neuropsychological deficits and their implications for treatment are major concerns of workers in the field. The neurologic evaluation of violent adolescents is an important issue facing clinicians. Data obtained in these evaluations demonstrate the seminal role of biological factors in many pathological presentations of violence.

A rational approach to novel drug development for the treatment of childhood disorders is clearly dependent on an appreciation of biological abnormalities in these conditions. Some of the recent directions pursued in the area of novel drug development are described.

Finally, as alluded to in the Introduction, many ethical issues are raised as investigators propose and implement research involving disturbed children. In fact, a resolution of these ethical issues is a necessary prerequisite for pursuing clinical investigations in this field.

21

Neuropsychological Evaluation of Children

Wilma G. Rosen

1. ISSUES IN CLINICAL CHILD NEUROPSYCHOLOGY

The major approaches to child clinical neuropsychology have used an adult model of neuropsychological functioning as an initial framework for analysis of cognitive dysfunction in childhood.[1–4] While this strategy seemed logical and initially proved useful as a model for the organization of ideas about brain–behavior relationships in childhood, the limitations of this approach, well articulated by Rudel,[5] have become increasingly apparent as brain-damaged children are evaluated more systematically in observational or experimental studies. Thus, child neuropsychologists began to accumulate evidence that countered the assumption that the developing organism suffers the same fate (both biologically and psychologically) following brain damage as does the mature adult. We know, from postnatal and prenatal studies of nonhuman primates and children, that focal brain damage produces different kinds of cognitive deficits and different degrees of recovery of function depending on the developmental stage or age at which the damage occurred.[6–13] Child neuropsychologists have also had to consider the logic of the assumption that abnormal development of a cognitive ability reflected pathology in the same brain structures that were determined from adult lesion studies to be necessary for the intact expression of that ability in adulthood.[14] In fact, Rudel[5] rejected that notion, stating, "Adult brain trauma studies shed light on what is essential for the *performance* of a function but not for its *development*" (p. 10). That is, during the acquisition of a skill, many parts of the brain may be involved; as proficiency increases, some structures may no longer participate in the mediation of this behavior. Thus, these brain structures, while crucial for the development of a function, may not be necessary for its expression once complete competency has been achieved.

Finally, it does not necessarily follow that a cognitive dysfunction or disturbance in childhood reflects the presence of a focal lesion or identifiable brain damage. Many professionals who deal with children displaying deviations from the norm that apparently cannot be

---

Wilma G. Rosen • Department of Psychiatry, Columbia University, and The Presbyterian Hospital, Neurological Institute, Columbia-Presbyterian Medical Center, New York, New York 10032.

attributed to functional factors have been accustomed to using a classification such as "minimal brain dysfunction" or "minimal brain damage." Invoking such a construct, in the absence of independent verification, conveys no more information than we already know, lumps together children with all kinds of difficulties into an all-inclusive category, provides little to no assistance for planning individual treatment strategies, and has the potential to do more harm than good. Thus, in several ways, clinical child neuropsychology cannot be considered a version of adult neuropsychology adapted for children.

2. DISTINGUISHING FEATURES OF CHILD NEUROPSYCHOLOGY

As a singular discipline, child clinical neuropsychology represents a unique blend of developmental psychology, pediatric neurology, pediatric psychiatry, educational assessment, and cognitive psychology. Part of what distinguishes child neuropsychology from adult neuropsychology is that the notion of development has to be considered in all child evaluations, while, in adults, a more static attained level of functioning is assumed. Thus, with child evaluations we require a more extensive set of "norms" for various stages of development and knowledge of the ways in which the manifestations of learning disabilities and other disorders change with development.[15,16]

The goals or purposes of child and adult assessments may differ as well. The adult evaluation is often performed to assist in diagnosis, to document current level of functioning in order to establish a baseline set of performances, or to assist in planning treatment strategies. The child evaluation, in addition to contributing to differential diagnosis or identification of the type(s) of learning disabilities or difficulties present, nearly always has the same pragmatic goals: recommendations for appropriate educational planning and services and/or techniques for behavioral management.

While the value of neuropsychological evaluation for children with identifiable neurological disorders, e.g., seizures, traumatic brain injury, tumors, and so on, seems self-evident, this kind of assessment also makes a significant contribution to the understanding and management of children with various disorders subsumed under the rubric of the DSM III-R,[17] including attention-deficit hyperactivity disorder, specific developmental disorders, such as a developmental reading disorder, Tourette's disorder, and so on. The neuropsychologist must understand how the symptoms of the disorder impact on the multiple demands of the educational process because schooling is a major aspect of children's lives, as well as how the symptoms affect socialization or may be a comcomitant of poor social skills.[18,19] More important, the neuropsychologist is in the unique position to analyze the interaction of the symptoms with the pattern of cognitive strengths and weaknesses revealed on testing. It is the consideration of the symptoms, cognitive functioning, and their interaction that allow the neuropsychologist to make a complete interpretation and to offer sensible recommendations.

3. THE EVALUATION PROCEDURE

Although the most common conceptualization of the procedure for a neuropsychological evaluation is simply to test and then interpret the results of testing, the procedure actually has several components designed to place the findings within a context. Since it is most important to remember that the testing reflects only a "slice" in the child's life, i.e., those few hours spent in the test situation, the findings are interpreted within contexts provided by both historical information and current observations. The historical context includes medical, genetic, behavioral, social, educational, and cognitive developmental information. The current context is the child's behavior during interview and testing.

3.1. Historical Context

This information is provided from interview with the parent(s) and all available records that document medical, genetic, developmental, and educational history. Each of these areas of inquiry can be addressed with both questionnaires[20,21] and on direct interview either informally or with a structured set of questions that directly assesses the presence or absence of diagnostic symptoms of particular disorders.[22]

The importance of medical history is the possible effects of a previous significant episode (e.g., head injury, high fever) on brain function and its short-term and long-term sequelae. Family history may be revealing because of the accumulating evidence of a possible heritable basis for learning disorders.[23,24]

The developmental history should include the ages at which motor and language milestones were attained. Of particular significance is the possible negative effect of delayed language development on the acquisition of reading skills.[5] Beginning with the infancy–toddler period, the developmental inquiry should also document early behavior problems in such areas as sleep patterns, activity level, impulsivity, attention span, and tolerance for frustration. The persistence of these early behavioral disturbances often represents the underpinnings of the formulation of the child's current difficulties.

Educational history is most easily attained from school records. Often, parents provide report cards, results of standardized testing, and, if the child has been evaluated before, reports of previous testing. This information represents additional valuable data, which is independent of parents' occasional faulty or biased recall of performances. Evaluation of school and test records allows for analysis of the pattern of performance over the years and comparison of the current test findings with previous findings to determine the stability of levels of functioning.

The child may also offer valuable information about school and family relationships, as well as their own explanation of the reasons for the evaluation. They may identify or describe school problems differently from the parents, may be unaware or unattuned to the concerns and complaints of others, or may have already erroneously concluded that the basis for their difficulties lies in a generalized incompetence, i.e., overall stupidity.

3.2. Behavioral Observations

Observation of the child occurs from the moment the child enters the office. One notes the child's physical appearance, interactions with the parent, and ability to handle being alone in an unfamiliar environment while the parent talks with the examiner. In the test situation, there are several behaviors of concern. These include how "nervous" the child appeared initially and the extent to which this persisted during the evaluation (i.e., anxiety), cooperativeness and persistence on difficult tasks which would reflect motivational status, levels of physical activity which may change depending upon task requirements, and social behaviors including eye contact, spontaneity of speech, dialogic interactions, and relatedness to the examiner.

The other kinds of behaviors that are observed during testing are actually the test-taking behaviors themselves. In particular, we are not only concerned with the actual scores that the child achieves, but, when possible, with assessing the way in which the child obtained that score. This technique, which is often referred to as the process approach,[25,26] stresses that the observed manner in which a task is executed permits a more complete analysis of the means by which a solution is achieved and potentially leads to a better understanding of the child's cognitive strengths and weaknesses. For example, on two different tasks, one requires copying a complex geometric form and the other requires learning names of many items, we observe the organizational strategies that the child uses in order to perform. While the outcome measure (score) actually is interpreted to reflect drawing accuracy or learning, the kinds of organizational strategies observed are used to provide additional but important information about how the child approaches unfamiliar material in order to maximize performance. In addition, the observed behavior may also give the examiner greater insight into the reasons for less than optimal performance. In this particular illustration, a low score on the learning task may not reflect impaired learning or memory skills *per se*, but lack of implementation of organization, a technique that enhances learning and recall.

3.3. Testing Procedures and Rationale

Just as in adult clinical neuropsychology, there are the two major lines of thought as to the most appropriate evaluation techniques, a set battery of tests and the hypothesis-testing approach; child clinical neuropsychology perpetuates this dichotomy. The battery approach is best represented by the Halstead-Reitan Neuropsychological Test Battery (HRB) in adults and its adaptations for children,[2,3,27,28] and similarly, the Luria-Nebraska Neuropsychological Battery (LNNB) and its version for children.[29,30] While the HRB was initially very useful for diagnosing the presence and location of brain damage in adults, this method is not especially helpful in children for whom there is no independent validation of the presence of brain damage, and the presence–absence differential is not critical for the assessment. In most cases, the child evaluation is directed toward documenting the child's cognitive strengths and weaknesses; localization of brain damage is not an issue.[5,31] The LNNB for children suffers from the same limitations. Consequently, modifications of the HRB for children have been made, and it should be noted that tests of intelligence and achievement, e.g., Wechsler Intelligence Scale for Children—Revised (WISC—R)[32] and the Wide Range Achievement Test—Revised (WRAT—R)[33] are also incorporated into the evaluation.

The hypothesis-testing approach has a few variations on its major thesis of beginning with a set of tests that samples a wide variety of behaviors, including other tests to target the *a priori* purported problems, and then as the deficits begin to emerge on testing, focusing on the problems in order to (1) verify their reliability by seeking redundancy in performance deficits, (2) identify the underlying causes of the deficit, and (3) determine related disorders.[5] This kind of approach requires that the examiner make decisions about performance using relative performance levels[34] or the capacity–achievement discrepancy as the criterion for deficit performance.[5] In both cases, there are no absolute quantitative criteria that specify the presence of a learning disorder. Sometimes, however, the discrepancy is expressed as the difference between chronological age and equivalent age level of performance, between standardized scores on two tests such as IQ and achievement, or between current grade level and equivalent grade level of performance.

The approach used in the cases presented below is the hypothesis-testing one. Strictly speaking, however, there is usually less flexibility in a child evaluation than in an adult assessment because of the need to sample a wider range of behaviors in children in order to rule out potential primary and secondary problems. For example, the detection of a reading disorder always requires a more extensive examination of language abilities.

4. CASE STUDIES

In the cases presented here, the evaluation included the WISC—R, achievement tests, and primarily tests of language, memory, and learning, attention, visuospatial and visuoconstructional abilities, and motor functioning. As will become apparent, there was a core set of tests used in all of the evaluations and then divergence in the remainder of the choice of tasks. The core set of tests included the WISC—R to determine the general level of intellectual functioning and to provide information about different kinds of functions and various tests of educational achievement in the areas of reading, spelling, and math. After those reference points had been established, the referral questions were addressed, which included reading difficulties, attentional problems, etc. As the expected difficulties were confirmed or disconfirmed, additional tests were administered according to the guidelines stated previously. The overlap among the cases in the tests administered is a function of the conceptualization of the disorders, which, in turn, suggests further examination using tasks that tap increasingly more basic functions.

4.1. Case 1: Reading Disability

4.1.1. Reason for Referral

D.Y. was a 12-year, 1-month-old, right-handed female who had been identified in grade two as having a reading problem due to "processing" difficulties. Her mother requested the current evaluation in order to delineate more explicitly the nature of the reading difficulty and to reassess plans for remediation and schooling because of anticipated entry into grade seven in junior high school.

4.1.2. Brief History

D.Y. was adopted at 4 days of age. She suffered from asthma, which required three hospitalizations during the previous eight years. Motor developmental milestones were achieved early. She spoke phrases at 2 years of age and sentences at 3 years; however, naming coins and repeating the alphabet reportedly occurred late in development. Psychoeducational evaluation at age 8, under the auspices of the school system, revealed that she was of average intelligence with a limited sight reading vocabulary, had difficulty with phonics, and evidenced sequencing errors. Her approach to learning new material was remarkable because of the resistance and anxiety noted by teachers and parents. Over the years, she had received reading remediation and classwork assistance for one period per day in school, had often voluntarily sought assistance from her teachers after school, and had a private tutor once per week.

4.1.3. Observations

D.Y. was an attractive, stylishly dressed early adolescent. At the first session she initially appeared shy and anxious, but became increasingly relaxed and spontaneously initiated conversation as the session proceeded. At the second session, she was quite at ease and socially confident. She worked willingly and diligently on all tasks, with the exception of story writing.

4.1.4. Test Findings

The major test scores are shown in Tables I and II.[32,33,35–42] This discussion centers on the major results, the referral problem, and related disorders; not all test findings are reported. On the WISC—R her overall level of intelligence was in the Average range, and her Verbal IQ score was nonsignificantly greater than her Performance IQ score. These scores were consistent with earlier test results.

The reading disability remained quite prominent as there were significant discrepancies between her average intellectual capacity and her borderline to low average range reading scores. She exhibited deficient decoding of nonsense words and single word reading. Passage comprehension ranged from average for factual material to low average for inferential questions. Oral reading revealed an unstable vowel-sound system, reordering of letter sequences,

Table I. Intelligence and Achievement Test Findings in a 12-Year-Old with a Reading Disorder

Intelligence

Wechsler Intelligence Scale for Children—Revised

Information	10	Picture Completion	12
Similarities	17	Picture Arrangement	11
Arithmetic	8	Block Design	11
Vocabulary	10	Object Assembly	7
Comprehension	13	Coding	8
Digit Span	7	Mazes	12
Verbal IQ	109		
Performance IQ	98		
Full-Scale IQ	104		

Achievement

Wide Range Achievement Test—Revised

	Standard score	*Percentile*	*Grade*
Reading	76	5	end of 3
Spelling	70	2	begin. of 3
Arithmetic	71	3	end of 4

Gilmore Oral Reading Test

Accuracy: 5.9 grade equivalent—average

Comprehension: 7.5 grade equivalent—above average

Rate: Very slow

Gray Oral Reading Test—Revised

	Standard score	*Percentile*
Comprehension	7	16
Time/Accuracy	6	9

Woodcock Reading Mastery Tests

Word Identification: 3.6 grade equivalent

Word Attack: 2.1 grade equivalent

Rapid Automatized Naming

Colors: 39 seconds—low average

Numbers: 43 seconds—impaired

Use Objects: 47 seconds—average

Small Letters: 33 seconds—impaired

and impaired consonant blending skills. Her very slowed reading speed was consistent with a lack of automaticity or fluidity of response on Rapid Automatized Naming.[38] Deficient spelling ability was manifested primarily in phonetic substitutions.

Examination of language abilities that might be related to the reading disorder revealed that verbal fluency, naming, repetition, and production and comprehension of syntax were in the low average to average range for her age. In particular, categorical fluency was variable, and when naming objects, there were several instances in which she recognized the item, described its function, but could not access the name. Story writing was laborious and effortful, as might be expected. In addition, examination of the WISC—R Coding and Digit Span scores suggests that these relatively lower scores are indicative of generalized deficits in symbolic representation and sequencing, respectively, both of which are consistent with her reading disorder. She also exhibited difficulties with mathematics, including the mental transformation of orally presented word problems and in written calculations.

Her strengths included good organizational skills during verbal learning and recall tasks, excellent visual memory abilities for geometric forms, and adequate visuoperceptual and visuoconstructional skills.

Table II. Language and Memory Test Findings in a 12-Year-Old with a Reading Disorder

Language
Verbal Fluency Tests
Semantic Categories
Food—average
Animals—low average
Words beginning with "sh" sound—low average
Boston Naming Test
Number correct: 48—average
NCCEA Sentence Repetition
Number correct: 12—average
Test of Written Language
Word Usage: Standard Score = 11, Percentile = 63
Story Writing: Insufficient amount written, unable to score
Neimark Memorization Strategies Test
Number of items recalled—maximum is 24
Trial 1: 14—average
Trial 2: 20—average
20-minute delay: 20—excellent
Benton Visual Retention Test
Number of forms correct: 9—above average
Number of errors: 1—above average

4.1.5. Conclusions and Recommendations

D.Y. had continued to exhibit a very significant dyslexia despite four years of remediation. The disorder appeared to be related to both sequencing difficulties and language problems. Interestingly, she exhibited quite adequate organizational abilities, which may reflect the study skills training she had received. With regard to her learning disability, the major recommendation was continued reading remediation with an emphasis placed upon deriving meaning from context, which will be the most needed skill as she proceeds through higher grades. Writing remediation should also be implemented. She will require support from the school system with possible adjustments to the curriculum and workload. If the school cannot provide the necessary changes, then a specialized high school should be sought.

4.2. Case 2: Attention-Deficit Hyperactivity Disorder (ADHD)

4.2.1. Reason for Referral

S.K. was referred for evaluation by his father and stepmother because of attentional problems and failure to perform up to expectations in school.

4.2.2. Brief History

S.K. was a right-handed, 11-year, 8-month-old male, who was in grade six and lived with his father, stepmother, and half-sister. Motor and language developmental milestones were achieved at the usual times, except for the later emergence of speaking in sentences and bicycle riding. His parents' descriptions of his behaviors and teachers' written comments on his behavior since grade one were remarkably consistent: S.K. exhibited a short attention span, low frustration tolerance, impulsivity, poor memory, and disorganization. In school he frequently failed to complete assignments, exhibited lack of self-control, appeared uncommitted to his schoolwork, and was socially immature. He explained that he began new assignments in school before finishing old ones because he was afraid that he would forget the instructions for the new tasks. Psychoeducational evaluation by the school system concluded that (unspecified) "emotional problems" hindered learning. Consequently, S.K. received two years of psychotherapy, but made little progress. S.K. enjoyed reading and reported that he had few friends.

4.2.3. Observations

S.K. is an attractive youngster with an engaging manner, whose overtalkativeness interfered with the progress of testing. He was quick-witted and enjoyed talking about his activities and his ideas. He showed signs of excessive motor activity as he frequently popped up from the chair or fidgeted. He was easily distracted by external noises.

4.2.4. Test Findings

The major findings are shown in Tables III and IV.[32,33,36,41–49] Again, this discussion centers on the most important findings, omitting less important results. On the WISC—R his overall level of intellectual functioning was in the Superior range. Both his Verbal and

Table III. Intelligence and Achievement Test Findings in a 11-Year-Old with Attention-Deficit Hyperactivity Disorder

Intelligence			
Wechsler Intelligence Scale for Children—Revised			
Information	14	Picture Completion	16
Similarities	14	Picture Arrangement	11
Arithmetic	11	Block Design	13
Vocabulary	12	Object Assembly	14
Comprehension	14	Coding	9
Digit Span	10	Mazes	11
Verbal IQ	118		
Performance IQ	118		
Full-Scale IQ	121		

Achievement			
Wide Range Achievement Test—Revised			
	Standard score	*Percentile*	*Grade*
Reading	102	55	begin. 7
Spelling	99	47	begin. 6
Arithmetic	101	53	begin. 6
Gray Oral Reading Test—Revised			
	Standard score	*Percentile*	
Comprehension	12	75	
Time/Accuracy	11	63	
Woodcock Reading Mastery Tests—Revised			
	Standard score	*Percentile*	
Visual–Auditory Learning	106	66	
Word Attack	106	65	
Word Comprehension	113	81	
Passage Comprehension	117	87	

Table IV. Learning, Memory, Attention, and Construction Test Findings in an 11-Year-Old with a Diagnosis of ADHD

Learning and Memory
  Neimark Memorization Strategies Test
    Number of items recalled—maximum of 24
    Trial 1: 10—impaired
    Trial 2: 19—average
    20-minute delay: 17—average
  Wechsler Memory Scale
    Paired-Associates Learning: 19—above average
  Benton Visual Retention Test
    Number of Forms Correct: 8—above average
    Number of errors: 3—average
Attention/Concentration
    Matching Familiar Figures Test
    Time to Completion: 37 percentile—average
    Number of Errors: 45 percentile—average
  Rey–Osterreith Complex Figure Test
    Copy: Fragmented
    Immediate Recall: Fragmented, diagonals incomplete details missing or misplaced
Construction
    Visual Motor Integration Test: 32–48 percentile for age
    Rosen Drawing Test: Within normal limits for age

Performance abilities were in the above average range. The pattern among the subtest scores reflected above average to superior problem solving abilities and fund of acquired information. The average range scores (9–11) reflected, primarily, a mild slowing in execution and some difficulties in the mental transformation of information.

Single word reading, decoding nonsense words and oral reading speed were all within the average range. Reading comprehension was above average and indicative of well-developed ability to derive meaning from context. His average spelling abilities were consistent with the limitations of his phonemic decoding ability apparent in reading. There was no evidence of associated language difficulties. Arithmetic calculations, while average, reflected carelessness and variable competency with fractions.

Verbal learning was excellent when the material was highly structured (Wechler Memory Scale, Paired Associates Learning[44]). On another verbal learning task (Neimark Memorization Strategies[41]), he failed to discern the categorical structure and organize the material accordingly both during studying and recall. His untenable explanation for his below average recall of trial 1 was his unfamiliarity with the items. Items he failed to recall included a sock, boat, table. On the second trial, he used some organizational strategy, and his recall improved to the average range.

This organizational difficulty was also apparent on the Rey-Osterrieth Complex Figure Test,[46,47] where his copy of a complex geometric form was fragmented and disjointed. Subsequent immediate recall of this form suffered for this initial lack of organization because his reproduction from memory included misplaced or omitted details. These inadequate drawings were primarily a function of failure to analyze the figure and plan a drawing strategy rather than a copying difficulty *per se* because his performances on other drawing tests were average for his age. Finally, on tasks usually interpreted as reflecting sustained attention and impulsivity (Matching Familiar Figures,[45] Coding, Digit Span, Mazes[32]) his performances were in the average range.

4.2.5. Conclusions and Recommendations

S.K. was a bright youngster with excellent expressive language capabilities and above average reading comprehension skills. The description and history provided by both his parents and teachers was highly characteristic of ADHD. On testing he exhibited significant difficulty in organizing material in order to learn or to reproduce it. This deficiency is often characteristic of children with ADHD, and it presents a significant problem in school as they enter the higher grades and must work more independently. S.K. also exhibited some relatively mild slowing on tasks, which might be interpreted as a compensatory strategy for impulsivity. He also had average phonetic decoding and encoding skills. The recommendations offered were (1) consultation with a pediatric psychiatrist or neurologist for a probable therapeutic trial with medication for ADHD, (2) structuring work schedules, organizing tasks, and maintaining consistency in expectations for levels of performance, (3) instruction in organizational and study skills, (4) additional instruction to increase his sight reading vocabulary and phonetic decoding and encoding abilities.

4.3. Case 3: Brain Damage and Rage Behavior

4.3.1. Reason for Referral

C.D. was referred for evaluation by a pediatric psychiatrist to assist in determining an appropriate academic setting.

4.3.2. Brief History

C.D. was a 13-year, 9-month-old boy from the Middle East who had lived in several English-speaking countries for most of his life. He had a left hemiparesis and bilateral sensorineural hearing loss. He was the product of a complicated delivery, which led to perinatal craniocerebral trauma, intracranial hemorrhage, and a three-week post-birth hospitalization. Developmental milestones were delayed as he walked at 18 months and night toilet training was successful at age 11. He received physical and speech therapy during early childhood, and was treated with Ritalin for ADHD from ages 3–5. Two years prior to evaluation, he was placed in a special classroom for learning disabled children, where he became physically aggressive with other children. He then transferred to a specialized school with initially good adjustment. After a few months, he became physically aggressive with his peers, teachers, and family. The intensification of this behavior led to a hospital inpatient evaluation, which was ongoing at the time of this testing.

4.3.3. Observations

C.D. acted much younger than his age, laughing or grinning as questions perceived as silly or "too easy." He was polite, cooperative, and apparently unaware of his errors. His speech was dysarthric.

4.3.4. Test Findings

The major test findings are shown in Table V.[32,33,41,50–52] On the WISC—R his level of general cognitive functioning was in the Borderline range, which was confirmed by his performance on the Raven Coloured Progressive Matrices. His Verbal abilities were in the Borderline range, while his Performance IQ was at the lower limit of the Low Average range.

Although single word reading was in the low average range for his age and at the beginning grade 6 level, reading rate and accuracy for stories was at the grade 2 level.[53]

Table V. Intelligence, Achievement, and Learning Test Findings in a Neurologically Impaired 13-Year-Old

Intelligence

Wechsler Adult Intelligence Scale—Revised

Information	6	Picture Completion	7
Similarities	7	Picture Arrangement	10
Arithmetic	7	Block Design	5
Vocabulary	5	Object Assembly	7
Comprehension	5	Coding	6
Digit Span	14	Mazes	6
Verbal IQ	75		
Performance IQ	80		
Full-Scale IQ	76		

Raven Coloured Progressive Matrices
Estimated IQ: 70
Percentile: 10

Achievement

Wide Range Achievement Test—Revised

	Standard score	*Percentile*	*Grade*
Reading	88	21	begin. 6
Spelling	68	2	begin. 3
Arithmetic	66	1	end of 4

Peabody Picture Vocabulary Test—Revised
Less than 1 percentile

Learning and Memory

Neimark Memorization Strategies Test
Number of items recalled—maximum is 24
Trial 1: 13—impaired
Trial 2: 13—impaired

Rey Auditory–Verbal Learning Test
Number of words recalled on 5 trials—maximum is 15
4,6,6,4,7

Spelling ability was significantly limited with weak phonological skills. Written arithmetic skills achieved included simple addition, subtraction, multiplication and division, including renaming (carrying and borrowing).

Expressive language skills appeared better developed than receptive language abilities, which were limited by difficulty in comprehension of complex syntax, temporal concepts, such as "before" and "after," and confusion of active and passive voice. Verbal fluency was above average for categorical retrieval, and sentence structure in oral language was moderately complex.

Immediate memory was above average, as he correctly repeated 8 digits forward. However, he was extremely limited in learning new material (Rey Auditory–Verbal Learning Test,[52] Neimark et al.[41]). On the latter test, he employed no strategy for studying the items and thought he had them all memorized in a short period of time. On maze tracing, he evidenced poor planning. Copy of geometric forms of increasing complexity was significantly impaired for his age but relatively consistent with his overall level of intellectual functioning.

4.3.5. Conclusions and Recommendations

C.D. exhibited a borderline range of intelligence functioning with diffuse dysfunction apparent on more extensive testing. He was considerably limited by receptive language diffi-

Table VI. Intelligence and Memory Retest Findings in a 13-Year-Old after 5 Months Treatment with Propranolol

Intelligence			
Wechsler Adult Intelligence Scale—Revised			
Information	6	Picture Completion	7
Similarities	7	Picture Arrangement	8
Arithmetic	3	Block Design	1
Vocabulary	3	Object Assembly	10
Comprehension	5	Coding	5
Digit Span	14	Mazes	7
Verbal IQ	66		
Performance IQ	65		
Full-Scale IQ	63		
Learning and Memory			
Neimark Memorization Strategies Test			
Number correct			
Trial 1: 12—impaired			
Trial 2: 14—impaired			
Rey Auditory–Verbal Learning Test			
Number of items recalled on 5 trials			
5,9,13,13,15			

culties, impaired learning, impulsivity, and lack of self-monitoring. Recommendations included (1) a special educational setting with a high teacher–student ratio, (2) language therapy, (3) remediation in reading, mathematics, and spelling.

4.3.6. Reevaluation of C.D. Post-Treatment with Propranolol

C.D. remained an inpatient and was reevaluated following a 5-month trial with propranolol for uncontrolled outbursts of rage[54] because of concerns about adverse effects of this agent on memory and intellectual functioning. At this second evaluation, C.D. recognized the examiner and was increasingly impulsive in performance. The brief test findings are shown in Table VI.[32,41,52] Although his IQ scores were in the mentally retarded range and were significantly lower than those obtained at the previous testing, the differences appeared to be primarily attributable to increases in impulsivity, inattention, and lack of self-evaluation of performance. His digit span remained similar to the previous testing; he showed a significant increase in the ability to learn a list of 15 words; organizational skills remained significantly impaired. The effects of propranolol on this youngster's cognitive functioning were inconsistent, that is, his intellectual functioning appeared to decline because of behavioral disturbances, but his verbal learning ability appeared improved.

5. CONCLUDING REMARKS

These cases serve as illustrations of some of the reasons for referral for a neuropsychological evaluation, the kinds of tests available for assessing various cognitive functions, the variations in interpretation of the data, the interfacing of the child's history with the test findings, and the range in recommendations based on the results. These cases are representative of the diagnostic categories that we most frequently encounter, that is, a primary learning disability with or without emotional problems, attention-deficit hyperactivity disorder with or without learning disabilities, and neurological impairment that produces significant compromise in many areas of cognitive, emotional, and behavioral functioning. The developmental nature of these evaluations, as contrasted with the adult assessment that is less subject to the effects of rapid changes with age and acquisition and assimilation of new knowledge and ideas, renders the child assessment more challenging.

REFERENCES

1. Denckla MB: Childhood learning disabilities, in Heilman KM, Valenstein E (eds), Clinical Neuropsychology. New York, Oxford University Press, 1979, pp 535–576
2. Reitan RM: Psychological effects of cerebral lesions in children in early school age, in Reitan RM, Davison, LA (eds), Clinical Neuropsychology: Current Status and Applications. Washington, D.C., V.H. Winston & Sons, 1974, pp 53–89
3. Boll TJ: Behavioral correlates of cerebral damage in children aged 9 through 14, in Reitan, RM, Davison, LA (eds), Clinical Neuropsychology: Current Status and Applications. Washington, D.C., V.H. Winston & Sons, 1974, pp 91–120
4. Rourke BP: Neuropsychological assessment of children with learning disabilities, in Filskov SB, Boll TJ (eds), Handbook of Clinical Neuropsychology. New York, Wiley–Interscience, 1981, pp 453–478
5. Rudel RG: Assessment of Developmental Learning Disorders. New York, Basic Books, 1988
6. Goldman PS, Galkin TW: Prenatal removal of frontal association cortex in the rhesus monkey: Anatomical and functional consequences in postnatal life. Brain Res 52:451–485, 1978
7. Teuber H-L, Rudel RG: Behavior after cerebral lesions in children and adults. Dev Child Neurol 4:3–20, 1962
8. Denckla MB: Learning for language and language for learning, in Kirk U (ed), Neuropsychology of Language, Reading, and Spelling. New York, Academic, 1983, pp 33–43
9. Goldman-Rakic PS: Development and plasticity of frontal association cortex in the infrahuman primate, in Ludlow CL, Doran-Quine ME (eds), The Neurological Bases of Language Disorders in Children: Methods and Directions for Research. Bethesda, NIH Publication No. 79-440, August 1979, p 1–16
10. Rudel RG: Neuroplasticity: Implications for development and education, in Chall JS, Mirsky AF (eds), Education and the Brain. Chicago, University of Chicago Press, 1978, pp 269–307
11. Rakic P, Goldman-Rakic P: Use of fetal neurosurgery for experimental studies of structural and functional brain development, in Thompson RT, Green OR (eds), Prenatal Neurology and Neurosurgery, Hampton, Virginia, Spectrum, 1983
12. Hecaen H: Acquired aphasia in children: Revisited. Neuropsychologia 21:581–587, 1983
13. Ewings-Cobb L, Levin HS, Eisenberg HM, Fletcher JM: Language functions following closed-head injury in children and adolescents. J Clin Exp Neuropsychol 9:575–592, 1987
14. Dennis M: The developmentally dyslexic brain and the written language skills of children with one hemisphere, in Kirk U (ed), Neuropsychology of Language, Reading, and Spelling. Orlando, Florida, Academic, 1983, pp 185–208
15. Fletcher JM, Satz P: Developmental changes in the neuropsychological correlates of reading achievement: A six-year longitudinal followup. J Clin Neuropsychol 2:23–37, 1980
16. Torgesen J: Problems and prospects in the study of learning disabilities, in Hetherington MG (ed), Review of Child Developmental Research, Vol. 5. Chicago, University of Chicago Press, 1975, pp 1–25
17. Diagnostic and Statistical Manual of Mental Disorders, ed 3, revised. Washington, D.C., American Psychiatric Association, 1987
18. Badian NA: Nonverbal disorders of learning: The reverse of dyslexia?. Ann Dyslexia 36:253–269, 1986
19. Rourke BP: Syndrome of nonverbal learning disabilities: The final common pathway of white-matter disease/dysfunction? Clin Neuropsychol 1:209–234, 1987
20. Gardner RA: The Psychotherapeutic Techniques of Richard A. Gardner. Cresskill, New Jersey, Creative Therapeutics, 1986
21. Levine MD, Brooks R, Shonkoff JD: A Pediatric Approach to Learning Disorders. New York, Wiley, 1980
22. Mannuzza S, Klein RG: Schedule for the Assessment of Conduct, Hyperactivity, Anxiety, Mood, and Psychoactive Substances (the CHAMPS). New York, Long Island Jewish Medical Center and New York State Psychiatric Institute, November 1987

23. Finucci JM, Isaacs SD, Whitehouse CC, et al: A quantitative index of reading disability for use in family studies. Dev Med Child Neurol 24:733–744, 1982
24. DeFries JC, Decker SN: Genetic aspects of reading disability: A family study, in Malatesha RN, Aaron PG (eds), Reading Disorders. Orlando, Florida, Academic, 1982, pp 255–259
25. Holmes JM: The context for assessment, in Rudel G (ed), Assessment of Developmental Learning Disorders. New York, Basic Books, 1988, pp 112–201
26. Milberg WP, Hebben N, Kaplan E: The Boston Process Approach to neuropsychological assessment, in Grant I, Adams KM (eds), Neuropsychological Assessment of Neuropsychiatric Disorders. New York, Oxford University Press, 1986, pp 65–86
27. Boll TJ: The Healstead-Reitan Neuropsychology Battery, in Filskov SB, Boll TJ (eds), Handbook of Clinical Neuropsychology. New York, Wiley, 1981, pp 607
28. Reitan RM: Manual for Administration of Neuropsychological Test Batteries for Adults and Children. Tucson, AZ, Reitan Neuropsychology Laboratories, 1979
29. Golden CJ: A standardized version of Luria's neuropsychological tests: A quantitative and qualitative approach to neuropsychological evaluation, in Filskov SB, Boll TJ (eds), Handbook of Clinical Neuropsychology. New York, Wiley, 1981, pp 642
30. Golden CJ: Luria-Nebraska Neuropsychological Battery: Children's Revision. Los Angeles, Western Psychological Services, 1988
31. Chadwick O, Rutter M: Neuropsychological assessment, in Rutter M (ed), Developmental Neuropsychiatry. New York, Guilford, 1983, pp 181–212
32. Wechsler D: Wechsler Intelligence Scale for Children—Revised. New York, The Psychological Corporation, 1974
33. Jastak JF, Jastak SR: The Wide Range Achievement Test Manual. Rev. ed. Wilmington DE, Jastak Associates, 1984
34. Rosen WG: Assessment of cognitive disorders in the elderly, in Perecman E (ed), Integrating Theory and Practice in Clinical Neuropsychology. Hillsdale, NJ, Lawrence Erlbaum, 1988, pp 381–394
35. Gilmore JV, Gilmore EC: Gilmore Oral Reading Test. New York, Harcourt Brace Jovanovich, 1968
36. Wiederholt JL, Bryant BR: The Gray Oral Reading Tests—Revised. Austin, Texas, Pro-Ed, 1986
37. Woodcock RW: Woodcock Reading Mastery Tests. Circle Pines, Minnesota, American Guidance Service, 1974
38. Denckla MB, Rudel RG: Rapid Automatized Naming (R.A.N.): Dyslexia differentiated from other learning disabilities. Neuropsychologia 14:41–79, 1976
39. Kaplan EF, Goodglass H, Weintraub S: The Boston Naming Test. Philadelphia, Lea & Febiger, 1983
40. Hammill DD, Larsen SC: The Test of Written Language. Austin, Texas, Pro-Ed, 1983
41. Neimark E, Slotnick NS, Ulrich T: Development of memorization strategies. Dev Psychol 5:427–432, 1971
42. Benton AL: The Revised Visual Retention Test, ed 4. New York, The Psychological Corporation, 1974
43. Woodcock RW: Woodcock Reading Mastery Tests—Revised. Circle Pines, Minnesota, American Guidance Service, 1987
44. Wechsler D: A standardized memory test for clinical use. J Psychol 19:87–95, 1945
45. Kagan J: Reflection-Impulsivity: The generality and dynamics of conceptual tempo. J Abnorm Psychol 1:17–24, 1966
46. Rey A: L'examen psychologique dans las cas d'encephalopathie traumatique. Arch Psychol 28:286–340, 1941
47. Osterrieth PA: Le test de copie d'une figure complex. Arch Psychol 30:206–356, 1944
48. Beery KE: Revised Manual for the Developmental Test of Visual Motor Integration. Cleveland, Ohio, Modern Curriculum Press, 1982
49. Goldberger E, Rosen WG, Gerstman L: The development of drawing skills in normal and learning-disabled children. J Clin Exp Neuropsychol 10:81, 1988
50. Raven JC: Coloured Progressive Matrices. New York, The Psychological Corporation, 1947
51. Dunn LM, Dunn LM: Peabody Picture Vocabulary Test—Revised. Circle Pines, Minnesota, American Guidance Service, 1981
52. Rey A: L'examen clinique en psychologie. Paris, Presses Universitaires de France, 1964
53. Durrell DD, Catterson JH: Durrell Analysis of Reading Difficulty, ed 3. New York, The Psychological Corporation, 1980
54. Williams DT, Mehl R, Yudofsky S, et al: The effect of propranolol on uncontrolled rage outbursts in children and adolescents with organic brain dysfunction. J Am Acad Child Psychiatry 21:129–135, 1982

22

Evaluation of the Violent Adolescent

Jonathan H. Pincus

1. INTRODUCTION

There is considerable disagreement regarding the prevalence and severity of neuropsychiatric impairment in the delinquent and criminal populations. Similar disagreement exists in the literature concerning the medical histories of delinquents. There is, however, a growing body of evidence that certain forms of delinquency are associated with disorders of the nervous system.

There is a consensus that delinquents tend to come from disturbed households. The degree of parental psychopathology however is still an area of controversy. The Gluecks[1] reported that parents failed to provide adequate affection and discipline. Patterson[2] described parents of delinquents as lacking household rules and predictable family routines. Others have called attention to more serious parental psychopathology. In spite of the commonly held notion that violence begets violence, the relationship of family violence, especially abuse, to delinquency has only recently received the attention it deserves though it would appear that most victims of abuse do not become violent themselves.

Much of the delinquency literature has emphasized individual factors, such as social deprivation, learning disabilities, and family characteristics, and has failed to consider the importance of the contribution of a variety of different biopsychosocial factors to delinquency. An exception is the study by Levine et al.,[3] in which delinquents were found to have a multiplicity of medical, neurodevelopmental, and educational vulnerability and to come from poorly educated nonintact families. To the best of our knowledge the only other studies to date integrating psychiatric and neurological factors, family psychopathology and a history of physical abuse and family violence have been our own.

In a recent paper Lewis et al.[4] compared matched samples of 31 incarcerated delinquents and 31 nondelinquents. The constellation of abuse, family violence, severe psychiatric symptomatology, cognitive impairment, minor neurological signs, and psychomotor symptoms correctly predicted group membership nearly 84% of the time. The most significant variable was abuse–family violence. The same constellation also distinguished the more aggressive

Jonathan H. Pincus • Department of Neurology, Georgetown University Hospital, Washington, D.C. 20007.

from the less aggressive subjects in each group. The existence of a syndrome characteristic of recurrently violent individuals composed of these variable was postulated.

This study was the outgrowth of earlier ones of 97 boys and 22 girls who were residents of the State Correctional School in Connecticut, which found, basically, that the more violent children were more likely to demonstrate psychotic symptomatology (paranoid ideation) and rambling associations and to have major and minor neurologic abnormalities. They were also more likely to have experienced and witnessed extreme physical abuse.[5]

The discovery of brain damage, especially epilepsy and a form of episodic limbic aggressive syndrome, as well as psychosis, is important because these conditions are potentially treatable. The danger of failing to diagnose and label correctly may be greater than the dangers that could result from accurate diagnosis because children and adults suffering from a wide variety of treatable psychiatric and neurologic disorders who come into conflict with the law and fail to receive an accurate diagnosis and appropriate therapeutic intervention are, by and large, eventually labeled "sociopathic." This label conjures up the image of unfeeling, amoral, impulsive individuals, who are responsible for their behavior, untreatable, and in need of incarceration or execution. There is also some danger inherent in correctly diagnosing a potentially treatable condition: treatment could fail and the subject might be prematurely freed from custody and repeat antisocial acts.

It is abundantly clear that the medical assessment of delinquent adolescents has seldom been well done. Lewis et al.[6] compared pediatric, psychiatric, neurologic, and hospital record data on the 20 adolescent boys of 97 seen in the reform school study who came from the New Haven area, all of whom had used the main general hospital there and had charts available for review. The three most striking findings were (1) the high prevalence of adverse medical events experienced by these 20 subjects—90% had major accidents, 85% severe head injuries, 75% were physically abused, and 50% had perinatal problems; (2) the failure of each medical specialty to identify important medical information that might have contributed to a youngster's adaptational difficulties, and (3) the fact that specialists often missed obtaining data traditionally considered relevant to their own specific areas of expertise.

It may be worthwhile to review some of the specific findings. The pediatricians or generalists significantly overlooked a history of perinatal problems, major accidents or injuries, severe head injuries, neurological abnormalities, epilepsy, blackouts, fainting, family psychiatric problems, and abuse. They were most likely to document major and minor medical illnesses and drug and alcohol abuse. The psychiatrists significantly overlooked major and minor medical illnesses, neurological abnormalities, and drug and alcohol abuse. The neurologist significantly overlooked a history of severe head injuries, major and minor medical illnesses, and family psychiatric problems. The hospital records were significantly deficient in documenting neurological abnormalities, family psychiatric problems, a history of physical abuse, and drug or alcohol abuse.

Although these findings reflect the clinical work of a small number of clinicians and cannot be generalized to all hospitals, pediatricians, psychiatrists, and neurologists, the findings were humbling for all these professions, which come into contact with delinquent adolescents. Several factors might have contributed to the inadequacy of the individual assessments. With the pediatricians, it is possible that a focus on immediate life-threatening medical problems may have diverted them from seeing the patient as the product of a lifetime of physical insults. In addition, the use of a "routine" medical form, rather than encouraging thoroughness, may have limited inquiry. In the interest of time, nurses often assisted in history taking, which may have limited the pediatrician's contact with the youngster.

It is likely that the main reason that the psychiatrists failed to document even one half the neurological abnormalities was because of the expectation that each child would be seen by the neurologist. The fallacy of this argument is that psychiatrists rarely have the collaboration of a neurologist, and even in this study the neurologist did not see all the patients. Psychiatrists are often reluctant to do a physical examination and the pediatrician cannot be relied on to perform more than a rudimentary neurological examination. The danger of maintaining this stance is amply illustrated by the number of disorders including current illnesses that may be overlooked.

Although it is not surprising that the neurologist was best at eliciting neurological abnormalities, it was also impressive that he was able to obtain good histories of physical abuse and drug and alcohol misuse by the subject and family psychopathology. This may be a reflection of the fact that he spent about an hour taking a history from each subject. However, he documented fewer than one half the instances of severe head injuries and only five of seven instances of epilepsy, blackouts or loss of consciousness. These oversights may have been a reflection of his inability to interview the parents and the unavailability to him of hospital chart information, but the amount of time spent with each person was proportional to the amount of relevant historical data that was obtained.

These data suggest that physicians evaluating delinquents must appreciate that most of their subjects will have histories of severe accidents or illnesses and will manifest neurological and psychiatric abnormalities that may contribute to their maladaptive behaviors. Community hospital records should be obtained whenever possible and family members interviewed, since both are vital sources of longitudinal medical information and each physician seeing a delinquent child should take the time to elicit a comprehensive medical history and not rely on a colleague to identify what he may have missed, as he may be the only interested, competent, thorough physician who will ever examine the youngster.

The major tests in the evaluation of the violent adolescent include medical history, physical examination, neurological history and examination, psychiatric history and examination, neuropsychological testing, electroencephalography (EEG), toxicology, serum prolactin levels, (in consideration of seizures), and imaging tests: computed tomography (CT) and magnetic resonance imaging (MRI) of the brain and, possibly, single photon emission computed tomography (SPECT) or positron-emission tomography (PET).

2. NEUROLOGIC HISTORY

A proper history and physical examination remain the essential ingredients of accurate diagnosis and therapy in the violent adolescent. It is essential to speak with the child's parent. A parent can give an account of the events of a child's history of which the child has no memory, and naturally a father or mother can relate information about his or her own past about which the child knows nothing. From them it is also possible to learn whether aunts, uncles, grandparents, and other members of the extended family have suffered from epilepsy, schizophrenia, depression, suicide attempts, drug or alcohol abuse, mental hospitalization, and a variety of illnesses thought to have some genetic components. Only a parent can describe a child's hand-flapping motions, his desire for sameness and fear of change, or his ability at an early age to sit for hours in one spot playing endlessly with a mechanical toy. A parent is also likely to be able to provide evidence of hyperkinesis with the child racing about frenetically unable to control his attention. The parent can also answer the question as to whether the child has ever received pharmacotherapy for behavior, including narcoleptics, anticonvulsants, or stimulants such as methylphenidate (Ritalin).

Families often have been in contact with a variety of different community resources; contact with these, with the parents' permission, provides valuable information. Schools, hospitals, child guidance clinics and welfare agencies contribute to the knowledge base about a given child. Reports of the clinic social worker who may have made a home visit and old hospital and school records all provide vital clues for understanding a child's behavior. A review of the hospital record showed that one child had made 72 visits to a hospital for a variety of complaints. The hospital record of another child was the only source of information concerning evidence of severe battering. Another record indicated that the mother had been treated for syphilis while the child was in utero. Another hospital record indicated that after a suicide attempt, a child had had a cardiac arrest for several minutes and had received closed-heart massage.

It is vital to get as complete a perinatal and subsequent medical history as possible. A history of prematurity, traumatic delivery, or severe early infection is not uncommon among violent delinquents, as are other events known to be associated with brain damage. Curiously, the standard developmental history has not been found to be of special help as developmental landmarks are often reported, accurately or not, as within normal limits or as precocious.

The seriousness of offenses with which a child has been charged may not necessarily indicate the severity of psychopathology. A child who had seriously injured another might have done so accidentally, whereas a mere truant might evidence psychotic symptomatology. Nonetheless, the quality of an offense sometimes gives a clue as to the nature of a given child's problems. For example, an impulsive or violent act, followed by no memory of it, suggests the need to investigate possible neurological vulnerabilities, such as seizures or intoxication. Bizarre acts may suggest psychosis or intoxication.

In an effort to minimize the threatening aspects of an evaluation, especially when legal proceedings are involved, it is important to establish the agreement that should issues arise that a child or the family wishes to discuss but not disclose, confidentiality can be guaranteed.

Reticence, suspiciousness, and frank paranoia can be alleviated to some degree by certain practices that facilitate the relationship with the child. A friendly greeting of the child in the waiting area rather than having him sent to the office can be effective. It is best not to have a desk come between the examiner and the child. If the child should be silent, one can begin by explaining the reasons for the evaluation and what is understood to be the activities that had brought him to medical attention. The interview technique has been codified in the revised Bellevue Adolescent Interview Schedule.

3. NEUROLOGIC EXAMINATION

The findings on neurologic examination encompass abnormalities of head circumference, motility, including inability to skip and motor incoordination, reflexes, visual acuity, eye movements, the retina, speech, and language. Special notation should be made concerning the presence of choreiform movements as these and poor coordination are the hallmarks of minor ("soft") signs of neurologic impairment. The physician is seeking evidence that either will point to a particular syndrome or specific diagnosis known to be associated with brain pathology or dysfunction, or the existence of a focal brain lesion. The medical evaluation of every child must include a screening evaluation of the mental status, i.e., alertness; orientation; interpersonal behavior; language; judgment; memory for digits (seven forward, four backward) words, and sentences, vocabulary; information; and drawing, and of scholastic skills, e.g., reading, writing, and arithmetic.

At this point in the investigation, the physician should have a relatively clear idea of the child's level of functioning, of the probability of detecting a specific deficit in brain function, and of the need for further tests and/or consultations with other specialists such as a neurologist, geneticist, neuropsychologist, audiologist, speech and language pathologist, or psychiatrist.

4. ELECTROENCEPHALOGRAM, EPILEPSY, COMPLEX PARTIAL SEIZURES, POSTICTAL STATE, AND VIOLENCE

Abnormalities of the EEG are quite prevalent among violent prisoners. In a study of 1250 persons in jail for crimes of aggression, Williams[7] found abnormal EEGs in 57% of the habitual aggressors and in only 12% of those who had committed a solitary aggressive crime, after prisoners who were mentally retarded or epileptic or who had sustained serious head injury were separated from the group. Of the habitual aggressors who had EEG abnormalities, the temporal lobe was involved in more than 80%. Unfortunately, inclusion of normal variants, such as the "14 and 6" pattern, among the "abnormals" may have exaggerated the estimate of the prevalence of EEG abnormalities in this paper.

In a study of more than 400 violent prisoners in a large penitentiary, it was discovered that one half had symptoms suggestive of epileptic phenomena and one third had abnormal EEGs, but fewer than 10% had frank temporal lobe epilepsy.[8] It was also apparent that these people had a characteristic social history, which included multiple physical assaults, aggressive sexual behavior including attempted rape, many traffic violations, serious automobile accidents, and "pathological intoxication."

A good deal of attention has been given to the question of whether patients with generalized and complex partial (psychomotor) seizures have a propensity toward violence before, during, after, or between seizures. One might predict that such an association exists, but it has been difficult to prove. Epileptiform changes on the EEG have been a large part of the standard by which behavior is judged to be epileptic or not.

The term epileptiform has been applied to any paroxysmal discharge containing spikes or sharp waves, either localized or generalized. Such discharges are usually but not always associated with epilepsy. In a study of 6497 unselected, nonepileptic patients, Zivin and Ajmone-Marsan[9] found spikes and sharp waves in only 2.2%. In this group of patients with paroxysmal discharges, seizures eventually developed in 15%. In another study, focal spikes were found in 1.5% of 1000 carefully selected normal children.[10] In 242 children whose EEGs showed spike foci, Trojaborg[11] found that 82% were epileptic, that another 13% without epilepsy had structural brain disease, and that only 5% were without any clear-cut evidence of brain disease, most of whom were diagnosed as having "behavior disorders." Clearly, epileptiform changes are usually associated with epilepsy, but it is also possible to have an epileptiform EEG and not have epilepsy.

If an epileptiform EEG cannot fully establish the diagnosis of epilepsy, neither does the finding of a normal interictal EEG rule out the possibility of epilepsy. Ajmone-Marsan and Zivin[12] reviewed 1824 EEGs from 308 epileptic patients who ranged in age from under 1 year to 64 years. Each patient had at least three recordings. Waking and sleep records were obtained for all, as well as responses to hyperventilation and photic stimulation. At the first examination, epileptiform activity was presented in only 56% of patients. In subsequent recordings an additional 26% of patients showed such activity. Nonetheless, 18% of patients had at least three consecutively negative recordings. Only 30% had epileptiform activity in all recordings.

The age of the patient at the time of the EEG and at the onset of seizures influenced the incidence of epileptiform tracings. When the EEG was recorded from children under 10 years of age, approximately 80% had positive first recordings, but during subsequent decades, there was a decline in the rate of positive recordings. Among those with seizures developing after the age of 30, there were several times the number of negative EEGs as repeatedly positive records. The type of epilepsy influenced the rate of positivity. Epileptiform EEGs were seen in almost 95 percent of patients with "absence" attacks. However, the investigators in this important study failed to state how they established the diagnosis of epilepsy and this failure probably underlies some of their more surprising findings: that 98% of patients with seizures arising in the temporal lobe had an epileptiform EEG at some time, and therefore that complex partial seizures correlate better with the EEG than any other clinical seizure type.

This conclusion is probably unwarranted. Even though a high correlation between psychomotor–temporal lobe epilepsy and positive EEG has been duplicated many times,[13–15] this flies in the face of the common clinical experience that epileptiform EEG abnormalities in complex partial seizures occur less often than in patients with most other clinical types of epilepsy. Indeed, it is known that a person can have symptomatic complex partial seizures with epileptiform activity recorded from the amygdala or hippocampus while the surface EEG is within normal limits during the attack. It would seem possible that in all these reports of the interictal EEG characteristics of complex partial seizures, the diagnosis of epilepsy was to some degree dependent on the EEG. In other words, the EEG was used as a diagnostic tool in establishing the diagnosis of complex partial seizures. If the EEG were not abnormal, the investigators might have believed they could not sustain the diagnosis of epilepsy. The high correlation of epileptiform patterns with complex partial seizures in these reports may reflect the great difficulty that clinicians face in establishing that diagnosis. Bolstering this interpretation, in a recent study of 87 simple partial seizures (focal motor or sensory) only 21% demonstrated ictal EEG changes. Thus a normal EEG is common during simple partial seizures and is probably more common in complex partial seizures and does not exclude the diagnosis.[16]

Because epileptiform features are especially critical for the diagnosis of complex partial seizures in the ordinary practice of neurology, but are difficult to uncover in patients suspected of having complex partial seizures, special recording and activating techniques have been devised. These include nasopharyngeal recordings, sphenoidal recordings, sleep deprivation, and sleep EEGs. The observation that serum prolactin levels rise immediately after generalized seizures and after complex partial seizures that involve the amygdala has provided another tool for the diagnosis of epilepsy.[17–19]

Unfortunately, nasopharyngeal electrodes are susceptible to artifacts that seriously compromise their utility. The most troublesome of these are the spikelike discharges produced by contractions of the nasopharyngeal muscles.[20] There is a difference of opinion concerning the value of nasopharyngeal leads; some experienced electroencephalographers hold them to be useful only rarely.[21] The basic question is: How often does one see epileptiform discharges in the nasopharyngeal electrodes in a patient suspected of complex partial seizures when such discharges are completely missing from the surface recordings? The answer to this question is not fully available for nasopharyngeal, sphenoidal, sleep-deprived, sleep, or even depth recording. The usual response of encephalographers is "occasionally."

Sphenoidal leads give essentially the same information that nasopharyngeal leads are supposed to provide, but with much less of a problem caused by artifactual activity. The difficulty of positioning them in the proper location and the need for a surgical procedure to insert them has resulted in their use only in a few specialized clinics.

Sleep activation is least complicated in its interpretation when natural sleep is recorded. Many laboratories encourage a patient to drift off into sleep after obtaining an initial recording in the waking state. Most EEG activation occurs in the drowsiness or slow-wave phases, during which it appears that there is a lower cerebral resistance to the synaptic spread of convulsive potentials. Thus, in slow-wave sleep, discharges limited during wakefulness to deep structures such as the limbic structures now spread to cortical regions. It is not clear how often sleep recording, or sleep recording following sleep deprivation, uncovers abnormalities that are not at all present in the waking record.

Sleep deprivation per se provides an additional degree of activation compared with sleep unrelated to deprivation. Mattson et al.[22] showed that an additional 34% of known epileptic patients activated significantly following sleep deprivation than with sleep alone. For this reason, sleep records are usually noted after a period of sleep deprivation. Metrazole and other convulsant drugs have been used to activate latent seizure propensities, but the fact that normals can be thus induced to generate EEG and clinical seizures has complicated the interpretation of such techniques. Convulsants are not used for this purpose in most centers any longer.

The sampling problem inherent in any standard EEG lasting 45–60 min is obvious. For this reason, 24-hr ambulatory cassette EEGs have been introduced. This is the EEG equivalent of Halter monitoring for cardiac diseases and has proved quite useful. The ambulatory EEG yield of evidence to support a diagnosis of epilepsy is 83% of that of intensive EEG/video monitoring, which is the "gold standard." EEG/video monitoring is more expensive and requires hospitalization. During recording the patient must be relatively immobilized. Ambulatory monitoring was more than 2.5 times more sensitive than routine EEG in yielding evidence supportive of a diagnosis of epilepsy, although episodes of aberrant behavior unaccompanied by ambulatory EEG change required direct behavioral observation for correct diagnosis.[23]

Complex partial seizures of frontal lobe origin were documented in 10 of 90 patients who were studied with depth electrodes. The constellation of symptoms characteristic of such seizures were bizarre and suggested hysteria. Without depth recordings, the epileptic origin of these symptoms could not be established but the use of such an invasive technique can only be justified in known epileptics, even though the diagnosis of epilepsy may be missed by more conservative recording techniques.[24]

There are several ways in which seizures and violence could be theoretically related:

1. *An episode of directed violence could be the automatism of a complex partial seizure:* There is considerable resistance to this concept. The directed aggression reported during epileptic attacks by Saint-Hilaire et al.,[25] Ashford et al.,[26] and Mark and Ervin[8] has been ascribed to "fear, defensive kicking and flailing" by Delgado-Escueta and colleagues,[27] who nonetheless reported seven patients who demonstrated aggression toward inanimate objects or another person during seizures recorded with scalp electrodes. Of these seven, "one had aggressive acts that could have resulted in serious harm to another person;" yet Delgado-Escueta and associates concluded that the commission of murder or manslaughter during psychomotor automatisms was "near impossibility." As the study of Delgado-Escueta did not select patients from a population with known aggressive behavior, violence, and psychosis and did not study any patients who were on trial for violent crimes during epileptic attacks and did not use depth electrodes for the most part, it left open the possibility that more harmful acts of aggression could characterize the automatisms of criminals with epilepsy or violence-prone patients with psychosis and epilepsy. One report of three excessively violent male patients, one of whom had experienced grand mal seizures, used depth electrodes implanted in and around the amygdala with continual recordings for 3–7 weeks. This study conclusively demonstrated that the episodic rage attacks of all three men were associated with spiking discharges in the amygdala and that the behavior could be reproduced by stimulation. In only one patient was there a clear indication of an ictal basis of violent behavior coming from the surface EEG recordings.[28]
2. *Directed violence could be an outgrowth of the encephalopathy associated with a seizure or the postictal state:* There have been many reports of status epilepticus presenting as prolonged confusional states[29] and at least one report of "a man who frequently enters a confused, paranoid psychotic state immediately after a (generalized) seizure and who killed his wife while in such a state".[30] This report is the more

impressive because its author has often written to the effect that there is "very little evidence of violent crimes being directly related to epileptic phenomena".[31]

3. *Anxiety, fear, or anger could precipitate a seizure, possibly by inducing hyperventilation:* In this way, aggressive actions could cause a seizure. The confusional state associated with such a seizure or its postictal period might allow continued aggression to occur unhindered by the inhibitions that intact cortical function might bring to bear on the situation. Even if such a seizure were brief, lasting less than half a minute,[27] the postictal period could be much longer.
4. *Brain damage that predisposes an individual to violence might also cause seizures:* This perspective, proposed forcefully by Stevens and Hermann,[32] regards epilepsy as an epiphenomenon in relation to violence or any other behavioral deviation but identifies limbic brain damage as a critical etiologic feature in the pathogenesis of psychopathology in epileptics. The preponderance of evidence points toward the conclusion that some patients with partial epilepsy and a focus of abnormality in the temporal lobe have a vulnerability to undergo personality change,[33,34] but the role of epilepsy per se in this vulnerability is moot. Possibly epileptic discharges in sensitive regions of the brain that do not result in clinical seizures can give rise to nonictal behavior disorders, that is, to the "interictal state."
5. *Violence and epilepsy may be only serendipitously related:* There have been many reports of an association between complex partial seizures and violence. Falconer et al.[35] reported that 38% of patients with temporal lobe epilepsy showed "pathological aggressiveness." Among 666 cases of temporal lobe epilepsy studied by Currie et al.,[13] 7% were found to be "aggressive." Glaser[14] reported "aggressive behavior" in 67 of 120 children with limbic epilepsy. Serafetinides[36] also found "aggressiveness" to be a characteristic of temporal lobe epilepsy."

Some seizure units with video/EEG-monitoring equipment discourage the admission of episodically violent individuals, since they are unprepared to handle violent patients and fear that their equipment may be damaged. This kind of selectivity has severely limited the number of prolonged EEG studies of violent individuals and probably has lowered the reported prevalence of violence among epileptics so monitored.[27]

How sure can one be that a violent prisoner does not in fact have complex partial seizures? The diagnosis of such seizures is not always easy to make. Among 400 prisoners with a documented history of violent assaults against other individuals, Mark and Ervin[8] found an obvious and known history of epilepsy in 38, an incidence more than 10 times that of the general population. Even more impressive was the fact that fully 50% of this group of 400 had experienced phenomena resembling epileptic symptoms, including altered states of consciousness, warning stages preceding their violent act and following it, sleep or drowsiness and/or lassitude. EEG abnormalities were seen in 50% of the group. How are we to decide whether the incidence of complex partial seizures in this group is less than 10% or approximately 50%?

The prevalence of symptoms suggesting complex partial seizures in violent individuals was shown in a study by Lewis,[37,38] who evaluated the psychiatric–psychological status of 285 children referred from the juvenile court and reviewed the psychiatric, medical, and EEG records of those who manifested psychomotor seizures symptoms. Of the 285 children, 18 experienced episodes of apparent loss of fully conscious contact with reality lasting from several seconds to several hours; on occasion, these episodes were observed during psychiatric interviewing or psychological testing. Four of the children had histories of the automatisms often associated with complex partial seizures, such as lip smacking and mouth movements, and four had experienced frequent episodes of *déjà vu*. Of the 18 children with psychomotor symptoms, eight (44%) had been arrested for crimes of extreme personal violence, including two who committed murder. Furthermore, six others who were referred for milder violations, such as truancy and property offenses, had, in fact, attacked other individuals at school or at home, but were charged with lesser offenses. All told, 14 of the 18 had engaged in serious acts of violence. It should be noted that violent attacks against person constitute 8% of the offenses for which children are referred to the juvenile courts and 9% of the juvenile offenders referred for psychiatric evaluation. The incidence of arrest for violence was significantly higher in the group with psychomotor symptoms ($x^2 = 11.4$), $p < 0.001$). In the group with psychomotor symptoms, 11 of 14 EEGs were abnormal, with three demonstrating temporal foci.

It is interesting that more than 75% of the prisoners studied by Mark and Ervin[8] had histories of significant periods of unconsciousness from head injury or disease and that 15 of the 18 reported by Lewis and Balla[37] had a similar history of events known to be associated with brain injury, such as perinatal problems including infection or prematurity, serious accidents with head trauma, or other disease. These studies strongly support the concept that violence has neurological determinants.

5. NEUROPSYCHOLOGICAL TESTING

Neuropsychological testing can provide critically important information. When cognitive deficits are moderate or severe, psychological testing usually serves to confirm the physician's clinical impression, but in cases of mild or questionable brain damage, psychological tests can provide clinical information that is not readily apparent to a careful interviewer.[39] Many of these tests can give information about the lateral site of a lesion and, when repeated at intervals, can provide precise information as to the progression or regression of signs of intellectual dysfunction. Psychological testing is valuable in determining whether to refer the patient to a neurologist or to a psychiatrist.

Too often, psychological testing is limited to the Wechsler Adult Intelligence Scale (WAIS) and projective tests. Personality inventories such as the Minnesota Multiphasic Personality Inventory (MMPI) are also extremely valuable in providing confirmatory data for certain psychiatric diagnoses, and in distinguishing true epilepsy from hystereoepilepsy.[40]

The MMPI has little use for the study of brain damage. Although intelligence scales do provide some useful information, assessment using the Halstead Reitan battery, or variants of it, is essential for the detection of subtle but neurologically significant deficits in cognitive functioning.[39]

Neuropsychological testing is a most important tool for determining deficits of brain function, as opposed to the imaging tests, which mainly reflect brain structure. The other tests currently available for assaying brain function are the EEG, which when used with topographical techniques can be valuable for determining the functional state of the brain. The EEG provides a tool for diagnosing epilepsy and for identifying focal or diffuse abnormalities. The EEG and neuropsychological testing are now standard. A newer technique for evaluating brain function is PET.

6. POSITRON-EMISSION TOMOGRAPHY

Positron-emission tomography scanning makes use of radioactive ligands, which are used by different portions of the brain. The most commonly used ligand is 18-F-2-fluoro-2-deoxy-D-glucose (FDG), which can give precise regional information concerning blood flow, glucose uptake, and glucose utilization. Sensory stimulation by various modalities causes functional activation and increases in the cerebral metabolic rate in particular cortical and basal gray matter structures. In many neurological disorders, the metabolic rate is altered in a disease-specific pattern. In dementia of the Alzheimer type, glucose utilization is impaired even in the early stages of disease, with particular accentuation of the abnormality in the parietotemporal cortex. In multi-infarct dementia, glucose utilization is mainly reduced in the multifocal small infarcts. In Huntington's disease, the most conspicuous changes are found in the caudate nucleus and putamen.

Additional applications of the PET technique include the determination of the metabolism of various substrates, of protein synthesis, of the function and distribution of receptors, of tumor growth, and of the distribution of drugs, as well as the measurement of oxygen consumption, blood flow, and blood volume. The technique has been used to distinguish Alzheimer's disease from normal pressure hydrocephalus and for localization of epileptic foci during the interictal state; it has just begun to be used in the investigation of psychiatric disorders. Changes in glucose metabolism have been described in the brains of patients who are depressed and in alcoholics and an increase in dopamine receptor sites has been noted in the brains of schizophrenics, although the latter finding has been questioned.

Positron-emission tomography scanning has not yet become a clinical tool for the investigation of violent adolescents, but the finding of hypometabolic regions in or around the temporal lobes in patients suspected of epilepsy whose EEGs were normal or nonspecific might, even at this early stage of the use of PET technology, point toward a diagnosis of complex partial seizures.

The expense of creating a PET unit has stimulated the development of another technique that measures regional blood flow and provides an indirect parameter of the rate of cerebral metabolism: single photon emission computer tomography (SPECT). There is as yet no published data on the use of this technique in brain-damaged and/or violent individuals.

Other tests that must be performed during or immediately after behavioral aberrations that concern the function of the brain are those of body fluids. Blood levels of alcohol and toxicology screening are extremely important tests that should be performed as soon as possible after a violent act has been committed. Alcohol has been the most completely studied of all drugs in relation to violence. It has been shown that 50–83% of murderers were drinking at the time of the murder.[41] Loberg[42] compared severely belligerent and nonbelligerent alcoholics. He found that paranoia and the experience of having been abused characterized the more belligerent alcoholics suggesting that personality defects and life-history qualities increase the risk of violent behavior under the influence of alcohol.

Holcomb and Anderson[43] studied 110 individuals charged with capital murder and divided them by drug history. The group who were not using alcohol or drugs at the time of the crime had the smallest percentage of offenders who "overkilled" the victim and this group contained the largest percentage who had a reason for killing the victim and who remembered the crime. The group that had taken illegal drugs, but not alcohol, was the most likely to overkill and have no apparent motive for murder; nearly one half this group could not remember the actual event. Those who had killed while under the influence of alcohol were intermediate. Paranoia was quite common to all groups. It would appear that alcohol and drugs at the scene of homicide enhance the likelihood of and the viciousness of a killing. The prevalence of paranoia and the fact that 45% of these accused murderers had received psychiatric treatment before committing homicide may suggest that alcohol and drug abuse cause problems primarily in the predisposed.

The main tests that provide information concerning brain *structure* are CT and MRI. Either one is extremely valuable, indeed indispensable, for the evaluation of patients suspected of brain disorders. There are few data concerning CT and/or MRI results in violent individuals but, with regard to complex partial seizures, MRI is much more sensitive than CT even with and without contrast, and it provides superior image quality and is more versatile in providing multiplanar imaging.[44] The invasive techniques for the study of the brain—lumbar puncture and angiography—are not to be performed routinely but may be indicated in certain cases.

7. SUMMARY

The neurological evaluation of the violent adolescent must focus on the history and neurological examination, as well as tests of brain function and structure. Tests of brain function include the standard EEG with sleep and sleep deprivation, ambulatory cassette monitoring, and topographical mapping. Special electrode placements may be helpful and include nasopharyngeal, sphenoidal and depth electrode studies. PET and SPECT scans may add important data in the future. During the immediate postepisode period, serum prolactin, alcohol, and drug toxicology screens are very helpful in resolving diagnostic questions concerning epilepsy and/or intoxication. Imaging tests of brain structure also can provide crucial information. These are MRI and CT scanning, the latter with and without contrast.

REFERENCES

1. Glueck S, Glueck E: Unraveling Juvenile Delinquency. New York, Commonwealth Fund, 1950
2. Patterson GR: Coercive Family Processes. Eugene, Oregon, Castalia, 1982
3. Levine MD, Karniski WM, Palfrey JS, et al: A study of risk factor complexes in early adolescent delinquency. Am J Dis Child 139:50–56, 1985
4. Lewis DO, Pincus JH, Lovely R, et al: Biopsychosocial characteristics of matched samples of delinquents and nondelinquents. J Am Acad Child Psychiatry 26:744–752, 1987
5. Lewis DO, Pincus JH, Glaser GH: Violent juvenile delinquents: Psychiatric, neurological, psychological and abuse factors. J Acad Child Psychiatry 18:307–319, 1979
6. Lewis DO, Shanok SS, Pincus JH, et al: The medical assessment of seriously delinquent boys: A comparison of pediatric, psychiatric, neurologic and hospital record data. J Adol Health Care 3:160–164, 1982
7. Williams DT: Neural factors related to habitual aggression. Brain 92:503–520, 1969
8. Mark VH, Ervin FR: Violence and the Brain. New York, Harper & Row, 1970
9. Zivin L, Ajmone-Marsan C: Incidence and prognostic significance of "epileptiform" activity in the EEG of non-epileptic subjects. Brain 91:751, 1968
10. Petersen I, Erg-Olofsson O, and Selder U: Paroxysmal activity in EEG of normal children, in Kellaway P, Petersen I (eds): Clinical Electroencephalography of children. New York, Grune & Stratton, 1968, p 167–187
11. Trojaborg W: Changes in spike foci in children, in Kellaway P (ed): Clinical Electroencephalography in Children. New York, Grune & Stratton, 1968, pp 213–226
12. Ajmone-Marsan C, Zivan LS: Factors related to the occurrence of typical paroxysmal abnormalities in the EEG records of epileptic patients. Epilepsia 11:361–381, 1970
13. Currie S, Heathfield KWG, Henson RA, et al: Clinical course and prognosis of temporal lobe epilepsy: A survey of 61 patients. Brain 94:173–190, 1971
14. Glaser GH: Limbic epilepsy in childhood. J Nerv Ment Dis 144:391–397, 1967
15. Gibbs FA, Gibbs EC: Atlas of Electroencephalography, 2nd ed. Reading, Massachusetts, Addison-Wesley, 1952
16. Devinsky O, Kelley K, Porter RJ, et al: Clinical and electroencephalographic features of simple partial seizures. Neurology (NY) 38:1347–1352, 1988
17. Yerby MS, Van Bell G, Friel PN, et al: Serum prolactin in the diagnosis of epilepsy. Neurology (NY) 37:1224–1226, 1987
18. Prichard PB III, Wannamaker BB, Sagel J, et al: Serum prolactin and cortisol levels in evaluation of pseudoepileptic seizures. Ann Neurol 18:87–89. 1985
19. Collins WC, Lanigan O, Callaghan, N: Plasma prolactin concentrations following epileptic and pseudoseizures. J Neurol Neurosurg Psychiatry 46:505–508, 1983

20. Bickford RG: Activation procedures and special electrodes, in Kass D, Daly DD (eds), Current Practice of Clinical Electroencephalography. New York, Raven, 1979, pp 269–305
21. Louis A, White JC, Langston JW: Nasopharyngeal Electrodes: Are they worth the trouble? Ann Neurol 12:95, 1982
22. Mattson RH, Pratt KL, Caverley JR: Electroencephalograms of epileptics following sleep deprivation. Arch Neurol 13:310–315, 1965
23. Bridgers SL, Ebersole JS: The clinical utility of ambulatory cassette EEG. Neurology (NY) 35:116–173, 1985
24. Williamson PD, Spencer DD, Spencer SS, et al: Complex partial seizures of frontal lobe origin. Ann Neurol 18:497–504, 1985
25. Saint-Hilaire JM, Gilbert M, Bouner G: Aggression as an epileptic manifestation: Two cases with depth electrode study. Epilepsia 21:184, 1980
26. Ashford JW, Schulz SC, Walsh FO: Violent automatism in a partial complex seizure. Arch Neurol 37:120–122, 1980
27. Delgado-Escueta AV, Mattson RH, King L, et al: The nature of aggression during epileptic seizures. N Engl J Med 305:711–716, 1981
28. Smith JS: Episodic rage, in Girgis M, Kiloh LG (eds), Limbic Epilepsy and the Dyscontrol System. Elsevier/North-Holland, Biomedical Press, 1980, pp 255–301
29. Somerville ER, Bruni J: Tonic status epilepticus presenting as confusional state. Ann Neurol 13:549–551, 1983
30. Gunn JC: Epileptic homicide: A case report. Br J Psychiatry 132:510, 1978
31. Gunn JC: Violence and epilepsy (Letter.) N Engl J Med 306:299, 1982
31. Stevens JR, Hermann BP: Temporal lobe epilepsy, psychopathology and violence: The state of the evidence. Neurology (NY) 31:1127–1132, 1981
32. Bear DM, Fedio P: Quantitative analysis of interictal behavior in temporal lobe epilepsy. Arch Neurol 34:454–467, 1977
33. Trimble MR: Personality disturbances in epilepsy. Neurology (NY) 33:1332–1334, 1983
35. Falconer MA, Hill D, and Wilson JL, et al: Clinical, radiological and EEG correlations with pathological changes in temporal lobe epilepsy and their significance in surgical treatment, in Baldwin M, Bailey P, (eds), Temporal Lobe Epilepsy. Springfield, Illinois, Charles C Thomas, 1958, p 396–410
36. Serafetinides EA: Aggressiveness in temporal lobe epileptics and its relation to cerebral dysfunction and environmental factors. Epilepsia 6:33–42, 1965
37. Lewis DO, Balla DA: Delinquency and Psychopathology. New York, Grune & Stratton, 1976
38. Lewis DO: Delinquency, psychomotor epileptic symptomatology and paranoid ideation: A triad. Am J Psychiatry 133:1395–1398, 1976
39. Lezak MD: Neuropsychological Assessment, ed. 2. New York, Oxford University Press, 1983
40. Wilkus RJ: Dodrill CB, Thompson PM: Intensive EEG Monitoring and psychological studies of patients with pseudoepileptic seizures. Epilepsia 25:100–107, 1984
41. Rosslund B, Larson CA: Crimes of violence and alcohol abuse in Sweden. Int J Addict 14:1103–1115, 1979
42. Loberg T: Belligerence in alcohol dependence. Scand J Psychol 24:285–292, 1983
43. Holcomb WR, Anderson WP: Alcohol and multiple drug use in accused murderers. Psychol Rep 52:159–164, 1983
44. Schorner W, Meencke HJ, Felix R: Temporal lobe epilepsy: Comparison of CT and MR imaging. AJR 149:1231–1239, 1987

23

Novel Drug Development in the Developmental Disorders

Frank J. Vocci, Jr. and Stephen I. Deutsch

1. INTRODUCTION

This chapter highlights the diagnostic concepts of developmental disorders, summarizes salient research activities directed toward understanding deficits in autism from the neuropsychological, neurophysiological, and neurobiological perspectives, develops a framework for drug development aimed at therapeutic intervention in these disorders, and discusses prototypic agents that may be of value.

2. DEVELOPMENTAL DISORDERS

A complete discussion of the differential diagnosis of developmental disorders is beyond the scope of this chapter. Briefly, developmental disorders of childhood may involve specific disturbances of skills development (e.g., cognitive, motoric, and social) or may be pervasive. Disturbances may be limited to delays in general intellectual functioning or may involve a more circumscribed deficit (e.g., an arithmetic skills disorder). Pervasive developmental disorders are manifested by disturbances in the development of reciprocal social interactions, language and communication skills, and imaginative activities. Frequently, children with pervasive development disorders exhibit a restricted behavioral repertoire, which is often repetitive and stereotypic. A variety of associated deficits and behaviors may also be present.

The American Psychiatric Association (APA) has subtyped pervasive developmental disorders into two categories: the Autistic subtype (299.00) and Pervasive Developmental Disorder Not Otherwise Specified (299.80).[1] In considering the diagnosis of a pervasive

---

Frank J. Vocci, Jr. • Medications Development Program, Division of Preclinical Research, National Institute on Drug Abuse, Rockville, Maryland 20857. *Stephen I. Deutsch* • Psychiatry Service, Veterans Administration Medical Center, and Department of Psychiatry, Georgetown University School of Medicine, Washington, D.C. 20007. The concepts, opinions, and conclusions expressed in this chapter are those of the authors. No official endorsement of the Food and Drug Administration is implied or intended.

developmental disorder of childhood, other disorders must be entertained in the differential diagnosis: mental retardation, schizophrenia, hearing impairments, specific language and speech disorders, some visual impairments, and schizoid and schizotypal disorders. For the diagnosis of autism to be made, the onset must be made before 30 months of age, and children must exhibit abnormalities in their development in three categories: reciprocal social interaction, verbal and nonverbal communication and imaginative play, and restrictions of activity or interests. Examples of Diagnostic and Statistical Manual (DSM-III-R) criteria of disturbed social interactions include a marked lack of awareness of the feelings of others, lack of or abnormal seeking of comfort in times of distress, lack of or impaired imitation of social behaviors, lack of or abnormal social play, and an impaired ability to make friendships. Examples of pathology of the second category include (1) lack of communication: no babbling, facial expression, gesture, mime, or spoken language; (2) markedly abnormal nonverbal communication: lack of eye-to-eye contact, lack of facial expression, and stiffening when held; (3) absence of imaginary activity in terms of role playing, fantasy characters, or animals; (4) markedly abnormal speech volume, pitch, rate, rhythm, or intonation; and (5) abnormalities in the form and content of speech: echolalia, pronoun switching, idiosyncratic phrasing, or conversational non sequitors, or impairment in the ability to sustain a conversation. Examples of the third category include stereotyped body movements, including self-injurious behavior; preoccupation with parts of objects or attachment to objects; distress over trivial changes in the environment; insistence on following routines in precise detail; or a restricted range of interest or preoccupation with a narrow interest (e.g., interested only in lining up objects or learning facts about train schedules). The symptom complex and severity of symptoms are variable in autistic children.

3. AUTISM: NEUROPSYCHOLOGICAL, NEUROPHYSIOLOGICAL, AND NEUROBIOLOGICAL CORRELATES

Research into the neuropsychological aspects of autism initially was concerned with description of abnormal behaviors. The emphasis then shifted to an attempt to characterize behaviors that are syndrome specific; more recently, there is an increased attempt at understanding the processes affected by autism.[2] This approach deals with phenomenological rather than etiological aspects and has led to the concept of a basic deficit in autism. Several constructs about this basic defect have been elucidated. This defect has been characterized as a cognitive deficit involving impairments in language, sequencing, abstracting, and coding functions.[2–4] It must be appreciated that the deficit in question should not be confused with etiology, may not have a single cause, may not necessarily be a unitary deficit, and may exhibit general syndromic qualities rather than (be) a specific abnormality.

Although this basic cognitive defect may be one of the core elements of the autistic syndrome, it is unclear as to what relationship this defect has to the psychological processes that subserve socialization.[2] The processes impaired that can adversely affect socialization are a failure to discriminate the emotional state of others, an inability to differentiate on the basis of age and gender,[5] and an inability to understand what others are thinking. It is not clear as to whether the last defect is different from the defects of lack of discriminability of emotional state and gender.

Neuropsychological efforts are currently being directed toward understanding the linkage between cognitive and socialization deficits, determining whether the cognitive deficits are necessary features of the disorder or may serve as a basis for subtyping, determining the relationship of these cognitive defects to the intellectual capacity of autistic subjects, and delineating the specificity of these defects to autism.[2]

Neurophysiological correlates of autism have been sought primarily in the electroencephalogram (EEG) and by analysis of evoked responses. The EEG data have been reviewed.[6] A wide range of abnormal EEGs have been reported, ranging from a low of 13% to a high of 83% in autistic/psychotic populations. A variety of abnormal waveforms were reported, both within and across studies. Thus, there are no dysrhythmias or paroxysmal activities that are pathognomonic for autism. Moreover, the literature is difficult to assess due to the lack of appropriate control groups, different criteria for diagnosis of both the clinical entity and the EEG abnormalities, and technical problems in the evaluation of the EEGs in autistic children. A second finding of heuristic importance is the report of low-voltage, fast (β) waves in EEG studies in autistic subjects,[7,8] although this observation has not been confirmed in all studies.[9] This has led to a hypothesis of faulty information processing due to a failure of the reticular activating system to screen out irrelevant stimuli. The information is passed on to the cortex, which attempts to process all the information. This leads to a sensory overload situation (hyperarousal), which is correlated to the low-voltage fast activity in the EEG. It also suggests a mechanism whereby novel stimuli could be lost in a "crowd" of irrelevant information. Thus, in certain autistic patients detection of novel signals could be problematic (i.e., a signal detection deficit). Moreover, it has been suggested that certain behaviors (i.e., ritualistic and stereotypic behaviors and social withdrawal) may serve to produce *dearousal.*[10]

Analysis of evoked responses has led to a modification of certain hypotheses of autism and the generation of others. It has been reported that autistic children habituated more rapidly to a photic stimulus than retarded or normal controls.[11] This suggests a modification of the hyperarousal hypothesis to accommodate the finding that hypoarousal may also be present.

In addition, several investigators have reported prolonged transmission of auditory evoked potentials,[12–14] which could result in abnormal or delayed maturation of cortical systems involved in processing of sensory input. This could affect language processing, initially in a receptive way, then later in an expressive form. Moreover, the delay in transmission may distort the signal presented to the areas in the cortex responsible for processing the information. Thus, signal delay and signal distortion may be present in autistic subjects.

A defect in information storage has also been reported.[15] These investigators used a stimulus presentation/stimulus deletion technique to test whether the stimulus information was available for transmission to long term memory stores. For example, visual stimuli of trains of 250 stroboscopic flashes (10-μsec duration) were presented at interstimulus intervals of 1 sec. A stimulus was randomly deleted at an average rate of 1 out of every 10 stimuli. In normal individuals this presentation elicits missing stimulus potentials (MSPs) in the associative cortex. The autistic subjects tested either has smaller MSPs or failed to generate MSPs. This suggests a defect in information storage that would possibly lead to defective or incomplete long-term memory. Thus, from the neurophysiological viewpoint, there is evidence of alterations in the processing of information in autistic patients. These differences may include, but not be limited to, altered stimulus transit time possibly leading to alterations in the quality of the stimulus, a high background level of sensory transmission reaching the cortex possibly leading to a problem in novel signal detection, a more rapid habituation of certain signals, and a defect of transmission of information to long-term memory stores. Further research should attempt to correlate alterations in neurophysiological parameters with specific autistic deficits and levels of intellectual functioning. The specificity of the deficits should be verified by comparing autistic groups with both age- and sex-matched controls and subnormal controls matched as closely as possible on age and intelligence criteria. If and when one of the neurophysiological abnormalities mentioned above is shown to be highly correlated to a defect in autistic patients, this could lead to the formation of testable hypotheses for pharmacological agents (e.g., enhanced learning of language as a consequence of improvement of novel stimulus detection, or reduction of background sensory input, increased habituation of constant information, and/or improved transmission of MSPs to long-term memory stores). The determination of mechanism may be secondary to the determination of a salutary effect but would be important for development of future drugs with both similar and dissimilar mechanisms as well as to aid in the correlation of neurophysiological and clinical events.

Several neurobiological approaches have attempted to elucidate abnormalities that could be associated with causes or deficits in autistic patients. Owing to space limitations, only three approaches are presented here. The first approach seeks biochemical differences in neurotransmitter systems as reflected by changes in levels of neurotransmitters or metabolites in body fluids (e.g., blood or CSF) or deduces putative abnormalities from the responses to pharmacological probes. Investigators have studied the serotoninergic system (see Section 5.2), the catecholamine system (see Section 5.4), and the endorphin system (see Section 5.3). Although specific findings are discussed under the appropriate subheadings later in the chapter, it is important to point out that the abnormalities noted are not present in all autistic patients. Thus, there is evidence of biochemical heterogeneity in this disorder. This point has implications for drug development and treatment strategies.

The second approach has analyzed autistic behaviors from the standpoint of behavioral neurology.[16] Damasio and Maurer[16] analogized autistic behaviors to neurological disorders with the same or similar clinical presentations and correlated these neurological disorders with the affected neuroanatomical sites and neurochemical systems. From this approach it was suggested that autism may result from bilateral damage to a phylogenetically distinct portion of cortex, known as periallocortex, and from damage to the neostriatum. Periallocortex is confined to the mesial portions of the frontal and temporal lobes. Interestingly, these regions of cortex receive the terminal portions of the mesolimbic dopaminergic system (projecting from the ventral tegmentum, area A10). It was suggested that disturbances in medial cortical areas are produced by or accompanied by alterations in this dopaminergic system.

The third approach has involved the study of possible immunological abnormalities in autistic patients (see Chapter 9, *this volume*).

4. FRAMEWORK FOR DRUG DEVELOPMENT IN THE DEVELOPMENTAL DISORDERS

The findings summarized in the previous section suggest several features of the pervasive developmental disorders that must be taken into account when devising drug treatment strategies. Currently, the major treatment for these disorders is behavioral/educational, and it is likely to remain that way for the foreseeable future. Drug development in these disorders should be viewed as an adjunct to the behavioral and educational techniques. Indeed, it would be a major advance to develop agents designed to increase the rate of learning and/or retention or to improve affect and/or interpersonal relations in the context of traditional education programs for children with specific or pervasive developmental disorders.

We have conceptualized general principles for drug development based on what is known in the literature and defined objectives for therapy with an ideal agent. We limit the scope of discussion mainly to the molecular concepts that would be of interest in the search for new agents. Comments regarding the totality of development are beyond the scope of this chapter, although some comments on preclinical issues and clinical trial considerations are appropriate.

An ideal agent would be one that aided appropriate unfolding of language, emotional, and social skills; reduced stereotypies/rigidities, and eliminated other maladaptive behaviors (e.g., self-injurious behaviors). Unfortunately, the fields of psychopharmacology or molecular biology do not hold promise for a single agent that would cure pervasive developmental disorders. Moreover, the search for such agents is hindered by the lack of an animal model of the disorder. What one must then consider is the development of multiple agents that may correct one or more defects. Furthermore, as the disorder is chronic, the goal would be to develop drugs for long-term therapy. Thus, agents should have low intrinsic toxicity, lack unacceptable side effects, and have salutary effects on one or more of the following: cognitive processes involved in acquisition and retention of new information, cognitive/emotional/psychological processes involved in social adaptation, reduction of stereotypies, and reduction of maladaptive behaviors (e.g., aggressive behaviors directed at self and others). Such agents will likely mimic, modulate, or antagonize neurotransmitter systems.

5. PROTOTYPIC AGENTS FOR THE DEVELOPMENTAL DISORDERS

5.1. Dopamine Antagonists

Neuroleptic drugs have been shown to have some beneficial effects in autistic patients. Studies using haloperidol have reported a salutary effect on a number of behavioral measures (withdrawal, stereotypies, hyperactivity, fidgetiness, and abnormal object relationships) and improved learning.[17–19]

Since agents in this drug category are already in clinical use, the problem in this case is to develop new agents with different mechanisms or different spectrums of action. Two approaches may yield new neuroleptic drugs in the treatment of developmental disorders. The first approach is based on an empirical analysis of the efficacy of certain neuroleptic drugs in the treatment of autism. Based on this type of analysis, the ideal neuroleptic agents was proposed to possess high affinity for dopamine and serotonin receptors and calcium channels and low affinity for adrenergic, histaminergic, and muscarinic receptors.[20]

The second approach to the selection of a novel neuroleptic agent involves the potential therapeutic manipulation of dopamine receptor subtypes. Evidence for two dopamine receptors, D-1 and D-2 has been reviewed.[21] Selective agonists and antagonists for the two receptor sites have been identified. There is evidence that D-1 receptor agonists act synergistically with dopamine at D-2 receptors[22] to increase the intensity of stereotypic behaviors. Moreover, SCH 23390, a selective D-1 antagonist,[23] blocked the effects of a selective D-2 agonist (quinpirole). This suggests that D-1 receptors play a necessary role in the expression of D-2 agonistic effects. A nonselective agonist, such as endogenous dopamine, would stimulate both receptor subtypes. Thus, a D-1 antagonist may be a novel neuroleptic.

It is not known whether a D-1 antagonist would be efficacious alone or would require the addition of a D-2 antagonist. In the first case, a D-1 antagonist could be tested as a single agent in unmedicated psychotic or autistic populations. Thereafter, fixed combinations of a D-1 and D-2 antagonist could be tested for the optimal D-1/D-2 antagonism ratio. Optimizing therapy on the basis of a fixed combination of D-1/D-2 antagonists or individual optimizations of D-1 and D-2 antagonists may be very important, as there may be a synergistic effect of two combined antagonists such that lower doses of the combination could be used with the anticipated corresponding reduction in side effects.

5.2. Serotonergic Agents

Fenfluramine, an anorexigenic agent and serotonin releaser, has been the subject of a major investigational effort in autistic patients. Increases were noted in the level of serotonin in the blood of a subgroup of autistic patients.[24] A second study noted autistic children with low IQ scores and more florid symptomatology had elevated serotonin levels.[25] Furthermore, in an open-label study of outpatients, it was reported that fenfluramine decreased behavioral symptoms and increased IQ scores.[26] Another open-label study in hospitalized autistic patients noted that fenfluramine increased relatedness and animated facial expression, and decreased irritability, temper tantrums, aggressiveness, self-mutilation, and hyperactivity.[27] Recently, a double-blind parallel-groups comparison of fenfluramine and placebo was conducted in 28 hospitalized autistic children.[28] After a 2-week placebo baseline period, fenfluramine (mean optimal dose 1.747 mg/kg per day) or matching placebo was administered for 8 weeks. In this study, fenfluramine was not superior to placebo in reducing the severity of several core symptoms of autism. Moreover, children receiving fenfluramine showed a higher rate of incorrect responding in an automated discrimination learning task. These data are not consistent with a beneficial therapeutic effect of fenfluramine in autism.

An additional caution must be raised about the possible use of fenfluramine in autistic children. Fenfluramine has been reported to alter the morphology of serotonergic cell bodies in the ventral tegmentum, to damage the presynaptic serotonergic nerve terminals of the cerebral cortex, and to deplete the serotonin content of the caudate, hippocampus, and cortex–limbic system of the rat brain.[29] Fenfluramine was administered twice daily (bid) for 4 days at doses ranging from 6.25 to 50 mg/kg subcutaneously (s.c.). Toxic effects persisted for at least 2 weeks after the last dose of fenfluramine. Although toxic effects were most often reversible with the lowest dosages, this study should caution clinicians against the indiscriminate or unmonitored administration of fenfluramine, especially for prolonged periods on young children.

The serotonergic alterations observed in autism require further exploration and, if applicable, extension to include an assessment of receptor subtypes. A possible approach to the problem is an assessment of the receptor specificity and functional activity in a model serotonin-mediated interaction, tryptophan-induced increases in plasma prolactin.[30] The tryptophan-induced increase in plasma prolactin is blocked by the nonselective antagonist, metergoline. Moreover, administration of the 5-hydroxytryptophan (5-HT_2) antagonist enhanced the response. This has been interpreted as suggesting that the response is mediated by 5-HT_1 receptors. (The data also suggest a functional interaction between 5-HT_1 and 5-HT_2 receptor systems.) This paradigm could be used to determine whether the 5-HT system in autistic patients is normal or abnormal. Appropriate serotoninergic agents and their receptor subtypes could be deduced from the responses in the tryptophan–prolactin model. For example, if there is an excessive release of prolactin associated with tryptophan challenge, this would suggest a serotonin antagonist be considered for future testing. The present understanding of this response would suggest that either a nonselective or 5-HT_1 antagonist could be tested.

5.3. Narcotic Antagonists

Endogenous opioids systems are characterized by a multiplicity of endogenous ligands and receptor populations, leading to a variety of possible actions within the CNS. The peptides and their binding sites will be described and then related to possible therapeutic interventions in developmental disorders.

Endogenous opioids are derived from three opioid gene families: the pro-opiomelanocortin family, the proenkephalin family, and the prodynorphin family. β-Endorphin is derived from the pro-opiomelanocortin family. Methionine enkephalin, leucine enkephalin, methionine enkephalin-8, methionine enkephalin-Arg^6-Phe^7, and peptide E are derived from the proenkephalin family. α-Neo-endorphin, β-neo-endorphin, dynorphin A-(1–8), dynorphin A-(1–17), and dynorphin B-(1–13) are derived from prodynorphin.[31] The three opioid systems have different distributions in the brain. The β-endorphin system is rather circumscribed, cell groups are found in the medial basal hypothalamus, and the nucleus of the solitary tract. The proenkephalin system is widely distributed in brain, with peptides derived from this group noted at all levels of the neuraxis. The dynorphin system is also widely distributed and, although it roughly parallels the enkephalin system, there are nuclei in which enkephalins are found that dynorphins are absent (e.g., habenular nuclei, interpeduncular nucleus, periventricular nucleus of the thalamus). In addition, there are several nuclei in which dynorphins are located and enkephalins are absent (e.g., the magnocellular nuclei of the hypothalamus).[32] The first differentiation of opioid receptors introduced three receptor subtypes: μ-, κ-, and σ-receptors.[33] Subsequently, the enkephalin preferring δ-receptors were discovered.[34] Shortly thereafter a β-endorphin preferring site, the ϵ-receptor, was characterized.[35] The proposed actions associated with stimulating these receptor systems has recently been reviewed.[36] Briefly, the supraspinal analgesic actions and respiratory depression activity of opioids are associated with stimulation of μ-receptors. Stimulation of κ- or δ-receptors is associated with analgesia at the spinal level. It is not known whether stimulation of σ- or ϵ-receptors occurs with endogenous effectors.

Other less classic actions of opioid drugs and endogenous opioid peptides have also been hypothesized. Endogenous opioids have been implicated in the regulation of attention, perception, affect, social behavior, and motor activity, as well as the pathogenesis of affective disorders, schizophrenia, and infantile autism.[37] The link of the endogenous opioid system to autism has been made on the basis of similarities between effects produced by opiates and autistic behaviors,[37,38] the effects of endogenous opioids,[39] or narcotic antagonists[40] on brain development, the analysis of endorphin/enkephalin levels in biological fluids of autistic patients,[41–43] and the effects of narcotic antagonists on self-injurious behavior in autistic patients.[44]

The behaviors common to both autistic subjects and morphine-treated animals and men include insensitivity to pain, diminished crying, poor clinging, reduced socialization, episodes of increased motor activity alternating with quiescence, affective lability, and stereotyped behaviors. This has led to a proposal that endorphin hyperactivity or hyperresponsiveness is responsible for social/emotional deficits seen in autism. Specifically, endorphins/enkephalins may subserve the drive to communicate and the normal reward system involved in positive reinforcement associated with social interactions. An overactivity of this system would lead to a reduced drive to communicate resulting in reduced social behavior and possibly an affective indifference toward social interactions. This latter effect may be the biochemical sybstrate for the lack of empathy noted in autistic patients. Thus, the need for social activity may be "short-circuited" by an excessive level of endorphins/enkephalins, dissociating social behavior and the positive affect generated by it.

The second link of endogenous opioid systems to autism derives from studies that have suggested opposite effects of opioid antagonists and agonists on brain development. Naltrexone increased the size of the brain and the number of neurons and glial cells in developing rats,[40] whereas exogenous β-endorphin administration to rats during the gestation period produced cognitive impairment and delayed development, two of the features of autism.[39] Assuming that these results are extrapolable to humans, it implies excessive endorphin/enkephalin activity during development may be one of the neurochemical imbalances responsible for autism. Moreover, as development does not stop at birth or even in early infancy in the human, this imbalance may persist. Administration of narcotic antagonists would then possibly have immediate effects on affective indifference to social activity (vide supra), and, conceivably, a long term effect to promote brain development, assuming this sort of plasticity still exists in the brain of the autistic child.

The third approach in this area has been to measure levels of endogenous opioids in autistic patients.[41–43] Measurement of a humoral endorphin fraction in the plasma of autistic patients showed a reduced level of this putative transmitter, suggesting a failure to validate the excessive endorphin hypothesis.[41] A second study that measured immunoreactive β-endorphin in the plasma of unmedicated autistic patients also noted reduced levels of this compound.[42] Although these data do not support an endorphin excess hypothesis, it must be considered that the relationship of central endorphin levels to circulating levels in the periphery is unknown. Thus, alterations in levels of other endogenous opioids are not ruled out. For example, a subset of autistic patients with pain insensitivity and self-destructive behavior had elevations of peptides in fractions of their CSF, notably methionine enkephalin-Lys.[6] These peptides are derived from proenkephalin A[43] and preferentially bind to delta receptors.

The fourth possible link of endogenous opioids to autism is the proposed role of endogenous opioids in the maintenance of self-injurious behavior and possibly stereotypies. The hypothesis has been successfully tested for self-injurious behavior using naltrexone.[44] There are two prevailing hypotheses on the role of this system in these behaviors. The first hypothesis is that there is an elevation of endogenous opioid tone in these patients, rendering them insensitive to pain. The second hypothesis is that self-injury is reinforced by the release of endogenous opioids during the behaviors. The hypotheses are not mutually exclusive.

What, then, are the implications of the current evidence and the hypotheses of endogenous opioid involvement in autism? The first implication is that there is sufficient rationale for further testing of both non-selective antagonists as well as receptor subtype selective antagonists. This rationale includes the possibly immediate therapeutic effects on self-injurious behaviors as well as the less likely effects on brain maturation. Naltrexone has been tested in a limited fashion (vide infra). It may well find a place in the treatment of autism although its potential hepatotoxic effect may limit its usefulness. Theoretically, the "problem" with non-selective antagonists is that they can block all the known and unknown actions of the endogenous opioid systems. By contrast, receptor subtype selective antagonists exert a beneficial effect by preserving sensitivity to specific classes of agonists (e.g., preserving sensitivity to mu agonists if strong analgesia was necessary).

The safety and putative salutary effects of naltrexone hydrochloride were evaluated in ten autistic boys (ages 3.42–6.5 years) who received single early morning doses (0.5, 1.0, and 2.0 mg/kg) in ascending order on three consecutive Wednesdays.[45] Drug was administered in an open-label fashion; multiple behavioral ratings were taken throughout the dosing day and the day after. Naltrexone was chosen for the study because of its oral potency and long duration of antagonist action. Children were observed in three settings (i.e., playroom, nurse's office, and classroom) by psychiatrists, nurses, and teachers. The severity of withdrawal on the Children's Psychiatric Rating Scale was the rating item most sensitive to change across the three dose levels studied. Dose-dependent reductions were observed in the severity of Underproductive Speech (moderately reduced at 0.5 mg/kg), and Stereotypies (greatly reduced at 2.0 mg/kg per day). Interindividual plasma levels of naltrexone varied greatly; there was no observable relationship between plasma level of naltrexone and response. Mild sedation was seen in seven children and was the most common side effect. The results support further investigation of naltrexone in autistic children under double-blind conditions.

The second implication of the data is that the involvement of the endogenous opioid system may extend to the peptides interacting with δ-receptors and possibly other receptor subtypes. Thus, a δ-antagonist could be the first subtype selective antagonist to be tested. This suggestion takes into account the neurochemical findings in the CSF of autistic children[43] and the wide distribution of the proenkephalin A system and δ-receptors (a 1 : 1 correspondence is not implied). δ-Antagonists with reasonably good selectivity; i.e., 10- to 30-fold delta/μ preference, have been reported (e.g., ICI 174864,[43] and 16-methylcyprenorphine[46,47]). The next type of antagonist we would select for testing in autistic patients is a κ-selective antagonist. Although specific dynorphin abnormalities have not been reported or hypothesized in autism, this system has the second largest distribution of the opioid peptide systems known to date. Two selective κ-antagonists, binaltorphimine and norbinaltorphimine, have shown κ/μ-receptor selectivity in the guinea pig ileum preparation of 52 and 19, respectively.[48] Of more theoretical interest would be the effect of a sigma receptor antagonist. No known prototype exists for this antagonist subtype.

The last implication of the data is that autistic patients may behave in way that increases their endogenous opioid tone through stereotypic and self-destructive behaviors. No pathophysiological mechanism is necessary for subserving these phenomena; i.e., pain and repetitive movement may be general conditions for endogenous peptide release. The autistic patient may "learn" that certain behaviors produce sought after changes in affective states. The administration of an antagonist in these circumstances may be beneficial to promote reduction of the stereotypies and the maladaptive behaviors. No preference for a selective versus nonselective antagonist can be suggested at this time.

5.4. Adrenergic Agents

There is some rationale for the investigational use of adrenergic blocking drugs in the treatment of autism. The use of β-blocking drugs has been proposed as a test of the hyperarousal hypothesis. One report of an open study of propranolol or nadolol in eight adult autistic patients described marked improvement in the subjects' aggressive and self-injurious behaviors.[49] Ratey et al.[49] suggested that the effects of the β-blocker were to reduce the level of arousal rather than to possibly correct a primary defect of autism. According to this hypothesis, one would also expect a reduction in secondary maladaptive behaviors (e.g., rituals, and the maintenance of sameness) as the secondary behaviors may serve to promote dearousal. The putative efficacy of the β-blocking drugs in autism needs to be confirmed in appropriately controlled and blinded studies.

The insistence on sameness, avoidance of novelty, and aggressive and explosive behaviors in autistic children may result from a disturbance of the locus ceruleus, which may function as a novelty detector[50] and differential amplification system for noxious stimuli.[51] In a normal person, low levels of arousal in this system may serve to detect novel stimuli and focus attention. At higher levels of activation, all novel stimuli are recorded and differentially amplified. In the case of noxious stimuli, cognitive and autonomic responses preparing for a fight-or-flight response may be activated. At the highest levels of activation, novel stimuli will be processed with activation for fight or flight recognized by the individual as somatic anxiety, fear, panic, or life-threatening terror. Under normal conditions, the system operates at the low level, detecting novel stimuli and processing them in a fashion that will focus attention to the stimulus. However, if stimuli that a normal individual would process as low-level novelty were processed pathologically as moderately noxious due to an abnormally sensitive differential amplifier, this could cause avoidance behaviors such as an insistence on sameness as well as avoidance of novelty and behavioral disturbances that appear disproportionate to the situation. Thus, many of the peculiar behaviors observed in autistic patients would be viewed as coping mechanisms to avoid activation of a pathological amplification system. We recognize that this is a variation of the hyperarousal hypothesis. However, it lends itself to pharmacological testing. We propose an α_2-agonist (clonidine) or partial agonist be tested in autistic patients for possible efficacy as a dampening agent of the differential amplifier system.

5.5. Putative Cognitive Enhancers

The efficacy of the various classes of agents that may have positive effects on cognitive processes remains to be established. We have identified four separate classes of agents that may affect cognition. We review the four classes of drugs that appear to be candidates for further development. Nootropic agents are a novel group of compounds that may have salutary effects in the presence of brain dysfunction. Their current mechanism of action is unknown; effects on cholinergic, dopaminergic, and adrenergic systems have been noted, as well as a selective increase of cerebral blood flow to an ischemic hemisphere after stroke has been reported. The prototype nootropic agent is 2-pyrollidoine acetamide (piracetam).

These agents are under active investigation for a possible indication in the treatment of specific developmental disorders, especially reading disability. In a double-blind placebo-controlled study, piracetam was administered to 257 boys (ages 8–13) with reading disability. A daily dose of 3.3 g was shown to increase reading speed.[52] Visual event related potentials (ERPs) were measured in a small subsample of the these children while performing a vigilance task. Effects on the ERPs were consistent with a facilitation of verbal processing mechanisms; i.e., the latency of a late component was increased in the left hemisphere in association with the correct detection of letter sequences, whereas the latency of the same component was increased in the right hemisphere in association with correct rejection of irrelevant stimuli.[53]

Piracetam has also been reported to increase short-term memory and retrieval from long-term memory stores in a sample of 60 dyslectic boys. The treated children showed an increased digit span, learned more objects on the first trial, and retained more after a delay in testing that the control group. A rapid automatized naming test was used to show that retrieval was enhanced after 12 weeks of treatment with piracetam.[54] A 36-week trial of the effect of piracetam in the treatment of dyslexic boys (ages 7–12) was conducted under double-blind placebo-controlled conditions. Treatment was associated with an increase in reading ability and comprehension at 12 weeks; the effect persisted for the remainder of the 36-week trial.[55]

A second possible treatment approach to enhancement of cognitive faculties is the use of agents that reduce γ-aminobutyric acid (GABA) transmission to the forebrain. The evidence for a GABAergic link to forebrain cholinergic transmission in the rat and the cognitive-enhancing effects of drugs that inhibit the effect of GABA in the forebrain has been summarized.[56] The prevailing hypothesis is that GABA decreases cholinergic turnover in the basal forebrain. Thus, agents that decrease GABAergic input may have a facilitatory effect on cholinergic transmission in an area linked to short-term memory. It is suggested inverse partial agonists and GABA antagonists may have the appropriate behavioral effects. Preclinical studies have identified three candidates for possible human study. The antagonist β-carboline ZK 93426 displayed salutary effects in aged rats in a spatial delayed alternation task, antagonized the scopolamine-induced impairment in a second study, and improved the performance of basal–forebrain lesioned rats in an automated radial tunnel maze learning paradigm. The partial inverse agonist FG7142 was shown to attenuate the disruptive effects of scopolamine in a passive avoidance task in mice. The inverse agonist β-carboline-3-carboxylate methylester (β-CCM) was shown to improve performance in a habituation test of mice to a novel environment, a memory test of a passive-avoidance behavior in mice, and imprinting in chicks. An effect on acquisition of new information through increased vigilance is suggested by the authors of these studies.

There is some evidence from preliminary studies in human volunteers that the antagonist ZK 93426 may have antiamnesic and promnesic effects. In a double-blind placebo-controlled study, this drug produced behavioral changes reported as stimulant and activating on self-report and visual analogue scales. Further evidence of effects on processes associated with memory was observed as an improvement in a logical reasoning and picture differences tests, which are thought to measure concentration and attention, respectively.[56] It was noted that these effects occurred at subanxiogenic doses. Improvement of recall was also noted in a picture-recall test in human volunteers administered ZK 93426. Assuming the effects of this compound are validated in appropriately controlled and blinded studies, this compound or one with a similar mechanism of action would be a candidate for testing in children with learning disabilities and pervasive developmental disorders.

A third approach to improvements in cognition involves the concept of stimulation of the glutaminergic system in the hippocampus. The response of the rabbit nictitating membrane/eyelid has been used as a model for evaluating the role of the hippocampus in classically conditioned learned responses. Long-term potentiation in hippocampal pyramidal neurons is thought to be the electrophysiological basis of this learned response. In this model, classic conditioning was associated with a 40% increase in sodium-independent [^{3}H]glutamate binding.[57] The response may be mediated by a subclass of glutamate receptor, the *N*-methyl D-aspartate (NMDA) receptor complex. This receptor subclass is localized to the terminal fields of major hippocampal pathways.

Experimental evidence suggests that glutaminergic agonists may enhance certain types of learning, whereas antagonists may disrupt learning. Glutamate administration to rats caused a more rapid extinction of a learned level-pressing response; i.e., it helped the animals learn to extinguish a response.[58] Administration of the NMDA agonist aminophosphonovaleric acid to rats resulted in the selective impairment of a spatial learning task.[59] Glutamic acid diethyl ester, a competitive antagonist of glutamate binding, interfered with acquisition of a reinforced behavior in rats.[60]

There is some clinical evidence for a positive effect of glutamate on cognitive function. A review of more than 50 clinical studies noted positive effects of administration of the free acid in some subjects.[61] Intellectually normal children and adults who were positive responders showed increased learning rates, superior scores on IQ subsets, and increased energy levels and motivation. In this review, 11 studies were noted in noninstitutionalized retarded children. Seven studies reported positive results on intellectual performance.

The experimental and clinical studies with glutamate suggest that stimulation of this system may be associated with a positive effect on acquisition. Appropriately controlled and blinded studies with glutamate or a glutaminergic agonist are needed to test the hypothesis of a glutaminergic component in certain types of learning. Assuming positive effects can be demonstrated in normal populations under controlled conditions, further trials should be considered in learning disabled and autistic populations. It is suggested that autistic populations be one of the last populations of children tested due to a concern that glutaminergic agonists may reduce the seizure threshold and autistic patients are at an increased risk of seizures.

The fourth possible group of agents that could influence learning are certain steroids, namely dehydroepiandrosterone (DHEA) and its sulfated derivative (DHEA-S). These steroids may be trophic factors that promote the survival and differentiation of neurons in cell culture and antagonize amnesia and promote learning in mice. In a tissue culture medium, DHEA and DHEA-S, in concentrations as low as 10^{-8} M, increased the number of neurons derived from 14-day mouse embryo brain. The increased number was not due to proliferation since there was no increase in [^{3}H]thymidine incorporation. Further, a reduction in the proliferation of astrocytes was noted.[62] Intracerebroventricular administration of DHEA in doses as low as 10^{-10} M improved retention of a learned avoidance task. Moreover, DHEA in a solution of dimethyl sulfoxide antagonized the amnestic effects of dimethyl sulfoxide administered alone.[63] It is not known whether conventional routes of administration are capable of increasing brain levels of these steroids. Moreover, it has yet to be shown that these compounds improve components of learning and memory in a clinical situation. Thus, this group of compounds remains the most speculative.

6. COMMENTS ON CLINICAL TRIALS IN AUTISM

A full discussion of all the ethical, scientific, medical, administrative, and regulatory issues involved in the conduct of clinical studies in autism is beyond the scope of this chapter. However, several comments that should be considered in future trial design are offered here:

First, a cognitive defect has been proposed as a core defect that may be responsible for abnormal language, coding, sequencing deficits; thus, clinical studies in autistic patients should either directly or indirectly measure aspects of the cognitive defect.

Second, there are no neurophysiological correlate(s) in autism that can be used in place of behavioral/educational rating scales although clinical studies utilizing a combined approach of psychopharmacological/ educational variables with concomitant measurement of neurophysiological parameters should be encouraged; neurophysiological measures should be made in two concurrent control groups, an intellectually normal age and sex matched control group and a group of retarded patients who are also age and sex matched to the autistic group.

Third, there is evidence for neurochemical heterogeneity of the disorder; thus, one would not expect uniform responses in autistic subjects in a clinical trial.

Fourth, the clinical presentation of the disorder is variable in severity, and the degree of mental retardation is variable across individuals, although deficits may be stable across time; the degree of mental retardation in an autistic population may be a critical variable as there may be differential effects of drugs based on intellect differences.

Fifth, substances with mild to moderate efficacious effects or effects in subpopulations of autistic patients may be missed in small clinical trials due to insufficient statistical power; a potential remedy for this problem is the conduct of multicentered trials.

ACKNOWLEDGMENTS. The authors would like to acknowledge the helpful discussions of Drs. Abraham and Ronit Weizman regarding the concepts in this chapter.

REFERENCES

1. American Psychiatric Association: Diagnostic and Statistical Manual of Mental Disorders, ed 3, revised. Washington, D.C., American Psychiatric Association, 1987
2. Rutter M, Schopler E: Autism and pervasive developmental disorders: Concepts and diagnostic issues. J Autism Dev Disord 17;159–187, 1987
3. Dawson G: Lateralized brain function in autism: Evidence from the Halstead–Reitan neuropsychological battery. J Autism Dev Disord 13:369–386, 1983
4. Rutter M: Cognitive deficits in the pathogenesis of autism. J Child Psychol Psychiatry 24:513–531, 1983
5. Hobson RP: The autistic child's recognition of age-related features of people, animals, and things. Br J Dev Psychol 4:343–352, 1983
6. James AL, Barry RJ: A review of psychophysiology in early onset psychosis. Schizophr Bull 6:506–525, 1980
7. Hutt C, Hutt SJ, Lee D, et al: Arousal and childhood psychosis. Nature (Lond) 204:908–909, 1964
8. Kolvin I, Ounsted C, Roth M: Cerebral dysfunction and childhood psychosis. Br J Psychiatry 118;407–414, 1971
9. Creak EM, Pampiglione G: Clinical and EEG studies on a group of 35 psychotic children. Dev Med Child Neurol 11:218–227, 1969
10. Kinsbourne M: Do repetitive movement patterns in children and animals serve a de-arousal function? J Dev Behav Pediatr 1:39–42, 1980
11. Hermelin B, O'Connor N: Psychological Experiments with Autistic Children. Oxford, Pergamon, 1970
12. Skoff BF, Mirsky AF, Turner D: Prolonged brainstem transmission time in autism. Psychiatry Res 2:157–166, 1980
13. Tanguay PE, Edward RE, Buchwald J, et al: Auditory brainstem evoked responses in autistic children. Arch Gen Psychiatry 39:174–180, 1972
14. Student M, Sohmer H: Evidence from auditory nerve and brainstem evoked responses for an organic brain lesion in children with autistic traits. J Aut Child Schizophr 8:13–20, 1978
15. Novick B, Kutzberg D, Vaughn HG: An electrophysiological indication of defective information storage in childhood autism. Psychiatry Res 1:101–108, 1979
16. Damasio AR, Maurer RG: A neurological model for childhood autism. Arch Neurol 35:777–786, 1978

17. Campbell M, Anderson LT, Meier DD et al: A comparison of haloperidol and behavior therapy and their interaction in autistic children. J Am Acad Child Psychiatry 17:640–655, 1978
18. Campbell M, Anderson LT, Small AM, et al: The effects of haloperidol on learning and behavior in autistic children. J Aut Dev Disord 12:167–175, 1982
19. Anderson LT, Campbell M, Grega DM, et al: Haloperidol in infantile autism: Effects on learning and behavioral symptoms. Am J Psychiatry 141:1195–1202, 1984
20. Deutsch SI, Campbell M: Relative affinities for different classes of neurotransmitter receptors predict neuroleptic efficacy in infantile autism: A hypothesis. Neuropsychobiology 15:160–164, 1986
21. Stoff JC, Kebabian JW: Two dopamine receptors: Biochemistry, physiology and pharmacology. Life Sci 35:2281–2296, 1984
22. Walters JW, Bergstrom DA, Carlson JH, et al: Dopamine receptor activation required for postsynaptic expression of D_2 agonist effects. Science 236:719–722, 1987
23. Hilditch A, Drew GM, Naylor RJ: SCH 23390 is a very potent and selective antagonist at vascular dopamine receptors. Eur J Pharmacol 97:333–334, 1984
24. Ritvo ER, Yuwiler A, Geller E, et al: Increased blood serotonin and platelets in early infantile autism. Arch Gen Psychiatry 23:556–572, 1970
25. Campbell M, Freidman E, Green WH, et al: Blood serotonin in schizophrenic children. Int Pharmacopsychiatry 10:213–221, 1975
26. Ritvo ER, Freeman BJ, Geller ED, et al: Effects of fenfluramine on 14 outpatients with the syndrome of autism. J Am Acad Child Psychiatry 22:549–558, 1983
27. Campbell M, Perry R, Polonsky BB, et al: An open study of fenfluramine in hospitalized young attitude children. J Aut Dev Disord 16:495–506, 1986
28. Campbell M, Adams P, Small AM, et al: Efficacy and safety of fenfluramine in autistic children. J Am Acad Child Adol Psychiatry 27:434–439, 1988
29. Schuster R, Lewis M, Seiden LS: Fenefluramine neurotoxicity. Psychopharm Bull 22(1):148–151, 1986
30. Cowen PJ: Psychotropic drugs and human 5-HT neuroendocrinology. Trends Pharmacol Sci 8:105–108, 1987
31. Akil H, Watson SJ, Young E, et al: Endogenous opioids: biology and function. Annu Rev Neurosci 7:223–255, 1982
32. Walker JM, Moises HC, Coy DH, et al: Comparison of the distribution of dynorphin system and enkephalin system in brain. Science 218:1134–1138, 1982
33. Martin WR, Eades CG, Thompson JA, et al: The effects of morphine and nalorphine-like drugs in nondependent and morphine-dependent chronic spinal dog. J Pharmacol Exp Ther 197:517–532, 1976
34. Lord JAH, Waterfield AA, Hughes J: Endogenous opioid peptides: Multiple agonists and receptors. Nature (Lond) 267:495–496, 1977
35. Akil H, Hewlett WA, Barchaw JD, et al: Binding of [^{3}H]Beta-endorphin to rat brain membranes: Characterization of opiate properties and interaction with ACTH. Eur J Pharmacol 64:1–8, 1980
36. Pasterniak GW: Multiple morphine and enkephalin receptors and the relief of pain. JAMA 225:1362–1367, 1988
37. Deutsch SI: Rationale for the administration of opiate antagonists in testing infantile autism. Am J Ment Defic 90:631–635, 1986
38. Sahley TL, Panksepp J: Brain opioids and autism: An updated analysis of possible linkages. J Autism Dev Disord 176;201–217, 1987
39. Sandman CA, Kastin AJ: The influence of fragments of the LPH chains on learning, memory, and attention in animals and man. Pharmacol Ther 13:39–60, 1981
40. Zagon IS, McLaughlin PJ: Naltrexone modulates body and brain development in rats: A role for endogenous opioid systems in growth. Life Sci 35:2057–2064, 1984
41. Weizman R, Weizman A, Tyano S, et al: Humoral–endorphin blood levels in autistic, schizophrenic, and healthy subjects. Psychopharmacology 82:368–370, 1984
42. Weizman R, Gil-ad I, Dick J, et al: Low plasma immunoreactive beta-endorphin levels in autism. J Am Acad Child Psychiatry (in press)
43. Gillberg C, Terenius L, Lonnerholm G: Endorphin activity in childhood psychosis. Arch Gen Psychiatry 42:780–783, 1985
44. Herman BH, Hammock MK, Arthur-Smith A, et al: Naltrexone decreases self-injurious behavior. Ann Neurol 22:550–552, 1987
45. Campbell M, Overall JE, Small AM, et al: Naltrexone in autistic children: an acute open dose range tolerance trial. J Am Acad Child Adol Psychiatry 28:200–206, 1989

46. Cotton R, Giles MG, Miller L, et al: ICI 174,864: A highly selective antagonist for the opioid–delta receptor. Eur J Pharmacol 97:331–332, 1984
47. Smith CFC: 16-Me Cyprenorphine (RX 8008M): A potent opioid antagonist with some selectivity. Life Sci 40:267–274, 1987
48. Portoghese, PS, Lipkowski AW, Takemori AE: Binaltorphimine and Nor-binaltorphimine, potent and selective K-opioid receptor antagonists. Life Sci 40:1287–1292, 1987
49. Ratey JJ, Mikkelsen E, Sorgi P, et al: Autism: The treatment of aggressive behaviors. J Clin Psychopharmacol 7:35–41, 1987
50. Mason ST, Fibiger HC: Current concepts. I. Anxiety: The locus coeruleus disconnection. Life Sci 25:2141–2147, 1979
51. Redmond DE, Huang YH: Current concepts. II. New evidence for a locus coeruleus–norepinephrine connection with anxiety. Life Sci 25:2145–2162, 1979
52. Wilsher CR, Bennett D, Chase CH, et al: Piracetam and dyslexia: Effects on reading test. J Clin Psychopharmacol 7:230–237, 1987
53. Conners CK, Blouin AG, Winglec M, et al: Piracetam and event-related potentials in dyslexic males. Int J Psychophysiol 4:19–27, 1986
54. Helfgott E, Rudel RG, Kairam R: The effects of piracetam on short and long-term verbal retrieval in dyslexic boys. Int J Psychophysiol 4:53–61, 1986
55. DiIanni M, Wilsher CR, Blank MS, et al: The effects of piracetam in children with dyslexia. J Clin Psychopharmacol 5:272–278, 1985
56. Sarter M, Schneider HH, Stephen DN: Treatment strategies for senile dementia: Antagonist beta-carbolines. Trends Neurosci 11(1):13–16, 1988
57. Mamounas LA, Thompson RF, Lynch G, et al: Classical conditioning of the rabbit eyelid response increases glutamate receptor binding in hippocampal synaptic membranes. Proc Natl Acad Sci USA 81:2548–2552, 1984
58. Agruso VM, Matthews MD: Effects of glutamic acid on operant behavior, activity and open field behavior in rats. Psychol Rep 57:1003–1004, 1985
59. Sahai S, Buselmaier W, Brussman A: 2-Aminophosphonobutyric acid selectivity blocks two-way avoidance learning in the mouse. Neurosci Lett 56:137–142, 1985
60. Freed WJ, Wyatt RJ: Impairment of instrumental learning in rats by glutamic acid diethyl ester. Pharmacol Biochem Behav 14:223–226, 1981
61. Vogel W, Broverman DM, Draguns JG, et al: The role of glutamic acid in cognitive behaviors. Psychol Bull 65:367–382, 1966
62. Bologa L, Sharma J, Roberts E: Dehydroepiandrosterone and its sulfated derivative reduce neuronal death and enhance astrocytic differentiation in brain cell culture. J Neurosci Res 17:225–334, 1987
63. Roberts E, Bologa L, Flood JF, et al: Effects of dehydroepiandrosterone and its sulfate on brain tissue culture and on memory on mice. Brain Res 406:357–362, 1987

24

Ethical Issues in Research in Child Psychiatry

Susan K. Theut and Arthur F. Kohrman

1. INTRODUCTION

In addition to the general concerns for protection of human subjects that govern all research, several special issues around research in children must be considered by investigators and Institutional Review Boards (IRBs). This chapter attempts to highlight those ethical questions that investigators in psychiatry and the neural sciences must consider in designing their projects to ensure the protection of children who are research subjects and to meet the scrutiny of the IRB process. All who contemplate research with children should be familiar with the Code of Federal Regulations[1] governing human research in the United States and the specific report and recommendations of the National Commission for the Protection of Human Subjects of Biomedical and Behavioral Research on Research Involving Children.[2] In this chapter, such familiarity is assumed as we discuss specific issues of importance to the expected readers of this volume.

Clearly, the first requirements of any research protocol are that the design of the project be rigorous, that a valid scientific question be asked clearly, that the methods are appropriate to the question pursued, and that the data be handled in a manner that will ensure significance and clarity of the results with the smallest possible number of research subjects. It is unethical and exploitative to perform research, whether with adults or children, that entails any risk in which the data derived or the handling of those data will not produce useful information for the benefit of the subjects themselves and/or the general class to which they belong.

Discussions and controversies about research in children as distinct from that in adults center about the following issues:

1. Justification for doing research with any significant risk that does not have at least the possibility of therapeutic benefit to the child subject or to children as a class.

Susan K. Theut • Psychiatry Service, Veteran's Administration Medical Center, and Department of Psychiatry, Georgetown University School of Medicine, Washington, D.C. 20007. *Arthur F. Kohrman* • Department of Pediatrics, Pritzker School of Medicine, University of Chicago, and La Rabida Children's Hospital and Research Center, Chicago, Illinois 60649.

2. Substantiation of the need to perform the research on a child population (as opposed to adults or animals), presumably to benefit the larger group of children to whom the knowledge gained will be usefully applied.
3. Definition of "minimal risk" to be applied in considering children as research subjects.
4. Moral boundaries of the legal empowerment of parents or guardians to give permission or substituted consent for research to be performed on minor children.
5. Indications for and methods of involving the child in the decision to participate in the research.

2. THERAPEUTIC AND NONTHERAPEUTIC RESEARCH

While it is generally considered acceptable to perform research with some degree of risk (and, presumably, commensurate potential benefit) or discomfort in fully informed consenting adults with competence to understand the hazards, the predominant sentiment of human research review groups in the United States is to be much more restrictive when such research is contemplated in children. Although there are those[3] who hold that children have a moral duty to participate in research which, while not benefiting them directly, will have benefit for the whole class of children, the general sentiment is to confine approval for such research to those studies in which risk or discomfort is extremely small or nonexistent. The principle of direct potential benefit to the subject or to other children with the child's condition or disease is closely adhered to in evaluating research proposals in which the child will be subject to significant risk or discomfort. The Federal Guidelines[1] provide a reasonable and generally workable definition of the acceptable type of interventions (§46.406b): "The intervention or procedure presents experiences to subjects that are reasonably commensurate with those inherent in their actual or expected medical, dental, psychological, social or educational situations." While this definition restricts the severity of those things that can be done to normal children, it also permits procedures to be done in the conduct of research which might be part of the standard diagnosis or therapy for a medical condition which preexisted the research intervention. Thus, it might be acceptable to perform an additional lumbar puncture on a child with a neurological problem in which lumbar puncture is an expected part of diagnosis or management, but unlikely that doing lumbar puncture in an otherwise well child would be sanctioned. The possible exception to these practices might be in those uncommon situations in which "The intervention or procedure is likely to yield generalizable knowledge about the subjects' disorder or condition which is of vital importance for the understanding or amelioration of the subjects' disorder or condition" (1, §46.406c). In practice, it appears that, at the present time, permission for nontherapeutic research on children, particularly those below the age of effective assent (see below), which has any significant degree of risk or discomfort will be difficult to obtain. While there will certainly be variations among IRBs at different institutions, the predominant sentiment is to apply stringent criteria to the approval of such research proposals.

3. ARE CHILDREN THE BEST OR ONLY POSSIBLE SUBJECTS?

Basic principles in the review of human research proposals are to ascertain that the research, indeed, must be performed in humans, as opposed to animals or other experimental systems (e.g., cell or tissue culture systems), that all alternatives have been exploited to their maximal yield of information, and that the performance of the research in humans is both the next logical step and will yield information of importance to humans in understanding of normal function or of disease or its cure. In contemplating research on children, additional criteria of appropriateness must be applied. Must children be used as subjects? If children below the age of effective assent are to be subjects, is there some characteristic of their unique physiology or of the disease process that makes it necessary to perform the research on this group of children as opposed to performing the research on older children or adults? Is the research of enough importance to the children who are subjects or to the class of children they represent to justify the risks of the specific research procedures?

4. WHAT ARE "MINIMAL RISKS" FOR CHILDREN?

The concept of minimal risk takes on special meaning when considering research involving children. Interventions that carry more than minimal risk are carefully separated for particular scrutiny in federal regulations and in reviews by IRBs. The National Commission for the Protection of Human Subjects'[2] Recommendations on Research Involving Children state: "Minimal risk is the probability and magnitude of physical or psychological harm that is normally encountered in their daily lives, or in the routine medical or psychological examination, of healthy children." The minimal risk standard further becomes the basis for establishment of those conditions in which research on children which has no direct benefit to the subjects may be carried out (1, §46.406). Here there is no better statement of intent or standards than the language of section 46.406 itself:

> HHS will conduct or fund research in which the IRB finds that more than minimal risk to children is presented by an intervention or procedure that does not hold out the prospect of direct benefit for the individual subject, or by a monitoring procedure which is not likely to contribute to the well-being of the subject, only if the IRB finds that:
>
> 1. The risk represents a minor increase over minimal risk.
> 2. The intervention or procedure presents experiences to subjects that are reasonably commensurate with those inherent in their actual or expected medical, dental, psychological, social, or educational situations.
> 3. The intervention or procedure is likely to yield generalizable knowledge about the subjects' disorder or condition that is of vital importance for the understanding or amelioration of the subjects' disorder or condition.

These are restrictive and specific statements of criteria for acceptability and expectations of outcomes of research on children; they leave little room for generous interpretations or "special cases." They do, however, permit judgment by IRBs of the child's specific medical condition and the usual therapies as a basis for approval of additional procedures to be done in the conduct of research. As an example, the IRB may permit additional bone marrow samples or lumbar punctures for research purposes in children who must undergo these inevitably as part of the standard diagnostic and therapeutic interventions for their medical conditions (e.g., leukemia or meningitis) but may not approve the same procedures in children who are well (e.g., to act as controls in clinical studies). The regulations and, indeed, the practices of individual IRBs permit some latitude in making determinations of suitability in particular circumstances or populations; however, procedures or interventions that carry with them more than minimal risk will necessarily undergo careful scrutiny and require clear and persuasive justifications to gain approval. The burden lies heavily on the investigator to be very clear about the need for and outcomes hoped for from research procedures on children which carry with them more than minimal risk; simple curiosity will not do.

5. LEGAL AND MORAL BOUNDARIES OF PARENTAL AUTHORITY

It is clear from the foregoing discussion and from practical experience that, for older children and adolescents, the ability to participate competently in decisions regarding research may be fully intact by any operational or comparative standard well before the legal age of emancipation from parental authority. Conflict between legal parental rights and the child's wishes may arise both in circumstances of the child's assent or of refusal. If we are to respect the child's separate identity as a moral agent, and the investigator has been satisfied that the child is competent to decide, a serious conflict may arise if the parent/guardian and child disagree about the child's participation. Great care must be taken that unwonted parental influence is not accepted when it would serve the investigator's needs (i.e., for the child to be ordered to participate), if the same degree of influence would be rejected in the opposite circumstances (i.e., if the parent ordered the child not to participate). The legal right of the parent/guardian to override the minor child's decision is uncontested, as long as the apparent motivation of the parent is the best interests of the child—all contemporary social and legal structures give the parent/guardian the greatest benefit of the doubt. It is more likely that serious conflicts of this sort will arise around questions of therapy, but they are becoming more common in issues of therapeutic research as well. Of particular recent prominence are decisions regarding children's participation in or continuation of risky and/or uncomfortable treatments as part of research protocols in serious or terminal illness.

6. ASSENT, CONSENT, AND PERMISSION: THE CHILD'S ROLE

The concept of the child's active involvement in decisions about research participation has been evolving in the last several years along with increasing concerns about paternalism in medicine and the growing national debate over the legal and moral limits of substituted or surrogate judgment in critical life and death decisions. These concerns resulted in establishment of the need to obtain the assent of the child to participate, where possible and appropriate, in the additions to federal (HHS) regulations in 1983.

In practice, there are two difficult judgments to be made by investigators and IRBs. The first of these is the determination of the criteria for the child's competence to participate intelligently in the decision at hand; the second is the appropriate response to the moral dilemma created when the child's participation is solicited in making a decision in which a refusal will not be honored. The latter situation is most likely to occur when the research involves a choice of therapies in a serious or potentially life-threatening situation. It is clearly within the legal and (probably) moral domain of a parent/guardian to make a judgment for a minor child, presumably in the child's best interest, which is not agreed to by the child; however, it is deceptive to offer choice when none really exists. Most who deal extensively with children agree that in these situations, explanation to the child of the reasons for the decision is mandatory. Failure to explain that decision shows a lack of respect for the child as a person. Moreover, "Mere failure to object, should not, absent affirmative agreement, be construed as assent." (1, §46.402). It is a principle of practice that all who deal with children, whether in research or in therapeutic situations, should disclose to the child all upcoming interventions or procedures, even when the seeking or giving of assent is not appropriate or possible.

McCormick[3] and others have presented the position that children should be educated to accept some discomfort or inconvenience on behalf of the greater social good, as part of their moral education. This principle, laudatory as it is at face, can be used to justify excessive paternalism; proposals to include otherwise normal children in research (especially as controls) that invoke the altruistic principle must be closely examined in the light of the child's best interests, both long and short term.[4] In these situations, the process for obtaining the child's assent may be as or even more important than the consent of the assent/consent document.

The language used by investigators and by IRBs should also reflect the realities of specific situations. Technically, parents are giving permission for children to participate in research (or for therapy); if a true choice exists for the child to participate after parental permission is obtained, then the child's positive response is assent. The concept of substituted consent implies that the parent (acting as surrogate for the child) is making a decision on the basis of what the child might have chosen if competent to make such a choice. This view of consent is invoked when very young children and infants are the potential research subjects, but even that situation is very different from the situation of obtaining substituted consent for an adult, where a reasonable presumption of the adult's wishes while competent can be made by the surrogate. There are important differences, often ignored, between making substituted judgment for a once competent person and for a never-competent person.[5] These differences put a special burden on those who would protect children from potential harm under the cover of a presumption of the child's best interests.

While involvement of the child in the assent process is a moral mandate, even the most well-intentioned professionals and IRBs must grapple with the problem of determining reasonable criteria of the child's competence to participate intelligently and voluntarily. Many attempts have been made to create guidelines for use by investigators and review boards in making determinations of competence of children. Clearly, there are wide variations between children of the same age in their development of cognitive and reasoning abilities. It must also be recognized that a given child may not have global competence (as is presumed in an adult) but may be very capable of understanding and making choices in a defined specific domain.

The concept of the ability of children to give consent to research at about age 7 can be understood through an exploration of what is known about the development of children's cognitive and reasoning skills. Piaget[6–8] described four stages leading to the development of adult thought: (1) sensorimotor stage (birth to 2 years); (2) preoperational thought (2–7); (3) concrete operations (ages 7–11); (4) formal operations (11 to end of adolescence). Individual children move through these stages at rates dependent on variables such as intellect, the stimulation of the family and significant others, and other environmental factors.

It is with good reason that children should not generally be presented with an option for informed consent until the age of 7, the stage of concrete operations. Before that time, in preoperational thought, the child does not understand cause and effect.[6–8] Just as an illness such as diabetes might be thought of as punishment for misbehavior, a research procedure might be understood in the same way. In the same sense, the preoperational child does not understand the concepts of conservation or of reversibility.[6–8] For instance, a child may be unduly frightened by a procedure that removes body fluids, fearing that they can't be replaced.

In the stage of concrete operations, the child engages the world through operational thought: the child begins to use reason to understand events in the world. Children in this stage also begin to understand rules and regulations. In this stage, normally developing children can understand the simple outlines of a research project ("we are studying how this medicine might change how your blood looks under a microscope") and can understand what the project entails if presented in a focused manner ("we will ask you to swallow this medicine every day with your juice for two weeks; every week we will take a small amount of blood out of your arm and look at it under the microscope"). Children in the concrete stage of operations can also understand the side effects of such a project ("the medicine may make you feel tired"; "when we draw your blood from your arm you will feel a pinch for a few seconds").

In the stage of formal operations, emerging adolescents begin to think abstractly, to reason in a logical manner and to understand probabilities and the conceptional underpinnings of a proposed procedure. Their thinking becomes more complex and abstract. At this developmental stage, the explanations of the broader or altruistic purposes of the projects and the expected side effects can and should be described in a much more detailed fashion.

The investigator needs to evaluate how to present the project not simply in reference to the age of the child, but also in recognition of the developmental stage of the child/adolescent. For various reasons (family background, intelligence, exposure), some early adolescents whom one would expect to think using formal operations still use concrete operations to a great extent.

The process of obtaining consent, assent, or permission has additional special aspects when dealing with certain groups of children. Among these are children who are mentally retarded; those who carry a psychiatric diagnosis such as autism, psychosis, or depression; children being treated in a culture (institutional or social) separate from their own; children whose parents are functionally incompetent or whose parental rights have been terminated or are in suspension; and children who are wards of the state, who reside in institutions for the mentally ill, or who are confined in correctional facilities.

The requirements for their protection are clearly outlined in federal regulations[1,2] and will not be restated here. However, the investigator who contemplates the necessity of involving such special populations of children in research must not only be familiar with the written regulations and the moral and ethical issues involved, but should also subject proposals to careful scrutiny before presenting them to the IRB. The use of expert neutral third parties in determining the child's competence to assent and in obtaining that assent as well as permission from the parent/guardian is becoming more common. There is much to recommend and practice in the pursuit of the greatest possible protection for the subject and for the investigator and institution in which the research is to be carried out.

7. CONCLUSIONS

Children who are the subjects of behavioral and neuroscience research necessitate special consideration. Such research can range from low invasiveness such as the child completing a questionnaire, to a venipuncture for a small amount of blood to a lumbar puncture. Few would argue that the first two should be permitted. However, a lumbar puncture should demand more rigorous consideration. Is the lumbar puncture beyond minimal risk as defined by the National Commission? If so, is it part of the usual treatment for the child or children who are to be the subjects of the proposed research? A continuing argument by neuroscientists who favor permitting lumbar punctures in normal children is that a lumbar puncture of normal children is often the only way to study normal amounts and types of neuropeptides. Others would argue that the same information might be obtained from the spinal fluid of children suspected to have neurological disorders but ultimately found to be normal, in whom the lumbar puncture would have been done in the course of standard diagnostic inquiry. Of such honest disagreements and difference of perspective are the discussions of IRBs made. A defensible position can be assembled for each position, and thoughtful ethicists and scientists will consider the merits of each individual research project and its probable scientific contribution.

The participation of children in research entails consideration of a variety of scientific and ethical principles. These include: the scientific validity of the question; the calculation of costs and benefits to the child and to society; the necessity for the project to be done in children; issues of assent, permission and consent; and consideration of the special status of children who are members of special groups (mentally retarded, mentally ill, wards of the court). As members of a society that strives to project its younger members who may be unable to effectively give consent or to protect themselves from authoritarianism or paternalism, all investigators have a responsibility to conduct research with children with a special sensitivity to the considerations we have discussed. Responsible scientists must themselves balance the research question with an obligation to protect children.

Institutional Review Boards provide a forum for discussion and critique, and are intended to establish and operate within a coherent set of standards; they thus serve as a means for insuring full airing of the investigator's motives and reasoning and uniformity of principles and processes in a given institution. However, the scientist who is the principal investigator understands in the most integral way the effects of the proposed research on the children who will serve as research subjects. The investigator should welcome the opportunity to submit the project to IRB review, in order that all parties in the institution and the public at large may have the greatest possible confidence that research in children is conducted with the highest ethical standards; but even after IRB review and approval, the primary responsibility for the conduct of ethical research with children remains with the scientist–investigator.

REFERENCES

1. Code of Federal Regulations: 45 CFR 46: Protection of Human Subjects. Department of Health and Human Services, NIH, Office for Protection from Research Risks, March 8, 1983
2. National Commission for the Protection of Human Subjects of Biomedical and Behavioral Research: Report and Recommendations on Research Involving Children. DHEW Publication No. (OS) 77-0005. Washington, D.C., U.S. Government Printing Office, 1977
3. McCormick RA: Experimentation in children: Sharing in sociality. Hastings Center Rep 6:41–46, 1976
4. Barondess JA, Kalb P, Weil WB, et al: Clinical decision-making in catastrophic situations: The relevance of age. J Am Geriatr Soc 36:919–937, 1988
5. Ramsey P: The enforcement of morals: nontherapeutic research on children. Hastings Center Rep 6:21–30, 1976
6. Piaget J, Inhelder B: The Psychology of the Child. New York, Basic Books, 1969
7. Piaget J: The Child and Reality. New York, Penguin, 1977
8. Piaget, J: The Language and Thought of the Child. London, Routledge and Kegan Paul, 1926

Index